MEMS and MOEMS Technology and Applications

MEMS and MOEMS Technology and Applications

P. Rai-Choudhury
Editor

SPIE PRESS
A Publication of SPIE—The International Society for Optical Engineering
Bellingham, Washington USA

Library of Congress Cataloging-in-Publication Data

Rai-Choudhury, P.
   MEMS & MOEMS: technology and applications / Prosenjit Rai-Choudhury.
   p. cm – (SPIE Press monograph ; volume PM85)
   Includes bibliographical references and index.
   ISBN 0-8194-3716-6
    1. Microelectromechanical systems. 2. Optoelectronic devices. I. Title.
   II. SPIE Press monograph ; PM85

TK7875 .R35 2001
621.381–dc21

                                                                                                   00-061209
                                                                                                          CIP

Published by

SPIE—The International Society for Optical Engineering
P.O. Box 10
Bellingham, Washington 98227-0010
Phone: 360/676-3290
Fax: 360/647-1445
Email: spie@spie.org
WWW: www.spie.org

Copyright © 2000 The Society of Photo-Optical Instrumentation Engineers

All rights reserved. No part of this publication may be reproduced or distributed
in any form or by any means without written permission of the publisher.

Printed in the United States of America.

Cover image courtesy of Sandia National Laboratories, www.mems.sandia.gov.

Contents

Preface / ix

Chapter 1. MEMS/MOEMS Technology Capabilities and Trends / 1
Patrick J. French

1.1 Introduction / 3
1.2 Basic Silicon Device Fabrication Steps / 8
1.3 Basic Micromachining Fabrication Processes / 14
1.4 Integration / 20
1.5 Micromachined devices / 20
1.6 Future Technologies / 33
1.7 Future Applications / 35
1.8 Conclusions / 39
    References / 40

Chapter 2. Operation and Design of MEMS/MOEMS Devices / 47
Douglas Sparks, Thomas Bifano and Dhiraj Malkani

2.1 Operation of MEMS/MOEMS Devices / 49
2.2 The Design of MEMS/MOEMS / 60
2.3 MEMS Design Examples / 75
2.4 Summary / 101
    References / 102

Chapter 3. Optical Microsystem Modeling and Simulation / 109
Jan G. Korvink, Arokia Nathan and Henry Baltes

3.1 Introduction / 111
3.2 Lagrangian Equations of Motion / 118
3.3 Discretization Methods / 136
3.4 Solution Methods / 152
3.5 Applications / 157
3.6 Discussion / 166
    Acknowledgments / 166
    References / 166

Chapter 4. The Digital Micromirror Device— A Micro-Optical Electromechanical Device for Display Applications / 169
Michael A. Mignardi, Richard O. Gale, David J. Dawson and Jack C. Smith

4.1 Introduction / 170
4.2 CMOS Fabrication / 171
4.3 Superstructure Fabrication / 172
4.4 Die Separation and Pixel Release / 176
4.5 Package Assembly / 179
4.6 Wafer, Package, and Projection Level Testing / 181
4.7 Device Reliability / 203
4.8 Summary / 203
References / 204
Appendix A: List of significant publications on DMD and DLP technology / 204

Chapter 5. Optical Waveguides and Silicon-Based Micromachined Architectures / 209
Christophe Gorecki

5.1 Introduction / 210
5.2 Status of Optical Waveguide Technologies on Silicon / 211
5.3 Silicon Micromechanics / 254
5.4 Examples of Micromachined Guided-Wave MOEMS / 260
5.5 New Trends and Conclusions / 295
References / 297

Chapter 6. Silicon Micromachines in Optical Communications Networks: Tiny Machines for Large Systems / 301
Randy Giles, Vladimir Aksyuk, Chris Bolle, Flavio Pardo and David J. Bishop

6.1 Introduction / 302
6.2 MEMS Devices / 302
6.3 MEMS in Lightwave Technology / 303
6.4 The Future / 327
References / 329

Chapter 7. Assembly and Test for MEMS and Optical MEMS / 331
Christopher Bang, Victor Bright, Michael A. Mignardi, Thomas Kocian and David J. Monk

7.1 Introduction / 333
7.2 Packaging Processes / 335
7.3 Packaging of Physical Sensors / 373
7.4 MEMS and MOEMS Testing / 427
7.5 Reliability of MEMS / 432

7.6 Cost Model and Summary / 449
     References / 455

Chapter 8. MEMS, Microsystems, Micromachines: Commercializing an Emergent Disruptive Technology / 479
Steven Walsh, Jonathan Linton, Roger Grace, Sid Marshall and Jim Knutti

8.1 Introduction / 481
8.2 Commercialization of Current M^3 Technology / 483
8.3 The M^3 Commercialization Process / 488
8.4 Market Studies: Their Problems and Value / 494
8.5 Technology Roadmap Development for an Emergent Industry / 506
8.6 Evolving M^3 Infrastructure / 510
8.7 Conclusions / 513
     References / 514

Index / 517

Preface

The silicon age that led the computer revolution has significantly changed the world. The invention of the transistor and the integrated circuit began a technological revolution—the miniaturization of electronics. Today, unprecedented increases in productivity are largely coming from the computer revolution. The next thirty years of the silicon revolution will incorporate new types of functionality onto the chip—structures that will enable the chip to reason, sense, communicate, and act. This new revolution will allow the silicon chip to affect their surroundings and to communicate the data to the world in new and dramatic ways. Besides processing data, the chips of tomorrow will process such things as chemicals, motions, light, and knowledge. This will be the basis of the next silicon revolution, and it will have a profound effect on the quality of life in this new century.

Mechanical systems are now poised for a similar revolution with its own miniaturization. The driving forces for the miniaturization of mechanical systems include cost, size, speed, weight, and precision while providing an effective interface between the macro and the microdynamic world. Micromachining technologies offer a wide range of possibilities for active and passive devices. Many of these technologies are based on surface micromachining evolving from the silicon integrated circuits technology. This book is written by experts in the field and it contains useful details in design and process, and can be utilized as a reference book or a textbook. It contains eight chapters covering all aspects of MEMS (microelectromechanical systems) and MOEMS (micro-optoelectromechanical systems) technologies.

Chapter 1 gives an overview of the subject and discusses the wide range of possibilities micromachining technology offers. Fabrication of three-dimensional structures by micromachining based on deposition and etching techniques is gaining momentum. The devices have been considerably scaled down in size, and many new applications from sensors to actuators to optical systems are emerging.

Conversion from one form of energy to another is the essential purpose of many micromachined devices. Chapter 2 reviews the operation and the design of micromachines using devices such as sensors and actuators. The design of these devices involves use of the basic principles of operation, microsystem and process modeling, circuit design, process technology, application environment, packaging, and manufacturing. For such multidisciplinary systems to work effectively, the design issues must be approached globally from the beginning, rather than trying to integrate the individual components at the end. Several examples of pressure sensors, accelerometers, resonant devices, and deformable mirrors are used to illustrate many of the factors encountered in MEMS design.

Chapter 3 reviews the growth of TCAD (Technology Computer-Aided Design) and modeling capabilities for MEMS and MOEMS. Commercial quality tools, similar to those available in the microelectronics industry, would aid in the rapid growth of the micromachining technology and microsystems. Further work is necessary to combine existing MOEMS simulation tools with an integrated TCAD simulation environment.

Chapter 4 describes the development of digital light processing (DLP) technology and production of digital micromirror devices (DMD) at Texas Instruments. The DMD is a semiconductor-based array of fast, reflective digital light switches that precisely controls a light source using a binary pulse-width modulation technique. Several DLP designs allow the technology to be used in compact and lightweight applications such as portable projectors, as well as in very high-brightness fixed installation applications such as digital cinema and boardroom projectors. Unlike most MEMS devices, the DMD is fully integrated and monolithically fabricated on a mature SRAM CMOS address circuitry. Thus, both the mirrors and the drive circuitry are integrated on a single silicon wafer.

Chapter 5 discusses the emerging technology of guided wave MOEMS and planar waveguides. For these applications silicon micromachining can be used to manufacture and assemble perfectly aligned optical components, as well as for the construction of optical modulation interfaces using the physical properties of thin-film multilayer structures.

In Chapter 6, interesting new micromachines that enable design and fabrication of optomechanically integrated circuits, and subsystems whose capabilities can facilitate the explosive growth in bandwidth and networking features of lightwave systems, are discussed. A variety of lightwave network applications benefit from the small size, scalability, low power consumption, and low cost of MEMS optical circuits. To date no optical devices in MEMS have been used in an active lightwave network, but many have potential for commercial applications.

Chapter 7 provides an in-depth treatment of the critical subject of assembly and test for MEMS and MOEMS. Packaging affects the operation of devices like pressure sensors and must be compatible with the design and operation of the system. Packaging and testing can represent over 50% of the product cost for microsystems. Die size and wafer cost are being reduced by the integration of control circuitry and MEMS/MOEMS. Custom packaging and testing will continue to keep the cost high for MEMS/MOEMS devices. Several examples of packaging for pressure sensors, accelerometers, micromirrors, other MOEMS, microfluidics, microvalves, and microswitches are discussed. This chapter provides an overview of several practical issues for the commercialization of microsystems.

MEMS and MOEMS technologies hold almost limitless possibilities for applications, including emerging fields such as biotechnology, medicine, telecommunications, and wireless RF applications. Recent systems estimates for year 2002 market valuations have ranged from $38 billion (US) to as high as $100 billion. Chapter

8 focuses on commercialization and discusses industry activities that could accelerate the commercialization of the technology. Some of the roadblocks to commercialization include industrywide differences in nomenclature, manufacturing and marketing infrastructure, and inherent problems that companies face trying to gain competitive advantage in such a broad interdisciplinary field. Despite many impediments, growing global interest will lead to widespread use of micromachining technology in the twenty-first century.

P. Rai-Choudhury
August 2000

MEMS and MOEMS Technology and Applications

CHAPTER 1

MEMS/MOEMS TECHNOLOGY CAPABILITIES AND TRENDS

Patrick J. French
Delft University of Technology

CONTENTS

1.1 Introduction / 3
    1.1.1 Origins of silicon processing / 3
    1.1.2 First steps in micromachining / 4
    1.1.3 Developing markets / 5

1.2 Basic Silicon Device Fabrication Steps / 8
    1.2.1 Oxidation / 9
    1.2.2 Patterning / 9
    1.2.3 Etching / 10
    1.2.4 Doping / 11
    1.2.5 Anneal/drive-in / 11
    1.2.6 Thin-film depositions / 11
    1.2.7 Bonding and packaging / 13

1.3 Basic Micromachining Fabrication Processes / 14
    1.3.1 Bulk micromachining / 14
    1.3.2 Surface micromachining / 15
    1.3.3 Miscellaneous silicon micromachining techniques / 17
    1.3.4 Nonsilicon technologies / 18

1.4 Integration / 20
    1.4.1 Compatibility issues / 20

1.5 Micromachined devices / 20
    1.5.1 Pressure sensors / 20
    1.5.2 Inertial sensors / 21
    1.5.3 Micromotors / 24
    1.5.4 Lateral resonators/comb drives / 26
    1.5.5 Micropumps / 26
    1.5.6 Digital light modulators / 28
    1.5.7 Microtip / 30
    1.5.8 Microgrippers / 31
    1.5.9 Microconveyer belt / 32

1.6 Future Technologies / 33
    1.6.1 Deep reactive ion etching / 33

       1.6.2   Stereolithography / 34
       1.6.3   Nanotechnology / 35
1.7   Future Applications / 35
1.8   Conclusions / 39
      References / 40

1.1 INTRODUCTION

Since the discovery of the transistor effect in semiconductors in 1947, the silicon industry has grown to proportions that nobody at the time would ever have dreamt of. The possibility of making mechanical structures in silicon came much later.

The development of silicon technology has provided a number of important advantages. Silicon is an extremely good mechanical material, as shown in Table 1.1. The micromechanical components can be integrated with the electronics to develop smart sensor and actuator systems with additional features such as self-test and self-calibration.

In Japan, long-term programs were set up to promote micromachining through MITI and with the establishment of the Micromachine Centre in Tokyo. In Europe a number of European Union programs were begun to cover this field and in the United States, the Defense Advanced Research Projects Agency (DARPA) has invested heavily in microelectromechanical systems (MEMS) technology. There are now a number of programs worldwide where micromachined structures can be ordered as part of multiproject designs.

1.1.1 ORIGINS OF SILICON PROCESSING

Silicon is the second most common element in the earth's crust. For centuries its oxide form, glass, has been used. However, it was not until the second half of the twentieth century that the electrical properties of silicon were fully realized and its potential in micromachining, was realized still later.

In the second half of the 1930s Mervin Kelly posed the question of whether mechanical relays in telephone systems could be replaced with electronic switches. He was at the time director of the Bell Telephone Laboratories. During the Second World War, considerable research effort was directed toward the properties of silicon and germanium, and some simple devices were made. It was in 1945 that William Schokley focused his semiconductor group on research in silicon and germanium. In his group were Walter Brittain and John Bardeen, and in 1947 they

Table 1.1 Properties of silicon-based materials compared with aluminum and steel.

	Melting point (°C)	Thermal expansion ($10^{-6}/°C$)	Density (g/cm^3)	Young's modulus (10^{11} Pa)	Yield strength (10^{11} Pa)
Si	1415	2.5	2.4	1.3–1.69	6.9
SiN	1900	2.8	1.48	2.43	14.0
SiO$_2$	1610	0.5	2.27	0.73	8.4
Al	660	25	2.70	0.70	0.17
Steel	1500–2000	12	7.9	2.1	4.2

discovered the transistor effect and successfully made the point contact transistor.[1] The name "transistor" came from the two words "transfer" and "resistor." These three researchers received the Nobel prize for physics in 1956 for the invention of the transistor.

In the early days, both silicon and germanium were used for the development of devices. In 1948 Teal and Little developed a technique for pulling high-quality germanium out of the molten material.[2] It was well known at the time that silicon, with its larger band gap, would be preferable for the transistor, but there were a number of practical problems in obtaining wafers of sufficient quality. An important breakthrough came in 1953 when Theuerer developed the floating zone method.[3] This technique yielded silicon of quality comparable to that of germanium. In 1954, Gordon Teal made the first silicon transistor grown with this new method at Texas Instruments.[4]

Jack Kilby joined Texas Instruments in 1958 and had soon made the first simple integrated circuits (ICs), and a patent application was made in February 1959. At the same time, Robert Noyce of Fairchild was working on planar technology and applied for a patent on his developments in July 1959. Other companies who wanted to develop ICs from this time on had to have a license from one of these two companies. Also in 1959, Atalla and Khang from Bell announced the first metal-oxide-silicon transistor (MOS). For many years MOS technology was not considered a serious competitor to bipolar technology and suffered from a number of technological problems. During the 1960s, further development was achieved by work at Hitachi and Toshiba in Japan and Motorola in the United States. In 1967 Texas Instruments open a new factory in Nice, France. From the end of the 1960s the momentum grew to create the industry we see today.

1.1.2 First Steps in Micromachining

Fine mechanics has been used for centuries. One only has to look at how the clock and watch industry has developed. The mechanical clock dates back to the 14th century. One of the earliest examples was built in 1386 and can still be seen today in Salisbury Cathedral, England. These early clocks used heavy weights and ropes and could certainly not be described as micro or even fine mechanics. The coiled spring was first used as a source of power in the late 15th century, probably in Nuremberg, and by the late 16th century the size of timepieces had been reduced sufficiently to enable a portable watch to be produced. In the 17th century, the Dutchman Christian Huygens invented the pendulum clock, which was an important development for the clock industry. The period from 1675 to 1800 saw a revolution in watch development, with further refinements and reductions in size. Interestingly, in the 18th century the problem of temperature compensation was confronted. Timepieces based on a steel balance spring had a temperature sensitivity of about 8 seconds per day per degree Celsius. The pressure for an improvement came from the shipping industry, and John Harrison responded by producing a number of highly accurate timekeepers in which temperature compensation was achieved by using bimetallic strips to adjust the effective length of the balance spring. The timekeeping industry has therefore traveled from macroengineering through to fine mechanics and today is even using micromachined wheels. During its development, the issue of cross-sensitivity also had to be confronted as it is now being confronted in MEMS devices.

There were limits in how small the structures could be made and further miniaturization would need new technologies. These new technologies are now being utilized by the watchmaking industry.

The term "micromachining" usually refers to the fabrication of micromechanical structures with the aid of etching techniques to remove part of the substrate or a thin film. Silicon has excellent mechanical properties,[5] making it an ideal material for machining. One of the first silicon (pressure) sensors was isotropically micromachined by Honeywell in 1962.[6] In 1966 Honeywell developed a technique to fabricate thin membranes using mechanical milling. Isotropic etching was used to produce membranes in 1970. Etchants dependent on crystal orientation led to more precise definition of structures and increased interest.[7] Anisotropic etching was introduced in 1976. An early anisotropically etched silicon pressure sensor was made by Greenwood in 1984.[8] In the 1960s the first surface micromachined structures using metal mechanical layers were presented.[9] Surface micromachining involves the formation of mechanical structures in thin films formed on the surface of the wafer. The early 1980s saw the growth of silicon-based surface micromachining using polysilicon as the mechanical layer.[10,11] In recent years a number of new technologies have been developed using both silicon and materials new to the semiconductor industry. The more recent developments in these technologies will be discussed in Sections 1.3.3 and 1.3.4.

With the ever-decreasing dimensions of the structures comes the need for more accurate tools. The development of tools and measurement equipment can be seen in Fig. 1.1.[12]

1.1.3 DEVELOPING MARKETS

In the early days there was no market for the new MEMS devices. Industry did not have sufficient trust in the new technologies and had to be convinced that these new devices were better than existing devices made using fine mechanics. The development of the pressure sensor is a good example of how sensors have developed. Pressure sensors based on stretched metal wires, such as that shown in Fig. 1.2, were found to be accurate and reliable and met the needs of industry. These devices used the change of resistance of the wires under mechanical stress. This change contains two components: the geometric effect and piezoresistive.

The piezoresistive effect was first discovered in the 19th century by Lord Kelvin, when he noticed that the change in resistance in metal rods could not be explained purely by dimensional changes.[13] Sensors based on the piezoresistive effect in metal alloys are still available today. The effect in silicon was report in

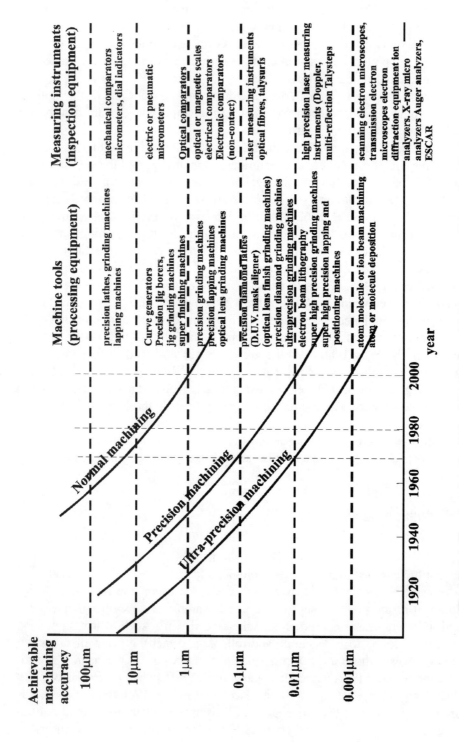

Figure 1.1 The development of achievable machining accuracy. Taken from Ref. 12.

MEMS/MOEMS Technology Capabilities and Trends

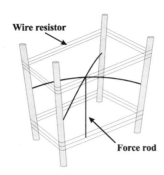

Figure 1.2 Early stretched-wire pressure sensor made by Bell & Howell.

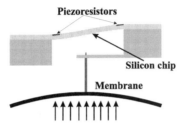

Figure 1.3 Pressure sensor based on the bending of a silicon chip.

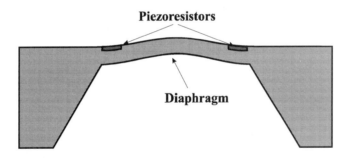

Figure 1.4 Pressure sensor based on a thinned silicon diaphragm.

the 1950s by Smith.[14] In 1958 the companies Kulite, Honeywell, and Microsystem introduced discrete silicon strain gauges. A silicon chip without any micromachining can be used as a pressure sensor. Diffused resistor-based pressure sensors were introduced by Kulite in 1961. An example of such a system, produced by Bell & Howell, is shown in Fig. 1.3.

Once micromachining techniques were available, a membrane could be formed in the silicon chip itself using the structure shown in Fig. 1.4.

The development of surface micromachining has resulted in a further reduction in size of the pressure sensor. Whereas the membrane in bulk micromachining, shown in Fig. 1.4, is typically 1–2 mm, a surface micromachined pressure sensor would have a membrane on the order of 100–200 μm. These devices usually use

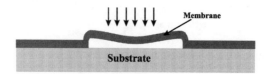

Figure 1.5 Pressure sensor fabricated using surface micromachining.

Table 1.2 Bryzek's MEMS market forecast.[15]

Year	1995	2005
Pressure sensor	1.0	2.5
Inertial sensors	0.4	0.8
Fluidic controls	0.01	0.1
Data storage	0.0	1.0
Display	0.0	1.0
Biochips	0.0	0.2
Communication	0.01	1.0
Misc.	0.03	0.1
Total MEMS	1.45	6.7
Total nonsensor MEMS	0.05	3.4

capacitive instead of piezoresistive output. A cross-section of such a device is given in Fig. 1.5.

It can be seen how the pressure sensor has developed. Other devices, such as accelerometers, have also gone through the same series of changes.

Through these advances the confidence of industry has increased and the market has continued to grow. The MEMS market in 1995 was US$1.45 billion, and is expected to grow steadily over the coming years. There have been wide-ranging estimates of the size of the market in 2005. One conservative estimate, made by Bryzek,[15] is given in Table 1.2. This predicts the continuing domination of the market by pressure sensors.

1.2 BASIC SILICON DEVICE FABRICATION STEPS

All silicon-based processes begin with the manufacture of the silicon wafers which generally come in sizes from 1 to 8 inches. It is important that the wafers can be grown with high purity and that precise levels of doping can be introduced. One important issue for bulk micromachining is the increasing thickness of the wafers. A 4-inch wafer, for example, has a thickness of 525 μm. For a 6-inch wafer this is increased to 750 μm. This increased thickness means longer etching for micromachining and larger dimensions for orientation-dependent etching.

The are a number of basic processing steps used in the electronics industry that which can be used in micromachining. These are standard processes and in many cases MEMS designers use variations on these to develop micromechanical

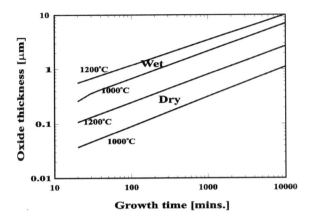

Figure 1.6 Oxide thickness as a function of growth time.

devices. These processing steps include oxidation, thermal applications and deposition.

1.2.1 OXIDATION

Oxidation is one of the basic steps in silicon processing. Silicon will oxidize at room temperature in a normal atmosphere. However, this oxidation is only a few atoms thick. Oxidation of silicon occurs at high temperature, usually above 800°C. The oxidation rate of silicon can be seen in Fig. 1.6. This figure gives two temperatures, 1000°C and 1100°C. As would be expected, the higher temperature yields a thicker oxide. This figure also shows two types of oxidation, wet and dry. With a wet oxidation, the gas is passed through a bubbler before entering the furnace. This added humidity results in a faster oxidation, although the density of the oxide is usually lower.

1.2.2 PATTERNING

In almost every step a pattern on the wafer must be defined. This is done by lithographic techniques. For many years it was thought that optical lithography could only be used down to 1 µm and that below this level electron beam or X-ray would have to be used. However, developments in photolithography continue to push the limits downwards. For MEMS technologies, optical lithography offers sufficient accuracy, although the future may show that new technologies for MEMS have to be developed. The basic steps for patterning the wafer are given in Fig. 1.7.

The resist is a photosensitive layer that is spun onto the wafer. The thickness of the resist is determined by its viscosity and the finning speed. After a light baking, the wafer is exposed to UV light through a mask. The resist becomes soft when exposed to UV light (positive resist) or hard (negative resist). A developer is used

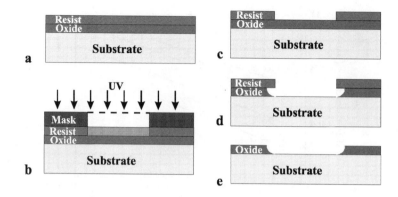

Figure 1.7 Basic steps for patterning the wafer.

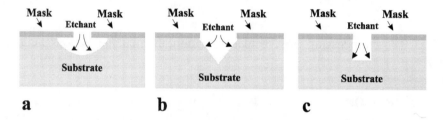

Figure 1.8 Three etching forms: (a) isotropic, (b) anisotropic-orientation dependent, and (c) anisotropic.

to remove the soft resist, leaving a pattern on the wafer. After baking, the resist is further hardened and can resist a number of etchants (but not all etchants).

1.2.3 Etching

Patterned wafers usually require etching. This may be the etching of thin films to form a mask for further processing or in the silicon itself to form 3D structures. Etching can be described as wet (chemical) or dry. Each process results in a defined shape of etch pit, depending upon the characteristics of the etching process. These can be classes in three groups as shown in Fig. 1.8.

Many wet chemical etchings are isotropic in nature, although as will be shown below, a number of etchants are available that are orientation dependent. Dry etching presents many new possibilities since the shape of the etched pit can be determined by the etching gases. New plasma etching systems are now on the market that will further expand the options for the MEMS designer. These enable pits to be fabricated with vertical walls. The different modes of plasma etching can be seen in Fig. 1.9. This shows that through combinations of ion bombardment and sidewall passivation, vertical walls can be achieved.

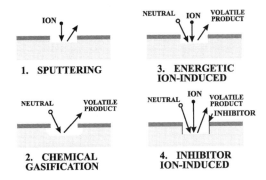

Figure 1.9 Four modes of plasma etching.

1.2.4 Doping

The basic silicon wafer is either *n* or *p* doped. Additional steps are used to define differently doped regions in the substrate. This can be performed by either diffusion or ion implantation. Basically, diffusion is a process in which the wafers are subjected to an atmosphere containing the desired dopant at a high temperature. The most commonly used dopants are $POCl_3$ for phosphorus doping (*n*-type) and boron nitride-based wafers for boron doping (*p*-type).

In the case of ion implantation, as the name suggests, the dopant ions are accelerated toward the wafers with sufficient energy that they enter the wafer. The depth of the implantation is dependent upon the ion type and the energy. The most commonly used dopants are boron for *p*-type doping and arsenic, phosphorus, and antimony for *n*-type material.

1.2.5 Anneal/drive-in

When doping atoms are introduced to the wafer, they have to be brought into the lattice and the desired doping profile reached. In the case of ion implantation, the crystal structure is severely damaged and this damage should also be repaired. This annealing does not need particularly high temperatures, and 800°C is sufficient to repair the damage. This step also ensures that the doping atoms take their position in the crystal structure, making them electrically active. If you wish to drive the doping atoms deeper into the substrate, higher temperatures (>1000°C) are required. These anneal and drive-in steps are often combined with an oxidation step.

1.2.6 Thin-film depositions

For a number of MEMS applications, in particular surface micromachining, additional thin films are required. Some of these films are available in standard IC processing, whereas others require the wafers to be removed from the clean room and processed elsewhere.

Chemical vapor deposition is widely used for the deposition of thin silicon-based films. Single-crystal silicon deposition is an essential part of many standard IC processes. This is known as epitaxial deposition (often called epi). Non-crystalline materials can also be deposited using chemical vapor deposition. These systems can be at atmospheric pressure (APCVD) and low pressure (LPCVD). For low-temperature depositions, a plasma is used to enhance the breakdown of the gases and this is known as plasma-enhanced CVD (PECVD).

Figure 1.10 gives an example of an epitaxial deposition system. The silicon is introduced in the form of SiH_2Cl_2 gas, known as dichlorosilane (DCS). Other gases that can be used include SiH_4, $SiHCl_3$, and $SiCl_4$. In general, the higher the chlorine content, the lower the deposition rate.

This system normally operates in the temperature range of 900°C and 1150°C and is used to form thin single-crystal layers of up to about 20 μm.

LPCVD is widely used for the deposition of thin films such as polysilicon, silicon nitride, and oxides. As with the epitaxial systems, there is a range of structures and layouts available. Figure 1.11 shows an example of a CVD tube.

A range of LPCVD processes are commonly used for micromachining, and some of these are listed in Table 1.3.

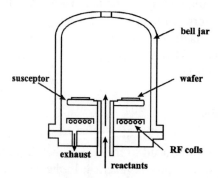

Figure 1.10 Example of an epitaxial deposition system. Based on the Gemini-1 pancake-type, cold-wall, induction-heated system.

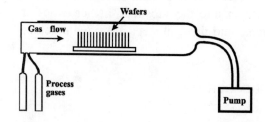

Figure 1.11 An example of an LPCVD system.

Since aluminum is commonly used for metallization in ICs, additional processing after aluminum deposition is limited to temperatures less than 400°C. In this case PECVD systems are usually used. The use of a plasma enhances the breakdown of the gases, and therefore temperatures in the range of 200–300°C can be used.

In addition to the silicon-based thin films, a range of metals have been used for micromachining purposes.[16,17] Aluminum has been successfully used for a number of surface micromachined structures.

In the field of actuators, piezoelectric activation is often used. Since silicon itself is not a piezoelectric material, alternative materials have to be used. These include quartz, polyvinylidene fluoride (PVDF), and zinc oxide.[18–20]

1.2.7 Bonding and packaging

In many applications, a sealed cavity or protection against excessive movement is required. In these applications it is important that the bond be mechanically strong and the seal be maintained. A range of techniques are available and these are listed in Table 1.4.[21] These techniques are further illustrated in Fig. 1.12.

Table 1.3 Commonly used gases and temperature ranges for CVD systems.

Layer	Gases	Temperature (°C)
Polysilicon	SiH_4	550–700
Silicon nitride	$SiH_2Cl_2 + NH_3$	750–900
	$SiH_4 + NH_3$	700–800
Silicon dioxide undoped	$SiH_4 + O_2$	400–500
PSG (phosphorus doped)	$SiH_4 + O_2 + PH_3$	400–500
BSG (boron doped)	$SiH_4 + O_2 + BCl_3$	400–500
BPSG (phosphorus/boron doped)	$SiH_4 + O_2 + PH_3 + BCl_3$	400–500

Table 1.4 Some examples of bonding techniques.[21]

Bonding method	Substrate	Intermediate layer	Temperature [°C]	Reference
Gluing	Si-Si, glass-Si, glass-glass	Spin-on adhesives	Room temp.	
Low-temp. glass bonding	Si-Si	Boron glass	450	22
Eutectic bonding	Si-Si	Au	370	23–25
Fusion bonding	Si-Si	—	>150	26,27
Anodic bonding	Glass-Si	—	>250	23,28–31
	Si-Si	Oxide	>850	
	glass-glass	—	<400	

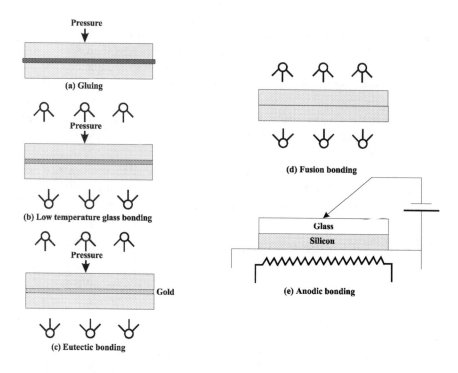

Figure 1.12 Five examples of bonding techniques.[21]

If bonding is to be performed at the end of the processing, the permitted temperature is extremely limited owing to the presence of aluminum. For many years this was a problem for fusion bonding, which used temperatures above 1000°C. However, more recent developments have brought the bonding temperature down to below 400°C, and thus compatible with aluminum.[26,27]

1.3 BASIC MICROMACHINING FABRICATION PROCESSES

As described in Section 1.1.2, the term "micromachining" usually refers to the fabrication of micromechanical structures with the aid of etching and/or deposition steps. The main silicon micromachining techniques can be divided into two groups bulk and surface micromachining, although there are a number of new technologies that will be discussed later.

1.3.1 BULK MICROMACHINING

Bulk micromachining usually refers to etching through the wafer from the back side to form the desired structures. A cross-section of a bulk micromachined membrane, integrated with bipolar electronics, is given in Fig. 1.13. These devices,

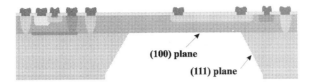

Figure 1.13 Cross-section of a simple membrane combined with bipolar electronics fabricated using anisotropic wet etching.

whether they be membranes or masses hanging on beams, usually have lateral dimensions on the order of 1–5 mm. The device shown here is fabricated using a standard (100) wafer. This structure makes use of wet anisotropic etchants. These etchants have the important property of an etch rate that is dependent upon crystal orientation. The low etch rate of the (111) plane yields the structure shown in Fig. 1.13. The main etchants used are ethylenediamine and pyrocatechol (EDP),[32] potassium hydroxide (KOH),[33] and tetramethyl ammonium hydroxide (TMAH),[34] although a number of alternatives are available.

In order to define structures such as a membrane, an etch stop should be defined. The most simple is a timed stop. The etch time is calculated from the wafer thickness and etch rate. This is extremely simple, but has limited accuracy. A more accurate etch stop is the so-called p^+ etch stop. EDP is particularly suitable for this application and has been used very successfully.[35] However, this techniques relies on a high boron-doped region, which can generate stress in the structure. Furthermore, such a high-doped region may not be compatible with the structure required. The final technique is the electrochemical etch stop.[36] In this case, the thickness of a membrane or beam is defined by the epitaxial thickness, which is extremely accurate. However, this is a more complicated process and difficult to batch fabricate. Furthermore, the holder required for the etching may introduce stress in the wafer, resulting in breakage. Recent developments in electrochemically controlled etching have removed the need for the external contact. The first of these is the photovoltaic etch stop.[37] As the name suggests, the voltage is generated using a strong light source. This yields a stop on a p-type layer on an n-type substrate. The second technique is the galvanic etch stop.[38] A gold/chromium layer deposited on the wafer creates a galvanic cell when the sample is immersed in tetramethylammoniumhydroxide. The voltage generated can be sufficient to create an etch stop on an n-type epitaxial layer on a p-type substrate.

1.3.2 SURFACE MICROMACHINING

Surface micromachining represents a totally different process from bulk micromachining. Instead of forming mechanical structures in the silicon substrates, devices are fabricated in thin films deposited on the surface. Early surface micromachined structures used metal mechanical layers, but polysilicon became dominant in the 1980s and today still represents an important material for surface micromachining.

The most important mechanical materials for silicon technology-based surface micromachining are polysilicon, silicon nitride, silicon dioxide, and aluminum. The basic surface micromachining process is given in Fig. 1.14. First a sacrificial layer is formed and patterned [Fig. 1.14(i–a)], followed by a similar process for the mechanical layer(s) [Fig. 1.14(i–b)]. Finally the sacrificial layer is removed by sacrificial etching to leave the free-standing structures [Fig. 1.14(i–b)]. As shown in Fig. 1.14(ii), the sacrificial layer is removed from under the mechanical structures either at the side of the structure or, in the case of membranes, through etch holes. An overview of surface micromachining techniques is given in Ref. 39. The dimensions of these devices are quite different from those of bulk micromachining. The lateral dimensions are usually between 100 µm and 1 mm. With thicknesses of the layers in the order of 0.5–2 µm.

The suitable sacrificial layer is dependent upon the mechanical layer used, with the important factor being the availability of an etchant that etches the sacrificial layer without significantly etching the mechanical layers or the substrate. A commonly used combination is an oxide sacrificial layer and a polysilicon or silicon nitride mechanical layer. Examples of these layer combinations can be seen in Table 1.5.

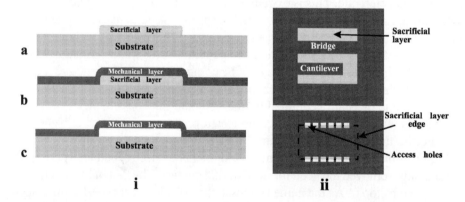

Figure 1.14 Basic surface micromachining process: (i) cross-sectional view, (ii) lateral view of a bridge, a cantilever, and a membrane.

Table 1.5 Layer combinations and etchants for surface micromachining.

Sacrificial layer	Mechanical layer	Sacrificial etchant
Oxide (PSG, LTO, etc.)	Polysilicon, silicon nitride, silicon carbide	HF
Oxide (PSG, LTO, etc.)	Aluminum	Pad etch, 73% HF
Polysilicon	Silicon nitride	KOH
Polysilicon	Silicon dioxide	TMAH
Resist	Aluminum	Acetone–oxygen plasma

1.3.3 MISCELLANEOUS SILICON MICROMACHINING TECHNIQUES

There are a number of silicon micromachining technologies that use the bulk material for mechanical structures but cannot be described as bulk micromachining in the traditional sense of the word.

A variation on surface micromachining is to use the epitaxial (epi) layer as the mechanical layer. A range of processes have been developed using both single-crystal epi [40–42] and epi-poly.[43,44] As with surface micromachining, epi-micromachining requires a sacrificial layer under the epi. For this purpose, oxide, high-doped buried layers and simply the substrate have been used. The basic process is given in Fig. 1.15.

Two examples of structures, fabricated at the Delft University of Technology in the Netherlands using epi-micromachining are given in Fig. 1.16. One has been fabricated using a sacrificial porous silicon process [41] and the other using epi-poly.[44]

Cornell University in the United States developed a process that used a combination of anisotropic (to etch the vertical pits) and isotropic (to underetch) plasma etching to obtain free-standing structures. This process was called SCREAM (single crystal silicon reactive etching and metallization). The basic process is given in Fig. 1.17.[45]

An alternative to the SCREAM process was developed at Twente University in the Netherlands.[46] This process is known as Black silicon. Basically the process

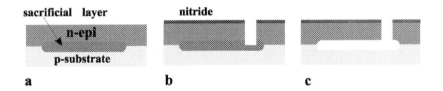

Figure 1.15 Basic epi-micromachining process.

Figure 1.16 Device fabricated using sacrificial porous silicon (left) and epi-poly (right).

uses a combination of different modes of plasma etching. The difference is the manner in which the two modes are achieved. In this case the second etch step deposits a fluorocarbon (FC) on the side wall of the etch pit. This protects the existing side wall against etching during the underetching. The basic etch sequence is given in Fig. 1.18.

1.3.4 NONSILICON TECHNOLOGIES

A number of nonsilicon-based technologies have emerged in recent years. Probably the most famous of these is LIGA.[47,48] This technology is a move away from the trend to make a devices smaller and smaller. The strength of the LIGA process is the ability to make a high aspect ratio. The LIGA process was first developed in Germany. Since then a micromachining version, known as S-LIGA, has been developed.[49] This basic process can be seen in Fig. 1.19.

An alternative to LIGA is the EPON SU-8. This is a special resist developed by IBM that has been applied to structures such as cogs for the watch industry at EPFL in Lausanne, Switzerland. Figure 1.20 shows a fabricated structure.[50,51]

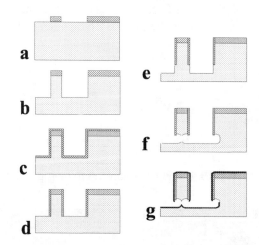

Figure 1.17 Basic sequence for the SCREAM process.

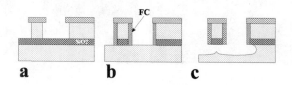

Figure 1.18 Basic process sequence for the Black silicon process.

MEMS/MOEMS Technology Capabilities and Trends

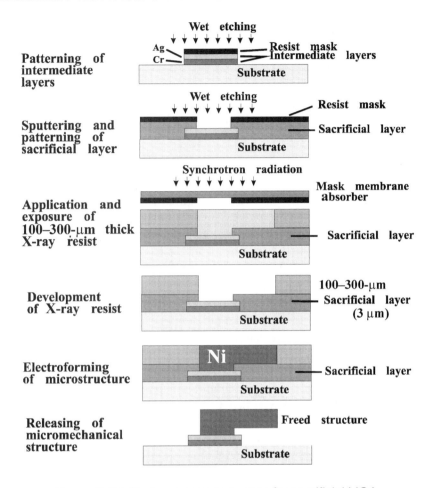

Figure 1.19 Basic process sequence for sacrificial LIGA.

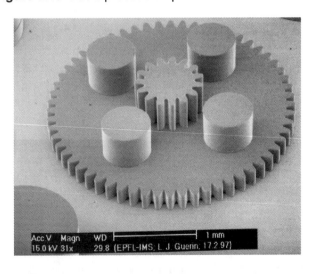

Figure 1.20 Cogs fabricated at EPFL using EPON SU-8 resist.[51]

1.4 INTEGRATION

Integration of MEMS devices refers to the fabrication of MEMS devices on the same substrate as the electronics. This requires the combination of two processes that have been optimized to give the required characteristics. Therefore any changes in the process should be made with care.

1.4.1 COMPATIBILITY ISSUES

If the micromachined structures are to be combined with the electronics on a single chip, the compatibility of the two processes (micromachining and electronics) must be considered. These considerations are quite different with different technologies. For bulk micromachining using anisotropic etchants, the compatibility problem is usually with the clean room. If KOH is used, contamination of the wafer surface limits the further processing, that can be performed. This can be avoided by using an alternative etchant, such as TMAH, or sodium hydroxide (NaOH), or by performing the etching at the end of the process. For surface micromachining, the main considerations are often the additional thermal budget of the depositions and annealing and masking during etching. The first of these problems can be avoided by either developing a deposition process that yields the desired mechanical properties without the need for high-temperature annealing [52–54] or by using layers already available in the standard process.[55,56] The second problems concerns the sacrificial etching. For example, if silicon dioxide is used as the sacrificial layer, the corresponding etchant is usually HF based. This creates problems for the aluminum used for metallization. Fortunately there are some solutions. One option is to use a metal layer that is not attacked by HF. An alternative solution is to use an etchant with a high selectivity of oxide etching over that of aluminum.[57–59]

Processes such as SCREAM and Black silicon do not suffer form these problems as long as a good selectivity of the plasma etch can be achieved. Processes such as LIGA and SU-8 can be built up on top of the substrate and are low-temperature processes.

1.5 MICROMACHINED DEVICES

Using a combination of standard and nonstandard processing techniques, a wide range of sensors and actuators have been fabricated. In the following sections examples of sensors and actuators will be given. However, with the rapidly expanding market for MEMS devices, this discussion is by no means complete.

1.5.1 PRESSURE SENSORS

Pressure sensors represent the largest market for silicon sensors and can be found over a wide range of applications. Particularly large markets are medical, automo-

Figure 1.21 A range of silicon pressure sensors fabricated by NovaSensors.[64]

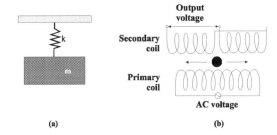

Figure 1.22 Two accelerometer structures: (a) mass and spring and (b) electromagnetic.

tive, and aeronautical. The applications for pressure sensors range from blood pressure to high pressures in diesel motors. With the introduction of micromachining techniques came a range of new sensor structures based on membranes. This enabled a reduction in size to such an extent that sensors could be fitted into a catheter and inserted into the body. A number of companies have entered this market.[60–63] Some pressure sensors fabricated by NovaSensors are given in Fig. 1.21.[64]

1.5.2 Inertial sensors

The first inertial sensors fabricated in silicon were accelerometers. Most of these devices were, and still are, based on a mass and a spring. Simple switch accelerometers can be made using a pendulum and a weight, or balls traveling in a coil. These principles can be seen in Fig. 1.22.

The first silicon devices on the market were based on bulk micromachining, using piezoresistive readout. Companies such as Silicon Microstructures and ICSensors have applied backside bulk micromachining to silicon vertical accelerometers.[65,66] A cross-section of the device fabricated by ICSensors is shown in Fig. 1.23.[66]

A feature of this device is the self-test electrode. This allows the sensors to be tested. When the device is activated, the self-test electrode will pull the mass upward electrostatically. The piezoresistors will then check that the mass moves as expected. Many modern cars that have airbags use a self-test each time the engine is started.

As with pressure sensors, a number of technologies have been used for the fabrication of accelerometers. Analog Devices developed a surface micromachined lateral accelerometer.[67] This device is integrated with the readout electronics in a complete system. Two photographs of one of these devices are given in Fig. 1.24.

The Analog Devices accelerometer uses polysilicon as a mechanical layer. A similar structure can be fabricated using the epitaxial layer as the mechanical layer, and an example of such a device can be seen in Fig. 1.25.[44]

The gyroscope is an important instrument for navigation. It was given its name by the French physicist Jean Foucault in the early 19th century and means "show-

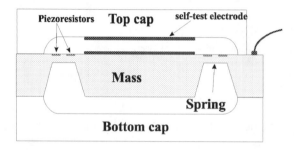

Figure 1.23 Cross-section of the accelerometer fabricated by ICSensors.[66]

Figure 1.24 Analog Devices surface micromachined accelerometer: (left) complete chip; (right) closeup of the sensing element.

ing turning" because Foucault used it to show the rotation of the earth. The gyroscope is based on interwoven spinning rings that can detect any rotation. Such a system is shown in Fig. 1.26.[68]

The introduction of micromachining to gyroscope manufacture is having an effect on size and price reduction similar to that found with accelerometers. One of the first of these micromachined devices used piezoelectric quartz in a tuning fork-type structure, such as that seen in Fig. 1.27.[69]

As can be seen in this figure, the rotation causes a vibration perpendicular to the input oscillation. Silicon gyroscopes have been developed using a range of techniques, including the tuning fork,[70] resonating plate,[71] and rotating rings.[72]

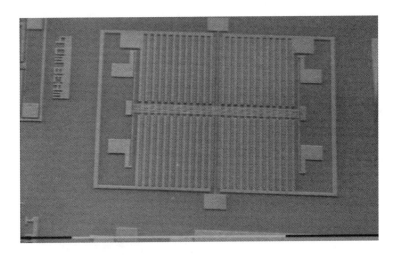

Figure 1.25 Accelerometer fabricated using epi-poly as the mechanical layer.

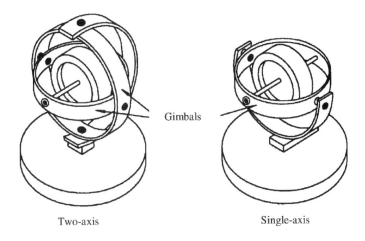

Two-axis   Single-axis

Figure 1.26 Basic gyroscope structure.

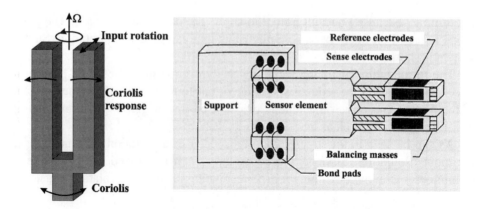

Figure 1.27 Principle and structure of a tuning fork-type gyroscope.

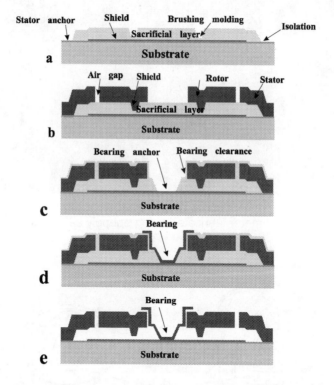

Figure 1.28 Basic process for the fabrication of micromotors.

1.5.3 MICROMOTORS

We often think of electric motors as large devices used for powering trains, washing machines, and vacuum cleaners, for example. Minimotors enabled us to make model railways that worked on electricity. These model trains were reduced in size

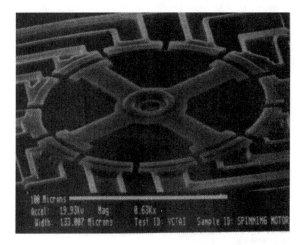

Figure 1.29 Electrostatic micromotor fabricated at Berkeley.[74]

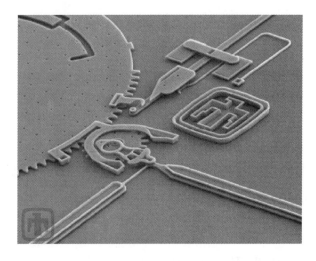

Figure 1.30 Locking system fabricated at Sandia National Labs.

over the years until we could barely see the mechanisms. In the 1980s the micromotor was born. This reduced the size of the motor from a few millimeters to hundreds of microns. Instead of being electromagnetic, these motors were electrostatic. In those early days, an attentive audience could be guaranteed for a video presentation on these motors.

The first polysilicon-based micromotors showed the potential of fabricating micromachined moving parts.[73,74] The devices required two polysilicon layers; the basic fabrication process is given in Fig. 1.28.

An example of one of these early devices is shown in Fig. 1.29. These devices showed how the technology had advanced, but at the time there were no applications to take up their development.

Figure 1.31 Closeup of the alignment clip.

In more recent years, Sandia National Laboratories has used these wheels to make micro safety locks using five polysilicon layers.[76] One of these devices is shown in Fig. 1.30. A closeup of the teeth of one of these devices is shown in Fig. 1.31.

The electrostatic devices have limited torque and therefore the applications will always remain limited. They have, however, found used in the field of optical choppers. As motors, they may never find true applications. However, technologies such as LIGA provide us with a compromise position. They are as small as surface micromachining, but are suitable for the integration of electromagnetic coils.

1.5.4 LATERAL RESONATORS/COMB DRIVES

Lateral resonators and comb drives make use of the flexibility of silicon and electrostatic application. An example of a resonator structure is shown in Fig. 1.32. Electrostatic attraction pulls the moveable plate from left to right. This basic idea has been applied to gyroscopes. It has also been used to drive rotors through a system similar to that of a steam engine.

1.5.5 MICROPUMPS

Pumping of fluids and gases can be traced back through the centuries. Pumps were used for applications such as lifting water out of wells or draining polders. The development of silicon micromachining has led to new micropumps capable of

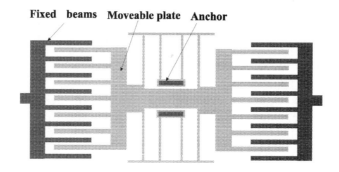

Figure 1.32 Lateral resonator structure.[77]

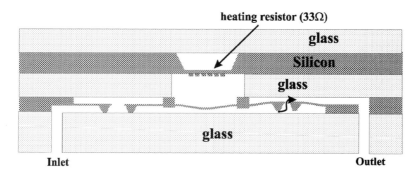

Figure 1.33 Cross-sectional view of a thermopneumatic pump.[78]

pumping rates in the range of nanoliters to milliliters per minute. Many macropumps use a rotating mechanism to force the fluid through the pump. This is much more difficult to achieve with micropumps. The pumping mechanism can use a number of techniques. Among these are electrostatic, piezoelectric, and thermal. These techniques operate by generating pressure differences in which the fluid is first sucked into a chamber and then forced out.

One thermal technique suitable for micropumps is the thermopneumatic technique. This pump consists of an additional chamber containing a heating resistor. This chamber may contain a gas or a gas–liquid mixture. Using a liquid such as trichloroethylene, sufficient pressure can be generated, with temperatures generated by the resistor of less than 100°C above ambient. The pumping pressure and rate are usually set off against each other to find the optimum characteristics. This is illustrated in Fig. 1.33.

Most pumps are fabricated using a number of wafers bonded together in order to achieve sufficient volumes. A further reduction in size will result in new fundamental problems, so that many fluids will simply not be able to pass through the pumps.

1.5.6 DIGITAL LIGHT MODULATORS

The deformation of membranes has been used for many years for projection displays. The Eidophor system used a thin oil film on a mirror. The membrane was scanned with an electric beam, which resulted in a pattern of charge that was then deformed using electrostatic attraction. A patent application for this technique was filed in November 1939. However, this technique could not be considered to be MEMS. An early MEMS version of this technique was the deformable mirror device.[79] The basic structure can be seen in Fig. 1.34.

In 1979 Texas Instruments produced a 16 × 16 array of membranes and this was increased to 128 × 128 in 1981.

Cantilever-type structures were made by Westinghouse in 1974.[80] These devices also used a scattering of the incoming light as shown in Fig. 1.35.

The digital micromirror device (DMD^{TM}), from Texas Instruments was first conceived in the late 1980s. The device is based on rotation about a diagonal axis [81] as shown in Fig. 1.36.

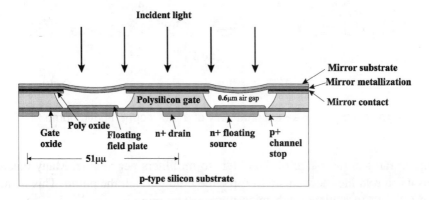

Figure 1.34 Deformable mirror device from Texas Instruments.[79]

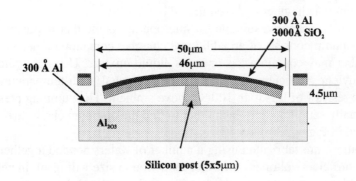

Figure 1.35 Mirror matrix element.

The device has seen a number of developments in which the area of the total chip covered by mirror has been increased. This high coverage can be seen in Fig. 1.37.[83]

These mirror arrays have been built into full-color projection systems as shown in Fig. 1.38. In this system, white light is split into three components and each color is given a pattern of light and dark using this chip. The three colors are then recombined to produce a full-color image.

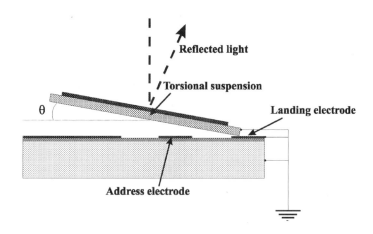

Figure 1.36 Basic deflection principle of a Texas Instruments-type device. Taken from Jaecklin et al.[82]

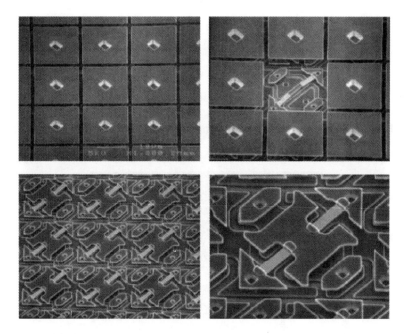

Figure 1.37 SEM photograph of a digital mirror array.

1.5.7 Microtip

Etching technologies have resulted in the ability to make extremely sharp tips. These tips have found a wide range of applications in displays,[85] scanning tunnelling microscopes (STMs),[86] atomic force microscopes (AFMs),[87] microvalves,[88] and mechanical sensors.[89] The microtip has also seen application in micromanipulation, where individual molecules can be arranged in a desired pattern.

The STM was invented by Gerd Binnig and Heinrich Rohrer, researchers at the IBM research labs in Zurich, Switzerland. Rohrer and Binnig worked through the 1970s and submitted their first patent on an STM in mid-1979. In 1982, they produced images of a silicon surface showing individual atoms. Microtips have also be used to move individual atoms. A famous example is the name IBM, which was written using 35 precisely placed atoms as in the STM picture in Fig. 1.39.[90]

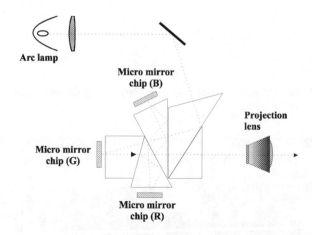

Figure 1.38 Projection system using three chips containing a micromirror array.[84]

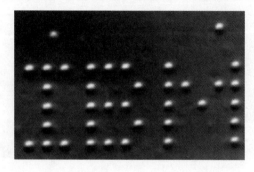

Figure 1.39 The name IBM written using individual atoms.[90]

MEMS/MOEMS TECHNOLOGY CAPABILITIES AND TRENDS

A number of etching technologies have been used to fabricate these tips. An example of a microtip fabricated by using underetching with reactive ion etching (RIE) followed by oxidation is shown in Fig. 1.40. The tip is on the order of 10 nm.

1.5.8 MICROGRIPPERS

Developments in surface micromachining have led to a range of microgrippers. These usually consist of electrostatic or thermal actuators to operate microtweezers. These may see future applications in microrobots and microassembly. Figure 1.41 shows a gripper fabricated by TIMA in France.[92]

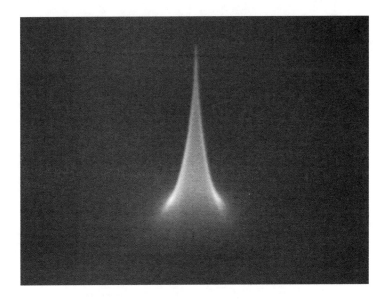

Figure 1.40 Example of a microtip.[88]

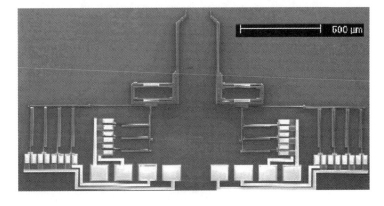

Figure 1.41 Microgripper fabricated by TIMA, France.

1.5.9 MICROCONVEYER BELT

A number of laboratories have developed microconveyer belts. These use tilting cantilevers or plates to transport objects placed on the chip. One system, developed at Tokyo University uses a bimetal effect to bend cantilevers.[92] An alternative developed at Delft University of Technology uses surface micromachining techniques to produce a tilting microplate.[93] A SEM of this system is shown in Fig. 1.42.

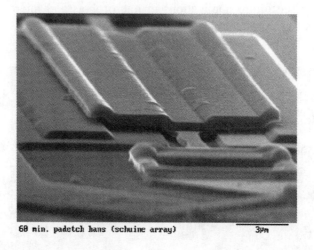

Figure 1.42 Microconveyer belt element fabricated using surface micromachining.[92]

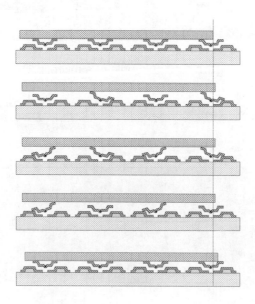

Figure 1.43 Basic action of the microconveyer belt.

The basic operation of this device uses these structures in pairs. These take turns picking up an object and moving it forward through a rotational movement. This basic mechanism is shown in Fig. 1.43. Although each of these components is extremely small, the high density enables large objects to be moved.

1.6 FUTURE TECHNOLOGIES

The technologies and devices described above are in a state of continual development. New devices and new applications are now becoming available with advances in the technology. A few examples of future trends are given below.

1.6.1 DEEP REACTIVE ION ETCHING

Deep reactive ion etching enables controlled etching deep into the wafer, giving full freedom to the designer since there are no crystal orientation considerations. Three companies that have been active in the development of etchers are STS, Alcatel, and Plasma Therm. Deep plasma etching can be achieved through low temperature or high plasma density. It is then possible to etch through the wafer using an oxide mask. As more research institutes obtain these etchers, an increasing number of structures are being developed.

One excellent example of a waveguide fabricated in Delft is shown in Fig. 1.44.[94] The etched strips make a channel for the wave to pass through. This channel turns the wave to the right and finally to a receiver.

Figure 1.44 Waveguide fabricated using deep RIE techniques.[94]

The plasma etching technique is being developed further and is leading to an increasing number of devices, such as lateral accelerometers and gyroscopes with straight side walls through the full thickness of the wafer.[95,96]

1.6.2 Stereolithography

Microstereolithography is a new microfabrication process that enables the fabrication of 3D objects that are complex in shape, such as the spring structure shown in Fig. 1.45.[97,98] In this process, the projections of an "active mask" induce a space-resolved photopolymerization of a liquid resin, allowing a complete layer of the object to be fabricated with only one irradiation. Layers of resin are irradiated in a pattern but are not developed, and are followed by further layer depositions and irradiation. At the end of the process, the total structure is developed. The superimposition of these multilayers (up to 1000 or more) results in the manufacture of real 3D objects.

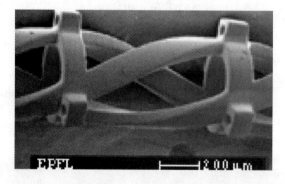

Figure 1.45 Example of a 3D structure made using stereolithography.[97]

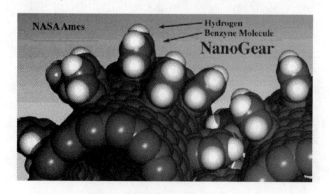

Figure 1.46 A nanogear.

1.6.3 Nanotechnology

Micromachining has seen a development from millimeter dimensions down to micron dimensions. The question remaining is how small we can go. Present-day micromachining generally involves a combination of depositions and etching. In nanotechnology, we are moving into submicron dimensions and eventually to molecular dimensions. Probably the inspiration for molecular nanotechnology came from the 1959 talk by Richard Feynman entitled "There's plenty of room at the bottom." The future is likely to see machining moving into atomic dimensions. One example of a proposed nanometer is shown in Fig. 1.46.[99]

1.7 Future Applications

As the reliability of MEMS devices improves along with industry's faith in the technology, these devices are finding increasing numbers of applications. Silicon sensors, including micromachined devices, are playing an increasingly important role in space exploration, where a reduction in size and weight is important.

New miniature satellites 10 cm across will make use of MEMS-based navigation and guidance systems.[100] MEMS devices will also travel deeper into space. The New Millennium Progam (NMP) is planning to use MEMS devices for a Mars microprobe.[101] This probe is illustrated in Fig. 1.47.

Perhaps an unexpected application for MEMS devices is in the field of spacecraft propulsion. As spacecraft are reduced in size, the propulsion system also needs to be miniaturized. Deep reactive ion etching allows chambers to be etched using the full thickness of the silicon wafer. This technique has the advantage of maintaining full freedom of the form of the side walls. This is particularly important in order to be able to optimize the thrust. The basic structure of the device is shown in Fig. 1.48.[102] The SEM photograph in Fig. 1.49 shows the advantage of deep RIE in the fabrication of straight side walls.

Optical MEMS are also finding their way into space applications. For example, the use of micromachined adaptive optics allows distortions to be corrected using compact and light devices. Figure 1.50 is a cross-section of a micromirror.

This mirror, which is based on an aluminum or gold-coated silicon nitride membrane, can be deflected electrostatically. The form of the mirror can therefore be adjusted to correct the optical wavefront. Figure 1.51 is a photograph of a 25-mm membrane.[104]

It is not only in space that MEMS devices are finding increasing applications. The modern automobile has a number of MEMS devices to increase safety and comfort and to reduce pollution and increase efficiency. Accelerometers are used to detect a collision and inflate an airbag. Tire pressure is constantly monitored and the brakes are adjusted on each wheel to prevent skidding. Initially, nonsilicon sensors were implemented, but the silicon solution is increasingly being used.[105]

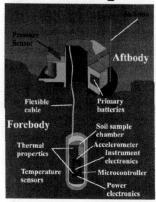

Figure 1.47 Structure of the Mars Microprobe.[101]

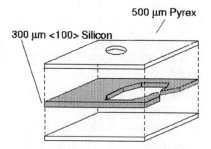

Figure 1.48 Exploded view of the microthruster.

Figure 1.49 SEM photograph of a fabricated thruster.[102]

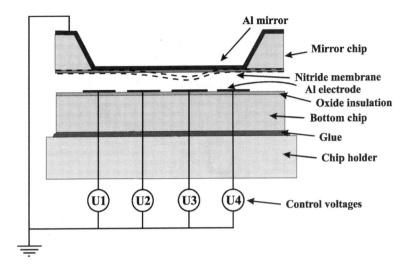

Figure 1.50 Cross-section of silicon-based adaptive mirror.[103]

Figure 1.51 Photograph of a 25-mm membrane. The left-hand membrane has been broken to reveal the under electrodes. Reproduced with kind permission of Gleb Vdovin.

The inclusion of sensors in automobiles is expected to continue to increase. An idea of the car of the future is given in Fig. 1.52, which shows the "advanced safety vehicle" being developed in Japan.[106]

Medicine is also awakening to the possibilities of MEMS devices, and a number are already in use. Silicon is an ideal material for measuring potential in the brain. Microprobes can be fabricated using bulk micromachining technology. Work performed at the University of Michigan in the 1980s used a p^+ etch stop to form microprobes. Recording sites could be incorporated on the tip for measuring biopotentials.[107] The basic structure can be seen in Fig. 1.53. These probes have been through a number of advances and refinements, and an example of a modern device is shown in Fig. 1.54.

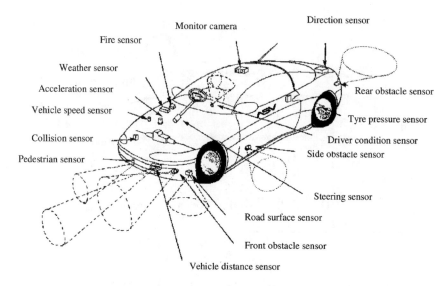

Figure 1.52 Advanced safety vehicle.[106]

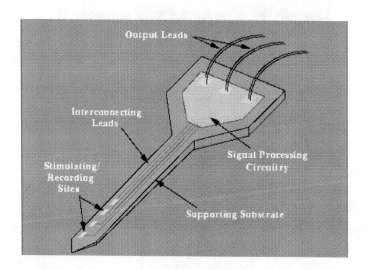

Figure 1.53 Silicon probe for biopotential measurements.

MEMS devices can also be used for external devices, such as the smart inhaler from Neuchâtel, Switzerland, which controls the number of droplets inhaled by a patient. The device can then be adjusted to optimize the droplet size. A cross-section of the device is shown in Fig. 1.55.[108]

Pressure sensors for medical applications have been in use for a number of years. Some of these are external, but many are now being used in the body itself. Smart sensors for catheters were proposed in 1990.[109] A system was presented for 16 or more sensors fabricated in CMOS technology. Parameters of interest include

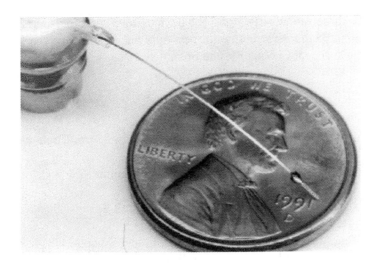

Figure 1.54 Photograph of the probe on an American penny.

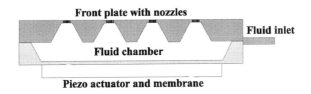

Figure 1.55 Cross-section of a smart inhaler from Neuchâtel, Switzerland.

temperature, pressure, electrocardiography, and oxygen or carbon dioxide content. Sensors in catheters will allow doctors to measure locally in some cases.

The future will see an increasing use of implanted systems. A number of implantable systems have already been developed, for both short- and long term use.[110] Pacemakers are incorporating increasing numbers of sensors to measure the activity of the user and adjust the rate of pulses appropriately.[111] Implantable micropumps will allow regular microdoses of medicine to be injected into the body.[112]

The future will see the use of nanorobots injected into the body. These could work in the blood system of the patient, which means that the total dimensions should be kept under about 3 µm. A number of structures have been proposed, but working devices have yet to be made.

1.8 CONCLUSIONS

After a number of years when industry mistrusted MEMS devices, there has been an explosion of interest. Micromachining technology offers a wide range of possibilities. Many of these technologies are based on deposition and etching techniques to make three-dimensional structures. The devices have been considerably reduced in size, bringing them into a number of new applications. Recent years have seen developments in sensors, actuators, and optical systems.

REFERENCES

1. J. Bardeen and W. H. Brattain, "The Transistor, A Semi-Conductor Triode," *Phys. Rev.* 74(2), 230–231 (1947).
2. G. K. Teal and J. B. Little, "Growth of Germanium Single Crystals," *Phys. Rev.* 78(5), 647 (1950).
3. A History of Engineering and Science in the Bell System, Physical Sciences (1925–1980), p. 582.
4. G. K. Teal, "Single Crystals of Germanium and Silicon—Basic to the Transistor and Integrated Circuit," *IEEE Trans. Electron Devices* ED-23(7), 621–639 (1976).
5. K. E. Petersen, "Silicon as a Mechanical Material," *Proc. IEEE* 70, 420–457 (1982).
6. O. N. Tufte and G. D. Long, "Silicon Diffused Element Piezoresistive Diaphragm," *J. Appl. Phys.* 33, 3322 (1962).
7. K. E. Bean, "Anisotropic Etching of Silicon," *IEEE Trans. Electron Devices* ED-25, 1185–1193 (1978).
8. J. C. Greenwood, "Etched Silicon Vibrating Sensor," *J. Phys. E. Sci. Instrum.* 17, 650–652 (1984).
9. H. C. Nathanson and R. A. Wickstrom, "A Resonant-Gate Silicon Surface Transistor with High-Q Band Pass Properties," *Appl. Phys. Lett.* 7, 84 (1965).
10. R. T. Howe and R. S. Muller, "Polycrystalline and Amorphous Silicon MicroMechanical Beams: Annealing and Mechanical Properties," *Sensors Actuators* 4, 447–454 (1983).
11. L.-S. Fan, Y.-C. Tai and R. S. Muller, "Pin Joints, Gears, Springs, Cranks and other Novel MicroMechanical Structures," in *Proc. Transducers 87*, pp. 849–852 (1987).
12. N. Taniguchi, *Annals of the CIRP* 32, 573 (1983).
13. W. Thompson, "On the Electrodynamics Qualities of Metals," *Proc. Roy. Soc. Lond.* 546–555 (1857).
14. C. S. Smith, "Piezoresistive Effect in Germanium and Silicon," *Phys. Rev.* 96, 42–49 (1954).
15. J. Bryzek, "Impact of MEMS Technology on Society," *Sensors Actuators* A56, 1–9 (1996).
16. H. C. Nathanson, W. E. Newell, R. A. Wickstrom and J. R. Davis, Jr., "The Resonant Gate Transistor," *IEEE Trans. Electron Devices* ED-14, 117–133 (1967).
17. D. D. L. Wijngaards, M. Bartek and R. F. Wolffenbuttel, "IC Process Compatible Metal Film Postprocessing Module," in *Proc. Eurosensors 97*, pp. 587–590 (1997).

18. R. Gerhard-Multhaupt, G. M. Sessler, J. E. West, K. Holdik, M. Haardt, and W. Eisenmenger, "Investigation of Piezoelectricity Distributions in Poly(vinylidene Fluoride) by Means of Quartz- or Laser-Generated Pressure Pulses," *J. Appl. Phys.* 55, 2769–2775 (1984).
19. W. C. Albert, "Vibrating Quartz Crystal Beam Accelerometer," in *Proc. ISA 28th International Symposium* 28(1), 33–44 (1982).
20. M. J. Vellekoop, C. C. G. Visser, P. M. Sarro and A. Venema, "Compatibility of Zinc Oxide with Silicon IC Processing," *Sensors Actuators* A21–23, 1027–1030 (1990).
21. S. Shuichi and M. Esashi, "Bonding and Assembling Methods for Realising a μTAS," in *Proc. Micro Total Analysis Systems*, pp. 165–179 (1994).
22. L. A. Field and R. S. Muller, "Fusing Silicon Wafers with Low Melting Temperature Glass," *Sensors Actuators* A21–23, 935–938 (1990).
23. C. D. Fung, P. W. Cheng, W. H. Ko and D. G. Flemming, *Micromachining and Micropackaging of Transducers*, pp. 41–61. Elsevier, Amsterdam (1985).
24. T. A. Knecht, "Bonding Techniques for Solid State Pressure Sensors," in *Proc. Transducers 87*, pp. 95–98 (1987).
25. R. F. Wolffenbuttel and K. D. Wise, "Low-Temperature Wafer-to-Wafer Bonding Using Gold at Eutectic Temperature," *Sensors Actuators* A43, 223–229 (1994).
26. J. Jiao, D. Lu, B. Xiong and W. Wang, "Low Temperature SDB and Interface Behaviours," in *Proc. MEMS 95*, pp. 394–397 (1995).
27. A. Berthold, B. Jakoby and M. J. Vellekoop, "Wafer-to-Wafer Fusion Bonding of Oxidized Silicon to Silicon at Low Temperature," *Sensors Actuators* A68(1–3), 410–413 (1998).
28. M. Harz, "Anodic Bonding for the 3rd Dimension," *J. Micromech. Microeng.* 2, 161–163 (1992).
29. T. Rogers, "Considerations for Anodic Bonding for Capacitive Type Silicon/Glass Sensor Fabrication," *J. Micromech. Microeng.* 2, 164–166 (1992).
30. T. A. Anthony, "Dielectric Isolation of Silicon by Anodic Bonding," *J. Appl. Phys.* 58, 1240–1247 (1985).
31. A. Berthold, L. Nicola, P. M. Sarro and M. J. Vellekoop, "A Novel Technological Process for Glass-to-Glass Anodic Bonding," in *Proc. Transducers '99*, pp. 1324–1327 (1999).
32. A. Bohg, "Ethylene-Diamine-Pyrocatechol-Water Mixture Shows Etching Anomaly in Boron Doped Silicon," *J. Electrochem. Soc.* 118, 401 (1970).
33. J. B. Price, "Anisotropic Etching of Silicon with KOH-H_2O-Isopropyl Alcohol," in *Semiconductor Silicon*, H. R. Huff and R. R. Burgess, eds., p. 339. Electrochemical Society, Pennington, NJ (1973).
34. O. Tabata, R. Asahi, H. Funabashi, K. Shimaoka and S. Sugiyama, "Anisotropic Etching of Silicon in TMAH Solutions," *Sensors Actuators* A34, 51–57 (1992).
35. H. Seidel, L. Csepregi, A. Heuberger and H. Baumgärtel, "Anisotropic Etching of Crystalline Silicon in Alkaline Solutions II. Influence of Dopants," *J. Electrochem. Soc.* 137, 3626–3632 (1990).

36. B. Kloek, S. D. Collins, N. F. de Rooij and R. L. Smith, "Study of Electrochemical Etch-Stop for High Precision Thickness Control of Silicon Membranes," *IEEE Trans. Electron Devices* 36, 663–669 (1989).
37. E. Peeters, D. Lapadatu, R. Puers and W. Sansen, "PHET, An Electrodeless Photovoltaic Electrochemical Etchstop Technique," *J. MEMS* 3, 113–123 (1994).
38. P. J. French, M. Nagao and M. Esashi, "Electrochemical Etch-Stop in TMAH without Externally Applied Bias," *Sensors Actuators* A56, 279–280 (1996).
39. P. J. French, "Development of Surface Micromachining Techniques Compatible with On-Chip Electronics," *J. Micromech. Microeng.* 6, 197–211 (1996).
40. Y. X. Li, P. J. French, P. M. Sarro and R. F. Wolffenbuttel, "Fabrication of a Single Crystalline Silicon Capacitive Lateral Accelerometer Using Micromachining Based on Single Step Plasma Etching," in *Proc. MEMS 95*, pp. 398–403 (1995).
41. T. Bell, P. T. J. Gennissen, D. de Munter and M. Kuhl, "Porous Silicon as a Sacrificial Layer," *J. Micromech. Microeng.* 6, 361–369 (1996).
42. M. Bartek, P. T. J. Gennissen, P. J. French, P. M. Sarro and R. F. Wolffenbuttel, "Study of Selective and Non-Selective Deposition of Single and Polycrystalline Silicon Layers in Epitaxial Reactor," in *Proc. Transducers 97*, pp. 1403–1406 (1997).
43. P. T. J. Gennissen, M. Bartek, P. M. Sarro and P. J. French, "Bipolar Compatible Epitaxial Poly for Smart Sensor-Stress Minimization and Applications," in *Proc. Eurosensors 1996*, pp. 187–190 (1996).
44. P. T. J. Gennissen and P. J. French, "Applications of Bipolar Compatible Epitaxial Polysilicon," in *Proc. SPIE Micromachining and Microfabrication Conference*, pp. 59–65 (1996).
45. K. A. Shaw and N. C. MacDonald, "Integrating SCREAM Micromachined Devices with Integrated Circuits," in *Proc. IEEE MEMS*, pp. 44–48 (1996).
46. M. de Boer, H. Jansen and M. Elwenspoek, "The Black Silicon Method V: A Study of the Fabricating of Moveable Structures for Micro Electromechanical Systems," in *Proc. Transducers 95*, pp. 565–568 (1995).
47. E. W. Becker, W. Erfeld, P. Hagmann, A. Maner and D. Munchmeyer, "Fabrication of Microstructures with High Aspect Ration and Great Heights by Synchrotron Radiation, Lithography, Galvanoforming and Plastic Molding (LIGA Process)," *Microelectron. Eng.* 4, 35–56 (1986).
48. H. Guckel, K. J. Skrobis, T. R. Christenson, J. Klein, S. Han, B. Choi and E. J. Lovell, "Fabrication of Assembled Micromachined Components via Deep X-Ray Lithography," in *Proc. IEEE MEMS 91*, pp. 74–79 (1991).
49. H. Guckel, K. J. Skrobis, T. R. Christenson, J. Klein, S. Han, B. Choi, E. G. Lovell and T. W. Chapman, "Fabrication and Testing of the Planar Magnetic Micromotor," *J. Micromech. Microeng.* 1, 135–138 (1991).
50. H. Lorenz, M. Despont, N. Fahrni, N. LaBianca, P. Renaud and P. Vettiger, "EPON SU-8: A Low-Cost Negative Resist for MEMS," in *Proc. MME '96*, (1996).

51. M. Despont, H. Lorenz, N. Fahrni, J. Brugger, P. Renaud and P. Vettiger, "High-Aspect-Ratio, Ultrathick, Negative-Tone Near UV Photoresist for MEMS Applications," in *Proc. IEEE MEMS 97* (1997).
52. H. Guckel, J. J. Sniegowski, T. R. Christenson, S. Mohney and T. F. Kelly, "Fabrication of Micromechanical Devices from Polysilicon Films with Smooth Surfaces," *Sensors Actuators* 20, 117–122 (1990).
53. P. J. French, B. P. Van Drieënhuizen, D. Poenar, J. F. L. Goosen, P. M. Sarro and R. F. Wolffenbuttel, "Low-Stress Polysilicon Process Compatible with Standard Device Processing," in *Sensors VI: Technology, Systems and Applications*, pp. 129–133. Manchester, UK (1993).
54. P. J. French, B. P. Van Drieënhuizen, D. Poenar, J. F. L. Goosen, R. Mallée, P. M. Sarro and R. F. Wolffenbuttel, "The Development of a Low-Stress Polysilicon Process Compatible with Standard Device Processing," *IEEE JMEMS* 5, 187–196 (1996).
55. G. K. Fedder, S. Santhanan, M. L. Read, S. C. Eagle, D. F. Guillou, M. S.-C. Lu and L. R. Carley, "Laminated High-Aspect Ratio Microstructures in a Conventional CMOS Process," in *Proc. MEMS 96*, pp. 13–18 (1996).
56. O. Paul and H. Baltes, "Novel Fully CMOS-Compatible Vacuum Sensor," *Sensors Actuators* A46–47, 143–146 (1995).
57. V. P. Jaecklin, C. Linder, N. F. de Rooij, J.-M. Moret and R. Vuilleumier, "Line Addressable Torsional Micromirrors for Light Modulator Arrays," *Sensors Actuators* A41–42, 324–329 (1994).
58. J. Bühler, J. Funk, F.-P. Steiner, P. M. Sarro and H. Baltes, "Double Pass Metallization for CMOS Aluminium Actuators," in *Proc. Transducers 95*, Vol. 2, pp. 360–363 (1995).
59. P. T. J. Gennissen and P. J. French, "Sacrificial Oxide Etching Compatible with Aluminium Metallization," in *Proc. Transducers 97*, pp. 225–228 (1997).
60. J. R. Mallon, Jr., F. Pourahmadi, K. Petersenm P. Barth, T. Vermeulen and J. Bryzek, "Low-Pressure Sensors Employing Bossed Diaphragms and Precision Etch-Stopping," *Sensors Actuators* A21–23, 89–95 (1990).
61. J. Knutti, "Manufacturing of Pressure Sensors, Accelerometers and other Microstructures," paper presented at *International Workshop on Microsystem Technologies (MST/MEMS)* in the USA and Canada (1995).
62. http://www.junctiontech.com/iegg1.html
63. http://www.melexis.com/
64. http://www.novasensor.com/tour/tour.html
65. T. Tschan, W. Urbanek, M. Lim, H. Allen and J. Knutti, "A High Performance, Low Cost Capacitive Accelerometer Trimmed Using Nonviolatile Potentiometers (EEPOT's)," in *Proc. 95*, pp. 601–603 (1995).
66. H. V. Allen, S. C. Terry and D. W. de Bruin, "Accelerometer Systems with Self-Testable Features," *Sensors Actuators* A20, 153–161 (1989).

67. W. Kuehnel and S. J. Sherman, "A Surface Micromachined Silicon Accelerometer with On-Chip Detection Circuitry," *Sensors Actuators* A45, 7–16 (1994).
68. A. Lawrence, *Modern Inertial Technology*. Springer-Verlag (1992).
69. J. Söderkvist, "Micromechanical Gyroscopes," *Sensors Actuators* A43, 65–71 (1994).
70. M. Weinberg, J. Bernstein, S. Cho, A. T. King, A. Koourepenis, P. Ward and J. Sohn, "A Micromachined Comb Drive Tuning Fork for Commercial Applications," in *Proc. Sensor Expo*, pp. 187–193 (1994).
71. Y. Mochida, M. Tamura and K. Ohwada, "A Micromachined Vibrating Rate Gyroscope with Independent Beams for the Drive and Detection Modes," in *Proc. MEMS 99*, pp. 618–623 (1999).
72. F. Ayazi and K. Najafi, "Design and Fabrication of a High-Performance Polysilicon Vibrating Ring Gyroscope," in *Proc. MEMS 98*, pp. 621–626 (1998).
73. L. S. Fan, Y. C. Tai and R. S. Muller, "IC-Processed Electrostatic Micromotors," in *Proc. IEEE Int. Electron Devices Meeting*, pp. 666–669 (1988).
74. M. Mehregany, S. F. Bart, L. S. Tavrow, J. H. lnags, S. D. Senturia and M. F. Schlecht, "A Study of Three Microfabricated Variable Capacitance Motors," in *Proc. Transducers 89*, pp. 106–107 (1989).
75. Y.-C. Tai and R. S. Muller, "Frictional Study of IC-Processed Micromotors," in *Proc. Transducers 89*, pp. 108–109 (1989).
76. J. J. Sniegowski and E. J. Garcia, "Surface Micromachined Gear Trains by an On-Chip Electrostatic Microengine," *IEEE Electron Dev. Lett.* 17, 366 (1996).
77. C. T.-C. Nguyen and R. T. Howe, "CMOS Micromechanical Resonator Oscillator," *Tech. Digest of IEDM*, pp. 199–202 (1993).
78. H. T. G. van Lintel, F. C. M. van de Pol and S. Bouwstra, "A Piezoelectric Micropump Based on Micromachining of Silicon," *Sensors Actuators* 15, 153–167 (1988).
79. L. J. Hornbeck, "128 × 128 Deformable Mirror Device," *Trans. IEEE Electron Devices* ED-30, 539–545 (1980).
80. R. N. Thomas, J. Guldberg, H. C. D. Nathanson and P. R. Malmberg, "The Mirror-Matrix Tube: A Novel Light Valve for Projection Displays," *Trans. IEEE Electron Devices* ED-30, 765–775 (1983).
81. P. F. van Kessel, L. J. Hornbeck, R. E. Meier and M. R. Douglass, "A MEMS-Based Projection Display," *Proc. IEEE* 86, 1687–1704 (1998).
82. V. P. Jaeklin, C. Linder, N. F. de Rooij, J.-M. Moret and R. Vuillermier, "Line Addressable Torsional Micromotors for Light Modulator Arrays," *Sensors Acuators* A41–42, 324–329 (1994).
83. L. J. Hornbeck, "Digital Light Processing: A New MEMS-Based Display Technology," in *Technical Digest of the 14th Sensor Symposium*, pp. 297–304 (1996).

84. L. J. Hornbeck, "Digital Light Processing for Projection Displays: A Progress Report," in *Proc. Euro Display 96*, pp. 67–71 (1996).
85. M. Tanaka, K. Takayama, A. Azeta, K. Yano and T. Kishino, "A New Structure for Driving System for Full-Color FEDs," in *Proc. SID 97*, pp. 47–51 (1997).
86. G. Binnig, H. Rohrer, C. Gerber and E. Weibel, *Phys. Rev. Lett.* 50, 120 (1983).
87. "AFM Fabricates A Tiny Transistor," *Science* 266(Oct.), 543 (1994).
88. J. A. Foerster, "Integrated Micro Vacuum Tubes in Silicon," Ph.D. Thesis, Delft University of Technology, Netherlands, 1996.
89. A. Hariz, H. G. Kim, M. R. Haskard and I. J. Chung, "Field-Emitter Cathode for Acceleration Sensors," *J. Micromech. Microeng.* 5, 282–288 (1995).
90. http://ww.foresight.org/UTF/Unbound_LBW/chapt_4.html
91. http://tima-cmp.imag.fr/tima/mcs/gripper.html
92. M. Ataka, A. Omodaka and H. Fujita, "A Biomimetric Micro Motion System," in *Proc. Transducers 93*, pp. 38–41 (1993).
93. J. F. L. Goosen and R. F. Wolffenbuttel, "Object Positioning Using a Surface Micromachining Distributed System," in *Proc. Transducers 95*, pp. 396–399 (1995).
94. T. Zijlstra, E. van Der Drift, M. J. A. de Dood, E. Snoeks and A. Polman, "Fabrication of Two-Dimensional Photonic Crystal Waveguides for 1.5 µm in Silicon by Deep Anisotropic Dry Etching," *J. Vac. Sci. Technol. B* (1999).
95. B. P. van Drieënhuizen, "Advances in MEMS Using SFB and DRIE Technology," in *Proc. SPIE Conference on Micromachined Devices and Components*, Vol. 3876, pp. 64–73 (1999).
96. S. S. Baek, Y. S. Oh, B. J. Ha, S. D. An, B. H. An, H. Song and C. M. Song, "A Symmetrical Z-Axis Gyroscope with a High Aspect Ratio Using Simple and New Process," in *Proc. IEEE MEMS 99*, pp. 612–617 (1999).
97. A. Bertsch, H. Lorenz and P. Renaud, "Combining Microstereolithography and Thick Resist UV Lithography for 3D Microfabrication," in *Proc. MEMS 98 Workshop*, pp. 18–23 (1998).
98. A. Bertsch, H. Lorenz and P. Renaud, "3D Microfabrication by Combining Microstereolithography and Thick Resist UV Lithography," *Sensors Actuators* A73, 14–23 (1999).
99. http://nanozine.com/nanogear.htm#motor
100. W. Tang, "Micromachines Blast off into Space," *Vacuum Solutions*, pp. 26–29 (1998).
101. http://www.fortunecity.com/roswell/avebury/97/nmp.htm
102. R. L. Bayt, K. S. Breuer and A. A. Ayon, "DRIE-Fabricated Nozzles for Generating Supersonic Flows in Micropropulsion Systems," in *Proc. Solid-State Sensor and Actuator Workshop* (1998).
103. G. Vdovin, P. M. Sarro and S. Middelhoek, "Technology and Applications of Micromachined Adaptive Mirrors," *J. Micromech. Microeng.* 9, R8–R20 (1999).

104. W. Holota, G. Vdovin, N. Collings, Z. Sodnik, R. Sesselmann and S. Manhart, "The Need of AO in Future Space Optical Instruments," in *Astronomy with Adaptive Optics, Proc. of the ESO*.
105. L. Halbo and P. Ochlckers, "'Electronics Components, Packaging and Production," in Chapter 9.
106. P. J. French, T. Mihara, H. Kaneko and T. Kita, "Smart Sensors in Automobiles," in *Silicon Sensors and Circuits: On-Chip Compatibility*, R. F. Wolffenbuttel, ed. Chapman & Hall, London (1996).
107. K. Najafi, J. Ji and K. D. Wise, "Scaling Limitations of Silicon Multichannel Recording Probes," *IEEE Trans. Biomed. Eng.* 37, 1–11 (1990).
108. http://www-samlab.unine.ch/projects/Smart_Inhaler/Smart_Inhaler.htm
109. Y. Manolo, J. Eichholz, M. Kandler, N. Kordas, A. Langerbein and W. Mokwa, "Smart Silicon System for Multisensor Catheter," in *Microsystems Technology*, H. Reichl, ed., pp. 710–715 (1990).
110. P. Dario, M. C. Carrozza, B. Allotta and E. Guglielmelli, "Micromechtronics in Medicine," *IEEE/ASME Trans. Mechtronics* 1, 137–148 (1996).
111. http://www.medtronic.com/brady/clinician/medtronicpacing/400intro.html
112. Dmaillefer, H. van Lintel, G. Rey-Mermet and R. Hirschi, "An High-Performance Silicon Micropump for an Implantable Drug Delivery System," in *Proc. MEMS 99*, pp. 541–546 (1999).

CHAPTER 2

OPERATION AND DESIGN OF MEMS/MOEMS DEVICES

Douglas Sparks
Delco Electronics

Thomas Bifano
Manufacturing Engineering, Boston University

Dhiraj Malkani
Lucent Technologies

CONTENTS

2.1 Operation of MEMS/MOEMS Devices / 49
    2.1.1 Bimetallic and thermal expansion actuators / 49
    2.1.2 Chemical and biological / 50
    2.1.3 Electrostatic / 51
    2.1.4 Magnetic sensors and actuation / 53
    2.1.5 Optical / 54
    2.1.6 Piezoelectric and acoustic / 56
    2.1.7 Piezoresistance / 57
    2.1.8 Resonant devices / 57
    2.1.9 Shape memory alloys / 59
    2.1.10 Tunneling / 59

2.2 The Design of MEMS/MOEMS / 60
    2.2.1 System and process modeling / 61
    2.2.2 Process design / 62
    2.2.3 Design rules / 63
    2.2.4 Scaling of MEMS devices / 64
    2.2.5 MEMS electronic design issues / 65
    2.2.6 Self-test design / 66
    2.2.7 Integration of MEMS/MOEMS devices and smart transducers / 66
    2.2.8 MEMS packaging and environmental exposure issues / 68
    2.2.9 Manufacturing-related design issues for high-volume production / 73
    2.2.10 Design reviews and DFMEA / 74

2.3 MEMS Design Examples / 75
    2.3.1 The design of pressure sensors / 75
    2.3.2 The design of accelerometers / 81

2.3.3 The design of resonant microsensors / 83
2.3.4 MOEMS design case study: deformable mirrors / 87

2.4 Summary / 101

References / 102

2.1 OPERATION OF MEMS/MOEMS DEVICES

Transduction, defined as conversion from one form of energy to another, is the essential purpose of many microelectromechanical/MEMS optical (MEMS/MOEMS devices). In this context, a transducer is a device that is actuated by energy of one form and supplies energy of another form. An example of a common MEMS transducer is the piezoresistive pressure sensor, in which variations in environmental pressure (fluidic energy) are transformed into the deflection of a micromachined membrane (mechanical energy). Using embedded resistors in the membrane, and an externally supplied current or voltage, a bridge circuit allows electronic sensing of a signal proportional to strain (electrical energy). Transducers encompass both actuators and sensors. More often than not, MEMS sensors convert a mechanical or optical signal into an electrical one. MEMS actuators do the converse, using electrical input to affect mechanical or optical devices, altering the environment. Although the number of phenomena and applications in the MEMS arena are vast, the initial sections of this chapter will review these most commonly used in MEMS/MOEMS. After this review of operational basics, the design of micromachines using these effects will be covered.

2.1.1 BIMETALLIC AND THERMAL EXPANSION ACTUATORS

When two thin layers of materials with different coefficients of thermal expansion (CTE) are bonded together, any change in temperature results in the generation of a bending force. This principle has been used for many years in thermostats and has been applied to MEMS as well. To generate the best temperature sensitivity or actuation force, alloys or materials with widely different thermal expansion coefficients are sought. In silicon integrated devices, existing materials such as aluminum and silicon are often employed.[1]

Bimetallic actuators are easy to produce in silicon-based microsystems since a variety of materials are already used in thin-film form in typical silicon processes. Unfortunately, the forces generated with silicon-compatible bimetallic films are often low for a given volume of material. Also, the only shape modification that can be obtained is flexural. High currents are also needed to generate a useful temperature change and hence motion for sensing or actuation.

In a thermal expansion device, an applied voltage forces a current through two legs of a silicon actuator. Electrical resistance is larger in the narrow leg than in the wide leg. The current through each leg is equivalent, but more electrical power ($\propto I^2 R$) is dissipated in the narrow leg, and it becomes hotter than the wider leg. Thermal expansion in each leg is directly proportional to the change in temperature, so the narrow leg expands more than the wide leg. Since the legs are anchored at one end and coupled at the other, this differential expansion results in mechanical bending of the structure. Actuation is limited by mechanical buckling in the narrow leg and by thermal damage to the structure. Nevertheless, tens of micrometers of in-plane actuation are achievable with this type of actuator.

The expansion of liquids or gases can be employed for actuation. A commercially available micromachined valve uses bulk thermal expansion of a liquid-filled chamber for actuation of the device. As the liquid expands through electrical resistive heating, it distorts the chamber walls, opening a valve port.

2.1.2 Chemical and Biological

A variety of microsystems have been designed for chemical and biological applications. Many of the underlying operational principles are covered in other sections of this chapter. This section will cover some of the microtechnologies used in these two fields.

Chemical microsystems extract data or interact with surrounding fluids. The miniaturization of chemical transducers has several advantages over traditional devices, including small sample volumes, high handling speed, fast response, and cost-effective multisampling. Thermal conductivity, density, and viscosity sensors have been produced in silicon and micromachined form.[2–4] Devices based on optical and electrical conductivity and capacitive measurements have been utilized.[5] Optical sensors include photodiodes, avalanche photodiodes, and optical filters. Surface acoustic wave (SAWs) devices have been used to measure the density and viscosity of fluids. These microsensors have been fabricated from silicon, quartz, and through the use of deposited ZnO, lead-zirconate-titanate (PZT), and AlN piezoelectric films. Resistive and capacitive chemical sensors have been made using deposited metal oxide films and cured polyimide. Galvanic cells have been fabricated from silicon/glass structures.[3]

Biological microsystems began with blood pressure monitoring, which is one of the highest volume applications for biological MEMS. Capacitive and piezoresistive pressure sensors have been used to monitor blood, urine, and intraocular pressure.[6] Blood flow and conductivity have been measured with microdevices. Retina implants made from CMOS silicon light sensors and RF links are under development.[7] Genetic diagnosis is a new application for micromachining. Microchannels for improved stretching and matching of DNA strands,[8] chemical amplification via microwell cavities,[9] electrophoresis using etched Pyrex wafers, and fluid processing with micromachined valves[10,11] are examples of microsystems used for genetic diagnostic applications.

Chemical amplification is necessary when concentration levels are too low to be practical for analysis. Amplification can be obtained by an enzyme-assisted chemical reaction that allows molecular duplication. Electrophoresis is a method used in separating DNA fragments of different sizes from a solution or mixture. The DNA segments are negatively charged and therefore drift under the presence of an electric field. Mobility will vary with segment size and so can be used to produce the well-known racetrack print. These genetic testing methods have enormous potential in diagnosing the hundreds of gene-related diseases, as well as in assisting with the human genome mapping project. Both of these chemical and biological analyses can be accomplished using microsystems.[12]

2.1.3 ELECTROSTATIC

Electrical actuation in MEMS is frequently achieved using Coulombic attraction between oppositely charged surfaces. This mode of actuation is particularly suited to micromachines, because the magnitude of actuation force scales inversely with the square of the separation between opposing surfaces. The electrostatic force (F) can be calculated from Coulomb's law, which gives the force between two point charges,

$$F_{\text{elec}} = \frac{1}{4\bar{n}\varepsilon_r \varepsilon_o} \frac{q_1 q_2}{X^2}, \qquad (2.1)$$

where q_1 and q_2 are two charges in Coulombs, ε is the permittivity, and X is the distance separating the charges. If there are more than two charges, it is necessary to determine the force between each charge pair and to superimpose these vectors to find the resultant force.

In many surface-normal electrostatic actuator applications, a movable conductive membrane or plate is positioned over a fixed electrode plate, with an air gap between them. In this parallel-plate configuration,

$$W = -\frac{1}{2}CV^2 = -\frac{1}{2}\frac{\varepsilon_o \varepsilon_r}{X} A V^2, \qquad (2.2)$$

where C is the capacitance, V the voltage, and the force (F) between the plates of area (A) is

$$F = \frac{dW}{dX} = \frac{1}{2}\frac{\varepsilon_r \varepsilon_o A V^2}{X^2}. \qquad (2.3)$$

In many cases, electrostatic attractive forces are balanced by mechanical restoring forces. A widely exploited electrostatic actuator geometry is the parallel-plate, surface-normal configuration in which one electrode is the rigid substrate while the other is a membrane suspended by a mechanical spring attachment. Such actuators balance applied electrostatic attractive forces with intrinsic elastic restoring forces due to mechanical suspension.

Both the electrostatic applied force and the mechanical restoring force act with almost no dissipation, making this actuation mechanism relatively efficient with respect to power consumption. Also, the fabrication technology used for MEMS allows for a wider design freedom in most of the relevant parameters required: electrode area, interelectrode gap, and mechanical stiffness of attachments. Electrostatic actuators based on Coulombic attraction are frequently incorporated into more complex transduction systems in MEMS applications, most notably in lateral comb drive devices.

Capacitance variation is a common operational concept used in MEMS sensors. Using a combination of processing methods, a MEMS structure is fabricated

Figure 2.1 Capacitive micromachine.

to form a capacitor structure on a silicon chip. One of the capacitor plates is exposed to the parameter to be measured and moves relative to the fixed plate. Figure 2.1 shows an example of an electroformed capacitive structure.[13,14] This device is composed of a number of rigid plates surrounding a flexible ring. As the ring moves in a lateral direction, the gap between this thin ring and the rigid outer plates varies. This variation in distance is sensed capacitively. Other capacitive micromachined devices work in the vertical direction with large plates. In the case of pressure sensors, a micromachined diaphragm bends relative to the fixed plate, thus giving a capacitance variation as a function of pressure over the diaphragm.[15–17] Signal conditioning circuitry detects this capacitance variation and converts it to a higher signal output. Micromachined sensors often have such small signals that on-chip amplification is often required. Further, electrostatic force rebalance measurement techniques can be used effectively with capacitive sensors. Force rebalancing is particularly effective when small plates or areas are employed. The sensing output of the capacitive ring shown in Fig. 2.1 operates in a force rebalance mode.

The capacitive signal (C) is proportional to the area (A) of opposing plates and inversely proportional to the distance between the plates (d):

$$C = \varepsilon A/d, \tag{2.4}$$

$$C = q/V. \tag{2.5}$$

As in any system, the larger the micromachined capacitive plate areas and closer they are placed, the larger the detectable signal change. Increasing the potential difference between the two plates (V) will allow smaller charge differences (q) to be detected with a capacitive microstructure.

2.1.4 MAGNETIC SENSORS AND ACTUATION

A variety of magnetic sensors and actuators have been employed in microsystems. Perhaps the simplest magnetic device is the inductor. An electroformed microcoil has been employed as part of a chip-size RF transmitter.[18]

Three-dimensional coils with and without center cores have been fabricated using MEMS technology for both sensors and motors.[19–23] The inductor can be best understood via Faraday's law. Faraday's law for a coil or toroid is:

$$\varepsilon = -\frac{d(N\Phi_B)}{dt}, \qquad (2.6)$$

where ε is the electromotive force, N the number of coil turns, Φ_B the magnetic flux, and t is time. Inductance can be defined as:

$$L = -\frac{\varepsilon}{di/dt}. \qquad (2.7)$$

Equation (2.6) is the magnetic analogy to Eq. (2.5). For inductance, the presence of a magnetic field is the significant feature, while for capacitance, the electric field is the significant feature. Adding a core to the toroid increases magnetic field in the volume of the toroidal.

Most magnetic core materials used in sensor, actuator, and signal-processing applications require soft magnetic properties. Such a material has low hysteresis when magnetic flux density is plotted as a function of magnetic field intensity (B–H curve). For high-frequency applications, such as RF MEMS, soft magnetic material with a high electrical resistivity is desired.

Magnetic materials for microsystems have been deposited using electroplating,[19] evaporation,[24] sputtering,[25] chemical vapor deposition,[26] and screen printing.[20] Alloys such as NiFe, NiFeMo, FeCoP, MnAs, CoFeCu, NiZn, and MnZn[19,24–26] have been deposited with these techniques. When designing microsystems for magnetic applications, it should be noted that the bulk properties cited in textbooks cannot be used. Thin films often have significantly different magnetic properties than their bulk counterparts.

Anisotropic magnetoresistive (MR) and giant magnetoresistive (GMR) films have been used in a number of sensor applications, including the high-volume field of read/write heads on computer hard drives.[27] Large changes in electrical resistance in these films are encountered when stimulated by an external magnetic field. The films used in these applications vary and are typically formed by evaporating or sputtering very thin alternating layers of Fe/Cr, Fe/FeMn, Ni/Cr, or Co/Cu/NiFe. Cr or FeMn layers are antiferromagnetic materials; Fe, Ni, Fe/Ni, and Co are ferromagnetic materials; and Cu is used as a conductive spacer layer.

Certain ferromagnetic materials exhibit a magnetostrictive effect when exposed to a magnetic field. Under an increasing magnetic field, these materials expand or contract. Rare-earth alloys can have a maximum elongation of 1000 to 2000 μm/m.

Elemental iron exhibits a contraction of -8 μm/m. While magnetostrictive devices have a relatively fast response time and actuation force, the large fields and hence electrical currents required, and hysteresis problems have prevented this technique from seeing widespread application.

Hall-effect devices are perhaps one of the most common magnetic sensors. Discovered in 1879 by E. H. Hall, this effect is observed when an electrical current is placed in a magnetic field. While not using moving parts, a deflecting force is exerted on the conductor perpendicular to the current flow and magnetic field. Hall-effect devices can be economically fabricated from semiconductors and so lend themselves to MEMS.

MEMS magnetic actuators make use of coupling between magnetic fields to generate motive forces or induce a change in a magnetic medium (e.g., writing magnetic storage disks). Most frequently this actuation employs electromagnetism based on the physical principle that a current-carrying coil induces a perpendicular magnetic field. In macroscopic systems, electromagnetic actuation has been exploited most frequently in motors. Thin-film planar fabrication used for MEMS is not particularly suited to fabrication of wrapped coils. Consequently, MEMS electromagnetic motors have not been widely fabricated. Some micromachined actuators have been reported that are based on magnetic flaps.

2.1.5 Optical

Optical microdevices are covered extensively in Chapters 4, 5 and 6 and include the application of the basic operating principles of reflection, refraction, diffraction, and attenuation. Gratings, scanners, displays, photodiodes, phototransistors, filters, mirrors, switches, optical/IR-based chemical sensors, and fiber optic components have all been fabricated using MOEMS technology.[28–32]

A free-rotating switch consisting of micromachined optical switches (FS-MOS) featuring free-rotating hinged micromirrors is illustrated in Fig. 2.2, which also shows a schematic of an FS-MOS incorporating a bridging function. The output is switched by rotating the desired mirror.

Figure 2.3 shows an 8 × 8 FS-MOS. The most important parameter that determines the scaling of these devices is loss. These devices have switching times of submilliseconds when actuated with a 100-V square wave at 500 kHz. The switching time is 500 μs for rotating the mirror from the "OFF" to the "ON" position. There are many losses involved; a few of them are insertion loss, path loss, and scattering and absorption loss.

Figure 2.4 shows a scanning electron microscope (SEM) photomicrograph of a micromachined fiber optic cross-connector. This switch basically separates the subscribing system from the main fiber networks. These switches have a delay time of 10–25 ms; an insertion loss of -2.5 and -0.84 dB for reflection and transmission, respectively; a crosstalk less than -45 dB; and a switching contrast of

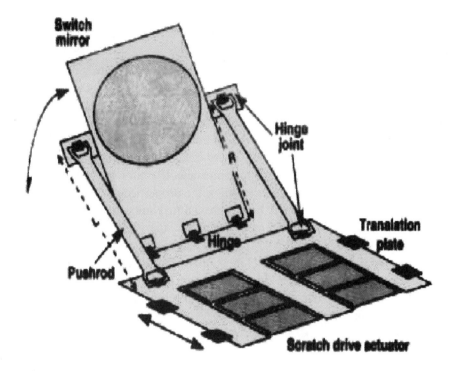

Figure 2.2 Schematic of a microactuated free-rotating mirror.[105]

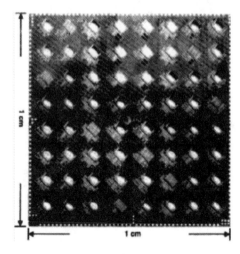

Figure 2.3 Top view photograph of an 8 × 8 FS-MOS.[105]

more than 45 dB. A few advantages of these switches are low loss, large switching contrast, and small crosstalk. This is one of many examples of micromachined optical devices that are discussed throughout this book.

2.1.6 Piezoelectric and acoustic

A piezoelectric material produces an electric voltage when mechanically stimulated. This comes about through asymmetric displacement in the crystal structure and the charge centers of the crystal ions. A mechanical effect can also be induced by the application of voltage to such a crystal.

The piezoelectric effect has been used in both sensors and actuators. Materials such as quartz, zinc oxide, and a variety of ceramics such as lead-zirconate-titanate are most commonly employed to produce piezoelectric microsystems. By varying the composition of multicomponent ceramics, the properties of the material can be modified. Traditional piezoelectric beams are 0.5 to 1 mm thick and are made up of alternating stacked ceramic and metal electrode layers. Piezoelectric micromachines typically employ thin films of sputtered material deposited onto a cantilever or diaphragm of silicon. A micromachined silicon cantilever that uses a thin film of sputtered, piezoelectric zinc oxide has been fabricated by Indermuhle et al.[33] This device uses the piezoelectric material to both sense and actuate the arms of an atomic force microscope (AFM).

Figure 2.4 Micromachined fiber optic crossconnector.[105]

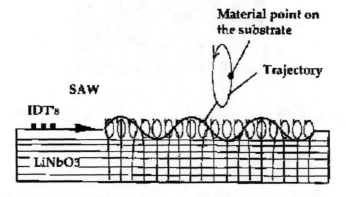

Figure 2.5 Illustration of SAW-based chemical sensor.[103]

Piezoelectric actuators produce a relatively high force or voltage-proportional elongation, have a high dynamic response, and do not consume electrical power in the "off" mode. Unfortunately, this sensing or actuation technique requires high voltages for operation, drifts with time, and exhibits hysteresis during shock or elongation. Most PZT materials are brittle in bulk form and so can be difficult to machine mechanically; however, this is not a problem for the thin-film materials used in silicon-integrated devices.

MEMS acoustic actuators are based on the principles of piezoelectricity. A surface acoustic wave can be used for actuation. Typically, a SAW device is composed[103] of a piezoelectric substrate, and by applying an electric field pattern to the surface of the substrate, we get a SAW wave.

Figure 2.5 shows the motion of points on the surface of a solid substrate excited by a SAW. These particles follow an elliptical trajectory governed by the boundary conditions. The elliptical motion has orthogonal vector components parallel and perpendicular to the surface. The component parallel to the surface is used for linear actuation.[103]

2.1.7 PIEZORESISTANCE

Piezoresistance is one of the most commonly employed micromachining sensing phenomena. The piezoresistive effect is the change in electrical resistance of a material in response to mechanical strain. The gauge factor of piezoresistive elements depends on material, grain size (for polycrystalline materials), doping level, crystallographic orientation, and temperature.[34,35] Single-crystal p-type resistors, formed in (100) wafers along the <110> direction are known to have the highest gauge factor of common piezoresistive material.[34] The piezoresistive effect in single-crystal silicon has been known since 1954.[35] Kanda[34] has explained the high sensitivity of single-crystal silicon piezoresistors using a carrier-transfer mechanism and the effective mass change associated with the k-space energy surfaces.

Figure 2.6 shows an example of an accelerometer fabricated with a Wheatstone bridge. Piezoresistors are most often used as part of a Wheatstone bridge configuration. Since CMOS and bipolar integrated circuits are fabricated from (100) silicon wafers, this transduction principle was adopted by the first microsystems. Piezoresistive devices have been and still are employed in millions of micromachined pressure sensors and accelerometers.[15,36,37]

2.1.8 RESONANT DEVICES

A number of microsystems have employed resonant structures, which are attractive because of the potential of high sensivitity and the option of a frequency shift output.[38] A natural resonant mode is utilized in these devices. A natural or normal mode of vibration is a specific pattern of a structure that in the absence of rotation is uncoupled. Each normal mode has a specific shape, or pattern of vibration, and a

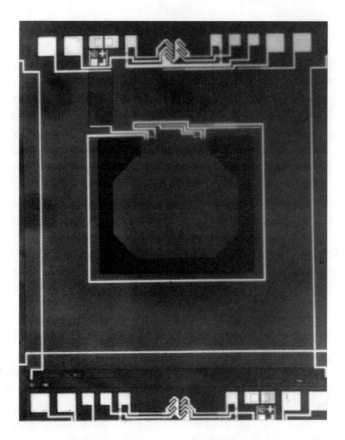

Figure 2.6 Piezoresistive accelerometer.

natural frequency. If a structure is caused to vibrate in one of the normal modes, the energy of vibration remains stored in that specific mode and the device continues to vibrate indefinitely in that shape, provided there is no damping or motion. Because of this stored energy, it is preferred to utilize natural resonant frequencies rather than forcing an arbitrary frequency on a structure. Computer modeling is an easy way of locating the resonant frequencies of any structure. Simulation is also useful in predicting how changes in mass or motion will affect the frequency output of the device.

Resonance in a microstucture is initiated and sustained by subjecting the device to an oscillating electrical input. This input can be generated through capacitive coupling; piezoelectric, electromagnetic, or surface acoustic wave oscillators, or other stimulation.[38-41] In the case of capacitive coupling, a square or sine wave is typically added on top of a larger d.c. voltage. Resonant motion has been employed to produce accelerometers, angular rate sensors, pressure sensors, and chemical sensors.[13,38-41]

2.1.9 Shape memory alloys

The use of shape memory alloys (SMA) offers a similar but much more promising actuation and sensing technique than that of bimetallic materials. When an SMA is heated past its martensitic transformation temperature, the material becomes austenitic. The crystal structure change that occurs at this point leads to a dramatic change in shape. The change is permanent until a temperature change takes the device down below the transformation point.

Like most other devices, SMAs were applied in the macro arena before seeing use in MEMS. SMA wires and ribbons have been used in the aerospace, automotive, medical, and consumer fields as triggering, drive, and final control elements. More recently, these devices have been used in microsystems for valves and injectors.[42]

Shape memory alloys have perhaps the best force output per unit volume of any reusable microactuator technology. By changing the alloy composition, the transformation temperature can be varied to give the designer additional flexibility. In microactuators, temperature change is most often induced by passing electrical current through the SMA element. A number of design techniques have been developed to optimize the performance and reliability of SMA microsystems.[42] These include the use of rounded springs to reduce fatigue problems.

2.1.10 Tunneling

According to quantum mechanics, an electron can pass through a finite potential barrier. The electron tunneling effect is limited to the atomic scale. For tunneling transducers, the location of specific atoms on opposite electrodes controls the behavior of the device. Tunneling has the potential to be an important displacement transduction method because of its high sensitivity and independence of scale. This type of microsensor measures a nanoampere current and can resolve a 0.01-nm displacement. The tunneling current (I) has the following dependence on the gap (G) between opposing metal electrodes:

$$I \propto V \exp(-\alpha\sqrt{\Phi G}), \tag{2.8}$$

where V is the bias voltage, Φ the tunneling barrier height, and α a constant.

While the atomic-level sensitivity of the tunneling sensor holds great promise, it also has some practical problems of reproducibility and stability. Oxidation and surface contaminants such as water and organic chemicals can cause errors in current level, which can change with time. To counter surface adsorption problems, extensive cleaning, avoidance of organic die attachment adhesives, and vacuum packaging are often employed.[43,44] The material used to form the tunneling tips can also migrate across the surface, resulting in output drift. This is a particular problem with gold, which while resistant to oxidation, is prone to surface rearrangement.

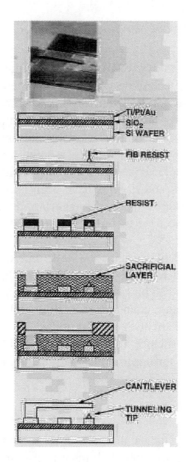

Figure 2.7 Tunneling accelerometer and process flow.[45]

Other developmental issues associated with tunneling devices include crashing together of the electrode tips, output errors due to mismatch of material expansion coefficients, and the wafer-level techniques of manufacturing many tunneling tips in large batches.[45,46] Micromachining has employed the tunneling phenomena to produce accelerometers, shown in Fig. 2.7,[45] thermometers,[47] pressure sensors,[48] IR detectors,[49] magnetometers,[50] and scanning tunneling microscopes.[33]

2.2 THE DESIGN OF MEMS/MOEMS

There are many technical issues that must be faced when designing and developing micromachined sensors and actuators. The design of a particular sensor or actuator must encompass several areas: a basic principle of operation, microsystem and process modeling, circuitry design, process technology, application environment, packaging, and a manufacturing system. Because of the multifaceted nature of sensor and actuator development, these aspects must be addressed globally in a common design at the start of development rather than summing the individual parts at the end. Too often, substantial effort is put into developing a device concept only to find later that the device is difficult or expensive to package suitably in a harsh environment. In some cases, the cost of materials, processing, and manufacturing make the sensor or actuator device impractical for low-cost applications. An economically successful consumer device technology is one that manages the proper balance between these interacting design and manufacturing aspects, while simultaneously achieving the required device performance at low unit cost.

2.2.1 SYSTEM AND PROCESS MODELING

In many ways computer modeling of MEMS is similar to that of integrated circuits. However, unlike ICs, both sensors and actuators must interact with their environments and often utilize moving parts. Interactions with the microsystem's environment may require predictions of fluid flow or the effect of thermal expansion coefficient differences between materials used to house the micromachine. Motion sensors such as accelerometers and angular rate sensors often move, or at least push against an electrostatic field if used in the force-rebalance mode. Pressure sensor diaphragms bend in response to pressure changes. Relays close and shut, as do valves. These represent some of the unique MEMS features that are often modeled. While the MEMS modeling tool set has been fragmented in the past, complete design software packages are being developed.

Mechanical modeling is often first employed when designing a micromachined device. Traditional mechanical computer modeling software such as ANSYS,[51,52] ABAQUCS, or NASTRAN is used, as are MEMS-specific programs such as IntelliCAD, Anise, MEMCAD, and SENSIM.[53–56] MEMS modeling is covered in great detail in Chapter 3. Mechanical modeling is utilized to determine the structural stability of a device as well as the resonant frequency and response to stress. Placement of electrical transducer elements can be optimized with this type of analysis. Prediction of transducer sensitivity, breakage, and susceptibility to mechanical shock and vibration can also be assessed in this manner. Possible sources of sensor error should be modeled and used to develop compensation methods where needed.

Flow sensors, nozzles, and valves interact with their surroundings directly and can also be modeled using programs such as FLOTRAN.[52] As in the case of other microdevices, performance, transducer sensitivity, breakage, and turbulence prevention can be better understood early on in the development cycle with computer modeling. Fluid damping is very important in the design of accelerometers, and so models have been developed for this application as well.[57]

Once a general micromachine structural model is complete, fabrication simulation can be undertaken. With silicon-based microsystems, traditional IC simulation programs are most often employed. These programs would include SUPREM,[58] DAVINCI, and PEPPER to determine the initial oxidation, diffusion, implant, and epitaxial parameters needed to form etch stops, diaphragms, piezoresistors, and other semiconductor and micromachine elements.

As in the case of integrated circuits, the process fabrication model can be linked to the electrical model of any circuit elements used by the microdevice. Just as IC simulation programs have been linked together, MEMS simulation software that joins mechanical process and electrical design has also been developed.[52–56] Extensive experimental work is required to optimize the accuracy of any model with results obtained in the factory. Uncalibrated modeling is useful for first-order development; however, confidence in any simulation package is only obtained with verification.

2.2.2 Process design

With MEMS technology, compatibility between the materials and operations in a production silicon-integrated circuit foundry is often a major issue. The application environment dictates what material must be used in the microsystem, while the processing technology and economics limits the materials that can be employed. It is possible that the application environment will limit what materials and hence processes are used in MEMS fabrication. Generally, the protection from the environment is accomplished via packaging.

Ideally, process modeling would be the first step in coming up with a MEMS wafer fabrication sequence. Practically, most MEMS processes start with an existing IC or micromachining process. Technology is reused to save time and development costs and to ensure reliability. Technology reuse takes advantage of both previously published work and available processes and equipment at the development and manufacturing site.

The piezoresistive pressure sensor processes used in the late 1970s were derivatives of the bipolar IC processes used at that time. The P+ piezoresistor layer came from the transistor base process, the N-type epitaxial layer came from the transistor collector process, and any N+ substrate tie came from the bipolar emitter process. In some cases even the P+ etch stop was derived from top-side/bottom-side junction isolation P+ buried-layer process modules. Metal and passivation layers were also borrowed from existing bipolar process flows. Twenty years ago, process modeling was rarely employed and processing experiments were used to adjust these existing bipolar IC processes into a manufacturable micromachining fabrication sequence. Most original work was required in the areas of silicon etching,[59–63] wafer bonding,[64] and micromachine packaging.[65–69]

In a similar manner, most piezoresistive accelerometers were derived from already existing micromachined pressure sensor processes.[37] Silicon doping, photolithography, oxidation, and etching processes were used virtually unchanged. Plasma or dry silicon etching was required to produce accelerometers, but again this had already been developed for IC manufacturing.[70–72] Capacitive accelerometers were borrowed from conventional CMOS or BICMOS processing in the same way.[73]

Figure 2.8 MEMS chips, fabricated using wafer bonding.

Examples of reuse of MEMS technology in production include the adoption of wafer-to-wafer bonding from pressure sensors and applied to accelerometers and angular rate sensors.[13,15,36] Figure 2.8 shows a photograph of pressure sensors, accelerometers, and gyroscope chips.

Radically new micromachined devices will still use existing IC and MEMS processes; however, they will rely more and more on computer modeling for quick implementation. This is especially true of items such as flow sensors, valves, nozzles, and ultrasonic and optical microdevices.

2.2.3 Design rules

Design rules are needed to produce robust microsystems. Often a developmental MEMS process will borrow many design rules from an existing bipolar or CMOS IC process. MEMS design rules should contain typical parametric targets for sheet resistivities, film thicknesses, and junction depths; they should also contain critical dimensions for line widths that are dependent on lithography and etch technology. It is recommended that experimental design rule verification test structures be run along with development wafer lots. These test dies consist of devices such as annular ring and interdigitated comb structures to determine minimum design spacing between diffused layers. Breakdown voltage measurements of such semiconductor layers with varied gaps are utilized to set the design rules used for the different levels. Piezoresisitive devices are typically diffused, single-crystal resistors and so are subject to spacing limits imposed by semiconducting breakdown voltages. For polysilicon and metal layers, the spacing between similar layers, which also pass

Table 2.1 Partial example of MEMS design rules.

Adjacent layers and openings	Spacing limit or minimum opening (μm)
P+ to P+	10
P+ to N+	8
N+ to N+	8
Metal to metal	8
Plasma trench	15
Pad	10

over topological steps created by previous layers, is used to generate rules. Photolithography and etching limits as well as shorts caused by stringers along steps often set the design guidelines of these deposited layers. Table 2.1 gives some typical design guidelines for spacings of various layers produced with proximity lithography etch processing.

Violations of design guidelines are typically checked via software prior to making photomasks.

For wet (100) silicon etching, the fabrication of convex corners can pose a problem. Convex corners are undercut at fast-etching silicon planes. Special sacrificial triangle or square structures must be designed into the corners to produce square convex corners. Proper corner compensation varies with the wet etching process and usually requires some trial-and-error experimentation to optimize the design and process.

Like wet etching, wafer-to-wafer bonding design rules are also arrived at through a combination of process and design layout development. For screen-printed bonding layers, such as epoxy or glass frit, the appropriate line width and spacings must be arrived at, as well as the screen mesh thickness and density. Minimum spacing and linewidth rules can also be important when using anodic and silicon fusion bonding on aggressive geometries. Narrow gaps will not bond properly or will have insufficient strength to be useful.

Test structures for detecting process-related defects such as pinholes in dielectric layers, or contamination-related junction leaks are also recommended for test dies. These devices can consist of large-area capacitors and diode strings. In development labs, many problems related, not to the design, but to contamination and defects, have the potential to delay progress. Without testing devices to detect defects, improper conclusions about new MEMS designs can be reached.

2.2.4 SCALING OF MEMS DEVICES

The miniaturization inherent in MEMS carries with it a number of scale-sensitive factors. Physical variables that dominate performance in macroscopic systems may become secondary at submillimeter scales (e.g., inertia), while those that are frequently neglected in macroscopic systems may be primary for other MEMS counterparts (e.g., surface tension). In addition, the laws of scaling for MEMS devices frequently do not permit the transducer designer to neglect as insignificant the effects of forces and fields that are secondary in macroscopic devices.

Important interactions occur with fluid dynamics and heat transfer at small scales. The geometric scaling laws for fluids are well understood, but the flow interactions with the surface pose a problem. At very small scale flows, the Reynolds number (Re) is often small, and the rarefied flow regime for gases is encountered.

2.2.5 MEMS ELECTRONIC DESIGN ISSUES

Circuit simulation using SPICE or other software is used to initiate the design of the microsystem electronics. Signal amplification, compensation over temperature, and decision making based on input are all circuit functions that have been included in existing MEMS circuitry.

After modeling, both historically and during the developmental process, the electronic hardware of a microsystem begins with discrete components on a circuit board. Discrete components make troubleshooting of the circuits inexpensive, fast, and flexible. Once system performance are optimized at both the simulation and circuit board level, the design and layout of a hybrid or silicon-integrated circuit can begin.

The choice of whether to use a hybrid approach, which includes a silicon circuit chip on a ceramic circuit board, often involves several considerations, including the decision as to what type of programming technique will be employed in the system and the level of confidence there is in the design of silicon circuits. If laser trim programming is the preferred technique for changing electronic output, and a large component size is required, a hybrid circuit approach may be chosen. Laser trimming of thermistors or resistors on ceramics has been done at the hybrid level on many high-volume micromachined sensors.[51] Laser trimming can be used at the chip level if the appropriate layers and masking levels have been included in the IC process.[73] Laser trimming is capital intensive and so has been avoided when possible.

One technique that has been used to replace laser trimming is that of fuse trimming. In a fuse programming design, a large number of bond pads are required to allow for wafer-level programming. Since bond pads are large in area on an IC, this technique takes up a significant amount of die area. Other potential problems with fuse programming include incomplete opening of a blown fuse. Residue from the blown fuse can leave behind a resistive electrical path. To overcome this residue problem, openings in the passivation layers above the metal or polysilicon fuse must be left behind. Unfortunately, openings in the passivation layer can lead to long-term reliability problems if the IC is used in a corrosive environment.

Other methods that have been employed to allow for silicon-level programming of microsystems include the blowing of Zener diodes or the inverting of MOS gate channels. Like the metal fuse, these programming elements do not require additional IC masking levels or processes. EPROM ICs require fewer bond pads than chips using fuse programming, and so can be smaller in area. They do require the design and development of a reliable programming element. If bipolar circuits are employed, a Zener zap programming element may be utilized. The Zener diode typically takes more than 100 mA to open up, while inverting an MOS channel requires an order of magnitude less current. Because of this programming current difference, considerably more IC area is often required for a Zener zap bit than is needed for an MOS EPROM bit. The Zener zap bit does have the advantage over the MOS EPROM of being able to withstand both extended time at temperatures >125°C and not being UV erasable during packaging.

EEPROM is another widely used microsystem programming method. These memory elements are electrically erasable, which allows for in-field recalibration. This could be a key advantage for smart sensor applications. The disadvantage to EEPROM is the requirement of a more complex and expensive IC process, and the possible loss of data at high temperatures.

2.2.6 SELF-TEST DESIGN

For moving transducers associated with safety-related systems such as air bag accelerometers and the angular rate sensors used in vehicle dynamic control, self-test features are important. Moving micromachine parts can be affected by particles. Particles can be dislodged from the interior walls of packages or the surface of the chip and prevent motion of the microstructures. An electrical self-test may be the only way to detect such a failure. Resonant micromachines in effect have a built-in self-test feature, their resonant motion. If the phase-lock-loop circuit fails to detect the primary resonant peak, the microsystem does not work and an error message can be sent to the output pin.[14] Cantilever devices operating on piezoresistive, capacitive, or piezoelectric principles must have a self-induced motion to provide self-test. For capacitive and piezoelectric devices this can be accomplished by applying a voltage. For a piezoresistive cantilever, an additional actuation layer or structure must be designed into the micromachine.[1] Electrostatic force, piezoelectric action, and bimetallic and shape memory alloy actuators can be employed to generate the self-test motion. Self-testing of MEMS-based accelerometers relies on exercising only a small portion of the operational bandwidth of the device, so specific self-test circuitry and/or software must be developed.

2.2.7 INTEGRATION OF MEMS/MOEMS DEVICES AND SMART TRANSDUCERS

Because the tools for MEMS fabrication and design evolved directly from semiconductor technology, there is a natural drive toward integration of MEMS devices with electronics on the same substrate using common or sequential processing. This transition and its challenges are illustrated in Table 2.2. Such integration has proven to be difficult and expensive in all but a small subset of applications. It is now generally accepted in the MEMS field that integration increases process

Table 2.2 Electronic challenges of MEMS integration.

Typical transducer characteristics		Desired system characteristics	Challenges
µV Signal outputs		Volt-level signals	Low-noise
Parts-per-million changes (mΩ, fF)		Amplified, buffered outputs	On-chip analog interface
Analog Signals	SIGNAL PROCESSING CIRCUITRY	Digital signals	In-module ADC
Cross parameter sensitivities		Pure parameters measured	Sensing of secondary variables
Individual device outputs		Multiplexed and addressable	Embedded microcontroller
Offset/slope/ linearity problems		Digitally compensated	Compensation standards
Outputs drift over time		Self-testing/ remote calibration	Stable references in-module actuators

complexity in comparison to a hybrid (side-by-side MEMS and electronics chips) approach and should only be considered when the advantages justify this added complexity. Two common types of MEMS devices that warrant consideration of integrated fabrication are those using capacitive sensors (where parasitic capacitance may make hybrid designs impractical) and those using large arrays of MEMS devices.

Putting both tranducers and circuits on the same piece of silicon presents a number of design issues. Process integration must not only combine micromachining and CMOS, bipolar, or BICMOS process steps in an efficient manner, it must also merge design rules. Certain micromachining processes cannot be performed at random in an integrated process flow. Thin silicon membranes and beams will undergo plastic deformation if inserted early into an IC process prior to long, high-temperature diffusion. Preventing plastic deformation can involve changing furnace processes and the design of micromachined diaphragms and beams. Material trapped in a microstructure also presents a contamination problem for traditional CMOS process lines. Surface micromachining cannot occur until most if not all photolithography processes have been completed. Spinning on photoresist can damage fragile microstructures.

The placement of IC elements on a chip can influence the output stability. Packaging interactions between a single-chip transducer and its environment are even more severe than those between a simple IC or sensor. In addition to the design constraints observed for a discrete transducer or IC, a one-chip sensor must take the layout of moving elements with respect to potentially stress-sensitive circuits. Mechanical modeling can indicate how far lateral stress gradients spread across a piece of silicon when a transduction element is deflected. Diffused resistors or transistors should be kept away from areas that are influenced by on-chip stress gradients. These type of stress-sensitive circuit elements must be placed away from

Figure 2.9 A MEMS-based, multisensor module.

the edges of the chip and bond pads and also from micromachined elements. If wafer-to-wafer bonding is employed, the bonding surfaces should be avoided since they could result in stress concentration and possibly even short metal runners if designed and bonded improperly. If the micromachine is used in a chemically aggressive environment, the protection of circuit elements must also be considered. Mobile ions such as hydrogen, sodium, potassium, and lithium can cause threshold voltage shifts in CMOS circuits, beta changes in bipolar transistors, and knee voltage shifts in diodes. Circuit elements must be protected by packaging, wafer bonding, or surface coatings. This can be very difficult or costly with a one-chip transducer design.

Future trends in the field of smart sensors include putting several sensors on one silicon chip. This has already been done in R&D labs.[74,75] These same sensor chips also have CMOS circuits on them, which could also lead to combining smarts with sensing capability. However, it remains to be seen if this can be accomplished in an economical manner with complex and varied sensors. Sensors that are made with the same process and in the same manner, such as optical arrays, will certainly be easier to include on a fully integrated chip. In the near term, discrete, dissimilar sensors and integrated circuits will be combined at the circuit board level, as shown in Fig. 2.9.

2.2.8 MEMS PACKAGING AND ENVIRONMENTAL EXPOSURE ISSUES

Package design is dictated by MEMS element technology and the application environment. The packaging operation can account for the largest fraction of MEMS

Table 2.3 The automotive, industrial, and aerospace MEMS environment.

Temperature:	−40°C to 85°C driver interior
	125°C near an engine or motor
	150°C on an engine or motor
	200 to >600°C in the exhaust and combustion areas
Mechanical shock:	3000g during assembly (drop test)
	50–500g in use
Mechanical vibration:	15g, 100 Hz to 2 kHz
Electromagnetic Impulses:	100 to 200 Volts/m, higher for military applications
Exposure to:	All: Humidity, salt spray
	Automotive, aerospace: fuel, oil, brake fluid, transmission fluid, ethylene glycol, Freon, or exhaust gases
	Industrial: solvents, acids, alkaline fluids

cost.[76] This is due to one-at-a-time packaging operations using highly specialized production equipment and end-of-line testing and calibration operations. Further, the package structure needs to be modeled concurrently with the transduction element to determine if there are any package-induced error sources, while protecting the sensor from the environment. Chapter 7 covers MEMS packaging in great detail.

The packaging of MEMS must begin with a consideration of the environment the microsystem will be exposed to.[51,77] Industrial (corrosives), medical (blood, antibodies), and aerospace (radiation, vibration, shock) applications have unique compatibility requirements. Table 2.3 lists the conditions that an automotive, industrial, or aerospace component is subject to.

Standardized testing of sensor components is partially covered in Society of Automotive Engineers standards via SAE J1221 and SAE J575G and for the military in Military Standards 750 and 883C. These standards detail accelerated testing procedures such as high- and low-temperature storage, temperature cycling, and thermal shock that are used in qualification testing of a packaged device. Specific sensors and actuators also have additional reliability tests.

As an example, pressure sensors are tested using pulsed pressure and temperature cycling while powered up. Actuators will have similar accelerated actuation testing. These extensive tests are required to ensure that a component will function over the 5- to 10-year, or 100,000- to 150,000-mile lifetime of an automobile in desert, tropical, and/or arctic locations. Commercial truck components require 10 years or 1 million miles of problem-free use.

To be an effective, a sensor or actuator must interact with the environment. Protecting a micromachined or electronic element from aggressive surroundings requires special attention to packaging. Protection starts at the die level. The same types of silicon nitride and doped oxide passivation used to prevent ionic contamination in integrated circuits are employed to protect most microsystem chips. Additional coatings such as parylene or passivating gels are next used to protect the circuit side of micromachined devices.[65] Chip-level encapsulation is also used on several types of micromachined sensors such as accelerometers and angular rate

Figure 2.10 Wafer-to-wafer bonded micromachines.

sensors to protect moving structures.[13,37] Chip-level packaging is accomplished by wafer-to-wafer bonding using a micromachined top cap, as shown in Fig. 2.10. In this capped slice, a cavity is etched to accommodate the moving element and through-holes are machined above the position of the bond pads to allow for wire bonding of the completed, stacked die. Since bonding can be done in a clean room, this type of wafer-level packaging is also an excellent method of keeping particles away from micromachined elements. Particles can lead to shorts or stop element motion. Figure 2.11 illustrates how the micromachined sensor is enclosed at the chip level.

Mechanical package design is the next level of protection for a microsystem. Hermetic packaging of the top side of differential pressure sensors and many motion sensors can be used. For lower cost, well-sealed plastic packages can be employed in conjunction with gel and/or chip-level packaging. Examples of these types of packages will be given in the sections on pressure sensors, accelerometers, and angular rate sensors. Solder sealing of motion sensors using ceramic side-brazed packages has also been employed to produce reliable devices.[67] Figure 2.12 shows such a ceramic package. This particular package was used to house a resonant micromachine. Early in the program, a Kovar lid was solder sealed under vacuum using gold/tin or silver/tin solder, as shown on the right side of Fig. 2.12. Wafer-to-wafer bonding under vacuum eventually replaced this expensive ceramic package. A number of other micromachined components need vacuum packaging for functionality or improved performance. Absolute vacuum reference pressure sensors,[15] tunneling devices,[43] and field emission displays all utilize vacuum seal-

OPERATION AND DESIGN OF MEMS/MOEMS DEVICES 71

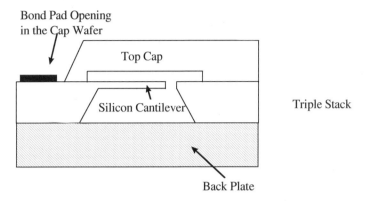

Figure 2.11 A cross-sectional view of a chip-level, packaged micromachined sensor.

Figure 2.12 Ceramic packaged MEMS device.

ing. This sealing can be done at the chip level using wafer-to-wafer bonding, or through a solder or weld hermetic seal performed under vacuum.

The package and assembly process can have a detrimental impact on microsystems. Since many micromachined sensors and actuators are fabricated from fragile semiconductor or ceramic material, breakage can occur.[36,78] Many sensors are essentially strain gauges and so can pick up packaging-induced stress as well as the phenomena they are intended to detect. Often packaging stress is not observed until the microsystem is exposed to temperature changes. Since automotive, industrial, and aerospace components see temperature changes of 125°C to over 400°C, this is a very critical source of stress. Differences in the thermal expansion coefficients of materials and in Young's modulus lead to these stresses. Table 2.4 shows a list

Table 2.4 MEMS material properties.

Material	CTE (ppm/K)	Young's modulus (MPa)	Poisson's ratio
Silicon	1.8 to 3.2	162,000	0.28
Alumina (aluminum oxide)	5.1 to 7.5	276,000	0.23
7740 glass	3.3	62,784	0.2
FR-4	15	18,200	0.25
Copper	17.6	117,200	0.33
Kovar	5.9	138,000	0.3
Silicone	800	6.9	0.4
Novolac epoxy	13	12,460	0.3
Silver filled epoxy	23 to 43	5200	0.34
Pb37-Sn63 solder	28	23,000	0.4

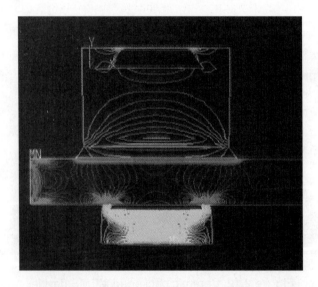

Figure 2.13 2D ANSYS model of a hybrid packaged silicon pressure sensor.

of these important material properties. Severe thermal stress can best be prevented early in the packaging design stage using computer modeling.[51,77,79,80]

Figure 2.13 shows an example of a simple 2-dimensional simulation of a silicon pressure sensor bonded to a glass pedestal that is glued to a ceramic substrate. On the opposite side of the ceramic substrate is a silicon integrated circuit, attached to the ceramic substrate via solder bumps. Material differences led to the stress gradients evident in the figure. Die attached materials with a low Young's modulus, as well as mounting sensor elements to glass pedestals using wafer bonding have been employed to minimize the package-induced stress that reaches the sensing elements in the silicon. This is a critical packaging area where component integra-

OPERATION AND DESIGN OF MEMS/MOEMS DEVICES

Table 2.5 Percentage of cracked pressure sensor dies for various substrate materials and adhesives.[62]

Adhesive	Plastic	Substrate material ceramic (Al_2O_3)	Glass (Pyrex)
Epoxy	100	11	0
Silicone	0	0	0

tion comes into play. Modeling gives the engineers an early indicator of potential problems. Table 2.5 shows some experimental data comparing die breakage for different adhesives and substrates. In this study a number of dies were thermally shocked 10 times from −40°C to 125°C to accelerate possible die cracking. While plastic substrates are less expensive than ceramic or glass (Pyrex), they can lead to die fracture when combined with hard (high Young's modulus) epoxy. Epoxy can only be employed when the substrate matches the thermal expansion coefficient of the silicon die.

Packaging of microsystems without breakage problems is just the first assembly barrier for many sensors. Lower levels of packaging stress can cause hysteresis and long-term output drift. Many customer specifications often call for less than 1% to 5% drift in output over time after prolonged operation or temperature cycling. Dies produced as indicated in Table 2.5, using hard epoxy and a plastic substrate, exhibited large amounts of hysteresis prior to fracture of the die. Using silicone die attachments or glass substrates not only prevented fracture but also reduced sensor error terms. This is typical of the types of package design improvements required for applications subjected to wide temperature variations.

The layout or design of silicon transducer elements and the integrated circuit used to support them should also take packaging stress into account. Die attachments, bond pad placement, and diffused resistor and transistor placement can affect the output, stability, and hysteresis observed. For (100) silicon, used in CMOS and BICMOS chips, diffused resistors are piezoresistive. While they are not intentionally placed over a flexible diaphragm or beam, they still can encounter enough stress if packaged improperly to affect the output of an IC. Designing with polysilicon resistors or orienting diffused resistors at a 45-degree angle to the chip perimeter are ways to reduce stress sensitivity. Placement of sensitive elements or circuits away from the chip corners, edges, and bond pads are other methods used in the design stage to prevent packaging stress from influencing the output of the transducer or IC.

2.2.9 MANUFACTURING-RELATED DESIGN ISSUES FOR HIGH-VOLUME PRODUCTION

The probability of successfully developing a new device technology is enhanced if the existing manufacturing infrastructure is taken into account when the device

Table 2.6 An example from a microsystem DFMEA.

Item	Potential failure mode	Potential effect of failure	SEV	Potential cause of failure	OCC	Current design controls	DET	RPN
Measure motion	Electrical short	Invalid sensor output	6	Bond pads too close. Layout error	3	Design verification software	4	72

concept is selected. It is much easier to bring a device technology into production when all that is needed is a modification of an existing manufacturing process.

High-volume customers expect component prices to be relatively low. This can only be accomplished using inexpensive materials and designing the MEMS package to be easily tested and calibrated. The requirement of reliable, stress-free packaging and system calibration often dominates the overall cost of the microsystem.[76,77] Micromachined devices will often be used to sense fluid pressure, flow rate, motion, or temperature. This requires custom test equipment. The variation in sensor output with temperature also requires calibration at different temperatures. These factors must be taken into account when designing a package in order to minimize the cost of the overall product.

While not always possible, a low-cost product can be obtained by employing production-proven techniques utilizing existing manufacturing equipment in order to minimize both development cost and time. One of the major capital investments in any sensor technology lies in the equipment needed to do automated testing and packaging. For high-volume production, this equipment is highly specialized and needs a large plant floor area. It is this aspect of production that comes into play when the next-generation sensors and new sensors are developed. Many times the manufacturing system has a direct bearing on the selection of a sensor concept.

2.2.10 DESIGN REVIEWS AND DFMEA

In industrial design, design reviews and design documentation, such as design failure mode and effect analysis (DFMEA), are employed to ensure high-quality products and prevent the repetition of previously observed problems. Design reviewers are a group of experienced engineers, not necessarily familiar with the current project, who go over the details of a design. These experts are looking for potential problems similar to those seen in the past. A DFMEA is a more formal method of documenting design issues. DFMEAs are often required and audited by customers and so can be of great importance in winning new business. Table 2.6 gives an example of the items typically found in a DFMEA.

The DFMEA format uses four numerical rankings. SEV stands for severity. The severity of a particular failure mode is gauged with this number. A 1 corresponds to an almost undectable impact; 2–3, a low severity that the end user will not notice; 4–6, a moderate failure causing some customer dissatisfaction; 7–8, a high degree of customer dissatisfaction resulting in an inoperable system; 9–10, a very high severity that may involve potential safety problems on failure. For frequency of occurrence, the OCC ranking, a 1 corresponds to a remote probability of occurrence; 2–3, a low failure rate; 4–6, a moderate failure rate; 7–8, a high failure rate, and 9–10, a very high failure rate. The chance of detecting a failure is gauged by the DET ranking. A DET ranking of 1 corresponds to a remote chance that this defect will reach the next operation or the customer without detection; 2–3, to a low (2–5%) chance the problem will escape detection; 4–6, to a moderate (10–20%) chance of avoiding detection; 7, to 25%; 8, to 40%, and 9, to a 50% chance; and 10, to a greater than 50% chance of escaping detection. The risk priority number (RPN) is found by multiplying the SEV, OCC, and DET numbers together. This allows the various potential design problems to be weighed against each other. The highest RPN numbers are assigned corrective actions to solve, prevent, or sort out these problem areas.

In the area of process development, a PFMEA (process failure mode and effect analysis) is also done to prevent process-related failures. The PFMEA tables look very similar to the DFMEA table already discussed. Design reviews and FMEAs should be used both early in the overall design process to save time and reduce errors, and as a final check before investing in photomasks and capital equipment.

2.3 MEMS Design Examples

Four examples of MEMS/MOEMS devices are next used to illustrate how many of the ideas discussed are applied. The pressure sensor, accelerometer, resonant micromachine, and deformable mirror will be covered in more detail.

2.3.1 The design of pressure sensors

Starting in 1979, the microprocessor-based automotive engine control module was phased in to control the engine air-to-fuel ratio fixed at the stoichiometric point. By doing this, the catalytic converter efficiently minimizes the tailpipe emissions, bringing them into compliance with federal regulation. With these engine control systems came the need for sensors on both the input and exhaust sides of the engine. The zirconia oxygen sensor filled the need on the exhaust side, giving a step function feedback signal at the stoichiometric point. A manifold absolute pressure (MAP) and a manifold air temperature (MAT) sensor were used to compute the density of air entering the engine.

Historically, it was this automotive system, requiring a MAP sensor, that introduced MEMS technology to the world in high-volume units[15,77,78] While several sensor technologies could satisfy the MAP sensor requirements, a silicon micromachined pressure sensor became the device of choice due to lower cost and smaller

Table 2.7 Vehicle applications for pressure sensors.

Air intake	Tire
Exhaust gas recirculation	Fuel rail
Fuel vapor pressure	Air conditioning fluid
Side impact detection	Brake
Turbo boost	Transmission
Barometer	Steering
Suspension	Oil

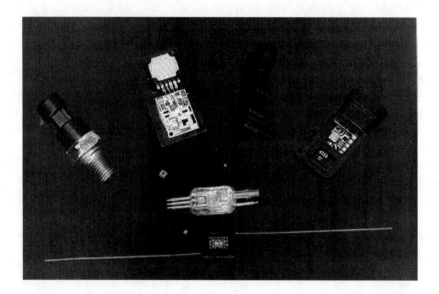

Figure 2.14 A variety of micromachined pressure sensor packages.

size. Today, the volumes of MAP sensors continue to grow and are measured in millions of units per year, virtually matching the total volume of cars and trucks produced by the automotive industry each year (Table 2.7). Further, newer versions of these MAP sensor devices have been designed that take advantage of the advances in the microsystem processing technology.

Figure 2.14 shows a variety of pressure sensor cells and other MEMS chips. Silicon piezoresistive pressure sensors were the first automotive micromachined products. These devices have been manufactured using standard IC processing along with wet silicon etching and wafer bonding. After MAP applications, this sensor technology was quickly applied to barometric and turbo boost monitoring in the automobile. Improvements in silicon etching technology have continued for these devices. Wet silicon etching is accomplished using timed etching, a P+ etch stop process and electrochemical etching. Silicon-to-silicon bonding is replacing anodic and glass frit bonding for forming the reference vacuum. The two dies on the left in Fig. 2.14 are piezoresistive pressure sensors. Die sizes have shrunk and

Table 2.8 Piezoresistive sensor design items.

Diaphragm thickness and area
Resistor placement with respect to cavity edge
Resistor junction depth and doping level
Wheatstone bridge offset voltage and resistance
Die attachment and packaging material

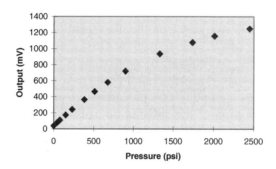

Figure 2.15 The output of a piezoresistive pressure sensor.

wafer diameters have been increased to lower the cost of these sensors. In addition to die size shrinkage, packages have changed with time. Figure 2.14 shows that the earliest MAP microsystems were silicon on ceramic hybrid products. As knowledge of the sensor performance improved, the ability to go to an all-silicon module, without the hybrid substrate, was developed. This type of module is shown on the top right of Fig. 2.14.

A more recently implemented micromachined automotive sensor that is being installed in millions of cars is the fuel vapor pressure sensor. This sensor is used to detect gasoline vapor leaks in the fuel tank in order to reduce raw fuel emissions into the environment. This device is similar to the 20-year-old absolute pressure sensor; however, it measures pressure down to the ± 10 kPa range and is differential. The extreme pressure sensitivity of the fuel vapor sensor makes stresses induced by the package a very important issue.

Table 2.8 gives a list of piezoresistive design issues that must be taken into account when developing a micromachined pressure sensor. The same issues are also relevant when designing piezoresistive devices such as accelerometers or flow sensors. The maximum stress (σ) and hence electrical output range with respect to pressure (P) is controlled by the ratio of diaphragm thickness (t) to the length (L) of a long, square cavity. Equation (2.9) shows how this relationship is expressed with diaphragms of uniform thickness:

$$\sigma = (0.3078 PL^2)/t^2. \quad (2.9)$$

The process used to etch out the cavity must be taken into account during the design process. The theoretical side-wall angle of etched (100) silicon is 54.7°.

Etch angle variation has been observed for different etchants. When etching from the back side of a typical 380-µm-thick wafer, switching from a KOH to EDP solution can cause a change of 20–30 µm in final cavity width. Experimental data are needed to accurately align the edge of the cavity to the placement of the piezoresistive or piezoelectric sensing elements, and so maximize the sensitivity of the device.

Central bosses, or thickened regions, have been employed to improve the performance and extend the range of pressure sensors. To optimize the sensitivity of a piezoresistive or piezoelectric sensor, the placement of the resistor with respect to the cavity edge is critical. The sensing resistor must be placed at this stress maximum, which in micromachined silicon devices is within 10 µm of the cavity edge. Diaphragm breakage is usually not a problem for silicon pressure sensors. Figure 2.15 shows how the output of a piezoresistive pressure sensor varies with pressure. The linear portion of the curve is limited to the lower pressure region. Breakage occurs at around 10 times past the useful range of the sensor. If employed in the linear output region, only during transient pressure spikes does the burst pressure of a silicon micromachined sensor come into play.

Modeling and experimental results have shown that it is important that the diffused resistor be in the top 10 to 15% of the deflecting diaphragm or cantilever. If the junction depth is much deeper than this, a portion of the resistor will not be exposed to the maximum stress induced by the deflection. Doping levels can be utilized to minimize temperature effects.[34] The offset voltage and resistance of the Wheatstone bridge formed on the piezoresistive micromachine must be well characterized prior to design of amplifiers and temperature compensation circuits. Certain circuits may only be able to tolerate positive offset voltages, and overall current draw of the microsystem will generally go up as the resistance of the Wheatstone bridge goes down. Finally, as has already been mentioned, the material used to package the silicon micromachine can have an important influence on the electrical output.

Capacitive pressure sensors have also been developed.[15–17] These devices have an advantage of lower power consumption then piezoresistive devices, and so have been used in remote tires and medical implants. The package shown at the bottom of Fig. 2.14 uses a capacitive pressure sensor for a tire application.

With the exception of consumer and medical systems, most microsystems must work in a range of 40 to 125°C. The output of piezoresistive devices is known to vary significantly with temperature. To overcome this, the amplification ICs used in automotive modules include temperature compensation circuits. Figure 2.16 shows the output of a touch-mode capacitive pressure sensor as a function of pressure and temperature.[15] A touch-mode device has four distinct output regions with respect to pressure. Initially the device acts as a traditional capacitive pressure sensor, showing a low sensitivity. The second region of Fig. 2.16 occurs when the top plate first touches the bottom plate, and is characterized by a very large jump in capacitance over a very small pressure range. The third region in the output slope is the useful region and is typically linear over pressure. The final region is encountered when the top diaphragm cannot deflect any further. In this final section of

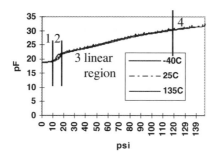

Figure 2.16 The output of a touch-mode capacitive pressure sensor.

the curve, the device is saturated and no longer responds to pressure. It can also be seen in Fig. 2.16 that there is very little change in output of the touch-mode capacitive pressure sensor with temperature. Using the linear portion of the curve can make system design simpler by reducing temperature compensation circuitry requirements.

Piezoresistive sensors also can have nonlinear outputs when operated at high pressures. This is shown in Fig. 2.15. Figure 2.15 is an output plot versus pressure for a 300 psi piezoresistive pressure sensor. Like the touch-mode capacitive pressure sensor, linearity is not maintained at all pressures. The slope of the output versus pressure for both types of sensors decreases at very high pressures. Active calibration data are typically not gathered in this high-pressure region, but the capability of surviving excursion there is needed for overpressure conditions.

Both piezoresistive and capacitive pressure sensors have undergone the standard automotive reliability tests. A difference did show up during electrostatic discharge (ESD) testing.[15] Piezoresistive cells can withstand an 8000-V ESD test, while the touch-mode capacitive pressure cells can only pass an 1800-V ESD test. ESD performance in the capacitive cells can be improved by increasing the cavity dielectric thickness. However, increasing the thickness of this layer decreases the sensitivity of the device. At the module level, the pressure cells are generally protected from ESD events by ICs or discrete components. ESD is still an issue that must be addressed at the capacitive cell level during die handling and assembly.

The silicon-to-silicon bonded touch-mode capacitive pressure sensor is much smaller and more reliable than the earlier silicon-to-glass bonded devices discussed in the literature.[81,82] The size reduction and process simplification leads to lower sensor cost. Using standard aluminum bond pads instead of heavily doped silicon or very thin Au/Cr wire-bonding sites leads to better reliability. Employing a production-proven silicon direct-bond process to form the sealed vacuum chamber instead of plugging a small hole with glass frit, or covering a thin metal crossunder is also a more reliable process. Using silicon-to-silicon bonding overcomes the thermal buckling problem that has been observed on capacitive pressure sensors made by bonding silicon to Pyrex glass.[81] These improvements are all part of the process of making a new MEMS product manufacturable.

Table 2.9 Comparison of piezoresistive and touch-mode capacitive pressure sensor technology.

Piezoresistive	Capacitive
High-volume experience	Low power usage
Small size	Small size
Zero pressure information and wider pressure range	Better temperature Performance
Better ESD resistance	Resistant to packaging stress

Table 2.9 presents a comparison of the two micromachined pressure sensor technologies. Capacitive technology certainly is superior with respect to electrical current use. This can be of utmost importance in remote battery-powered applications. The improved performance over temperature is another advantage for the capacitive sensor, resulting in simpler circuitry. Piezoresistive pressure sensors have been in use for over 20 years, and more than 250 million such devices have been utilized by the automotive industry. High-volume experience is certainly an advantage for this type of product. Piezoresistive sensors have a proven track record and can be made at a low cost. Temperature-compensation ICs do exist for piezoresistive sensors, but their use results in a slightly larger overall system than required for a capacitive module. Calibration over temperature is an expensive process. Unlike a touch-mode capacitive sensor, the piezoresistive device works well from a zero pressure point through the range of interest. While this is important for some applications, in others, such as tire, oil, and braking, it is only necessary to know when pressure levels are getting low. Since capacitive touch-mode pressure sensors use thin dielectric layers, they can be more sensitive to failure from electrostatic discharge or electrical overstress than sensors based on a Wheatstone bridge.

Additional areas of growth for future micromachined pressure sensors include monitoring suspension fluid, oil, fuel, air conditioning fluid, transmission fluid, and steering fluid.[15,68,69] Because oil and brake fluid dissolve or cause swelling of protective gels and parylene, stainless steel diaphragms have been used to fabricate high-pressure sensors with micromachined silicon. The sensor in the top left of Fig. 2.14 is a high-pressure sensor. Two silicon-based sensors have emerged for these applications. In one type, a standard silicon element is immersed in silicone oil and covered with a corrugated stainless steel diaphragm. Pressure on the steel diaphragm is transmitted through the incompressible oil to the silicon diaphragm. Both absolute and differential silicon cells have been used to make this type of package. The other type of silicon high-pressure sensor in use mounts or deposits a thin silicon strain gauge on the back side of a glass-coated steel diaphragm. This silicon strain gauge is wire bonded to a circuit board containing integrated circuits

OPERATION AND DESIGN OF MEMS/MOEMS DEVICES   81

and other components. In both packages, a steel diaphragm protects the micromachined silicon from the corrosive environment.

2.3.2 THE DESIGN OF ACCELEROMETERS

Prior to the 1990s, non-MEMS, ball-in-tube acceleration sensors were used for relatively high-cost, inflatable restraint systems when these were vehicle options. When mandated, the need for a lower cost motion sensor became apparent. Micromachined accelerometers were able to fill this need, and have replaced most original impact sensing technologies in the automobile.[36,37,73]

In the mid-1990s, micromachined accelerometers began to appear in cars for detecting frontal impacts and were associated with airbag deployment. Initially the same type of bulk micromachining processes were used to produce these devices as had been used to fabricate pressure sensors. Currently a mix of bulk-etched sensors and surface micromachined devices are being used in the automobile for crash detection. Similar micromachined devices are now being used to detect side impacts. High-g accelerometers for side impact[37] and low-g accelerometers for ride control applications are areas of future growth for this type of micromachined motion sensor.

The piezoresistive acceleration sensor consists of a thin paddle connected in a cantilever fashion to a supporting frame, as shown in Fig. 2.6. The paddle is free to move, and the mass of the paddle responds to acceleration. Piezoresistors are formed in the thin beam close to the point of attachment. As the paddle is deflected, the resistance value changes in response to the strain. The resistors are arranged in a Wheatstone bridge, which translates the paddle deflection into a differential voltage.

Figure 2.17 shows the process flow and a cross-sectional schematic of the triple wafer stack employed to make these sensors. Standard integrated circuit processes are first used to create piezoresistors and the sensor circuit, as shown in Fig. 2.17(a). An electrochemical etching process is used to etch a cavity into the back side of the wafer [Fig. 2.17(b)], forming a thin diaphragm of the epitaxial layer. At this point, the device is essentially a pressure sensor.

A silicon backplate wafer is bonded to the back side of the sensing slice, forming a two-wafer stack. Virtually any kind of wafer-to-wafer bonding process can be used with the smooth surfaces.

After bonding, a dry or plasma etching process is used to cut through the diaphragm from the front side, forming a cantilevered beam of a paddle shape. The piezoresistive sensing element is now free to move.

A separate cap wafer is made using wet etching processes, and is bonded to the two-wafer stack, as shown in Fig. 2.17(f). Since silicon steps are to be covered, traditional anodic and silicon fusion bonding cannot be employed. This top cap protects the accelerometer during the harsh wafer dicing process. Without this top cap, the silicon cantilevers may be broken off under the high-pressure water spray

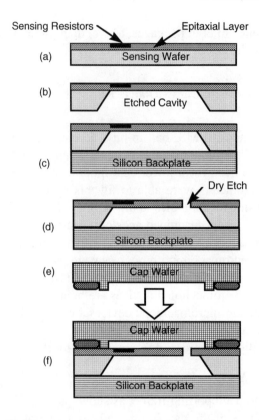

Figure 2.17 The process flow of a piezoresistive accelerometer.

used during wafer sawing.[36] Reducing the sawing water flow can prevent breakage; however, it leads to silicon particle adhesion to the exposed aluminum bond pads.[83] Any kind of particles near micromachined sensors can present a quality problem for unprotected devices in cavity packages. Exposure to particles can continue during packaging and assembly. The complete packaging of a micromachine at the wafer level in a clean room minimizes the chances of particle contamination. Not only does the top cap prevent particle contamination and breakage, it also is used for damping cantilever motion to improve sensor performance, and enables the die to be mounted directly to a circuit board.

Accelerometers of this type are used in two automotive passive-restraint control systems. In the first, the accelerometer is mated with a signal conditioning and temperature compensation integrated circuit, and assembled in a custom multichip package suitable for mounting on a circuit board. After the accelerometer assembly is tested and calibrated over the temperature range, it is assembled into a side-impact passive restraint control module. The control module contains an accelerometer, microcomputer, and software algorithm to determine when airbag deployment is required.

In more advanced systems, the accelerometer is teamed with a custom integrated circuit that contains both the signal conditioning and crash detection func-

Table 2.10 Accelerometer design items.

Cantilever thickness, length, and width
Mass
Cross-axis sensitivity
Damping
Cantilever breakage or motion impediments
Warpage due to film stress
Fundamental resonant frequencies

tions. The sensor, custom IC, and off-chip components are assembled into a rugged hybrid module that is mounted directly to the vehicle structure. The system is unique in that it contains all of the functions necessary to determine when airbag deployment is required. The unit can be calibrated for either side- or front-impact applications for a wide variety of vehicles. The system performs continuous diagnostics and communicates with the central airbag control module via a two-wire, current loop interface.

Key accelerometer design issues are listed in Table 2.10. Micromachine mass and cantilever dimension are to a first order the primary design parameters. In addition to controlling sensitivity, these items prevent breakage. Micromachined wafer capping or other protective measures are needed during wafer sawing and die handling to eliminate fracture during packaging. Secondary but important design issues are damping, particularly for microstructures with a large area mass and cross-axis sensitivity, the resonant frequency of the device, and squeeze film damping effects. Cross-axis sensitivity can lead to error when unwanted accelerations are detected from the wrong directions. For industrial, aerospace, and automotive applications, system vibrations under 2000 Hz are common. The microstructure resonant frequency should be higher than this value or extensive package dampening employed to keep resonant vibrations from being interpreted as false accelerations.

Finally, film stress on or in the thin cantilever and loose particles should be considered when designing an accelerometer. If film stress is high enough, the cantilever can be deflected up or down while at rest. In both cases this can lead to cross-axis sensitivity. If it is deflected down, surface micromachined structures can touch the surface of the silicon chip. This contact, like the presence of particles, can impede motion and again lead to error.

2.3.3 THE DESIGN OF RESONANT MICROSENSORS

The next motion sensor that is seeing high-volume application in the automotive industry is the angular rate sensor.[13] These micromachined sensing devices have been applied to vehicle control systems in both the yaw and roll axis. Improved braking, safety, and navigational assist are areas in which these devices are being employed. Angular rate sensors are also used in camcorders, robotic control, virtual

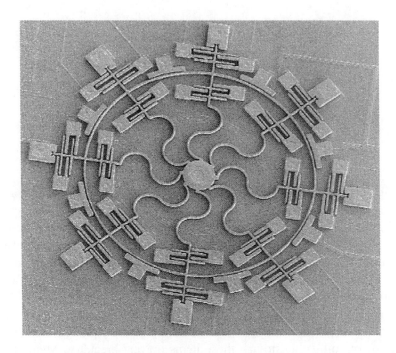

Figure 2.18 A resonant angular rate sensor.[14]

Table 2.11 Design issues of resonant micromachines.

Resonant frequencies
Vacuum level and/ or Q
Cyclic fatigue, material, and drive level

reality headsets, and various aerospace applications. Various MEMS angular rate sensors utilize tuning forks, combs, and rings.

Figure 2.18 shows a CMOS integrated, electroformed ring device used for angular rate sensing. The metal micromachine is electroformed on top of a standard CMOS wafer as illustrated in Fig. 2.19. Once the electroformed micromachine is completed, wafer-to-wafer bonding is employed to seal the resonant stucture in a vacuum. Vacuum sealing is required to obtain a high signal or Q value. Design issues associated with this product include planarization and placement of the underlying CMOS metal runners,[84] prevention of stiction with microstructures, wafer alignment between the cap slice and micromachined CMOS wafer, and the overall integration of the process and design.

The 1-mm-diameter micromachined ring gyroscope shown in Fig. 2.18 is driven into resonance. Nodal points are monitored capacitively for any change. When the chip is rotated, the Coriolis effect causes a change in the standing-wave pattern. This pattern shift leads to a signal change at the nodes, which is used to

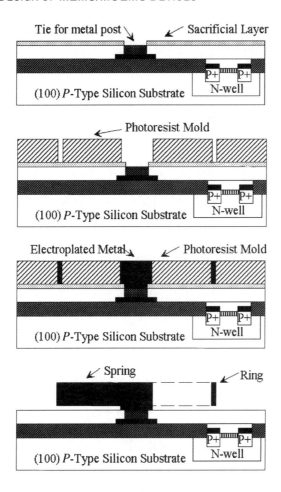

Figure 2.19 The process flow for making a resonant angular rate sensor.

Table 2.12 X–Y vibrational modes for a 1-mm-diameter ring.

Mode	Typical frequency (Hz)
Torsional	8,000
X and Y translational	13,500
X–Y compression	20,000
Tri-modal compression	30,000

sense angular rate. The symmetrical ring structure provides excellent linear vibration rejection. Vibration and shock due to rock impingement and power train vibration are a real issue in vehicular use. Cantilever angular rate sensors have had performance problems caused by such mechanical stimulus. It was necessary to CMOS amplifiers to the MEMS element because of the small signal size produced.

Generic design issues that must be taken into account when developing a resonant microstructure must start with the resonant frequencies (Table 2.11). In general, only one resonant mode is used for sensing. Other resonant peaks can lead to error and should be located as far away from the desired resonant peak as possible. This can be accomplished efficiently through the use of computer simulation. Table 2.12 shows the type of resonant modes and their frequencies for a 1-mm-diameter, 15-μm electroformed ring.[13] The X–Y compressive mode is utilized by the sensor.

Spring stiffness, length, and shape and well as mass and mass distribution can all be used to control the distribution of resonant frequencies. Resonant frequency shifts as chemicals are absorbed onto a film attached to a resonanting beam have been employed to produce chemical sensors.[39] The effect of mass changes on resonant frequency can be modeled prior to fabrication. Process variations that lead to linewidth and thickness variations can also be simulated to obtain an indication of how robust a resonant sensor design is. Mass changes due to linewidth variation in typical photolithography processes can have a significant effect on frequency.

The design and material used for a micromachine influences the Q of the resonating device. The Q is calculated by dividing the resonant frequency by the frequency delta at 3 dB below the peak. The damping and/or vacuum level of the package has a big influence on the Q of the micromachine. Low Q due to viscous damping has restricted the application of resonant devices to chemical sensing.[38] In air without a vacuum, Q values typically range from 100 to 500, depending on damping design parameters. Under vacuum, Q values can be increased to 10 K–100 K for silicon, and 1 K to 3 K for nickel. Cyclic fatigue can be a problem with micromachined resonant structures,[85] just as it is with macromachined structures.[86] Drive levels, mechanical design, and material selection can control this problem. Materials selection for resonant microstructures has trade-offs. Silicon is ductile, deforming through dislocation motion above 700°C.[87,88] At room temperature silicon fails via fracture. Complex resonant polysilicon micromachines have had fracture[89] and cracking[85] problems during normal handling or resonant operation. Electroformed microsystems have not exhibited these failure modes because of the more ductile room temperature nature of the metals employed.

Stiction occurs when two surfaces stick together. This has been a problem with light, large-area resonant and capacitive microstructures, such as those used for motion sensors.[84,90–92] There are examples of stiction in the literature for silicon,[90] dielectric,[91] and electroformed[84,92] micromachines. Design rules limiting the proximity of vertical and horizontal features are required. Special microbumps can be used to prevent stiction.[84] Coatings during processing have also been employed to keep pieces from sticking together.[93]

Other problems associated with resonant sensor design include proper substrate connection design to optimize Q by reducing mechanical loss. Residual film stress on resonant beams has been a source of performance degradation with time.[38]

2.3.4 MOEMS DESIGN CASE STUDY: DEFORMABLE MIRRORS

Replacement of macroscopic optical processing with MEMS-based alternatives offers higher bandwidth, lower cost, smaller size, and easier integration. The silicon substrate itself provides an optically smooth, flat platform that can be built up through surface micromachining or etched away through bulk micromachining to manufacture optomechanical systems. The design process for MOEMS devices couples the traditional challenges associated with MEMS electromechanical design to additional constraints imposed by the required optical performance. The severity of such additional design challenges is strongly affected by the optical function of the device.

A microsystem that offers a rich matrix of optical, mechanical, and electronic design issues is the micromachined deformable mirror (μDM). Such devices consist of a membrane mirror shaped by an array of surface-normal actuators. Most deformable mirrors (DMs) that are commercially available today are macroscopic devices made with flat glass mirror plates supported by an array of piezoelectric actuators. Deformable mirrors based on micromachining technology promise to alter the field of adaptive optics (AO) by providing a low-power, compact, high-performance, and economical alternative to existing mirror systems. The emergence of μDMs is likely to extend the field of adaptive optics from its roots in astronomical imaging systems to commercially important emerging areas such as laser-based communications, biomedical imaging, laser welding, and terrestrial imaging.[29,94–98]

The principles of resolution enhancement by adaptive optics have been established for several decades, and have been applied successfully to a number of large-aperture ground-based telescope systems where the large cost of deformable mirror systems is not prohibitive.[99] Put simply, adaptive optics is a way of improving optical resolution by compensating for fabrication errors (e.g., misshapen or thermally deformed mirrors) or optical path aberrations (e.g., turbulent atmosphere effects). This requires that a compensating mirror be deformed in such a way that unwanted aberrations are measured and then cancelled (usually through a process called "phase conjugation"). In many visible wavelength imaging applications, system requirements include nanometer-scale precision, several micrometers of stroke, hundreds of hertz of bandwidth, and tens to hundreds of actuation zones to achieve diffraction-limited performance.

In the following MOEMS design case study, a recently developed μDM is described. Relevant specifications of the required microsystem are explored, leading to a set of performance goals. Microfabricated mirrors have proven to be considerably less costly, smaller, and faster than competing macroscopic systems, although their optical performance is not yet comparable.

2.3.4.1 Deformable Mirror Specifications

The primary variables defining the performance of a deformable mirror in an AO system are the number of actuators, the control bandwidth, the maximum actuator stroke, and the actuator resolution. For astronomical observations, these design parameters can be estimated based on the desired Strehl ratio, the optical wavelength of interest, the characteristics of the optical disturbance (e.g., intensity of atmospheric turbulence), and the system aperture, using theoretical turbulence and photonic models. The Strehl ratio, S, is a common measure of imaging performance; it is the ratio of the on-axis intensity of an aberrated image of a point-source to the on-axis intensity of a unaberrated image.

Based on assumptions of moderate atmospheric turbulence visible light, and a 1-m diameter aperture, a deformable mirror would require 400 actuators, a closed-loop bandwidth of 100 Hz, a stroke of 2 µm, and a resolution of 10 nm to achieve diffraction-limited performance. These parameters were used as initial design goals for the MEMS deformable mirror detailed in this case study.

2.3.4.2 Advantages of MEMS Deformable Mirrors

Three alternative approaches to the conceptual design of µDMs, each of which has been recently fabricated and tested with some success, are illustrated in Figs 2.20–2.22.[29,96,98] All three designs employ electrostatic actuation to achieve mirror deflection. Electrostatic actuation is preferred in this application because it uses very little power, requires no exotic fabrication steps or nonstandard thin-film materials, and introduces no inherent hysteresis. The topmost figure shows a two-level surface-micromachined structure in which an array of identical actuators is independently attached to a single mirror membrane. This design will be explored in greater depth in the following sections. It offers local mirror shaping, smooth phase continuity, and is closely analogous in concept to the most common macroscopic DMs. Recent work at Stanford[100] has extended this architecture to bulk-micromachined devices, allowing longer stroke and flatter single-crystal mirrors.

The middle figure shows a bulk micromachined membrane supported over an array of electrodes. Because the mirror membrane is clamped along its edges but otherwise is not mechanically attached to the actuator array, the electromechanical

Figure 2.20 Continuous MEMS mirror architecture with post-like attachments.

Figure 2.21 Continuous MEMS mirror architecture with edge-clamped attachment.

behavior of this mirror is quite different from that of the first design. The actuators in the array are highly coupled, and the range of contours achievable with this mirror is more limited (e.g., no points of inflection, no edge deflection or slope). However, an advantage is that the membrane is fabricated separately from the electrode array through bulk micromachining. As a result, it does not suffer from print through or pattern transfer effects that occur in μDMs made by surface micromachining.

The third design is a segmented mirror array in which each mirror element is attached to an actuator that allows piston-motion displacement. Such a mirror array can be made to approximate the conjugate shapes achievable on continuous-membrane mirrors. Also, because there is no mechanical coupling between adjacent mirror segments, the actuator's mechanical stiffness can be reduced sufficiently to allow signal-level (0–5 V) control of mirror position. However, more actuators are needed to achieve comparable wavefront compensation with this mirror array than with either of the continuous mirror devices.

The deformable mirror system to be described in more detailed here incorporates a continuous mirror sheet actuated at discrete points, the deformation being normal to the surface and continuous over a desired range. Figure 2.23 is a schematic showing a cross section through three actuators in a row with a mirror sheet over them. The actuators are fixed-end beams, a structurally robust design. When voltage is applied between an actuator and the substrate or a ground pad, the electrostatic force developed deflects the actuator toward the substrate. The mirror sheet, which is attached to the actuator at midspan by a post, also deflects as a result. The amount of deflection depends monotonically but nonlinearly on the

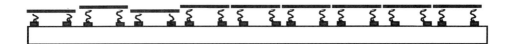

Figure 2.22 Segmented mirror architecture.

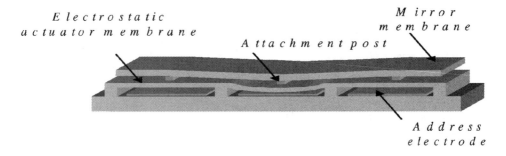

Figure 2.23 Cross-sectional schematic of three actuators in a zonal micromachined deformable mirror. A membrane mirror is attached by posts to an array of parallel-plate electrostatic actuators.

magnitude of the applied voltage. In the schematic, the actuator in the center is deflected, while the others are undeflected.

2.3.4.3 DM Modeling

A number of numerical approaches to modeling electromechanical behavior have been developed. For the mirror being described, the method of finite differences as applied to plate theory was used to numerically simulate the electrostatic actuator mechanics and the continuous deformable mirror mechanics. The general partial differential equation for plate bending is

$$\nabla\nabla w(x, y) = \frac{q}{D}, \qquad (2.10)$$

where w is the deflection of the plate, q is the distributed loading, and D is the flexural rigidity of the plate. The continuous problem was discretized into equally spaced arrays for the actuators and the continuous mirror as shown in Fig. 2.24. The finite-difference equivalent for this equation was then applied to the arrays. The boundary conditions for the actuators consisted of two clamped and two free

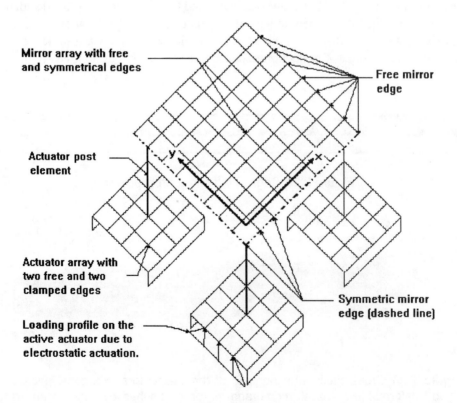

Figure 2.24 Schematic for numerical model of mirror using finite differences.

edges. The clamped-edge boundary conditions are obtained by applying the zero-slope condition at the edges. The boundary conditions on the membrane mirror required that all edges be free. The free edge boundaries yield two new differential equations,

$$\frac{\partial^2 w}{\partial x^2} + v\frac{\partial^2 w}{\partial y^2} = 0, \qquad \frac{\partial^3 w}{\partial x^2} + (2-v)\frac{\partial^3 w}{\partial x \partial y^2} = 0, \qquad (2.11)$$

which represent zero moment and zero shear on the free edge when the edge is normal to the x-axis. These equations are solved simultaneously with the homogeneous plate-bending equation to obtain the simulated deflections.

The loading conditions on the electrostatically activated actuator were determined using a deflection-dependent load formula,

$$q_{i,j} = \frac{k^2 \varepsilon V^2}{2(q - y_{i,j})^2}, \qquad (2.12)$$

where k is the dielectric constant for air, ε is the permittivity of free space, V is the voltage, g is the gap between the actuator and the electrostatic pad, y is the deflection, and i, j are the discretized locations in the x, y domain of the actuator array. Since the loading on the actuator varies with the inverse square of the deflection, extreme loading conditions (high voltage on the actuator) result in solutions that converge to deflections that are beyond geometrically constrained limits. These solutions represent loading conditions that are beyond a critical voltage and for which solutions cannot be obtained with the finite-difference method.

The loading condition on the mirror as a result of the electrostatic actuation was obtained by matching deflections of the mirror and the actuator at the location where the two are attached by a post. An iterative process was used to find the correct load that resulted in matched deflections between each actuator in the array.

A relaxation method was used to minimize the finite-difference residuals during iteration. Figure 2.25 shows simulated deflections of a nine-actuator mirror with free edges. These compare favorably with the deflections measured for a mirror subjected to the same conditions.[101]

2.3.4.4 Design of a Fabrication Process

As with most MEMS devices, electromechanical and functional design are strongly coupled to fabrication processes. "Design for manufacturing," a concept that has recently emerged as an important consideration in macroscale production systems, has always been essential for microsystem fabrication.

For this device, surface micromachining was used. Alternating thin films of structural material (polycrystalline silicon), and sacrificial material phosphorsilica glass (PSG) were deposited, patterned, and etched to create the mirror and actuator.

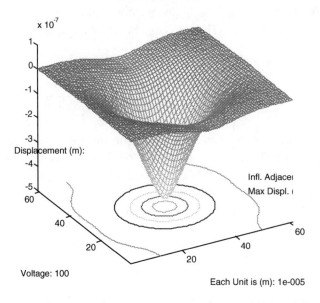

Figure 2.25 Theoretical deflections for a continuous mirror with free edges over a 3 by 3 array of actuators with voltage applied to the center actuator.

A 0.5-μm-thick base nitride layer insulates the surface-micromachined layers from the underlying silicon wafer. The first polysilicon layer, measuring 0.5 μm-thick, is used to create an array of electrode pads for the electrostatic actuators. It is also used for routing polysilicon wires on the substrate surface. The first PSG layer defines the space that eventually becomes the air gap for the parallel-plate surface-normal electrostatic actuators. This layer was made 5 μm thick to allow a useful actuation stroke of approximately 2 μm before electrostatic instability. On top of the first PSG, a 2-μm-thick polysilicon layer is patterned into an array of fixed-end beams that act as the movable electrodes for each of the electrostatic actuators. A second PSG layer, 2 μm thick, creates a gap between the electrostatic actuators and a second polysilicon layer, measuring 2 μm thick, which serves as the mirror membrane. Anchor holes in the second PSG allow attachment of this polysilicon mirror to the center of each electrostatic actuator.

2.3.4.5 Mirror Planarity

The most significant source of contour or figure error* in the micromachined optical devices produced on a smooth, flat substrate can be attributed to "print through" of multiple conformal, patterned thin films. When a cut is made in one film to define the desired two-dimensional pattern, subsequent films are unintentionally

---

* Contour error and figure error, used interchangeably in this document, refer to the root-mean-square deviation of the actual surface from the intended surface (in the surface normal direction).

given additional topography as they conform to the underlying contour. For this type of contour error, the amplitude of the error on the uppermost thin film can be (and generally is) as large as the cumulative thickness of all underlying layers. This is a prominent design and fabrication issue in optical applications requiring smooth surfaces.

Early versions of the device described here exhibited excessive nonplanarity on the mirror surface, owing the conformal growth processes inherent in surface micromachining. For the μDM, contour variations (e.g., nonplanarity) on the mirror surface generally added up to about 5 μm. Since the mirror itself has a thickness of less than 2 μm, postfabrication planarization through polishing or lapping is not possible.

A novel design strategy to generate acceptable planar surfaces was developed in collaboration with MCNC.[102] Briefly, this design concept is based on the idea that very narrow cuts in any layer get filled up by subsequent layers, resulting in greater planarity in the uppermost layers.

Acceptably planar surfaces were obtained by adopting this more restrictive design and layout rule for the device: All cuts in polysilicon and oxide layers were restricted to widths no larger than 1.5 μm. With pattern sizes restricted to this width, the conformal deposition processes in micromachining rapidly fill in cuts from previously deposited layers, largely attenuating the magnitude of their "print-through" to subsequent layers. As a result, greater planarity in the uppermost layers is achievable. Anchoring sites, which would be structurally weak if they were only 1.5 μm in width, are fabricated as honeycombs of thin polysilicon walls encapsulating a thicker structure of oxide. Figure 2.26 is a schematic illustrating this design concept.

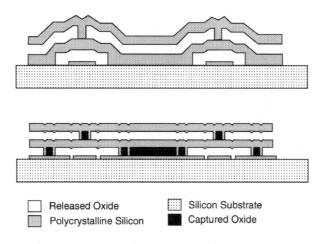

Figure 2.26 Top: Conventional surface micromachined structure crosssection, illustrating nonplanarity. Bottom: Crosssection of new design for continuous deformable mirror to achieve planar mirror surface.

The success of this processing strategy for planarization is illustrated in Fig. 2.27. The bottom SEM photo is of a mirror supported on sixteen actuators, fabricated without design rules favoring planarization. The underlying features

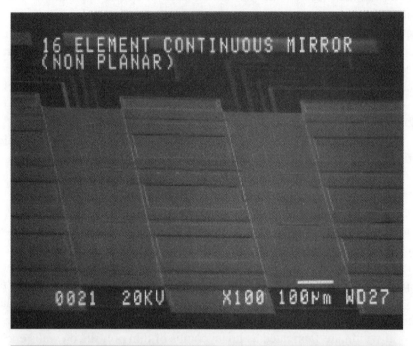

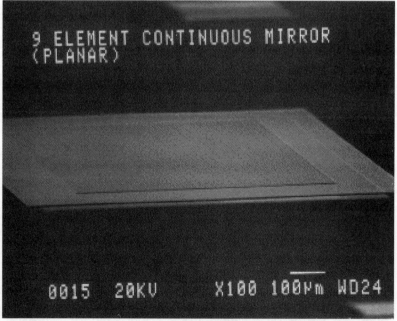

Figure 2.27 Planarization results for surface-micromachined mirrors.[102]

(electrostatic actuators and polysilicon electrodes) emboss the mirror membrane, leading to poor planarity. At the top is a mirror supported over a nine-element array of actuators. In the planarized mirror, no "print through" of underlying structures is observed. Examination using scanning microscopy indicates significant gains in surface planarity achieved through this technique.

A geometric model was developed to calculate the topography generated by features in the underlying layers. The model assumes sharp corners and vertical side walls for all features created by reactive ion etching. This is a reasonable assumption except in the case of very thick layers, where the side wall angle may become significant. At every point on the underlying surface, the film grows at the same rate normal to the local tangent to the surface at that point. The development of an actual profile depends on the material of the film being deposited, the process parameters, the edge profile of the underlying features, and the loading. For the purposes of a mathematical model, an index c is defined which characterizes the extent of conformal deposition. This index describes the ratio of film growth rate in the vertical direction to the film growth rate normal to the side wall at every point.

The effect of a decreasing gap (from 5 μm to 1.5 μm) has been fabricated as well as modeled. SEM photos and corresponding model results are detailed in Fig. 2.28 for topography generated by a gap in the first PSG layer. To generate this topography, a gap g was patterned into PSG1, and all the other films (Poly1, PSG2, Poly2) were deposited without any patterning over the gap. The oxides were released using hydrofluoric acid. The nonplanarity h from the final layer, Poly2, is included with the computer-generated topography for each case. A good qualitative match can be seen between the experimental and computed results.

Systematic avoidance of this nonplanarity is possible through the design-based algorithm detailed in the previous section. Using this algorithm, with cuts restricted to no wider than 1.5 μm, nonplanarity due to printthrough has been reduced to less than 50 nm over large areas of a complex, seven-layer device.

Additional reductions in printthrough can be achieved with is the addition of a processing step to physically flatten films at critical stages of production, thereby reducing or eliminating nonplanarity from all previous layers. The standard technique, chemomechanical polishing (CMP), has a long history of success in CMOS fabrication, and has recently found use in the development of MEMS micromirrors. Of course, CMP is only useful if the cumulative nonplanarity to be eliminated is significantly smaller than the thickness of the film being polished. An advantage of CMP for planarization is the improvement in surface smoothness that accompanies this process.

2.3.4.6 Actuator Design and Characterization

The heart of the actuation system for these devices is an array of parallel plate electrostatic actuators, consisting of 250-μm square diaphragms suspended several micrometers above similar size electrodes fixed to the substrate. These diaphragms are anchored to the substrate along two opposing edges. The critical issues relating to device performance are yield (indicating process reliability), position repeatability, and frequency response. In a systematic evaluation of these parameters, hundreds of individual actuators were tested. The results were promising. Fabrication through conventional surface micromachining yielded 95% useful, working actuators. Electromechanical performance was evaluated by driving the actuators with a high-speed voltage controller and measuring the dynamic motion response. It was found that actuators exhibited no hysteresis (when not driven beyond their insta-

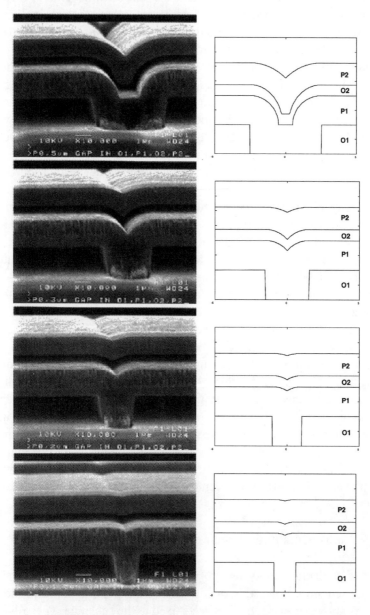

Figure 2.28 SEM micrographs and predicted topographies for gaps in oxide1.

bility voltage), and that they could be repeatably positioned to a precision of 10 nm over a 2-μm stroke. Moreover, their frequency response was measured to be at least 7 kHz for actuators beneath a membrane mirror.[103]

The actuators exerted influence functions (i.e., coupling through the mirror) that resulted in approximately 10% deflection at adjacent actuator locations. The magnitude of this influence is appropriate for adaptive optics applications. Larger influences would introduce excessive crosstalk between adjacent actuators, and further complicate dynamic control of the mirror. On the other hand, small influences would limit smoothness on the mirror surface, degrading optical performance.

Position repeatability was measured for several actuators through ensemble averaging of the data from a series of ramp actuation tests. A typical data set consisted of 100 sample runs to produce an ensemble of records of deflection as a function of voltage. Experimental conditions were kept as statistically similar as possible over the course of acquiring a set of data. The applied voltage was in the form of a ramp from 0 to 50 V with deflection data measured at 0.5-V increments, followed by a 50–0 V ramp. Since no hysteresis was observed, displacement data corresponding to the rampup and the rampdown were treated as identical, resulting in 200 sample data points at each voltage value. The data were analyzed using standard statistical analysis procedures for random data.

The results are presented for an ensemble of 200 records, with a measurement resolution of 2.5 nm. Figure 2.29 shows the ensemble-averaged displacement curve with the sigma limits corresponding to 99% probability. The limits were calculated based on the normal curve, using the unbiased ensemble mean and standard deviation estimates. The average standard deviation along the curve is 1.89 nm, resulting in an average position repeatability of 9.76 nm (for 99% confidence levels).

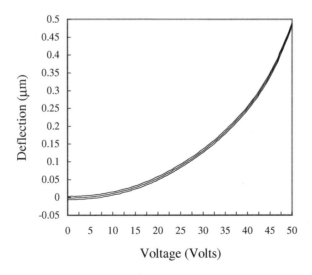

Figure 2.29 Position repeatability of actuator (ensemble averaged mean curve with limits corresponding to 99% probability).

2.3.4.7 Stress and Strain in MOEMS Thin Films

The performance of a microoptoelectromechanical system is often limited by the optical quality of the micromachined devices that are used to reflect or refract light. The processes used to make micrometer-scale structures typically result in figure errors and optical scattering that exceed the tolerances used for producing larger-scale components. While the principal sources of contour errors, and optical absorption in surface micromachined devices are well understood, particularly for silicon and aluminum mirror elements. Compensating for these errors is a sugnificant challenge.

In the last section, gross contour errors in MOEMS devices due to print-through were described, along with a design-based algorithm for mitigating them. A second source of contour errors in micromachined structures is the residual stress that results from layering many dissimilar films in a composite sandwich to achieve complex mechanical function. These films are deposited at temperatures significantly above ambient, and they are frequently doped to modify their electrical behavior. When the sacrificial layers of the device are dissolved, residual stresses in the elastic structural layers are partially relieved by deformation of those structures. Without proper annealing and processing, such stresses can be sufficient to cause fracture in the structural film, or to buckle a device, or to cause dramatic out-of-plane bending of released structures. The presence of residual stresses and strains probably represents the single most ubiquitous problem associated with surface micromachining for optical applications.

Even with careful consideration paid to annealing and process control, residual stresses can vary dramatically from batch to batch with the same process, and significantly from wafer to wafer in the same batch. Variations in stress through the thickness of the film are particularly troublesome from an optical standpoint because they cause curvature of the membrane making it difficult to obtain an optically flat mirror.

The residual stresses in micromachined silicon thin films are caused by the deposition (e.g., LPCVD) process. The steps responsible for stress generation vary considerably from one fabrication facility to the next, or for different devices made within the same facility. For the purposes of discussion, nominal process steps for a common foundry process (MCNC multiuser-micromachining program-MUMPS) are detailed here. MUMPS represents a time-tested relatively repeatable process with reasonably low-stress polycrystalline silicon films. The silicon is vapor deposited, typically on top of a sacrificial phosphorus silica glass (PSG) film, at a temperature of ~550°C. After cooling, the biaxial tensile stresses in the film are typically more than 100 MPa in magnitude owing to a mismatch of the thermal expansion coefficient between the silicon and PSG films. A PSG film is then deposited on the top of the silicon before annealing at 1050°C for 1 hr. Postdeposition annealing serves two purposes: It dopes the polycrystalline silicon by diffusing phosphorus from the PSG layers and it reduces this tensile stress in the as-deposited polycrystalline silicon film. Typically, this results in a net compressive

OPERATION AND DESIGN OF MEMS/MOEMS DEVICES

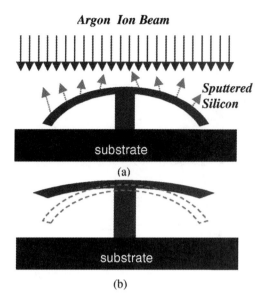

Figure 2.30 Accelerated argon ions uniformly etch the upper surface of the test structure (a) thereby modifying the residual stress gradient profile and changing the curvature of the membrane (b).

biaxial stress in the film. Average biaxial stresses in a film deposited on a wafer can be measured indirectly by destructive testing; the film is etched away in a reactive ion etcher. Wafer bow is recorded before and after this process. Through an elastic plate-bending formula, one can infer the average biaxial stress that was present in the layer that was removed.

It is difficult to reduce residual stress gradients (and consequently their effects on deformation) in the course of processing itself since relatively significant variations in these gradients occur within a single batch of wafers that are processed simultaneously.

Recently, an attempt was made to modify the residual stress profile, and consequently the shape, of test mirrors by removing thin layers of the top of the membrane. This was achieved through neutral ion machining with a uniform argon-ion beam. A schematic of the process is depicted in Fig. 2.30.[104,105]

Test mirrors were machined for a period of time, then removed from the ion machining system and measured using the interferometric microscope. The inverse of the radius of curvature was plotted as a function of the silicon film thickness, as measured periodically during the ion machining test. An important result was that the test mirrors did in fact change shape repeatably and controllably as a result of the ion beam machining. More important, the effect was such that nearly perfect planarity was achieved.

The ion beam immediately (within the first 10 s) affected the mirror curvature, driving it into a more convex shape. This was caused by the *creation* of a thin compressive layer as a result of ion bombardment.

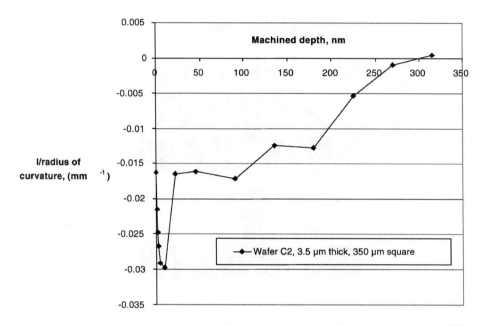

Figure 2.31 Graph of experimental results.

After this initial effect, a more gradual change in mirror shape accompanied additional ion machining time. In total, 300 nm was removed from the initially 3.5-μm-thick silicon mirrors. The mirror became flatter with continued machining until it was essentially planar (within ±10 nm). Upon continued machining, the mirror edges began to curl progressively into a concave shape. Surface roughness of the mirrors was unaffected by this process.

Figure 2.31 is a graph of the results on one test mirror. The rapid formation of a compressive layer on the top surface of the membrane due to ion bombardment can be inferred by the immediate and steep change in curvature upon commencement of machining. As expected, this additional compressive stress causes the membrane to become more convex. After several minutes of machining, the effect of thinning the membrane and thereby removing layers of residual stress due to fabrication begins to cause a reduction in membrane curvature. Because continued machining causes the membrane to become progressively less convex, it can be concluded that the layers removed by ion machining on this upper layer of the film were predominantly compressive.

Interferometric surface maps of one test mirror before machining (radius of curvature: 60 mm; total sag: 160 nm peak-to-valley) and after machining to the point of optimum flatness (radius of curvature: 1400 mm; total sag: 18 nm peak-to-valley) demonstrate that this process has the potential to significantly affect postrelease deformation in mirror structures.

The electromechanical performance of the μDM described here is detailed in Table 2.13.

Table 2.13 Micromachined deformable mirror characteristic.

Optical Characteristic	
Clear aperture of mirror	3.3 mm
Effective fill factor	99.95%
Unpowered mirror surface figure	50 nm rms, 470 nm PV
Unpowered segmented surface figure	6.7 nm rms, 36 nm PV
Electro-mechanical Characteristic	
Number of actuators	140
Actuator configuration	12 × 12 Square grid (w/o corners)
Actuator spacing	300 μm
Inter-actuator coupling	15%
Stroke	2 μm
Hysteresis	0%
Drive voltage	240 V max
Lifetime	500 M cycles @ 1/2 full stroke
Temporal open loop bandwidth	6.6 kHz (−3 dB @ 50% full stroke)
Inpulse response	Critically damped, 25 μs time constant
Bandwidth	DC-6.7 kHz

2.3.4.8 Case study summary

A design for a continuous-membrane micromachined deformable mirror has been detailed. It was found that in addition to the conventional design criteria that pertain to electromechanical microsystems, MOEMS devices such as this deformable mirror require more elaborate consideration of print-through, planarity and thin film stress gradients. However, when these additional design constraints are met, the resulting optical microsystems can provide opto-electromechanical performance that enable unprecedented optomechanical device performance.

2.4 SUMMARY

There are many technical issues that must be faced when designing and developing micromachined sensors and actuators. The design of a particular sensor or actuator must take into amount several areas: the basic principle of operation, microsystem and process modeling, circuitry design, process technology, application environment, packaging, and a manufacturing system. Because of the multifaceted nature of sensor and actuator development, these aspects must be addressed globally in a common design at the start of development, rather than summing the individual parts at the end. Examples for micromachined pressure sensors, accelerometers, resonant devices, and deformable mirrors have been used to illustrate many of the factors encountered in MEMS design.

REFERENCES

1. F. Pourahmadi, L. Christel and K. Petersen, "Silicon Accelerometer with New Thermal Self-Test Mechanism," in *Technical Digest of the IEEE Solid State Sensor and Actuator Workshop*, pp. 122–125 (1992).
2. R. Srinivasan, I. Hsing, J. Ryley, M. Harold, K. Jensen and M. Schmidt, "Micromachined Chemical Reactors for Surface-Catalyzed Oxidation Reactions," in *Technical Digest IEEE Solid State Sensor and Actuator Workshop*, pp. 15–18 (1996).
3. M. J. Vellekoop, "Integration of Physical Chemosensors," in *Proceedings of 1998 Microsystems Symposium*, p. 117 (1998).
4. C. Von Benken, "Next Generation IR Gas Measurement," *Sensors* 15, 49 (1998).
5. J. Lin and E. Obermeier, "Capacitive Thin Film Gas Sensor with Signal Processing System for Determination of SO_2," *Sensors Actuators B* 15–16, 319–322 (1993).
6. W. Mokwa and U. Schnakenberg, "On-Chip Microsystems for Medical Applications," in *Proceedings of 1998 Microsystems Symposium*, p. 69 (1998).
7. B. van der Schoot, M. Boilat and N. de Rooij, "Micro-Instruments for Life Science Research," in *Proceedings of 1998 Microsystems Symposium*, p. 111 (1998).
8. P. Shrewsbury, S. Muller and D. Liepmann, "Characterization of DNA Flow Through Microchannels," in *Modeling and Simulation of Microsystems*, pp. 578–580 (1999).
9. J. Harrison, K. Fluri, K. Seiler, Z. Fan, C. Effenhauser and A. Manz, "Micromachining a Miniaturized Capillary Electrophoresis-Based Chemical Analysis System on a Chip," *Science* 261, 898–897 (1993).
10. P. Wilding, M. Shoffer and L. Kricka, "PCR in a Silicon Microstructure," *Clin. Chem.* 40, 1815–1818 (1994).
11. A. van den Berg and P. Bergveld, *Micro Total Analysis Systems*. Kluwer, Boston, MA (1994).
12. C. Mastrangelo, M. Burns and D. Burke, "Microfabricated Devices for Genetic Diagnostics," *Proc. IEEE* 86(8), 1769–1787 (1998).
13. D. Sparks, S. Zarabadi, J. Johnson, Q. Jiang, M. Chia, O. Larsen, W. Higdon and P. Castillo-Borelley, "A CMOS Integrated Surface Micromachined Angular Rate Sensor: Its Automotive Applications," in *Transducer '97*, pp. 851–854 (1997).
14. S. Zarabadi, T. Vas, D. Sparks, J. Johnson, Q. Jiang, M. Chia and E. Borzabadi, "A Resonant Comb/Ring Angular Rate Sensor Vacuum Packaged via Wafer Bonding," *SAE Technical Proceedings*, No. 1999-01-1043, p. 95 (1999).
15. W. Baney, D. Chilcott, X. Huang, S. Long, J. Siekkinen, D. Sparks and S. Staller, "A Comparison between Micromachined Piezoresistive and Ca-

pacitive Pressure Sensors, *SAE Technical Proceedings*, No. 973241, pp. 61–64 (1997).
16. Y. Lee and K. Wise, "A Batch-Fabricated Silicon Capacitive Presssure Transducer with Low Temperature Sensitivity," *IEEE Trans. Elect. Dev.* ED-29, 42–48 (1982).
17. Y. Zhang and K. Wise, "An Ultra-Sensitive Capacitive Pressure Sensor with a Bossed Dielectric Diaphragm," in *Technical Digest of the IEEE Solid State Sensor and Actuator Workshop*, pp. 205–209 (1994).
18. B. Ziaie, N. Kocaman and K. Najafi, "A Generic Micromachined Silicon Platform for Low-Power, Low-Loss Miniature Transceivers," in *Transducers '97*, No. 1D4.01, pp. 257–260 (1997).
19. J. Park and M. Allen, "Development of Magnetic Materials and Processing Techniques Applicable to Integrated Micromagnetic Devices," *J. Micromech. Microeng.* 8, 307–316 (1998).
20. H. Baltes and R. Popovic, "Integrated Semiconductor Magnetic Field Sensors," *Proc. IEEE* 74, 1107–1132 (1986).
21. C. Ahn, Y. Kim and M. Allen, "A Planar Variable Reluctance Magnetic Micromotor with Fully Integrated Stator and Coil," *J. MEMS* 2, 165–173 (1993).
22. Y. Sugiyama, "Recent Progress on Magnetic Sensors with Nanostructures and Applications," *J. Vac. Sci. Technol.* B13, 1075 (1995).
23. M. Paranjape, I. Filanovsky and L. Ristic, "A 3-D Vertical Hall Magnetic-Field Sensor in CMOS Technology," *Sensors Actuators A* 34, 9–14 (1992).
24. T. Hasiwa and M. Matsumoto, "Fe-Co-P Thin Films Prepared with Reactive Evaporation Method," *IEEE Trans. Magn.* 24, 2055–2058 (1988).
25. M. Hanazono, S. Narishige and K. Kawakami, "Fabrication of a Thin Film Head Using Polyimide Resin and Sputtered Ni-Fe Films," *J. Appl. Phys.* 64, 2608–2610 (1982).
26. P. Lane, B. Cockayne, P. Wright and P. Oliver, "Metallorganic CVD of MnAs for Thin Film Applications," *J. Cryst. Growth* 143, 237–242 (1994).
27. K. Derbyshire and E. Korczynski, "Giant Magnetoresistance for Tomorrows Hard Drives," *Solid State Tech.* (Sept.), 57–66 (1995).
28. M. C. Wu, N. F. De Rooij and H. Fujita, "Introduction to the Issue on Microoptoelectromechanical Systems," *IEEE J. Selected Top. Quant. Electron.* 5, 2 (1999).
29. T. Bifano, J. Perreault, R. Mali, and M. Horenstein, "Microelectromechanical Deformable Mirrors," *IEEE J. Selected Top. Quant. Electron.* 4, 83–89 (1999).
30. A. A. Yasseen, J. N. Mitchell, J. F. Klemic, D. A. Smith and M. Mehregany, "A Rotary Electrostatic Micromotor 1 × 8 Optical Switch," *IEEE J. Selected Top. Quant. Electron.* 5, 26 (1999).
31. S. Nagaoka, "Compact Latching-Type Single-Mode-Fiber Switches Fabricated by a Fiber-Micromachining Technique and their Practical Applications," *IEEE J. Selected Top. Quant. Electron.* 5, 36 (1999).

32. A. A. Yasseen, S. W. Smith, F. L. Merat and M. Mehregany, "Diffraction Grating Scanners Using Polysilicon Micromotors," *IEEE J. Selected Top. Quant. Electron.* 5, 75 (1999).
33. P. Indermuhle, G. Schurman, G. Racine and N. de Rooij, "Tip Integration on Arrays of Silicon Microcantilevers with Self-Exciting Piezoelectric Sensor for Parallel Atomic Force Microscopy Applications," in *Transducers '97*, No. 2B3.02, pp. 451–453; Fig. 2, p. 452 (1997).
34. Y. Kanda, "A Graphical Representation of the Piezoresistance Coefficients in Silicon," *IEEE Trans. Electron Dev.* ED-29, 64–70 (1982).
35. C. Smith, "Piezoresistance Effect in Germanium and Silicon," *Phys. Rev.* 94, 42–45 (1954).
36. D. Sparks, D. Rich, C. Gerhart and J. Frazee, "A Bi-Directional Accelerometer and Flow Sensor Made Using a Piezoresistive Cantilever," in *Proc. EAEC Conf.*, No. 97A2IV40, pp. 1119–1125 (1997).
37. D. Rich, W. Kosiak, G. Manlove and D. Schwarz, "A Remotely Mounted Crash Detection System," *SAE Technical Proceedings*, No. 973240, pp. 53–59 (1997).
38. R. Howe, "Resonant Microsensors," in *Transducers '87*, pp. 843–848 (1987).
39. R. Howe and R. Muller, "Resonant-Microbridge Vapor Sensor," *IEEE Trans. Electron Dev.* ED-33, 499–506 (1986).
40. E. EerNisse, "Quartz Resonator Frequency Shifts Arising from Electrode Stress," in *Proc. 29th Annual Symp. on Frequency Control*, pp. 1–4 (1976).
41. T. Lammerink and W. Wlodarski, "Integrated Thermally Excited Resonant Diaphragm Pressure Sensor," in *Technical Digest Third Intl. Conf. on Solid-State Sensors and Actuators*, pp. 97–100 (1985).
42. M. Kohl and K. Skrobanek, "Linear Microactuators Based on the Shape Memory Effect," in *Transducers '97*, No. 3A2.04, pp. 785–788; Fig. 6, p. 787 (1997).
43. S. Waltman and W. Kaiser, "An Electron Tunneling Sensor," *Sensors Actuators A* 44, 201–210 (1989).
44. J. Grade et al., "Progress in Tunnel Sensors," in *Technical Digest of the IEEE Solid State Sensor and Actuator Workshop*, pp. 80–84 (1994).
45. R. Kubena, G. Atkinson, W. Robinson and F. Stratton, "A New Miniaturized Surface Micromachined Tunneling Accelerometer," *IEEE Elect. Dev. Lett.* 17, 306–308 (1996).
46. T. Kenny, W. Kaiser, J. Reynolds, J. Podosek, H. Rockstad, E. Vote and S. Waltman, "Electron Tunnel Sensors," *J. Vac. Sci. Technol.* 10, 2114–2118 (1992).
47. M. Chmielowski and A. Witek, "Tunneling Thermometer," *Sensors Actuators A* 45, 145–151 (1994).
48. C. Yeh and K. Najafi, "A Bulk-Silicon Tunneling-Based Pressure Sensor," in *Technical Digest of the IEEE Solid State Sensor and Actuator Workshop*, pp. 123–126 (1994).

49. T. Kenny, J. Reynolds, J. Podosek, E. Vote, L. Miller, H. Rockstad and W. Kaiser, "Micromachined Infrared Sensors Using Tunneling Displacement Transducers," *Rev. Sci. Instrum.* 67, 112–114 (1996).
50. L. Miller, J. Podosek, E. Kruglick, T. Kenny, J. Kovacich and W. Kaiser, "A Micro-Magnetometer Based on Electron Tunneling," in *Proceedings 1996 IEEE-MEMS Workshop*, pp. 467–471 (1996).
51. D. Sparks, "Component Integration and Packaging of Automotive Microsystems," in *Proceedings of 1998 Microsystems Symposium*, p. 37 (1998).
52. L. Castaner et al., "Design and Fabrication of a Low Cost Water Flowmeter," in *Transducers '97*, No. 1B4.08, pp. 159–162 (1997).
53. K. Lee and K. Wise, "SENSIM: A Simulation Program for Solid-State Pressure Sensors," *IEEE Trans. Electron Dev.* ED-29, 34–41 (1982).
54. T. Kurzweg, S. Levitan, P. Marchand, K. Prough and D. Chiarulli, "CAD for Opto-Electronic Microsystems," in *Modeling and Simulation of Microsystems*, p. 687 (1999).
55. Y. Zhang, S. Crary and K. Wise, "Pressure Sensor Design and Simulation Using CAEMEMS-D Module," in *Technical Digest of the IEEE Solid-State Sensor and Actuator Workshop*, pp. 32–35 (1990).
56. R. Harris, F. Maseeh and S. Senturia, "Atomatic Generation of 3-D Solid Model of a Microfabricated Structure," in *Technical Digest of the IEEE Solid-State Sensor and Actuator Workshop*, pp. 36–41 (1990).
57. Y. Yang and S. Senturia, "Numerical Simulation of Compressible Squeezed-Film Damping," in *Technical Digest of the IEEE Solid-State Sensor and Actuator Workshop*, pp. 76–79 (1996).
58. P. Griffin and J. Plummer, "Advanced Diffusion Models for VLSI," *Solid State Tech.* (May), 171–176 (1988).
59. K. Petersen, "Silicon as a Mechanical Material," *Proc. IEEE* 70, 420–457 (1982).
60. W. Kern, "Chemical Etching of Silicon, Germanium Gallium Arsenide and Gallium Phosphide," *RCA Review* 29, 278–285 (1978).
61. K. Bean, "Anisotropic Etching of Silicon," *IEEE Trans. Elect. Dev.* ED-25, 1185–1190 (1978).
62. L. Smith and A. Soderbarg, "Electrochemical Etch Stop Obtained by Accumulation of Free Carriers without PN Junction," *J. Electrochem. Soc.* 140, 271–275 (1993).
63. K. Lee, "The Fabrication of Thin, Freestanding, Single-Crystal Semiconductor Membranes," *J. Electrochem. Soc.* 137, 2556–2574 (1990).
64. A. Brooks and R. Donovan, "Low Temperature Electrostatic Silicon-to-Silicon Seals Using Sputtered Borosilicate Glass," *J. Electrochem. Soc.* 119, 119–124 (1972).
65. D. Monk et al., "Media Compatible Packaging and Environmental Testing of Barrier Coating Encapsulated Pressure Sensors," in *Technical Digest of the IEEE Solid-State Sensor and Actuator Workshop*, pp. 36–41 (1996).

66. J. Butler, V. Bright and J. Comtois, "Advanced Multichip Module Packaging of MEMS," in *Transducers '97*, No.1D4.02, pp. 261–264 (1997).
67. D. Sparks, J. Frazee and L. Jordan "Flexible Vacuum-Packaging Method for Resonating Micromachines," *Sensors Actuators A* 55, 179–183 (1996).
68. K. Korane, "Silicon for High-Pressure Sensing," *Machine Design* (May), 64–70 (1989).
69. K. Sidhu, "Packaging Very High Pressure Transducers for Common Rail Diesel Injection Systems," in *Proceedings SensorExpo*, pp. 29–32 (1997).
70. N. Lu et al., "A Substrate-Plate Trench-Capacitor (SPT) Memory Cell for DRAMs," *J. Solid-State Circuits* SC-21, 627–633 (1986).
71. P. Karulkand and M. Wirzbicki, "Plasma Etching of Ion Implanted Polysilicon," *J. Electrochem. Soc.* 136, 2716–2720 (1989).
72. D. Sparks "Plasma Etching of Silicon, Silicon Oxide, Silicon Nitride and Resist with Fluorine, Chlorine and Bromine Compounds," *J. Electrochem. Soc.* 139, 1736–1741 (1992).
73. T. Core, W. Tang and S. Sherman, "Fabrication Technology for an Integrated Surface-Micromachined Sensor," *Solid-State Tech.* (Oct.), 39–47 (1993).
74. D. Polla, R. Muller and R. White, "Integrated Multisensor Chip," *IEEE Elect. Dev. Lett.* 7, 254–256 (1986).
75. D. Sparks, X. Huang, W. Higdon and J. Johnson, "Angular Rate Sensor and Accelerometer Combined on the Same Micromachined CMOS Chip," *Microsystem Tech.* 4, 139–142 (1998).
76. R. Grace, "Automotive Applications of MEMS," in *Proceedings SensorExpo*, pp. 299–307 (1997).
77. D. Eddy and D. Sparks, "Applications of MEMS Technology in Automotive Sensors and Actuators," *Proc. IEEE* 86(8), 1747–1755 (1998).
78. D. Sparks and R. Brown, "Buying Micromachined Sensors in Large Volumes," *Sensors* 12(Feb.), 53–57 (1995).
79. Y. Lin, P. Hesketh and J. Schuster, "Finite-Element Analysis of Thermal Stresses in a Silicon Pressure Sensor for Various Die-Mount Materials," *Sensors Actuators A* 44, 145–149 (1994).
80. B. Hong and L. Burrell, "Modeling Thermal Induced Viscoplastic Deformation and Low Cycle Fatigue of CBGA Solder Joints in a Surface Mount Package," *IEEE Trans. Components, Pack. Manufact. Tech.* 20, 280–284 (1997).
81. M. Habibi, E. Lueder, T. Kallfass and D. Horst, "A Surface Micromachined Capacitive Absolute Pressure Sensor Array on A Glass Substrate," *Sensors Actuators A* 46, 125–128 (1995).
82. W. Ko, Q. Wang and Y. Wang, "Touch Mode Capacitive Pressure Sensors for Industrial Use," in *Technical Digest of the IEEE Solid-State Sensor and Actuator Workshop*, pp. 24–248 (1996).
83. D. Sparks, "Chemically Accelerated Corrosion Test for Aluminum Metallized ICs," *Thin Solid Films* 235, 108 (1993).
84. D. Sparks, "Method for Forming a Micromachined Motion Sensor," U.S. Patent, No. 5,547,093, Aug. (1996).

85. C. Muhlstein, "Fatigue Crack Initiation and Growth Testing of MEMS and Small Structures," in *JPL MEMS Reliability and Qualification Workshop*, pp. 98–121 (1997).
86. R. Reed-Hill, *Physical Metallurgy Principles*. Van Nostrand, New York, pp. 749–826 (1973).
87. L. Dyer, H. Huff and W. Boyd, "Plastic Deformation in Central Regions of Epitaxial Silicon Slices," *J. Appl. Phys.* 42, 5680–5688 (1971).
88. D. Sparks, "Dislocation and Stacking Fault Interactions in Silicon," *J.Electron. Mat.* 16, 119–122 (1987).
89. J. Geen, "A Path to Low Cost Gyroscopy," in *Technical Digest of the IEEE Solid State Sensor and Actuator Workshop*, pp. 51–54 (1998).
90. C. Linder and N. der Rooij, "Investigation of Freestanding Polysilicon Beams in View of their Application as Transducers," *Sensors Actuators A* 21–23, 1053–1059 (1990).
91. P. Scheeper, J. Voorthuyzen, W. Olthuis and P. Bergveld, "Investigation of Attractive Forces Between PECVD Silicon Nitride Microstructures and an Oxidized Silicon Surface," *Sensors Actuators A* 30, 231–239 (1992).
92. S. Chang, M. Chia, P. Castillo-Borelly, W. Higdon, Q. Jiang, J. Johnson, L. Obedier, M. Putty, Q. Shi, D. Sparks and S. Zarabadi, "An Electroformed CMOS Integrated Angular Rate Sensor," *Sensors Actuators* 66, 138–143 (1998).
93. R. Legtenberg, H. Tilmans, J. Elders and M. Elwenspoek, "Stiction of Surface Micromachined Structures after Rinsing and Drying: Model and Investigation of Adhesion Mechanisms," *Sensors Actuators* 43, 230–238 (1994).
94. M. Horenstein, T. Bifano, R. Mali and N. Vandelli, "Electrostatic Effects in Micromachined Actuators for Adaptive Optics," *J. Electrostatics* 42, 69–82 (1997).
95. R. Krishnamoorthy, T. Bifano, N.Vandelli and M. Horenstein, "Development of MEMS Deformable Mirrors for Phase Modulation of Light," *Opt. Eng.* 36, 542–548 (1997).
96. G. V. Vdovin and P. M. Sarro, "Flexible Mirror Micromachined in Silicon," *Appl. Opt.* 34, 2968–2972 (1995).
97. M. K. Lee, W. D. Cowan, B. M. Welsh, V. M. Bright and M. C. Roggemann, "Aberration-Correction Results From a Segmented Microelectromechanical Deformable Mirror and a Refractive Lenslet Array," *Opt. Lett.* 23, 645–647 (1998).
98. J. H. Comtois and V. M. Bright, "Applications for Surface Micromachined Polysilicon Thermal Actuators and Arrays," *Sensors Actuators A* 58, 19–25 (1997).
99. R. L. Clark, "Adaptive Optics Solves Real-World Problems," *Photonics Spectra* (April), 101–106 (1997).
100. T. Mansell, P. Catrysse, E. Gustafson and R. Byer, "Silicon Deformable Mirrors and CMOS-Based Wavefront Sensors," presented at *Society of Photooptical Instrumentation Engineers (SPIE) 45th Annual Meeting*, San Diego, 1–24 August, 2000. To appear in *Proc. SPIE* Vol. 4124.

101. T. Bifano, R. Mali, J. Perreault, K. Dorton, N. Vandelli, M. Horenstein and D. Castanon, "Continuous Membrane, Surface Micromachined Silicon Deformable Mirror," *Opt. Eng.* 36, 1354–1360 (1997).
102. R. K. Mali, T. Bifano and D. A. Koester, "Design-Based Approach to Planarization in Multilayer Surface Micromachining," *J. Micromech. Microeng.* 9, 294–299 (1999).
103. T. G. Bifano, J. Perreaolt, R. K. Mali and M. N. Horenstein, "Microelectromechanical Deformable Mirrors," *J. Selected Topics in Quantum Elec.* 5, 83–90 (1999).
104. H. T. Jonhson, P. A. Bierden and T. G. Bifano, "Strain and Curvature in an Ion-beam Machined Free-standing Thin-film Structure," submitted to *Appl. Phys. Lett.*, (2000).
105. T. G. Bifano, H. T. Jonhson and P. A. Bierden, "Elimination of Stress-induced Curvature in Thin-film Structures," submitted to *J. of MEMS*, (2000).

CHAPTER 3

OPTICAL MICROSYSTEM MODELING AND SIMULATION

Jan G. Korvink
The Albert Ludwig University

Arokia Nathan
University of Waterloo

Henry Baltes
ETH

CONTENTS

3.1 Introduction / 111
    3.1.1 Progress in microsystem microsystem technology / 111
    3.1.2 Definitions / 111
    3.1.3 Benefits of CAD / 112
    3.1.4 Organization of chapter, approaches followed, and effects considered / 115

3.2 Lagrangian Equations of Motion / 118
    3.2.1 Lagrangian for crystalline continua / 121
    3.2.2 Material interface / 124
    3.2.3 Kinematics, constitutive models, material data, and energy expressions / 125
    3.2.4 Mechanical effects / 127
    3.2.5 Dielectric effects / 132

3.3 Discretization Methods / 136
    3.3.1 Weighted residuals and discretization techniques / 137
    3.3.2 Strong form / 137
    3.3.3 Weighted residual statement / 138
    3.3.4 Finite-difference method / 138
    3.3.5 Control volume method / 141
    3.3.6 The weak form / 144
    3.3.7 Bubnov–Galerkin finite-element method / 145
    3.3.8 Inverse statement / 148
    3.3.9 Boundary-element method / 149
    3.3.10 Other integral equation methods / 150

3.4 Solution Methods / 152
    3.4.1 Algebraic equation systems / 153
    3.4.2 Nonlinearities / 153

       3.4.3   Time integration / 154
       3.4.4   Linear equation systems / 155
       3.4.5   Eigensolutions / 156
       3.4.6   Coupling strategies / 157

3.5 Applications / 157
       3.5.1   Thermal imager / 157
       3.5.2   Optomechanical pressure sensor / 158
       3.5.3   The shape of microlenses / 164

3.6 Discussion / 166

     Acknowledgments / 166

     References / 166

The field of optical microsystem modeling is reviewed with special emphasis on the equations of motion, constitutive equations, solution methods, and applications. The phenomena covered include acoustics, heat transport, optics, and couplings such as thermoelasticity, piezoelectricity and photoelasticity in a single, unified formalism.

3.1 INTRODUCTION

Progress in microsystem technology is echoed by the availability of production-quality computer-aided design (CAD) tools. In this section we review the growth of this field. Next we make some clarifying definitions before progressing to an example of micro-optoelectromechanical systems (MOEMS) simulation, the digital micromirror. The last subsection introduces the rest of the chapter.

3.1.1 PROGRESS IN MICROSYSTEM MICROSYSTEM TECHNOLOGY

Microsystems are rapidly progressing from being the exclusive focus of research teams to becoming viable commercial products,[1] as confirmed by such industrial examples as the airbag accelerometer,[2] the widely available atomic force and scanning tunneling microscopes,[3] inkjet printer heads as consumer items, magnetic field sensors and hard-disk read-write heads, and the digital micromirror display projection device.[4] This trend is accompanied by a rapid growth in specialized computer-aided design support, currently manifested by the recent emergence of a number of microsystem CAD companies,[5–8] by two international conferences dedicated to the field, the CAD for MEMS[9] and the MSM,[10] by a number of review articles,[11] by the appearance of a journal special issue,[12] and by the recent announcement of a comprehensive monograph.[13] The development of microsystem CAD tools closely follows the trend set by the more mature field of microelectronic technology CAD tools. From this perspective, commercial CAD support for MEMS is still in its infancy.

3.1.2 DEFINITIONS

To narrow the topic of MOEMS modeling to what appears most relevant and applicable to date, we first make the following operating definitions:[14]

- A *microsystem* is a combination of packaged microelectronic sensor and actuator transducers and possibly, circuits, at the micrometer to millimeter scale, suitably integrated to perform a variety of functions. The microsystem converts an input physical (such as radiant, thermal, magnetic, mechanical) or chemical signal into an output electrical signal, which if so desired, can be processed and converted into a suitable form. For our discussions we only consider single-chip microsystems manufactured using industrial complementary metal oxide semiconductor (CMOS), bipolar or BiCMOS, combined with additional postprocessing steps that add (through deposition, coating, or electro-plating) functional materials or remove (through etching) materials in selective regions.[15] In this spirit, the following commercial products fit our definition for general microsystems: the Analog Devices airbag accelerometer,[2] the Honeywell magnetic field sensor, inkjet printer heads, and hard-disk storage media read-write heads.

- MEMS, which stands for microelectromechanical systems, is a popular acronym for microsystems in general. MOEMS, as indicated, is an acronym for micro-optoelectromechanical systems, i.e., microsystems with a strong optical application or component. The Texas Instruments digital mirror array chip[4] is a good example. In this chapter we use the terms microsystems, MEMS, and MOEMS interchangeably.
- *Microsystem modeling* is the synthesis (or tailoring) of a set of model equations whose solution is representative of the physical response of certain microsystems.
- *Microsystem simulation* is the solution, with a computer program, of a set of model equations using an implemented numerical method.
- A *CAD tool* implements the discretized model equations and provides additional geometry specification and analysis capabilities, which enable the program user to simulate aspects of the physical behavior of the microsystem.
- A *technology CAD* (TCAD) system combines CAD tools to simulate the manufacturing, device behavior, and embedded response of microsystems in one environment.

3.1.3 BENEFITS OF CAD

Microsystem development is a broad field that combines a large range of technical disciplines. Microsystem modeling suffers from this lack of focus, which strongly contrasts with modeling in the digital and analog microelectronic fields. CAD tools are clearly needed to reduce the consumption of development resources, and frequently help provide insight into complex physical processes. Whenever they are applicable and useful, software modeling tools rapidly gain acceptance by the design community.

The benefits of CAD are best illustrated in terms of an example. Here, we consider the electrostatically deflectable digital micromirror. Electrostatically driven microactuators have low power consumption and relatively short response times, making them ideally suited to large-array imaging devices. Examples of other actuators can be found in Section 3.5.

Figure 3.1 illustrates an electron micrograph and the CMOS mask layout of the micromirror. It is free standing, supported by two thin Al beam hinges, and suspended over four independently addressable electrodes embedded in the chip

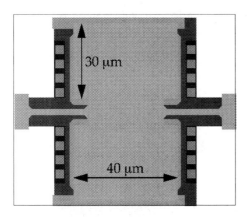 

Figure 3.1 Micromirror mask (left) and scanning electron micrograph (right) showing device details. Device features: mirror size, 30 × 40 μm; mirror thickness, 1.4 μm; gap width beneath mirror, 1.4 μm; torsional hinge size, 15 × 2.4 μm; torsional hinge thickness, 0.3 μm; max. angular deflection, 4.2°; driving voltage, 14 V. (From Ref. 17.)

surface and formed by n-well diffusion tubs. Mirror deflection is controlled by the charge on the electrodes. Critical design requirements for efficient, stable, and reliable micromirror operation include:

- Highly smooth mirror surface to minimize divergence of reflected optical beam.
- High chip-surface fill factor.
- Low-voltage operation.
- High deflection angle.
- Low deflection threshold voltage.
- Control of maximum stresses in the structure so as not to exceed the yield stress limit or the material fatigue limit.
- Fast dynamic response, implying control of squeezed-film air damping.

Micromirror design and optimization with respect to the above requirements can only be performed by numerical simulation. In addition, numerical simulations enable extraction of compact model parameters for synthesis of compact circuit models for SPICE-type simulations, needed for cointegration of mirror arrays with circuitry.

BEMMODULE[16] was used for characterizing the electrostatic behavior of the aluminum micromirror device.[17] For the numerical simulation, the geometry of the torsional micromirror was generated from the original mask layout using numerical emulation of the double metal CMOS process DIMOS01 at Delft University of Technology, the Netherlands. In Fig. 3.1, the mask layout and a surface electron micrograph (SEM) of the fabricated device are shown.

Process emulation by PROSIT-ISE produced the device geometry, including the sacrificial first metallization and the intermetal silicon dioxide layer. The geometry of the generated device is densely discretized, featuring 57,693 surface panels [see Fig. 3.2(I)]. The REMESH utility was used to simplify the initial mesh. Sub-

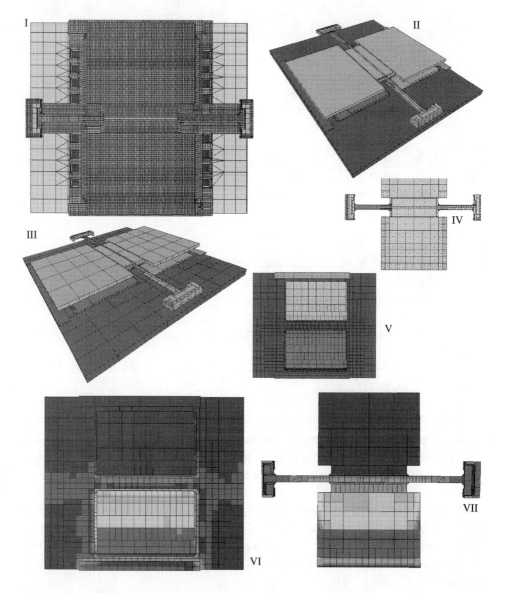

Figure 3.2 CMOS micromirror process and device simulation shown in a sequence of images. From the mask and process description, the (overmeshed) geometry was obtained using PROSIT (I). Excessive surface panels are removed with REMESH (II). Surface mesh is refined based on geometrical shape and arrangement (III). Mesh is adapted according to numerical solution (IV and V). Typical field solution for a mirror bias of 14 V (VI and VII).

sequently, the sacrificial layers were stripped to yield the geometry for electrostatic simulation, using a minimal description of 386 interface panels [see Fig. 3.2(II)]. The minimal mesh created by REMESH is not suited as an initial discretization for adaptive refinement, since it is too coarse to allow computing a valid distribution of the boundary element error indicator. BEMMODULE allows mesh refinement, based on the boundary element shapes, without having to perform any electrostatic computations. Thus a suitable initial mesh for adaptive refinement was generated from the minimal geometric description, in a fully automated manner [see Fig. 3.2(IV), (V)].[16]

The electrostatic solution for the mirror, for a range of mirror deflection angles, was computed using the multipole accelerated-boundary element method. For each angle, the electrode charge was computed by integrating the charge density over the electrode surface. In addition, the electrostatic force and torque was obtained by evaluating and integrating the electrostatic pressure's zeroth and first moments over the electrode surfaces. The results for charge and torque are shown in Fig. 3.3.

3.1.4 Organization of Chapter, Approaches Followed, and Effects Considered

In this chapter we have selected a subset of physical effects that lie at the heart of many current optical microsystems. We demonstrate the modeling process by obtaining, for these microsystem applications, a set of modeling equations, constitutive data, a discretization scheme, and a solution strategy. The connection between the various processes is illustrated in Fig. 3.4, which also provides an overview of the organization of this chapter.

For MOEMS it is essential to recognize that the cost and effort to produce specialized CAD tools is much larger than for microelectronics. This is because of the larger variety of important physical effects, available technologies used, and optimization goals pursued. We believe that much can be gained from recognizing, as soon as possible, general principles in the equations and software components that may be reused. This fact has guided us in writing this chapter. In fact, we exclusively use a Lagrangian description for the physical phenomena. This choice of method tremendously simplifies and structures the discussion:

- The Lagrangian method allows the modeler to concentrate on describing the required detailed behavior by formulating individual terms for stored and dissipated energy densities. We are required to follow a standardized yet familiar format for these energy terms.
- The energy terms are then simply plugged into the Lagrange equation to obtain the so-called "equations of motion." Without additional effort, we obtain the correct description for the coupled energy domains relevant here.
- If we require more detail on material behavior, all that is required is to update the energy terms to reflect this detail. In this way we can go so far as to

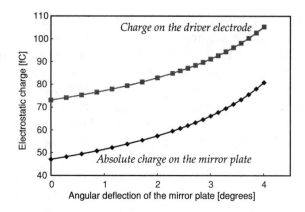

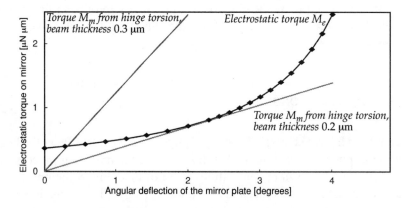

Figure 3.3 Top: Electrostatic charge on the polysilicon driver electrode of the micromirror device and induced charge on the mirror plate as a function of the applied deflection angle. The driving potential is 14.0 V. Bottom: Torque resulting from electrostatic pressure and from hinge torsion acting on the mirror as a function of the angular deflection. The driving potential is 9 V.

include quantum phenomena in the resulting "semiclassical" model without changing the method.
- A very important additional benefit exists for the developer of simulation programs. Lagrange-like formalisms translate directly into the finite-element method. Thus it is possible to plan ahead and create software that can grow with added requirements, keeping up with the tough demands of creative device invention.

Next we discuss the "kinematic" requirements of the "motion," as well as the constitutive models needed, to obtain expressions for energy storage and work increment for the materials under consideration. We also present the currently most important discretization methods used in the simulation of MEMS and MOEMS.

OPTICAL MICROSYSTEM MODELING AND SIMULATION

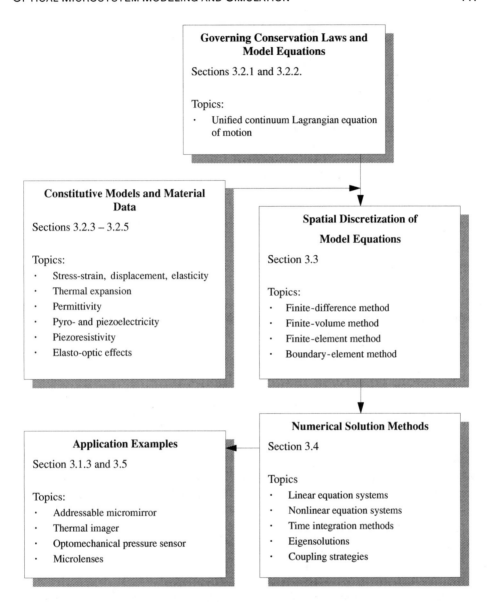

Figure 3.4 The organization of this chapter is guided by the interaction between various modeling and simulation processes.

With the important constituents prepared, we model three MOEMS. The examples are chosen to demonstrate a variety of simulation approaches in use today, and include lumped SPICE modeling of the control volume method, a semianalytical method, and a panel-based method.

The ability to convert optical signals into and out of the electrical domain, together with the large set of modulation mechanisms available for an optical signal, are the main motivations for the development of MOEMS. For example, physical

Table 3.1 Matrix of physical effects available to MOEMS.

Primary signal	Optical effect	Device
Mechanical	Photoelastic effect	Strain sensor, pressure sensor
	Sagnac effect	Gyroscopes
	Doppler effect	Velocimeters
	Movement	Moving optical elements
Thermal	Thermo-optic effect	Temperature sensors, switches
	Radiation	Thermal scene generators
Electrical	Electro-optic effects: Pockels, Kerr	Modulators, switches
Magnetic	Magneto-optical effects: Faraday	Electrical current sensors

Source: Adapted from Ref. 18.

and chemical signals may be used to modulate the intensity, phase, frequency, and polarization of the optical signal, as illustrated in Table 3.1.[18] In this chapter we consider a subset of these effects. We do not consider the modeling of the optics of microlenses in this article, but touch upon the shaping of microlenses.

This text updates a preceding article dealing mainly with MEMS.[14] In particular, its content is based in part on two short courses presented at the CAD for MEMS '97 workshop in Zurich, in part on a graduate-level course held by Nathan while on sabbatical in Zurich, and in part on various senior undergraduate courses taught by the authors at the University of Freiburg, the University of Waterloo, and at the ETH Zurich. A comprehensive review of transducer CAD by two of us is found in Ref. 13, and contains comprehensive tables of measured material properties as required for accurate simulation. The symbols used are summarized in Table 3.2.

3.2 LAGRANGIAN EQUATIONS OF MOTION

By "equations of motion" we refer to the combination of conservation laws and constitutive equations for a selected subsystem. Many methods exist to derive the equations. Usually, following a "Lagrangian" rather than a "Newtonian" method, we obtain the equations of motion through the satisfaction of the:[19]

- *Geometric requirements*: We will select a complete and independent set of generalized coordinates for the system. The coordinates are kinematically admissible.
- *Constitutive requirements*: These will enter the model through energy density and co-energy density expressions for the inertial elements and the conservative forces that make up the Lagrangian density, and through nonconservative work increments for the dissipative processes.

Table 3.2 Symbols used in the chapter. Symbols may have more than one meaning, depending on context.

Symbol	Units	Description
A	m^2	Surface area
$\mathbf{A}$	$V\,s\,m^{-1}$	Vector potential field
$\mathbf{B}$	$V\,s\,m^{-2}$	Magnetic induction field
C	$N\,m^{-2}$	Elasticity tensor
C_P	$m^{-2}\,N^{-1}$	Photoelasticity tensor
c	$m\,s^{-1}$	Speed of light in vacuum
$\mathbf{D}$	$A\,s\,m^{-2}$	Electric displacement field
d	$A\,s\,N^{-1}$	Piezoelectric tensor
$\mathbf{E}$	$V\,m^{-1}$	Electric field strength
F	N	Force
e	$A\,s$	Electronic charge
f	$N\,m^{-3}$	Force density field
$\mathbf{H}$	$A\,m^{-1}$	Magnetic field strength
$\mathbf{J}$	$A\,m^{-2}$	Current density field
k	m^{-1}	Wave number
$\mathfrak{L}$	J	Lagrangian
L	$J\,m^{-3}$	Lagrangian density
L_{ij}		Onsager kinetic coefficients
ΔL	m	Optical pathlength difference
M	–	Crystal configuration constant
$\mathbf{M}$	$A\,m^{-1}$	Magnetic dipole moment density
m	kg	Mass
n	–	Refractive index
$\mathbf{n}$	–	Unit vector
$\mathbf{P}$	$A\,s\,m^{-2}$	Polarization density
p	$A\,s\,m^{-2}\,K^{-1}$	Pyroelectric coefficients
p	$V\,m\,A^{-1}\,s^{-1}$	Pockels elasto-optic tensor
Q	V	Electric quadrupole moment density
Q	–	Quality factor
R	–	Rotation matrix of direction cosines
$\mathbf{R}$	m	Position vector
s	$A\,V^{-1}\,m^{-1}$	Electrical conductivity tensor
T	J	Kinetic energy
T	K	Temperature
t	s	Time
U	J	Electric and magnetic field energy
$\mathbf{u}$	m	Displacement field
V	m^3	Volume
V	J	Potential energy
v	$J\,m^{-3}$	Potential energy density
$\mathbf{v}$	$m\,s^{-1}$	Velocity field
W	J	Mechanical work
w	$J\,m^{-3}$	Energy density
x	m	Displacement in spatial coordinates
X	m	Displacement in material coordinates
∇_x	–	Gradient operator with respect to x

Table 3.2 (Continued).

Symbol	Units	Description
α	K^{-1}	Thermal expansion coefficient
α	$V K^{-1}$	Seebeck coefficient
α_0	m	Constant length
β	m^{-1}	Propagation constant of the guided mode
γ	$N m^{-1}$	Surface tension
δ	–	Identity tensor (Levi-Civita)
δ		Variation operator
$\varepsilon, \varepsilon_T$	–	Strain tensor, thermal strain tensor
ε_0	$A s V^{-1} m^{-1}$	Permittivity of free space
κ	$A s V^{-1} m^{-1}$	Dielectric permittivity tensor
κ_{th}	$W K^{-1} m^{-1}$	Thermal conductivity tensor
λ	m	Wavelength
ζ_j	Arb.	Generalized displacement
ζ	Pa s	Viscosity parameter
$\dot{\zeta}_j$	Arb.	Generalized velocity
η	Pa s	Viscosity parameter
Ξ_j	Arb.	Generalized force
ξ	Arb.	Generalized force density
Π	–	Internal coordinate measure
π	$m^2 N^{-1}$	Piezoresistive coefficients
ρ	$kg m^{-3}$	Mass density
σ	$N m^{-2}$	Stress
$\Delta \Phi$	–	Relative optical phase difference
χ	–	Dielectric susceptibility tensor
χ_0	–	Electro-optic susceptibility tensor
ψ	V	Electric potential
Ω		Integration domain

- *Dynamic requirements*: We will use the Hamilton principle to obtain the Lagrange equations for our system.

A classical Lagrangian description appears to be a suitable choice because it:

- Provides a means to deal with multiple energy domains of interacting subsystems, such as optical, magnetic, thermal, mechanical, and chemical subsystems.
- Unites reversible (e.g., piezoelectric or elasto-optic) and irreversible (photoelastic) phenomena with one formalism.
- Is compact and efficient in its description, yielding all observable effects.

The method that we have adopted here largely follows the detailed exposition presented in Ref. 20, with some concepts from Refs. 21, 22, and 23.

3.2.1 LAGRANGIAN FOR CRYSTALLINE CONTINUA

For a given Lagrangian $\mathfrak{L}(\zeta_j, \dot{\zeta}_j, t)$ of a physical system, defined with the generalized scalar variables ζ_j, their scalar velocities $\dot{\zeta}_j$ and the time t, the requirement that the variation of the time integral of L, the action, should vanish, i.e.,

$$\delta \int_{t_1}^{t_2} \mathfrak{L} dt = 0, \tag{3.1}$$

which is also known as Hamilton's principle, leads to Lagrange's continuum equations:

$$\frac{d}{dt}(\nabla_{\dot{\zeta}_j} L) - \nabla_{\zeta_j} L + \nabla_{Z_K}(\nabla_{\nabla_{Z_K}\zeta_j} L) = \sum_n \xi_n, \tag{3.2}$$

for the Lagrangian density L, with $\mathfrak{L} = \int_V L dV$, and the generalized scalar forces $\Xi_j = \int_V \xi_j dV$. It is the generalized force densities that take into account the irreversible, dissipative effects in the system, defined through the nonconservative work expression:

$$W^{nc} = \sum_{i=1}^N \rho f_i^{nc} \cdot R_i = \sum_{j=1}^n \left(\sum_{i=1}^N \rho f_i^{nc} \cdot \nabla_{\zeta_j} R_i \right) \zeta_j = \sum_{j=1}^n \rho \xi_j \zeta_j. \tag{3.3}$$

Lagrangian mechanics renders the procedure of formulating suitable modeling equations to the formulation of a suitable Lagrangian density and nonconservative work expressions. Constitutive equations enter the method in the expressions for energy storage and in the terms for the energy dissipation. For the dielectric crystal, we use a Lagrangian made up of three terms, the matter Lagrangian $\mathfrak{L}_M$, the external electromagnetic field Lagrangian $\mathfrak{L}_F$, and the field–matter interaction Lagrangian $\mathfrak{L}_I$:[20]

$$\mathfrak{L} = \mathfrak{L}_M + \mathfrak{L}_F + \mathfrak{L}_I. \tag{3.4}$$

The matter Lagrangian is the familiar difference between the kinetic co-energy and the potential energy of the individual particles that make up the crystal;

$$\mathfrak{L}_M = T^* - V = \frac{1}{2} \sum_n m_n (\dot{x}_n)^2 - V(x). \tag{3.5}$$

The matter Lagrangian's terms must be formulated so as to represent all phenomena of interest in the crystalline materials making up the microsystem components. The phenomena of interest include acoustics, heat transport, optics, and couplings such as thermoelasticity, piezoelectricity, and photoelasticity, but see Table 3.1. In forming the Lagrangian we control the detail that the model considers, typically to describe semiclassical models that mix quantum-mechanical and continuum-mechanical approaches.

For our purposes dielectric crystals are considered as periodic Bravais lattices. We formulate the Lagrangian terms for a Wigner–Seitz unit cell [see Fig. 3.5] of the crystal. The unit cell for a monoatomic silicon diamondlike lattice and the diatomic quartz lattice is exemplified in Fig. 3.5(g). As constituent particles we consider the ions present in the unit cell, together with a single independent electron. The total kinetic co-energy is obtained from a sum over the N particles per unit cell and over

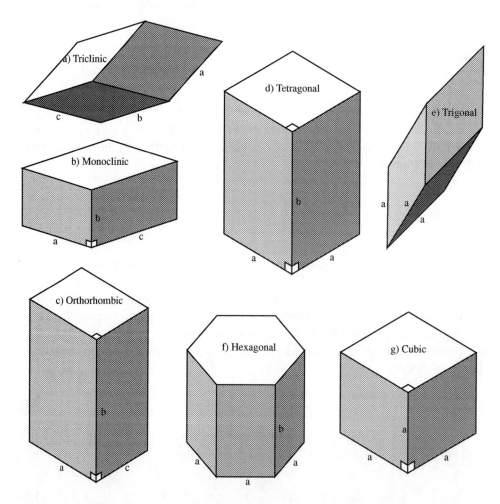

Figure 3.5 The seven basic crystal systems. Each object describes a unit cell for that crystal class. The non-equal edge dimensions a, b, and c indicate the extent of the crystal cell along an axis parallel to that edge. Perpendicular sides are indicated. The illustration starts with the least symmetric crystal system, triclinic, and ends with the most symmetric crystal, the cubic system.

the M cells in the crystal:

$$T^* = \frac{1}{2}\sum_i^M \sum_j^N m^j \left(\dot{x}^{ji}\right)^2 = \frac{1}{2}\sum_j^N \rho^j \int_V \left[\dot{x}^j(X,t)\right]^2 dV. \qquad (3.6)$$

On the right-hand-side of Eq. (3.6) we have also gone over to the continuum limit, with $\rho^j = m^j/V$ giving the per particle mass density. The material coordinate X points to the unit cell center of mass for the undeformed crystal. We will consider these position vectors more precisely when we define the mechanical deformation. In particular, we associate the momentum with the center of mass, and employ internal coordinates to take care of the deformation modes of the crystal. The kinetic co-energy is an approximation due to the presence of the single electron, resulting in a necessary simplification of the lattice's dynamics. To compensate for this simplification, a suitable choice for the periodic potential is required. Fortunately we hardly ever require a complete description of the energy band structure of a crystal, for the dynamics are usually due to a small perturbation in the vicinity of a stable configuration. Thus, e.g., for silicon, conduction electrons are mainly found at the <100> conduction band minima, where the potential energy may be expanded as a quadratic function of the wave vector, yielding ellipsoidal equipotentials with the major axis along the <100> axis. Dynamics are thus described using an effective mass tensor for the electrons to take into account crystal anisotropy. Similarly, conduction holes are found at the valence band maxima at the <000> positions—the high symmetry results in a diagonal mass tensor, but band degeneracy results in heavy and light holes.[24] In dielectric insulators such as undoped silicon, SiO_2 and Si_3N_4, we can often neglect the dynamics of the electrons.

Again drawing on the assumption that the lattice ions are only slightly displaced from their equilibrium positions in the course of any interaction, we may use a periodic potential for the lattice with the following characteristics:

- The potential must show a functional dependence on the relative position of the particles over the entire lattice—not just over the unit cell. In effect, it is not the center of mass position of the unit cell (because of the inherent translational invariance of the crystal), but only the internal coordinates and the derivatives of the spatial positions with respect to the material positions that are relevant.
- The potential must permit conservation of energy, momentum, and angular momentum.

These requirements can be combined to arrive at a dependence of the potential on only the Green finite strain tensor ε and an invariant measure of the internal coordinates Π. In this case the potential density is most conveniently expressed as a deviation from the natural state of the crystal by using a series expansion,[20]

$$\left.\begin{aligned}v(\varepsilon, \Pi) &= {}^{01}M\varepsilon + {}^{02}M\varepsilon\varepsilon + {}^{03}M\varepsilon\varepsilon\varepsilon + \cdots \\ &+ \sum_{\nu}{}^{10}M^{\nu}\Pi^{\nu} + \sum_{\nu\mu}{}^{20}M^{\nu\mu}\Pi^{\nu}\Pi^{\mu} + \sum_{\nu\mu\lambda}{}^{30}M^{\nu\mu\lambda}\Pi^{\nu}\Pi^{\mu}\Pi^{\lambda} + \cdots \\ &+ \sum_{\nu}{}^{11}M^{\nu}\Pi^{\nu}\varepsilon + \sum_{\nu}{}^{12}M^{\nu}\Pi^{\nu}\varepsilon\varepsilon + \sum_{\nu\mu}{}^{21}M^{\nu\mu}\Pi^{\nu}\Pi^{\mu}\varepsilon + \cdots\end{aligned}\right\}. \quad (3.7)$$

In Nelson's notation the coefficients ${}^{ab}M^{\nu\mu\cdots}$ are matrices (indicated by the superscript $\nu\mu\ldots$) of tensors, with a indicating the number of Π factors and b the number of ε factors, so that $\text{rank}({}^{ab}M^{\nu\mu\cdots}) = a + 2b$. The entries of ${}^{ab}M^{\nu\mu\cdots}$ are numerical constants that only depend on the mechanical configuration of the crystal in its natural state;[20] these coefficient values must either be measured or derived using the methods of quantum mechanics [see, e.g., Ref. 25, Chapters 11 and 17].

The electromagnetic field Lagrangian is written as the difference between the electrical coenergy and the magnetic energy:[20]

$$\mathcal{L}_F = U_E^* - U_B = \int_V L_F dV = \int_V \frac{\varepsilon_0}{2}(E^2 - c^2 B^2) dV, \quad (3.8)$$

where E is the vector-valued electric field strength in V m^{-1}, c is the speed of light in a vacuum in m s^{-1}, B is the vector-valued magnetic induction in V s m^{-1}, and ε_0 is the permittivity of free space. The field–matter interaction Lagrangian is:[20]

$$\mathcal{L}_I = (e\dot{x} \cdot A) - (e\psi), \quad (3.9)$$

where e is the electronic charge and $\dot{x}$ is the vector-valued velocity field. ψ is the scalar potential in volts, and A is the vector potential in V s m^{-1}, defined by the following two relations;

$$\left.\begin{aligned} E &= -\nabla\psi - \frac{\partial A}{\partial t} \\ B &= \nabla \times A \end{aligned}\right\}. \quad (3.10)$$

The two potentials ψ and A and the position x of the charge are selected as generalized displacements of the formulation.

3.2.2 MATERIAL INTERFACE

The dimension of microsystem features, in the range of micrometers, implies that the surface tension forces that arise at liquid/gas interfaces are sufficiently large relative to bulk forces to require consideration. Often considered a disadvantage, it is, however, possible to exploit capillary forces usefully during manufacture. It has, for example, been shown that microlenses may be conveniently formed by reflowing patterned photoresist.[44] It is necessary to know the resultant shape of the droplet to predict the optical quality of the lenses.

When we introduce a new interface into a material system, we expect the energy of the system to increase proportionally to the amount A of area created. It is argued that this holds because the bulk material does not create area spontaneously.[26] The surface energy density that is required as a proportionality parameter is called the surface tension γ, so that the surface energy associated with the extra area is $V_s = \gamma A$. It is instructive to look at the surface energy when the surface area is changed by a small amount. We may write that

$$\left. \begin{array}{l} dV_s = A \dfrac{\partial V_T}{\partial \varepsilon} d\varepsilon \\ = A\sigma {:} d\varepsilon \end{array} \right\}, \qquad (3.11)$$

where V_T is the total energy for the system, and σ and ε are the stresses and strains associated with the surface. From thermodynamic considerations, we find that the stress is defined in terms of the surface tension as

$$\sigma = \gamma \delta + \left. \dfrac{\partial \gamma}{\partial \varepsilon} \right|_T. \qquad (3.12)$$

If the surface tension γ depends on the strain state of a material surface, the second term is important and is an explanation for surface buckling. For liquids, where the atoms are less constrained, the second term may be neglected and the familiar result of "surface tension equals surface stress" results. Combining Eqs. (3.11) and (3.12) yields

$$V_s = A\left(\gamma \delta + \left. \dfrac{\partial \gamma}{\partial \varepsilon} \right|_T \right) {:} \varepsilon, \qquad (3.13)$$

relative to some arbitrary datum, and for liquid/gas interfaces becomes $V_s = A\gamma \operatorname{tr}(\varepsilon)$. Caution is necessary in using this expression other than for describing liquid stiction, because a full treatment of the surface state includes entropy considerations, anisotropy, and other effects.[26]

3.2.3 KINEMATICS, CONSTITUTIVE MODELS, MATERIAL DATA, AND ENERGY EXPRESSIONS

For the Lagrangian description of solids we require expressions for the quantity of energy stored by component materials through various mechanisms, as well as the

Table 3.3 Selected data for some materials at room temperature.

Material		Dielectric permittivity (–)	Thermal conductivity (W c m^{-1} K^{-1})	Elastic modulus (GPa)	Expansion coefficient (10^{-6} K^{-1})	Piezoelectric coefficient (p C N^{-1})
Aluminum			0.73–0.95	42.7	23–25	
Silicon	Crystalline Si (*n*-type)	11.7	1.56	$C_{11} = -97.1$ $C_{12} = -54.8$ $C_{44} = -172.0$	2.33	
	Crystalline Si (*p*-type)			$C_{11} = -80.5$ $C_{12} = -115.0$ $C_{44} = -52.8$		
	Poly-Si		0.2–0.4	130–150		
SiO$_2$	Non-crystalline	3.9		72–75		
	α–SiO$_2$	$\varepsilon_{r11} = 4.5$ $\varepsilon_{r22} = 4.5$ $\varepsilon_{r33} = 4.6$	0.014	$C_{11} = 86.80$ $C_{33} = 106.2$ $C_{12} = 7.10$ $C_{13} = 11.91$ $C_{44} = 58.17$ $C_{66} = 39.85$ $C_{14} = -18.02$	0.40–0.55	$d_{11} = -2.31$ $d_{14} = -0.727$
Si$_3$N$_4$		7.0	0.032	260	1.1–3.8	

quantity of energy dissipated through work done. Energy expressions require constitutive equations because it is the choice of material that determines the quantity of energy that a field can store. In this section we detail the constitutive equations and the material parameter data for a variety of material systems. If an effect represents the coupling of two or more energy domains, we will list it under the subsection we consider most useful for the exposition. Table 3.1 details the energy domains considered in this text. We will often speak of typical complementary metal oxide semiconductor (CMOS) materials. Currently this means the *n*- and *p*-doped Si, SiO$_2$, Si$_3$N$_4$, polycrystalline Si and Al, as available from most multiproject wafer CMOS foundries. The list of materials that are compatibly added to a CMOS wafer after the foundry process is growing rapidly. Typical materials are the polyimides through spin coating, magnetic nickel-iron by electroplating, and piezoelectric zincoxide by radiofrequency (RF) sputtering.[27] Such materials are meant to contribute to the functioning of a microsystem; we therefore include more general material considerations in the text. Table 3.3 provides a compilation of selected material data, for a small choice of metals, semiconductors, and dielectrics.

3.2.4 MECHANICAL EFFECTS

Almost all microsystem components are subject to stress, and hence experience some strain. More often than not, a large portion of the stress is caused by the packaging acting on the microsystem chip. Microsystems with movable parts also experience stress. Their sensors are often designed to indirectly measure a quantity through its modulation of the stress distribution. In other cases, an electrical quantity is modulated by the stress state. Examples of both are pressure sensors, accelerometers, deflectable micromirrors, and even thermometers. Mechanical resonators have finite quality factors Q. These are determined by the dissipation of energy away from the resonator, perhaps acoustically into the air, or as heat. In this subsection we look at stress and strain and their relation to each other. Then we add the coupling of the mechanical to the thermal domain, so as to describe useful and parasitic effects. Other mechanical cross-couplings are dealt with in the following subsections.

3.2.4.1 Kinematics of Deformation

The local deformation gradient in a material body may be described by investigating the environment about two points, **P** and **Q**, separated by a distance dx as the body undergoes a deformation. By deformation we refer to processes that change the shape of a material body reversibly, without introducing discontinuities in the body. The vectors required for the discussion are shown in Fig. 3.6. We assume a general linear transformation for the deformation of an arbitrary vector $\boldsymbol{P}'$ from its original state $\boldsymbol{P}$

$$\boldsymbol{P}' = \alpha_0 + (\delta + \alpha) \cdot \boldsymbol{P}, \tag{3.14}$$

so that for the differential line element $P'Q'$

$$d\boldsymbol{x}' = \boldsymbol{Q}' - \boldsymbol{P}' = \alpha_0 + (\delta+\alpha)^T \cdot \boldsymbol{Q} - \left(\alpha_0 + (\delta+\alpha)^T \cdot \boldsymbol{P}\right) = d\boldsymbol{x} + \alpha^T \cdot d\boldsymbol{x}. \tag{3.15}$$

If we now write $d\boldsymbol{x}' = d\boldsymbol{x} + \nabla \boldsymbol{u}^T \cdot d\boldsymbol{x}$, using $d\boldsymbol{u} = \nabla \boldsymbol{u}^T \cdot d\boldsymbol{x}$, then writing out the squared length of the deformed line element gives

$$\begin{aligned} dl'^2 &= d\boldsymbol{x}^T \cdot d\boldsymbol{x} + \left(\nabla \boldsymbol{u}^T \cdot d\boldsymbol{x}\right)^T \cdot d\boldsymbol{x} + d\boldsymbol{x}^T \cdot \left(\nabla \boldsymbol{u}^T \cdot d\boldsymbol{x}\right) \\ &\quad + \left(\nabla \boldsymbol{u}^T \cdot d\boldsymbol{x}\right)^T \cdot \left(\nabla \boldsymbol{u}^T \cdot d\boldsymbol{x}\right) \\ &= d\boldsymbol{x}^T \cdot \left(\delta + \nabla \boldsymbol{u} + \nabla \boldsymbol{u}^T + \nabla \boldsymbol{u} \cdot \nabla \boldsymbol{u}^T\right) \cdot d\boldsymbol{x}, \end{aligned} \tag{3.16}$$

so that we can define the Lagrangian strain tensor and its first-order small-strain approximation, using a first-order Taylor series to obtain the square root of Eq. (3.16), as

$$\varepsilon = \frac{1}{2}\left(\nabla \boldsymbol{u} + \nabla \boldsymbol{u}^T + \nabla \boldsymbol{u} \cdot \nabla \boldsymbol{u}^T\right) \cong \frac{1}{2}\left(\nabla \boldsymbol{u} + \nabla \boldsymbol{u}^T\right). \tag{3.17}$$

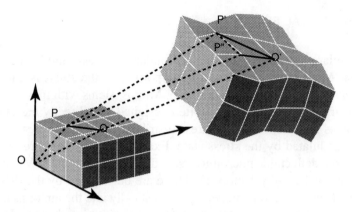

Figure 3.6 The vectors describing the relation between the points P and Q in an undeformed and deformed body.

The fact that it is possible to refer all deformation quantities to positions in the undeformed body using material coordinates, as well as in general spatial coordinates, leads to a duality of descriptions and a blossoming of terminology. Computationally it is often more useful to use spatial coordinates (where we assume that material and spatial coordinate systems coincide), a choice we adopt in this chapter.

The strain describes the instantaneous state of deformation in a body. It is a second-rank symmetric tensor. It has a set of principal axes and when its component matrix is referred to these axes, the matrix is diagonalized and characterized by three principal strain component values. The strain tensor components obey the rule of tensor transformation for a second-rank tensor:

$$\varepsilon' = R\varepsilon R^T, \tag{3.18}$$

where R is the "rotation" tensor of direction cosines. Under dynamic loading conditions, the strain rate tensor may be important as well. It is defined by

$$\dot{\varepsilon} = \frac{1}{2}(\nabla v + \nabla v^T), \tag{3.19}$$

where we have introduced the time rate of change of strain and the velocity field of material points from the relation $v = \partial u/\partial t$ so that

$$\frac{\partial}{\partial t}(\nabla u) = \nabla\left(\frac{\partial u}{\partial t}\right) = \nabla v. \tag{3.20}$$

3.2.4.2 Stress

The stress σ is a rank two tensor field, observed in a material body in reaction to external forces acting on the body, with the units of a surface force density.

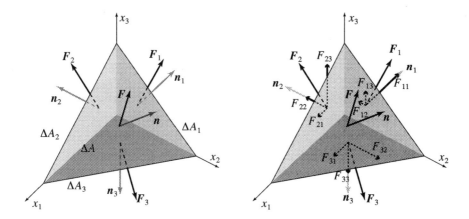

Figure 3.7 The stress at a point; given w.r.t. a set of coordinates <x,y,z> may be related to any other direction by requiring equilibrium for an elemental volume and taking the limit of the dimensions to zero.

The stress is defined as $\sigma = \lim_{\Delta A \to 0} F/\Delta A$. The relation of stress tensor components to force components is derived with the aid of the infinitesimal tetrahedron of Fig. 3.7. We associate index j with the face along the negative cartesian axis direction j, and denote the infinitesimal surface area of the tetrahedron face in this direction $\Delta A_j = -\Delta A(\boldsymbol{n} \cdot \boldsymbol{n}_j)$. We associate the force acting on face i in direction j as $F_{ij} = \boldsymbol{F}_i \cdot \boldsymbol{n}_j$. Then we define the stress tensor component σ_{ij} as

$$\sigma_{ij} = \lim_{\Delta A_j \to 0} \frac{F_{ij}}{\Delta A_j}. \tag{3.21}$$

Enforcing force and moment equilibrium for the infinitesimal tetrahedron relates the stress components on the diagonal face to those on the coordinate faces through the usual tensor transformation rule.

3.2.4.3 Thermal Expansion

When heat energy is added to a material body, its temperature increases, and hence the molecules or ions that make up the body tend to increase their separation because they are more energetic, which in turn leads to an expansion of the dimensions of the body. This is termed thermal expansion, and the inverse effect is also observed; removing heat causes shrinking. For small changes in temperature the effect is linear, and may be described by a linear relation between temperature and the induced "thermal" strains:

$$\varepsilon_T = \alpha \Delta T, \tag{3.22}$$

where α is a symmetric second-rank tensor of linear thermal expansion coefficients. Its principal values are usually specified; these must then be referred to the actual direction required.

3.2.4.4 Crystal Elasticity

In the elastic range, the dependence of stress on strain may be expressed in a Taylor expansion of the stress about a default state. Since the strain is already "small," we need only retain the terms linear in strain:

$$\sigma = \sigma_0 + C\varepsilon + O(\varepsilon^2). \quad (3.23)$$

Because both the stress σ and the strain ε are symmetric second-rank tensors, the linear elastic tensor C must be a fourth-rank tensor that reflects these symmetries. Only 36 of the possible 81 coefficients can be unique. It is usual to introduce an engineering or matrix notation that reflects this fact. We associate the indices of the tensor component matrices with the new compact engineering matrices in the following manner: $11 \to 1, 22 \to 2, 33 \to 3, 23 \to 4, 13 \to 5, 12 \to 6$. The character of C is determined by the crystal symmetries that the material possesses. The least symmetric crystal structure is the triclinic system, for which only 21 of the 36 coefficients are unique. Cubic crystals such as silicon require three elastic constants, and isotropic materials only two. The full set is shown in Fig. 3.8. The elastic energy density at a specific location in a crystal is

$$v_\varepsilon = \int_0^\varepsilon \sigma\, de = \int_0^\varepsilon (\sigma_0 + Ce)\, de. \quad (3.24)$$

3.2.4.5 Mechanical Dissipation

The strain rate defined earlier is needed to describe dissipative processes that take place during viscoelastic or plastic deformation. For a viscoelastic model we obtain the following contribution to the total stress in an isotropic body:

$$\sigma_D = \eta\,\text{trace}(\dot\varepsilon) + 2\zeta\dot\varepsilon. \quad (3.25)$$

In the analysis of harmonically excited structures, Eq. (3.25) gives a good two-parameter approximation of the internal dissipation of energy and can be well correlated with quality factor measurements. The work done by these dissipative forces in the body is then

$$W = \int_\Omega \left(\int_0^\varepsilon \sigma_D\, d\varepsilon \right) d\Omega. \quad (3.26)$$

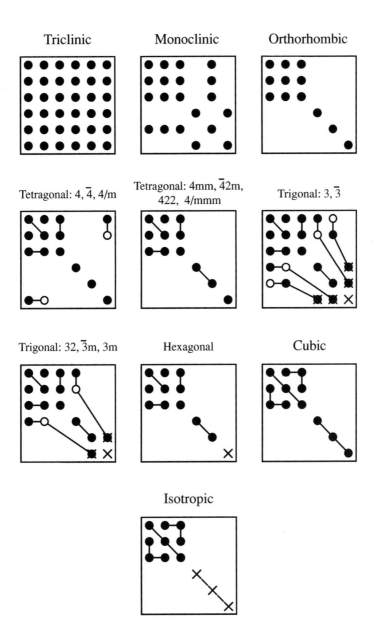

Figure 3.8 The structure of the symmetric fourth-rank elastic and photoelastic tensor coefficient matrices, written in compact engineering notation, for different crystal classes. A solid dot indicates a nonzero entry. A line indicates numerical equality. A hollow dot indicates a negative entry. A cross indicates $(s_{11} - s_{12})/2$. A crossed dot indicates that the entry is twice the value of the solid dot connected to it. The crystal classes are indicated using the Hermann–Mauguin notation. Adapted from Refs. 21 and 20.

3.2.5 Dielectric effects

Using a statistical-mechanical averaging of microscopic quantities, it is possible to show[28] that for general macroscopic media, the electric displacement D can be written as

$$D = \varepsilon_0 E + (P - \nabla \cdot Q' + \cdots), \qquad (3.27)$$

and the magnetic field H as

$$H = B - (M + \cdots). \qquad (3.28)$$

The symbols have the following usual meaning. E is the applied electric field and B is the applied magnetic induction. In the presence of the applied E and B fields, P is the polarization density, Q' the electric quadrupole moment density, and M the magnetic dipole moment density of the material. The ellipses indicate higher order moment densities, which, together with the electric quadrupole moments, vanish for most materials. The exact form of the two constitutive equations, Eqs. (3.27) and (3.28), is determined by a material's atomic structure and by the interactions between the constituent atoms or molecules.

3.2.5.1 Permittivity, Pyroelectricity, and Piezoelectricity

Unless the applied electric field E is close to the breakdown field of the material, we may assume that the intrinsic polarization density is proportional to the electric field, so that

$$P = \chi \kappa E \quad \text{and} \quad D = \kappa E. \qquad (3.29)$$

It is usually the dielectric permittivity rank two tensor κ, rather than the dielectric susceptibility tensor χ, that is measured and tabled. The structure of the dielectric permittivities is illustrated for all crystal classes in Fig. 3.9. The polarization is influenced by other state variables as well. Thus pyroelectricity is an electrothermal effect by which a local change in temperature leads to a local increase in polarization density:

$$(\delta P)_T = p \Delta T, \qquad (3.30)$$

with p the pyroelectric coefficient vector. Pyroelectricity is not observed in all crystal classes, as shown in Fig. 3.10. It does not appear as an effect in typical CMOS materials. The increase in polarization due to mechanical stress is called piezoelectricity, and is written as

$$(\delta P)_E = d : \sigma, \qquad (3.31)$$

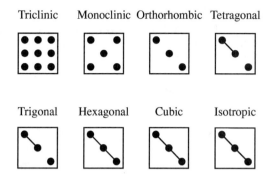

Figure 3.9 The structure of the symmetric second-rank dielectric tensor coefficient matrix for different crystal classes. The same structure holds for the electric susceptibility tensor and the thermal conductivity tensor. A solid dot indicates a nonzero entry. A line indicates numerical equality. The crystal classes are indicated using the Hermann–Mauguin notation. Adapted from Refs. 21 and 20.

where d is the rank three tensor of piezoelectric coefficients, and σ is the rank two stress tensor. The opposite effect of electrostriction is observed when an applied polarization P causes a change in the strain state. It involves the same property tensor d, so that

$$(\delta \varepsilon)_\sigma = d^T \cdot P. \tag{3.32}$$

For crystalline materials, the tensor d has a structure dictated by crystal symmetry. The possible cases are illustrated in Fig. 3.10 using an engineering matrix representation. Silicon's cubic structure prevents piezoelectric or electrostrictive effects. This is unfortunate, because these effects are very useful for a large range of sensor and actuator applications. Materials must be added to silicon microsystems, and typical examples are PZT and ZnO.[27] The energy density stored in the electric field is:[28]

$$w_e = \int_0^D E \cdot \delta D, \tag{3.33}$$

and if the medium is linear, then $E \cdot \delta D = \delta(E \cdot D)/2$, so that the total electrostatic energy is

$$w_e = \frac{1}{2}(E \cdot D). \tag{3.34}$$

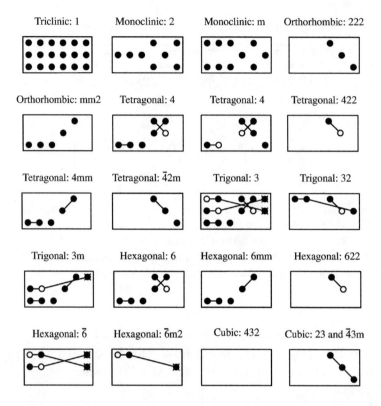

Figure 3.10 The structure of the partially symmetric third-rank pyroelectric and piezoelectric tensor coefficient matrix for different crystal classes, written in compact engineering notation. The same structure holds for the piezoelectric strain tensor, the electro-optic susceptibility tensor, and the Pockels strain-free electro-optic tensor. A solid dot indicates a nonzero entry. A line indicates numerical equality. The crystal classes are indicated using the Hermann–Mauguin notation. Adapted from Refs. 21 and 20.

The electrostatic co-energy density is the Legendre transform of the electrostatic energy w.r.t. E and D, $w_e^* = E \cdot D - w_e$, so that

$$w_e^* = \int_0^E \delta E \cdot D. \tag{3.35}$$

In a linear medium, Eqs. (3.33) and (3.35) are equal.

3.2.5.2 Charge Transport: Conductivity and Piezoresistivity

The flow of electric charge in a semiconductor may also be modulated by other fields. A full discussion is beyond the scope of this chapter, and the interested reader is referred to specialized texts dealing specifically with this issue.[13,24] Because of its huge significance in many MEMS, we briefly state the results of piezoresistivity. The deformation field experienced by a semiconductor tends to slightly deform the band structure, causing a small perturbation of the effective mass tensor of the charge carriers (holes and electrons) and hence the charge mobilities.[13] In nonjunction resistive regions of a doped semiconductor, we may express the current density dependence on the electric field in a point as

$$\boldsymbol{J} = -s\nabla\psi, \quad (3.36)$$

where s is the conductivity tensor of the material at that point. The perturbation of the conductivity tensor by an imposed stress field σ may be written as

$$\Delta s = -\pi\sigma s, \quad (3.37)$$

with π a rank four tensor of piezoresistive coefficients. Since the values of the coefficients of π are dependent on the doping level of the semiconductor, these are best measured for a given technology. The structure of π follows that of Fig. 3.8. We insert Eq. (3.37) into Eq. (3.36) to get the modified constitutive equation for conduction:

$$\boldsymbol{J}(\sigma) = -(\delta - \pi\sigma)s\nabla\psi. \quad (3.38)$$

Thus we see that for materials with isotropic conductivity s in the absense of stress, piezoresistivity can be the cause of the effective conductivity $(\delta - \pi\sigma)s$ becoming anisotropic.

3.2.5.3 Elasto-Optic Effects

The elastic deformation of a crystal can interact with a traversing electromagnetic wave to produce coupling in both directions: the modulation of the electromagnetic wave, and a modification of the deformation state of the crystal. Many interactions have been discovered, and a full treatment is beyond the scope of this chapter. The description starts with a constitutive equation that relates the polarization $\boldsymbol{P}$ to the electric field $\boldsymbol{E}$ and deformation gradient $\nabla_x \boldsymbol{u}$ through the rank four elasto-optic susceptibility tensor χ_0:

$$\boldsymbol{P} = \varepsilon_0 \chi_0 (\boldsymbol{E} \nabla_x \boldsymbol{u}). \quad (3.39)$$

The elasto-optic susceptibility has three terms: the direct elasto-optic susceptibility arising from strain, the direct elasto-optic susceptibility arising from rotation, and the indirect elasto-optic susceptibility arising from a piezoelectrically generated electric field and the electro-optic effect.[20] The rank four Pockels elasto-optic tensor $\boldsymbol{p}$ is related to the rank four elasto-optic susceptibility tensor χ_0 through the rank two inverse dielectric tensor κ^{-1}:

$$\boldsymbol{p} = -2\kappa^{-1}\kappa^{-1}\chi. \quad (3.40)$$

The rank two elastic strain tensor ε causes a perturbation of the rank two inverse dielectric tensor κ^{-1} according to

$$\Delta\kappa^{-1} = p\varepsilon = -2\kappa^{-1}\kappa^{-1}\chi_0\varepsilon. \tag{3.41}$$

Since $\Delta\kappa^{-1}$ can be linearly related to the perturbation of the rank two refractive index tensor Δn, and strain linearly to stress, we are able to define a further relation:

$$\Delta n = C_P\sigma, \tag{3.42}$$

where C_P is a rank four constitutive tensor called the photoelastic tensor.

3.2.5.4 Heat Transport and Thermoelectricity

Using the thermodynamics of irreversible processes, the energy fluxes J_i and their driving forces F_j are linearly related through the symmetrical Onsager kinetic coefficients L_{ij} through:[13,24]

$$J_i = \sum_j L_{ij} F_j. \tag{3.43}$$

The choice of a unipolar charge conduction model and some algebraic manipulation leads to the following dynamic constitutive equation:

$$\begin{bmatrix} J_{\text{el}} \\ J_{\text{th}} \end{bmatrix} = \begin{bmatrix} -s & (\alpha s) \\ (\alpha s T) & -(T s \alpha^2 + \kappa_{\text{th}}) \end{bmatrix} \begin{bmatrix} \nabla\psi \\ \nabla T \end{bmatrix}, \tag{3.44}$$

where α is the thermoelectric power or Seebeck coefficient and κ_{th} the rank two thermal conductivity tensor [also see Fig. 3.9] of the material.[14,29]

3.3 DISCRETIZATION METHODS

The model equations, i.e., a set made up of the combination of equations of motion together with constitutive relations, are now cast into an appropriate form, for which special solution methods and strategies exist. We consider the differential forms of the model equations, the partial differential equations (PDEs), as starting point. We use the weighted residual method to derive the discrete equations that form the basis of a variety of established numerical methods, including the finite-difference method (FDM), the control-volume method (CVM, alias the box method), the finite-element method (FEM), and the boundary-element method (BEM, alias the panel method). Under special conditions, certain problems may be cast into the form of integral equations of the second kind.[30] These lead to numerically stable solution algorithms.

3.3.1 WEIGHTED RESIDUALS AND DISCRETIZATION TECHNIQUES

We use the weighted residual method to obtain expressions that are suitable as the starting point for the most common PDE discretization methods. The basic idea is to project the residual by means of an inner product with a weighting function (also defined on a space of approximation functions), and to require this inner product to be zero.[31] The discretization error is therefore orthogonal to the approximate solution. The result is an expression that achieves an approximation error that vanishes in an integral sense. The method is useful when a variational principle for the PDE does not exist, and equivalent when one is known. We demonstrate the method for the anisotropic diffusion equation, and tabulate results for the mechanical equations of motion.

3.3.2 STRONG FORM

The residual of the anisotropic diffusion equation, $R_I = \nabla \cdot \boldsymbol{q} - s + c\dot{f}$ is defined on the geometric domain $\Omega \subset R^n, n \in \{1, 2, 3\}$ [see Fig. 3.11] and vanishes if the flux density $\boldsymbol{q} = a \cdot \nabla f$, and hence the potential $f(\boldsymbol{x}, t)$, is the exact solution of the diffusion equation. The tensor a is a symmetric positive-definite rank two material property tensor such as the dielectric tensor. The scalar source density field s is likewise defined on Ω. The diffusion equation is subject to initial conditions where $f(\boldsymbol{x}, t_0)$ and $\dot{f}(\boldsymbol{x}, t_0)$ is specified, and Dirichlet and Neumann boundary conditions. On the Neumann segment of the domain boundary $\Gamma_N = \bigcup_b \Gamma_{Nb}, \Gamma_N \subset \Gamma \equiv \partial\Omega$, the flux density is specified, and on the Dirichlet segment of the boundary $\Gamma_D = \bigcup_b \Gamma_{Db}, \Gamma_D \subset \Gamma \equiv \partial\Omega$, the potential is specified. Note that the Dirichlet and Neumann boundary segments are disjoint so that $\Gamma_N \cap \Gamma_D = \emptyset$. Furthermore, the boundary is everywhere specified by either a Neumann or a Dirichlet boundary condition, so that $\Gamma = \partial\Omega = \Gamma_N \cup \Gamma_D$. With each boundary segment we associate a separate residual.[32] The Dirichlet boundary residual $R_{II} = (\bar{f} - f)$ vanishes if the specified potential $\bar{f}$ and its approximation

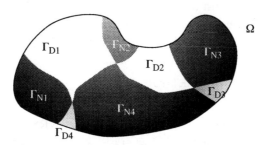

Figure 3.11 The simulation domain $\Omega \in R^n$ and its boundary $\Gamma \equiv \partial\Omega$. The boundary of the domain may be the union of many disjoint segments, each of type Dirichlet or Neumann.

Table 3.4 The weighted residual statements for the Poisson equation and the mechanical equation of motion.

Poisson equation
WRS: $\int_\Omega (\nabla \cdot \boldsymbol{q} - s) f^* d\Omega - \int_{\Gamma_N} (\boldsymbol{n}^T \cdot (\boldsymbol{q} - \bar{\boldsymbol{q}})) f^* d\Gamma - \int_{\Gamma_D} (\bar{f} - f)(\boldsymbol{n}^T \cdot \boldsymbol{q}^*) d\Gamma = 0$
WF: $\int_\Omega ((\nabla f)^T \cdot \boldsymbol{q}^*) d\Omega + \int_\Omega s f^* d\Omega - \int_\Gamma (\boldsymbol{n}^T \cdot \boldsymbol{q}) f^* d\Gamma + \int_{\Gamma_D} (\bar{f} - f)(\boldsymbol{n}^T \cdot \boldsymbol{q}^*) d\Gamma = 0$
IS: $\int_\Omega (\nabla \cdot \boldsymbol{q}^*) f d\Omega - \int_\Omega s f^* d\Omega + \int_\Gamma (\boldsymbol{n}^T \cdot \boldsymbol{q}) f^* d\Gamma - \int_\Gamma (\boldsymbol{n}^T \cdot \boldsymbol{q}^*) f d\Gamma = 0$
Mechanical equation of motion
WRS: $\int_\Omega ((\nabla \cdot \sigma - \boldsymbol{b})^T \cdot \boldsymbol{u}^*) d\Omega - \int_{\Gamma_N} ((\boldsymbol{n} \cdot (\sigma - \bar{\sigma})) \cdot \boldsymbol{u}^*) d\Gamma - \int_{\Gamma_D} ((\bar{\boldsymbol{u}} - \boldsymbol{u}) \cdot (\boldsymbol{n} \cdot \sigma^*)) d\Omega = 0$
WF: $\int_\Omega \left(\frac{1}{2}(\nabla \boldsymbol{u}^* + \nabla^T \boldsymbol{u}^*) \cdot \sigma\right) d\Omega + \int_\Omega (\boldsymbol{b} \cdot \boldsymbol{u}^*) d\Omega - \int_\Gamma ((\boldsymbol{n} \cdot \sigma) \cdot \boldsymbol{u}^*) d\Gamma - \int_{\Gamma_D} (\boldsymbol{u} \cdot (\boldsymbol{n} \cdot \sigma^*)) d\Omega = 0$
IS: $\int_\Omega ((\nabla \cdot \sigma^*) \cdot \boldsymbol{u}) d\Omega - \int_\Omega (\boldsymbol{b} \cdot \boldsymbol{u}^*) d\Omega + \int_\Gamma ((\boldsymbol{n} \cdot \sigma) \cdot \boldsymbol{u}^*) d\Gamma - \int_\Gamma (\boldsymbol{u} \cdot (\boldsymbol{n} \cdot \sigma^*)) d\Omega = 0$

f are equal, and the Neumann boundary residual $R_{III} = (\boldsymbol{q} - \bar{\boldsymbol{q}})$ vanishes if the specified flux density $\bar{\boldsymbol{q}}$ and its approximation $\boldsymbol{q}$ are equal.

3.3.3 WEIGHTED RESIDUAL STATEMENT

The *weighted residual statement* (WRS) of the diffusion equation is constructed from the first residual term R_I:

$$\int_\Omega R_I f^* d\Omega = \int_\Omega (\nabla \cdot \boldsymbol{q} - s + c\dot{f}) f^* d\Omega = 0. \tag{3.45}$$

The quantities marked by an asterisk refer to the weighting function field and its associated flux $\boldsymbol{q}^* = a \cdot \nabla f^*$. The weighting function f^* is any arbitrary scalar field. The WRS of the strong form of the PDE is the starting point for finite-difference and control-volume methods. We summarize the WRS for the Poisson equation as well as the mechanical equation of motion in Table 3.4.

3.3.4 FINITE-DIFFERENCE METHOD

The finite-difference method is generated from the WRS by colocation with a weighting function constructed as $f^* = \delta(\boldsymbol{x} - \boldsymbol{x}^k), \boldsymbol{x}^k \in \{\text{Grid}\}$. This is a Dirac delta function that evaluates to unity at the grid points, for example on the rectangular $n \times m$ tensor-product array of points spaced as $\boldsymbol{x}^k = \boldsymbol{x}^{n,m} \in \{(x_i^0 + n\nabla x_i) \times (x_j^0 + m\Delta x_j)\}$ shown in Fig. 3.12. Inserting the weighting function into Eq. (3.45)

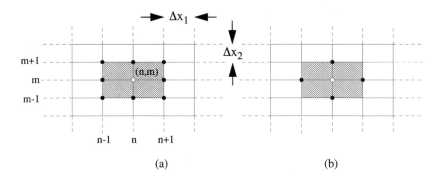

Figure 3.12 The finite-difference grid. To the left in (a) we show the nine-point scheme required for a general, anisotropic Poisson equation. To the right in (b) we show the familiar five-point scheme that results when the material property in the Poisson equation is isotropic.

implies that the residual (and hence any differential operator involved) is to be computed using data at the grid points only:

$$\int_\Omega (\nabla \cdot \mathbf{q} - s + c\dot{f})\delta(\mathbf{x} - \mathbf{x}^{n,m})d\Omega \Rightarrow (\nabla \cdot \mathbf{q} - s + c\dot{f})|_{\mathbf{x}^{n,m}} = 0. \quad (3.46)$$

The next step is to obtain expressions for the discrete differential operators, which is done by appropriately combining finite-difference operators,[33] some of which are listed in Table 3.5. The differential operator D^2 is written,[33] using the forward difference operator Δ_0 [see Table 3.5], as

$$D^2 = \frac{1}{h^2}(\Delta_0^2) + O(h^2) \quad (3.47)$$

so that we obtain the following difference formula when it is applied to the potential at grid point (n, m):

$$(D_{x_1}^2 f)^{n,m} = \left.\frac{\partial^2 f}{\partial x_1^2}\right|_{\mathbf{x}^{n,m}} = \frac{1}{(\Delta x_1)^2}\Delta_{0,1}^2 f^{n,m} + O[(\Delta x_1)^2]. \quad (3.48)$$

We now construct the differential operator D in terms of the averaging operator Υ_0 and the central difference operator Δ_0 (both defined in Table 3.5) using[33]

$$D = \frac{1}{h}(\Upsilon_0\Delta_0) + O(h^2), \quad (3.49)$$

Table 3.5 Some useful linear finite-difference operators.

Operator	Expression
Shift operator	$(Ez)^k = z^{k+1}$
Forward difference operator	$(\Delta_+ z)^k = z^{k+1} + z^k$
Backward difference operator	$(\Delta_- z)^k = z^k - z^{k-1}$
Central difference operator	$(\Delta_0 z)^k = z^{k+\frac{1}{2}} - z^{k-\frac{1}{2}}$
Identity	$(Iz)^k = z^k$
Average operator	$(\Upsilon_0 z)^k = z^{k+\frac{1}{2}} + z^{k-\frac{1}{2}}$
Differential operator	$(Dz)^k = \frac{1}{h}(\Upsilon_0 \Delta_0 z)^k = z'(kh)$

Source: Ref. 33.

so that when approximating the first partial derivative of f with respect to x_1 at the grid point (n, m), we obtain

$$(D_{x_1} f)^{n,m} = \left.\frac{\partial f}{\partial x_1}\right|_{x^{n,m}} = \frac{1}{\Delta x_1} \Upsilon_{0,1} \Delta_{0,1} f^{n,m} + O\left[(\Delta x_1)^2\right]. \tag{3.50}$$

Applying Eq. (3.48) and Eq. (3.50), we now approximate Eq. (3.46):

$$\left.\begin{aligned}
\left(\nabla \cdot \mathbf{q} - s + c\dot{f}\right)\big|_{x^{n,m}} &= \left(\nabla \cdot (a \cdot \nabla f) - s + c\dot{f}\right)\big|_{x^{n,m}} \\
&= \left(a_{11}\frac{\partial^2 f}{\partial x_1^2}\right)\bigg|_{x^{n,m}} + \left(2a_{12}\frac{\partial^2 f}{\partial x_1 \partial x_2}\right)\bigg|_{x^{n,m}} \\
&\quad + \left(a_{22}\frac{\partial^2 f}{\partial x_2^2}\right)\bigg|_{x^{n,m}} - s\big|_{x^{n,m}} + (c\dot{f})\big|_{x^{n,m}}
\end{aligned}\right\}, \tag{3.51}$$

at grid point $x^{n,m}$ using finite differences to obtain for the diffusion equation the finite-difference formula:

$$\left.\begin{aligned}
&\frac{a_{11}}{(\Delta x_1)^2}\Delta_{0,1}^2 f^{n,m} + \frac{2a_{12}}{\Delta x_1 \Delta x_2}\Upsilon_{0,1}\Delta_{0,1}\Upsilon_{0,2}\Delta_{0,2} f^{n,m} \\
&+ \frac{a_{22}}{(\Delta x_2)^2}\Delta_{0,2}^2 f^{n,m} - s^{n,m} = \frac{a_{11}}{(\Delta x_1)^2}\left(f^{n-1,m} - 2f^{n,m} + f^{n+1,m}\right) \\
&+ \frac{a_{12}}{2\Delta x_1 \Delta x_2}\left(f^{n+1,m+1} + f^{n-1,m-1} + f^{n+1,m-1} + f^{n-1,m+1}\right) \\
&+ \frac{a_{22}}{(\Delta x_2)^2}\left(f^{n,m-1} - 2f^{n,m} + f^{n,m+1}\right) - s^{n,m} + c\dot{f}^{n,m}
\end{aligned}\right\}. \tag{3.52}$$

Note that this nine-point formula, illustrated in Fig. 3.12(a), reduces to the familiar five-point formula, illustrated in Fig. 1.12(b), when the material property is isotropic and hence $a_{12} = a_{21} = 0$. The time derivative $\dot{f}$ and initial conditions are dealt with in Section 3.4.3.

For non-Dirichlet grid points that lie on the boundary of the computational domain, not all neighbors of the difference formula are available, and hence we have to use either a backward difference or forward difference scheme as is appropriate. Furthermore, if general, curved boundaries are allowed, care must be taken in forming the difference operator for near-boundary points, because accuracy could be lost otherwise. Another point to note is that when each grid point is made to lie within a single material (and not on a material interface), it greatly simplifies the formalism.

Requiring that Eq. (3.52) be satisfied at each of the k grid points results in k simultaneous equations that may be grouped into a matrix equation $Ax = b$, where x is a vector of the grid point unknowns f^k, and b contains the corresponding grid point source terms s^k and Neumann boundary contributions b_N^k.

3.3.5 CONTROL VOLUME METHOD

The control volume method is also generated from the WRS. It is sometimes called the box method, notably by the electronic device simulation community. The idea is straightforward and intuitively satisfying because flux conservation is immediately apparent. The WRS of Eq. (3.45) is used with the weighting function chosen as $f^* = 1$. Its first term is now transformed using Gauss's divergence theorem so that

$$\left. \begin{array}{l} \int_\Omega (\nabla \cdot \mathbf{q} - s + c\dot{f}) d\Omega = \int_\Gamma (\mathbf{n} \cdot \mathbf{q}) d\Gamma - \int_\Omega s d\Omega + \int_\Omega c\dot{f} d\Omega \\ \Leftrightarrow \int_\Gamma (\mathbf{n} \cdot a\nabla f) d\Gamma - \int_\Omega s d\Omega + \int_\Omega c\dot{f} d\Omega = 0 \end{array} \right\}, \quad (3.53)$$

for the unit outward normal to the boundary $\mathbf{n}$. Equation (3.53) holds for the entire domain of interest, and therefore also for any subdomain. We now require the simulation domain to be meshed as shown in Fig. 3.13(a) and (b), and apply Eq. (3.53) to each Voronoi box [shown in Fig. 3.13(c)]. To evaluate the boundary integral, we require an expression for ∇f at the box edges along which we will integrate. This is easily computed if we refer all computations to the Delaunay elements [shown in Fig. 3.13(d)]. Since the element is a simplex, we may use simplex shape functions $N^p(\mathbf{x})$ to approximate $f(\mathbf{x})$ and hence $\mathbf{q}$ from the discrete values of f at the mesh nodes:

$$f = \sum_{p=i,j,k} f^p N^p(\mathbf{x}) \quad \text{and} \quad \mathbf{q} = \sum_{p=i,j,k} f^p \nabla N^p(\mathbf{x}). \quad (3.54)$$

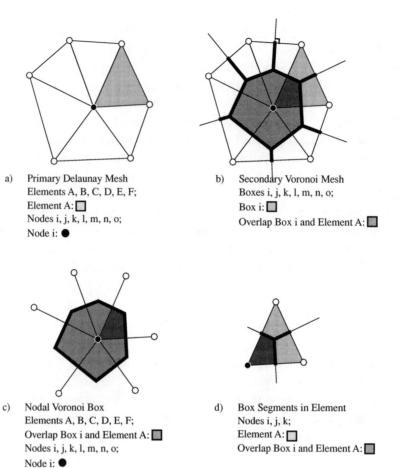

Figure 3.13 The construction of a Voronoi box around the vertex node of a 2D Delaunay ( ) simplex mesh. In a 2D Delaunay mesh, no triangle has inner angles greater than 90 degrees. The Voronoi box edges are the perpendicular bisectors of the Delaunay triangle edges. Each vertex of the Delaunay mesh is contained in a unique Voronoi box. Since the meshing presupposes simplices, it is possible to apply these ideas to 1D and 3D meshes as well.

The shape functions are linear polynomials if we restrict the positions of the nodal unknowns to the Delaunay vertex positions, so that

$$\left.\begin{array}{l} N^i(x_1, x_2) = \left(x_1^j x_2^k - x_1^k x_2^j\right) + \left(x_2^j - x_2^k\right)x_1 + \left(x_1^k - x_1^j\right)x_2/(2A) \\ N^j(x_1, x_2) = \left(x_1^k x_2^i - x_1^i x_2^k\right) + \left(x_2^k - x_2^i\right)x_1 + \left(x_1^i - x_1^k\right)x_2/(2A) \\ N^k(x_1, x_2) = 1 - N^i(x_1, x_2) - N^j(x_1, x_2) \end{array}\right\}, \quad (3.55)$$

for the triangle with area $A = (x_1^i x_2^j + x_1^j x_2^k + x_1^k x_2^i - (x_1^i x_2^k + x_1^j x_2^i + x_1^k x_2^j))/2$. A mesh vertex j has spatial coordinates x_i^j in the Cartesian coordinate direction i.

The following properties must hold for the shape function set (N^i, N^j, N^k) to be a valid interpolator over a Delaunay triangle's domain Ω^e

$$\left. \begin{array}{c} \sum_{l=i,j,k} N^l(x_1, x_2) = 1 \\ N^j(x_1^i, x_2^i) = \delta^{ij} \end{array} \right\}, \quad \forall x \in \Omega^e. \tag{3.56}$$

An element uniquely overlaps the three boxes associated with each of its vertices, and contributes two boundary terms to each box [see Fig. 3.13(d)]. The gradient of the potential inside an element is a constant vector. It is usual to assume that the material property is piecewise (element-wise) constant, a reasonable assumption for many problems. The contribution for element e written out in matrix format then gives:

$$\left. \begin{array}{c} \int_{\Gamma_e} (\mathbf{n}^T \cdot \mathbf{q}) d\Gamma = \begin{bmatrix} A^{ij}[n_1^{ij} \ n_2^{ij}] + A^{ik}[n_1^{ik} \ n_2^{ik}] \\ A^{ji}[n_1^{ji} \ n_2^{ji}] + A^{jk}[n_1^{jk} \ n_2^{jk}] \\ A^{ki}[n_1^{ki} \ n_2^{ki}] + A^{kj}[n_1^{kj} \ n_2^{kj}] \end{bmatrix} \begin{bmatrix} a_{11} & a_{12} \\ a_{21} & a_{22} \end{bmatrix} \\ \times \begin{bmatrix} \frac{\partial N^i}{\partial x_1} & \frac{\partial N^j}{\partial x_1} & \frac{\partial N^k}{\partial x_1} \\ \frac{\partial N^i}{\partial x_2} & \frac{\partial N^j}{\partial x_2} & \frac{\partial N^k}{\partial x_2} \end{bmatrix} \begin{bmatrix} f^i \\ f^j \\ f^k \end{bmatrix} = \begin{bmatrix} K_e^{ii} & K_e^{ij} & K_e^{ik} \\ K_e^{ji} & K_e^{jj} & K_e^{jk} \\ K_e^{ik} & K_e^{jk} & K_e^{kk} \end{bmatrix} \begin{bmatrix} f^i \\ f^j \\ f^k \end{bmatrix} \end{array} \right\}, \tag{3.57}$$

for the partial box boundary area A^{ij} separating vertex i and j, the boundary's normal vector pointing toward vertex j. The matrix K is unsymmetric because of the anisotropy of a. We may make K symmetric by choosing the box edge normals to be orthogonal to each other w.r.t. a using $n^\alpha \cdot a \cdot n^\beta = \delta^{\alpha\beta}$; in effect, defining an anisotropic Delaunay criterion. Caution is necessary, because this may lead to rank-deficient formulations. The source term and capacity terms are also integrated over each element. Numerical Gaussian quadrature is efficient and is arbitrarily accurate when the spatial distribution of s or c is polynomial in the spatial coordinates. A simpler approximation is to lump the distribution by using the nodal values for s and for c. Assuming a linear distribution value over the element e gives

$$R_e = \sum_{l=i,j,k} \int s^l N^l d\Omega = \sum_{l=i,j,k} R_e^i, \tag{3.58}$$

for the contributions to the source vector. Similarly, the consistent capacity entries for the matrix are

$$\int_\Omega c\dot{f} d\Omega = \sum_{m=i,j,k} \sum_{l=i,j,k} \int c^l N^l N^m d\Omega \dot{f}^m = \sum_{m=i,j,k} \sum_{l=i,j,k} C_e^{lm} \dot{f}^m. \tag{3.59}$$

Repeatedly building the matrix and vector contributions for each mesh element and assembling these to form the global K and C matrix and R vector results in the algebraic system of equations

$$\left.\begin{array}{r}\sum_{j=1}^{N} K^{ij} f^j + \sum_{j=1}^{N} C^{ij} \dot{f}^j = R^i \\ \sum_{j=1}^{N} \sum_{e=1}^{Ne} \sum_{m=1,n=1}^{3} P_e^{im} K_e^{mn} P_e^{nj} f^j + \sum_{j=1}^{N} \sum_{e=1}^{Ne} \sum_{m=1,n=1}^{3} P_e^{im} C_e^{mn} P_e^{nj} \dot{f}^j \\ = \sum_{e=1}^{Ne} \sum_{m=1}^{3} P_e^{im} R_e^m \end{array}\right\}, \quad (3.60)$$

with P_e^{im} the eth Delaunay element's Boolean incidence matrix that identifies the local element vertex node m with global node i. Boundary conditions are treated as follows: Dirichlet conditions, which specify the values of f at certain nodes f^d and hence zero the corresponding velocities $\dot{f}^d$, require that all individual matrix-vector product terms involving these known values be moved to the right-hand side, thereby "eliminating" the corresponding equations and columns and thus reducing the dimension of K, viz.,

$$\left.\begin{array}{r} \begin{bmatrix} K^{ff} & K^{fd} \\ K^{df} & K^{dd} \end{bmatrix} \begin{bmatrix} f^f \\ f^d \end{bmatrix} + \begin{bmatrix} C^{ff} & C^{fd} \\ C^{df} & C^{dd} \end{bmatrix} \begin{bmatrix} \dot{f}^f \\ 0 \end{bmatrix} = \begin{bmatrix} R^f \\ R^f \end{bmatrix} \Rightarrow \\ \begin{bmatrix} K^{ff} \\ K^{df} \end{bmatrix} \begin{bmatrix} f^f \\ f^d \end{bmatrix} + \begin{bmatrix} C^{ff} \\ C^{df} \end{bmatrix} \begin{bmatrix} \dot{f}^f \\ 0 \end{bmatrix} = \begin{bmatrix} R^f - K^{fd} f^d \\ R^f - K^{dd} f^d \end{bmatrix} \Rightarrow \\ K^{ff} f^f + C^{ff} \dot{f}^f = R^f - K^{fd} f^d \end{array}\right\}. \quad (3.61)$$

A Neumann condition describes a flux source density at a boundary node. This is achieved by adding the surface-integrated value of the source density to the corresponding nodal position in the right-hand side vector of Eq. (3.60). The time derivative of f and initial conditions are considered in Section 3.4.3.

3.3.6 THE WEAK FORM

We integrate the first term of the WRS, Eq. (3.45), by parts and obtain

$$\left.\begin{array}{r} \int_\Omega (\nabla \cdot \boldsymbol{q}) f^* d\Omega = \int_\Gamma (\boldsymbol{n} \cdot \boldsymbol{q}) f^* d\Gamma - \int_\Omega (\boldsymbol{q} \cdot (\nabla f^*)) d\Omega \\ = \int_\Gamma (\boldsymbol{n} \cdot \boldsymbol{q}) f^* d\Gamma - \int_\Omega (\nabla f) \cdot \boldsymbol{q}^* d\Omega \end{array}\right\}, \quad (3.62)$$

where we have used the complementarity and symmetry

$$\left.\begin{array}{r} \boldsymbol{q} \cdot (\nabla f^*) = (a \cdot \nabla f) \cdot (\nabla f^*) = (\nabla f) \cdot (a \cdot \nabla f^*) = (\nabla f) \cdot \boldsymbol{q}^* \\ a = a^T \end{array}\right\}. \quad (3.63)$$

Equation (3.62) is substituted into the WRS, Eq. (3.45), to obtain the *weak form* (WF) of the diffusion equation:

$$\int_\Gamma (\mathbf{n} \cdot \mathbf{q}) f^* d\Gamma - \int_\Omega ((\nabla f) \cdot \mathbf{q}^*) d\Omega - \int_\Omega s f^* d\Omega + \int_\Omega c \dot{f} f^* d\Omega. \qquad (3.64)$$

Note that even though the WF is algebraically equivalent to the WRS, it has relaxed the conditions on the required differentiability of approximations to f, from functions that are twice differentiable, to only once differentiable.[31] Also, the Neumann boundary condition is now removed into the governing equation through the first term of Eq. (3.64). The weak form is the starting point for the finite-element method, where this simplification is exploited.

3.3.7 BUBNOV–GALERKIN FINITE-ELEMENT METHOD

As for the CVM, the finite-element method requires a mesh of elements. One mesh condition, however, is relaxed. Elements must still be convex polytopes (lines, triangles, quadrilaterals, tetrahedra, and hexahedra are typical 1D, 2D, and 3D continuum finite elements), but an acceptable mesh need not be a Delaunay mesh; the finite-element method handles considerably less regular tessellations than the CVM. The field value f and its time derivative are represented using suitable shape functions defined for each mesh node, thus

$$f = \sum_p f^p N^p(\mathbf{x}), \quad \dot{f} = \sum_p \dot{f}^p N^p(\mathbf{x}) \quad \text{and} \quad \nabla f = \sum_p f^p \nabla N^p(\mathbf{x}), \quad (3.65)$$

where the index p runs over the number of nodes in the mesh. The Galerkin finite-element method now makes a particular choice for the weighting functions f^*. They are chosen from the same space of functions as the weighting functions, and one convenient choice is

$$f^* = \sum_p N^p, \quad \text{so that} \quad \nabla f^* = \sum_p \nabla N^p(\mathbf{x}), \qquad (3.66)$$

with the restriction that f^* vanishes at the Dirichlet boundaries. We need two more terms written out before forming the finite-element equations,

$$\mathbf{q} = a \cdot \nabla f = a \cdot \sum_p f^p \nabla N^p(\mathbf{x}) \quad \text{and} \quad (\mathbf{n} \cdot \mathbf{q})|_{\Gamma_N} = q_{\perp N}. \qquad (3.67)$$

Inserting Eqs. (3.65), (3.66), and (3.67) into Eq. (3.64) yields the finite-element equation system for the anisotropic diffusion equation in a global formulation:

$$\left. \begin{array}{l} \sum_p \sum_q \int_\Omega (\nabla N^p) \cdot (a \cdot \nabla N^q) f^q d\Omega - \sum_p \int_\Gamma q_{\perp N} N^p d\Gamma \\ - \sum_p \sum_q \int_\Omega N^p c N^q \dot{f}^q d\Omega + \sum_p \int_\Omega s N^p d\Omega = 0 \end{array} \right\}. \quad (3.68)$$

In matrix format, using $B = \nabla N$, Eq. (3.68) compacts to

$$\left(\int_\Omega B^T \cdot a \cdot B d\Omega\right) f - \left(\int_\Omega cNN d\Omega\right) \dot{f} + \int_\Omega sN d\Omega - \int_\Gamma q_{\perp N} N d\Gamma = 0. \quad (3.69)$$

The power of the finite-element method lies in its ability to construct and solve variants of Eq. (3.69) with enormous flexibility for the program user; both the choice of N and the positions of the nodes need only be fixed at run time. This is partly achieved by building Eq. (3.69) from the mesh on an element-by-element basis. This requires an assembly operation as described for the control volume method, and an element formulation of Eq. (3.69) which turns out to be rather straightforward because we may always write that $\int_\Omega I d\Omega = \sum_e \int_{\Omega_e} I d\Omega_e$, since $\Omega = \bigcup_e \Omega_e$. Furthermore, since the definition of the nodal shape functions is now moved to an element viewpoint, we only consider the overlap of each shape function with the element's domain. This fact can be exploited to simplify the formulation and hence the computer implementation; actually, it is now a straightforward matter to generate elements of any shape function order. We conclude this brief description of the finite-element method with a widely used variant.

Much simplification for implementation in a computer program is gained by using the isoparametric finite-element concept. Now the discretization is performed on a reference geometry, followed by a mapping of the discretized equation system to the actual element geometry in the mesh. The term isoparametric refers to the fact that the finite-element shape functions also serve to map the reference element's geometry to the mesh. For the four-noded isoparametric bilinear quadrilateral, the (invertible) geometrical mapping function from the reference coordinates ζ to the mesh coordinates x is

$$x = \sum_{p=i,j,k,l} x^p \overline{N}^p(\zeta). \quad (3.70)$$

We use an upper bar to indicate quantities defined in the coordinates of the reference finite element (also see Fig. 3.14). Using an element Jacobian:

$$J = (\nabla \zeta)^{-1} = \nabla_\zeta x = \sum_{p=i,j,k,l} x^p \nabla_\zeta \overline{N}^p(\zeta), \quad (3.71)$$

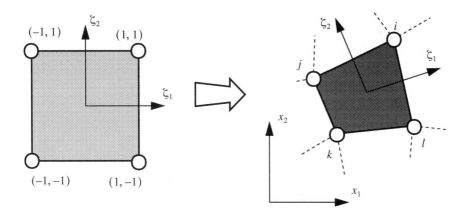

Figure 3.14 The isoparametric formulation is illustrated here for a 2D quadrilateral element. Left is the square reference element geometry which is mapped to the actual element geometry in the mesh. The element's shape functions are used to perform the geometrical mapping.

we now proceed to form the geometric matrices of the mesh coordinates in the reference coordinates so that

$$B^p = \nabla N^p = \nabla \zeta \cdot \nabla_\zeta \overline{N}^p = \nabla \zeta \cdot \overline{B}^p \quad \text{or} \quad B = J^{-1}\overline{B}. \tag{3.72}$$

The integral must also be transformed, and the differential area integral becomes

$$dx_1 dx_2 = |\det(J)| d\zeta_1 d\zeta_2, \tag{3.73}$$

where $\det(J)$ is the local element Jacobian determinant, defined using Eq. (3.71) as

$$\det(J(\zeta)) = \det\left(\begin{bmatrix} \frac{\partial \overline{N}^i}{\partial \zeta_1} & \frac{\partial \overline{N}^j}{\partial \zeta_1} & \frac{\partial \overline{N}^k}{\partial \zeta_1} & \frac{\partial \overline{N}^l}{\partial \zeta_1} \\ \frac{\partial \overline{N}^i}{\partial \zeta_2} & \frac{\partial \overline{N}^j}{\partial \zeta_2} & \frac{\partial \overline{N}^k}{\partial \zeta_2} & \frac{\partial \overline{N}^l}{\partial \zeta_2} \end{bmatrix} \begin{bmatrix} x_1^i & x_2^i \\ x_1^j & x_2^j \\ x_1^k & x_2^k \\ x_1^l & x_2^l \end{bmatrix}\right). \tag{3.74}$$

Now substitute Eqs. (3.72), (3.73), and (3.74) into Eq. (3.69), to get the isoparametric finite-element formulation:

$$\left.\begin{aligned} &\left(\int_{\overline{\Omega}} (J^{-1}\overline{B})^T a J^{-1}\overline{B}|\det(J)| d\overline{\Omega}\right)\{f\} - \int_{\overline{\Gamma}} q_{\perp N} \overline{N} |\det(J)| d\overline{\Gamma} \\ &- \left(\int_{\overline{\Omega}} c \overline{N}\,\overline{N} |\det(J)| d\overline{\Omega}\right)\{\dot{f}\} + \int_{\overline{\Omega}} s\overline{N} |\det(J)| d\overline{\Omega} = 0 \end{aligned}\right\}. \tag{3.75}$$

Table 3.6 Abscissae and weights for 1D Gauss quadrature using $\int_{-1}^{1} I(\eta)d\eta = \sum_{i}^{N} \omega^{i} I(\eta^{i})$.

$\pm\eta$	ω
$N=1$	
0.00000000	2.00000000
$N=2$	
±0.57735027	1.00000000
$N=3$	
±0.77459667	0.55555556
0.00000000	0.88888889
$N=4$	
±0.86113631	0.34785485
±0.33998104	0.65214515
$N=5$	
±0.90617985	0.23692689
±0.53846931	0.47862867
0.00000000	0.56888889

The integration limits are now the same for all elements in the mesh, and three integrand quantities are user-defined; a, s, and q. This makes numerical quadrature particularly suitable to perform the integration in Eq. (3.75). In fact, for many applications, Gauss quadrature can be made exact (within the limits of machine precision) with moderate effort. The essential idea is to replace each integral with a sum of the form

$$\int_{\eta} I(\eta)d\eta = \sum_{i} \omega^{i} I(\eta^{i}), \qquad (3.76)$$

where $I(\eta)$ is the integrand depending on the variable of integration η, and ω^{i} and η^{i} are quadrature weights and variable positions to be chosen. For Gauss integration rules, the values are specified in Table 3.6.

3.3.8 Inverse Statement

Proceeding as for the weak form, we now integrate the second term of the WF, Eq. (3.64), by parts

$$\int_{\Omega} (\nabla f) \cdot \boldsymbol{q}^{*} d\Omega = \int_{\Gamma} (\boldsymbol{n} \cdot \boldsymbol{q}^{*}) f d\Gamma - \int_{\Gamma} (\nabla \cdot \boldsymbol{q}^{*}) f d\Omega, \qquad (3.77)$$

and substitute this result into the WF to get the *inverse statement* (IS):

$$\int_\Gamma (\mathbf{n}\cdot\mathbf{q})f^* d\Gamma - \int_\Gamma (\mathbf{n}\cdot\mathbf{q}^*)f d\Gamma + \int_\Omega (\nabla\cdot\mathbf{q}^*)f d\Omega - \int_\Omega sf^* d\Omega + \int_\Omega c\dot{f}f^* d\Omega = 0. \tag{3.78}$$

We have used partial integration to swap the role of f and f^*, notably helped along by a convenient choice of f^* and, using hindsight, the choice for the signs of the terms in the original WRS. In this way we have also obtained two additional boundary integrals. The inverse statement is the starting point for the boundary-element method presented next.

3.3.9 Boundary-Element Method

The boundary-element method is a discretization of the inverse statement of Eq. (3.78). The method eliminates as many volume integrals as possible, so that only the material interfaces of the simulation domain need be discretized. This is done for the first volume integral by choosing the weighting function as the fundamental solution (the Green function) of the Laplace operator:

$$\left.\begin{array}{l} f^*(\mathbf{r}) = \dfrac{1}{2\pi\sqrt{|a|}} \ln\left(\dfrac{1}{\sqrt{\mathbf{r}^T a^{-1} \mathbf{r}}}\right) \quad \text{in 2D} \\[2ex] f^*(\mathbf{r}) = \dfrac{1}{4\pi\sqrt{|a|}} \left(\dfrac{1}{\sqrt{\mathbf{r}^T a^{-1} \mathbf{r}}}\right) \quad \text{in 3D} \end{array}\right\}, \tag{3.79}$$

so that

$$(\nabla\cdot\mathbf{q}^*) = \nabla\cdot(a\nabla f^*) = -\delta(\mathbf{r}). \tag{3.80}$$

When Eq. (3.80) is inserted into Eq. (3.78), we obtain the boundary integral equation (BIE):

$$\left.\begin{array}{l} 2\alpha\pi f(\zeta) + \int_\Omega sf^* d\Omega - \int_\Omega c\dot{f}f^* d\Omega \\[1ex] -\int_\Gamma (\mathbf{n}\cdot\mathbf{q})f^* d\Gamma + \int_\Gamma (\mathbf{n}\cdot\mathbf{q}^*)f d\Gamma = 0 \end{array}\right\}, \tag{3.81}$$

where $\alpha = 1$ in 2D and $\alpha = 2$ in 3D. We also require expressions for the flux associated with the fundamental solution, which are given by

$$\left.\begin{array}{l} \mathbf{q}^*(\mathbf{r}) = a\nabla f^*(\mathbf{r}) = \dfrac{a}{2\pi\sqrt{|a|}} \nabla\left(\ln\left(\dfrac{1}{\sqrt{\mathbf{r}^T a^{-1} \mathbf{r}}}\right)\right) \quad \text{in 2D} \\[2ex] \mathbf{q}^*(\mathbf{r}) = a\nabla f^*(\mathbf{r}) = \dfrac{a}{4\pi\sqrt{|a|}} \nabla\left(\dfrac{1}{\sqrt{\mathbf{r}^T a^{-1} \mathbf{r}}}\right) \quad \text{in 3D} \end{array}\right\}, \tag{3.82}$$

Unless the source s and the capacity c can each be represented by the divergence of a vector field, there is no obvious method by which to transform their volume integrals into surface integrals. The volume integrals should be evaluated using the techniques shown for the finite element or control-volume methods. This is clearly a disadvantage of our particular choice for the fundamental solution. For the remainder of this discussion, we consider only the sourceless stationary boundary integral equation:

$$c(\zeta)f(\zeta) - \int_\Gamma q_\perp f^* d\Gamma + \int_\Gamma f q_\perp^* d\Gamma = 0, \tag{3.83}$$

where we also set $q_\perp = (\boldsymbol{n}^T \cdot \boldsymbol{q})$ and $q_\perp^* = (\boldsymbol{n}^T \cdot \boldsymbol{q}^*)$, the surface normal flux components. The first integral in Eq. (3.83) is called the surface potential integral, the second the surface flux integral. The variable $c(\zeta)$ is a factor that depends only on the smoothness of the boundary at the evaluation position ζ. For smooth boundary points $c(\zeta) = 0.5$. The two integrals are evaluated by discretizing the boundaries into boundary elements, much like 2D finite-elements embedded in a manifold in 3D space. The field and surface normal flux are interpolated separately using standard shape functions $f(\boldsymbol{x}) = \sum_j N^j(\boldsymbol{x}) f^j$ and $q(\boldsymbol{x}) = \sum_j N^j(\boldsymbol{x}) q^j$.

3.3.10 OTHER INTEGRAL EQUATION METHODS

When a partial differential equation can be written as a Fredholm integral equation of the second kind, i.e., as $y(x) - \lambda \cdot \int_a^b K(x,t) y(t) dt = f(x)$, with λ a known constant factor, $K(x,t)$ a known kernel function, $y(x)$ the unknown function, and $f(x)$ a given right-hand-side function, the discretized version (based, for example, on the boundary element method) leads to a system of algebraic equations whose coefficient matrix has an inverse with eigenvalues bounded from below by 1.0. Such matrix systems are amenable to rapid and stable solution using iterative methods. A case at hand is the electrostatic problem described by the Laplace equation:

$$\nabla \cdot (\varepsilon \nabla \psi) = 0, \tag{3.84}$$

in the domain Ω for the case of regions of piecewise constant dielectric permittivity ε and continuously variable electrostatic potential ψ. We follow the compact presentation in Ref. 34, which is in turn based on the integral equation solution presented in Ref. 30. We denote by Γ the interdielectric interfaces, by $\partial\Omega$ the boundary of the domain Ω, and by $\Omega\backslash\Gamma$ the dielectric domains excluding Γ.

Everywhere within $\Omega\backslash\Gamma$ we have that $\nabla^2 \psi = 0$, because ε is constant. On Γ the normal component of the electric field must be continuous, so that

$$[[\varepsilon \nabla \psi]]_\Gamma = 0, \tag{3.85}$$

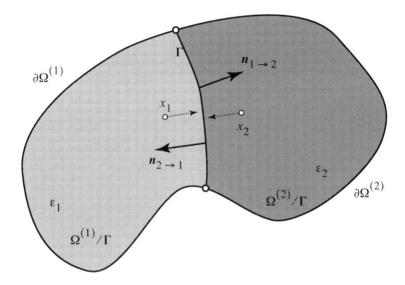

Figure 3.15 Nomenclature and geometrical concepts for the definition of the jump operator as required for the derivation of an integral equation formulation of the second kind.

where $[[.]]_\Gamma$ is the (linear) jump operator on Γ defined by

$$[[T]]_\Gamma = \lim_{x_1 \to \Gamma} T(x_1) \cdot \boldsymbol{n}_{1 \to 2} + \lim_{x_2 \to \Gamma} T(x_2) \cdot \boldsymbol{n}_{2 \to 1}, \qquad (3.86)$$

for a unit normal $\boldsymbol{n}_{i \to j}$ on Γ pointing from region i to region j (see Fig. 3.15). The dielectric permittivity can be written in terms of the permeability χ as $\varepsilon = \varepsilon_0(1+\chi)$, so that

$$[[\varepsilon \nabla \psi_\Gamma]] = 0 \Rightarrow [[\nabla \psi]]_\Gamma + [[\chi \nabla \psi]]_\Gamma = 0. \qquad (3.87)$$

We now introduce the field q defined on $\partial \Omega$ as $q = \partial_n \psi$ and on Γ as $q = [[\chi \nabla \psi]]_\Gamma$, so that

$$-\nabla^2 \psi = q \delta_\Gamma, \qquad (3.88)$$

where $\delta_\Gamma = 1$ on Γ and zero elsewhere. Equation (3.88) is a Poisson equation (with a fundamental solution satisfying $\nabla^2 \Phi - \delta = 0$) with the general solution

$$\psi(x) = \int_{\partial \Omega \cup \Gamma} \Phi(x-y) q(y) dS(y) = \text{slp}(\partial \Omega \cup \Gamma) q(x), \qquad (3.89)$$

in terms of the single-layer potential operator slp. The solution is defined everywhere except on $\partial \Omega$. The gradient of ψ is defined everywhere but on Γ. Approaching Γ from one side, we have that

$$\lim_{\theta \to 0^+} (\nabla \psi)(x - \theta n(x)) = \mathbf{slf}(\partial\Omega \cup \Gamma)q(x) + q(x)n(x)/2, \tag{3.90}$$

where the single-layer field (**slf**) is defined as

$$\mathbf{slf}(\partial\Omega \cup \Gamma)f(x) = \int_{\partial\Omega\cup\Gamma} \nabla\Phi(x - y)f(y)dS(y). \tag{3.91}$$

Combining Eqs. (3.85) and (3.90) gives

$$[[\varepsilon]]_\Gamma \, \mathbf{slf}(\partial\Omega \cup \Gamma)q + [[\varepsilon n]]_\Gamma q/2 = 0. \tag{3.92}$$

We can now define a vector c as

$$c = \frac{[[\varepsilon]]_\Gamma}{[[\varepsilon n]]_\Gamma} = \frac{(\varepsilon_1 - \varepsilon_2)n_{1\to 2}}{\varepsilon_1 + \varepsilon_2}, \tag{3.93}$$

so that the equation set to be solved becomes

$$\left.\begin{aligned} \mathrm{slp}(\partial\Omega \cup \Gamma)q &= u \quad \text{on} \quad \partial\Omega \\ q + 2c \cdot \mathbf{slf}(\partial\Omega \cup \Gamma)q &= 0 \quad \text{on} \quad \Gamma \end{aligned}\right\}. \tag{3.94}$$

3.4 SOLUTION METHODS

Through modeling we determined a set of PDEs and associated BCs from transport and constitutive equations. Using one of the spatial discretization schemes of the previous section transforms the modeling equations into a set of ordinary differential equations (ODEs). Typically we obtain one of the following equation forms

$$Ax = b, \tag{3.95}$$

$$C\dot{x} + Kx = f, \tag{3.96}$$

$$M\ddot{x} + C\dot{x} + Kx = f. \tag{3.97}$$

The first equation arises from stationary, quasi-stationary, and steady-state equations and is time-independent and therefore already algebraic. The second and third equations (ODEs) appear in connection with transient or time-dependent phenomena; the second is a generalized diffusion ODE, the third a generalized ODE of motion.

3.4.1 ALGEBRAIC EQUATION SYSTEMS

We need to transform the ODEs [Eqs. (3.96) and (3.97)] into algebraic systems of equations. Using one of a variety of assumptions, this results in equations of one of the following three forms, each of which has special solution algorithms available:

$$Ax = b, \tag{3.98}$$

$$(K + \omega^2 I)x = 0, \tag{3.99}$$

$$(K + \omega^2 M)x = 0. \tag{3.100}$$

The first equation is the standard algebraic equation system, and if linear, it can be solved immediately using the algorithms discussed in Section 3.4.4. The second and third equations represent the basic and generalized eigenvalue algebraic systems. They arise, for example, from harmonic time-dependent excitation of Eqs. (3.96) and (3.97), and from stability analysis (such as buckling).

3.4.2 NONLINEARITIES

If Eq. (3.98) is nonlinear, we may form the discrete nonlinear residual equation:

$$R(x) = A(x) \cdot x - b(x) = 0, \tag{3.101}$$

which may be turned into an iterative (fixed-point) scheme involving linear equations that yield iterative incremental solutions, by applying the generalized Taylor expansion to the residual equation:

$$R(x+d) = R(x) + d^T \cdot \nabla_x R(x) + \frac{1}{2}(d^T \cdot \nabla_x)^2 R(x) + \cdots = 0, \tag{3.102}$$

so that when we substitute Eq. (3.101) into Eq. (3.102) and truncate after the second term on the right-hand-side, we get the full Newton–Raphson scheme for the ith update using data from iteration $i-1$:

$$\left.\begin{array}{c}[A(x_i) \cdot x_i - b(x_i)] + d_i^T \cdot [\nabla_x A(x_i) \cdot x_i + A(x_i) - \nabla_x b(x_i)] = 0 \\ x_i = x + d_i\end{array}\right\}. \tag{3.103}$$

This algorithm is convergent only if the ith approximant x_i is sufficiently close to the required solution and is therefore not considered to be "globally" convergent. Note that the Jacobian matrix $J_i(x_i) = \nabla_x A(x_i) \cdot x_i + A(x_i) - \nabla_x b(x_i)$ is computed new for each iteration, making the Newton–Raphson algorithm time-consuming. Various schemes exist to make this algorithm stable, or to speed it up, or to ensure "global" convergence.[35] A popular method used to speed up the Newton–Raphson update, especially for "mild" nonlinearities, is the quasi-Newton update. Retaining the initially computed Jacobian matrix $J_0(x_0)$ for all updates saves computational resources and is often sufficiently accurate. In addition, in conjunction with direct linear solvers, the large effort of the LU-decomposition of the Jacobian is then only performed once. A step toward global convergence of the Newton–Raphson algorithm is to use a line search and/or an arclength constraint to optimize the incremental update.[35]

3.4.3 TIME INTEGRATION

The spatial discretization methods presented in the previous sections produce systems of ODEs. To solve the ODEs numerically, we have to discretize the time coordinate. Two broad methods are popular.

- Transform methods remove the time coordinate by making an ansatz for the time dependence in the frequency domain. Assume that we may write that $f(y,t) = \sum_j F_j(y)e^{-i\omega_j t}$ and $x(y,t) = \sum_j Y_j(y)e^{-i\omega_j t}$. Inserting this ansatz into Eq. (3.97) yields

$$\sum_j [-M\omega_j^2 Y_j(y) + C i\omega_j Y_j(y) + K Y_j(y)] e^{-i\omega_j t} = \sum_j F(y) e^{-i\omega_j t}. \quad (3.104)$$

By factoring out the exponential term $e^{-i\omega_j t}$, we obtain a complex algebraic equation to be solved for different parameters ω_j. If dissipative effects can be neglected (i.e., $C=0$), the equation can be solved as an eigenvalue problem (see Section 3.4.5).

- Difference methods discretize the velocity and acceleration terms of Eq. (3.97) directly, for discrete time steps Δt where the values of x, $\dot{x}$, and $\ddot{x}$ are known, and require an interpolation scheme for the intervals. We distinguish two types of difference schemes: implicit methods and explicit methods. Implicit methods require the solution of a linear equation system; explicit methods avoid this burden at the expense of accuracy and possibly stability, for which the only remedy is the use of excessively small time steps.

A popular implicit method for the second-order system of Eq. (3.97) is the Newmark single-step method,[36] by which the following time-discretized equation system is solved:

$$\left. \begin{array}{l} M\ddot{x}_{n+1} + C\dot{x}_{n+1} + Kx_{n+1} = f_{n+1} \\ x_{n+1} = x_n + \Delta t \dot{x}_n + \dfrac{\Delta t^2}{2}((1-2\beta)\ddot{x}_n + 2\beta\ddot{x}_{n+1}) \\ \dot{x}_{n+1} = \dot{x}_n + \Delta t((1-\gamma)\ddot{x}_n + \gamma\ddot{x}_{n+1}) \end{array} \right\} . \quad (3.105)$$

The parameters β and γ are used to adjust the stability and accuracy of the algorithm. For example, setting $\beta = 0.25$ and $\gamma = 0.5$ achieves an unconditionally stable (trapezoidal) method.[36] A second-order accurate explicit integration method may also be generated from Eq. (3.105) by setting $\beta = 0$ and $\gamma = 0.5$, and by requiring both M and C to be of a diagonal formulation, usually a reasonable physical assumption, so that their inverses are trivially computable:

$$\left.\begin{array}{l} M\ddot{x}_{n+1} + C\dot{x}_{n+1} + Kx_{n+1} = f_{n+1} \\[4pt] x_{n+1} = x_n + \Delta t \dot{x}_n + \dfrac{\Delta t^2}{2}\ddot{x}_n \\[4pt] \dot{x}_{n+1} = \dot{x}_n + \dfrac{\Delta t}{2}(\ddot{x}_n + \ddot{x}_{n+1}) \end{array}\right\}. \qquad (3.106)$$

Many excellent texts discuss the issues of time integration. In the context of finite-element methods, Ref. 36 is thorough. A more current text with a mathematical basis is Ref. 33.

3.4.4 LINEAR EQUATION SYSTEMS

The basic linear algebra problem, as far as simulation is concerned, is the solution of the simultaneous linear algebraic Eq. (3.95) for the unknown vector x with matrix A with known numerical coefficients and right-hand-side vector b with known numerical coefficients. We need only consider nonsingular square matrices A. Depending on the structure of the matrix A, and the cost of building it, different methods are appropriate.

Direct methods loosely based on Gaussian elimination are best when the matrix is highly sparse and its structure is easily computed.[35,37] In a sparse matrix, most of the matrix elements are identically zero. The finite-difference method on a regular grid with a regular grid equation numbering scheme produces a banded matrix A that can be stored in few vectors (one per matrix band), each no longer than the number of equations in the system. For this structure, the solution process may be made especially efficient; see, for example, the Mathematica[38] function TridiagonalSolve "tridiag" of Press et al.[35] The finite-element and finite-volume methods allow the discretization of unstructured grids, and here the structure of the matrix depends sensitively on the ordering of the grid unknowns. The matrix rows and columns must often be reordered to reduce the fill-in that the algorithm will produce. Fill-in causes initially zero matrix entries to assume nonzero values. Reordering is an NP-hard problem, but very efficient schemes exist, notably the Cuthill–McGee and the minimum degree algorithms.[37] Owing to the need to solve for multiple right-hand-side vectors b, the equation is then solved using LU decomposition.[35] The matrix condition, combined with the finite size of floating-point words on a digital computer, can lead to an accumulation of roundoff errors. Pivoting is essential to reduce part of this effect.[35] Another useful technique is called iterative improvement, and involves one additional solution step.[35] An excellent review of direct solution methods is found in Refs. 39 and 40.

As the effective matrix bandwidth increases, iterative methods may become more effective. Iterative solution methods include the conjugent gradient method and its variants. A search for the solution uses iterative updates based only on multiplying the matrix A or its transpose by a vector. This matrix-vector product can usually be obtained quite efficiently, especially on dedicated vector-processing computers. Iterative methods suffer bad convergence rates when the matrix A is badly conditioned. Conditioning may be improved through the use of matrix preconditioners. Iterative methods are available as C++ templates;[40] Ref. 40 also presents a handy decision diagram to aid in the selection of the "best" solver, depending on known linear system parameters, such as sparseness and symmetry.

3.4.5 Eigensolutions

The vectors x that satisfy each of the following two equations are called the eigenvectors:

$$\left(K + \omega^2 I\right)x = 0, \tag{3.107}$$

$$\left(K + \omega^2 M\right)x = 0. \tag{3.108}$$

With each eigenvector x we associate an eigenvalue $\lambda = -\omega^2$. Formally, as long as M has an inverse, it is possible to transform the generalized eigenproblem of Eq. (3.108) into the standard eigenproblem of Eq. (3.107). We now restrict our discussion to Eq. (3.107). Finding the eigenvalues of K is equivalent to diagonalizing K, and we may write that for X the matrix of eigenvectors of K

$$X^{-1}KX = \mathrm{diag}(\lambda_1 \ldots \lambda_N). \tag{3.109}$$

"The grand strategy of virtually all eigensystem routines is to nudge the matrix K towards diagonal form by a sequence of similarity transforms.[35]" This is usually done with the combination of two techniques. First, the matrix K is reduced to tridiagonal (rather than diagonal) form, using Givens rotations or Householder reductions, and then an iterative technique such as the QL (QR) algorithm is used to extract the eigenvalues. The QL algorithm costs $O(N)$ per iteration for a tridiagonal matrix [$O(N^3)$ for a general matrix], which is why this strategy is so popular.[35] The current state of the art in techniques is usually available in the NETLIB package, which is also available in MATLAB and Mathematica[38] implementations. An excellent overview of the eigenproblem and its numerical solution is contained in Ref. 41.

3.4.6 COUPLING STRATEGIES

Consider a coupled problem comprising n residual equations G_i in n generalized displacements ζ_i, which we write as

$$G_i(\zeta_1,\ldots,\zeta_n) = 0, \quad i = 1,\ldots,n. \tag{3.110}$$

In general we can form the Taylor expansion of Eq. (3.110) with respect to a reference state so that we obtain

$$G_i(\zeta_1,\ldots,\zeta_n) = 0 = G_i^0(\zeta_1^0,\ldots,\zeta_n^0) + \sum_{j=1}^{n} \nabla_{\zeta_j} G_i(\zeta_1,\ldots,\zeta_n)\delta\zeta_j + \cdots. \tag{3.111}$$

The first term in the sum of Eq. (3.111) is known as the Jacobian of the coupling, and is nonzero if there is a coupling term between equation i and generalized variable j. The next higher Taylor term (omitted above) is the Hessian, and it is usually not considered in practical coupled simulation algorithms. Many numerical schemes exist to efficiently implement this coupling algorithm. The most general and hence computationally most expensive scheme calculates the Jacobian matrix of the coupling explicitly.[36] Other schemes have borrowed from nonlinear solvers, since these are based on the same starting point, namely a Taylor expansion. Very often knowledge from the problem domain allows the simplification of the coupled problem. An extensive treatment of the issue from the viewpoint of the finite-element method can be found in Ref. 36.

3.5 APPLICATIONS

In this section we discuss three optical microsystem modeling applications that have led to special numerical solutions.

3.5.1 THERMAL IMAGER

Micro-hotplates are integrated thermoelectric microactuators that are used as dynamic thermal scene simulators (DTSS), arrays of thermal pixels used to generate thermal images and mainly applied to the calibration of infrared (IR) detectors and imaging systems.[42] In these devices, an electric current is dissipated in a thermally isolated resistor so that its temperature rises and it is caused to radiate heat. The device was manufactured using the commercial 1.2 µm CMOS process of Northern Telecom in Canada together with a front-side anisotropic etch to release the hotplate and isolate it from the silicon wafer. The hotplate consists of a polysilicon heater embedded in an SiO_2 sandwich [see Fig. 3.16]. Using a two-and-a-half-dimensional model for the sandwich, including in-plane conduction and

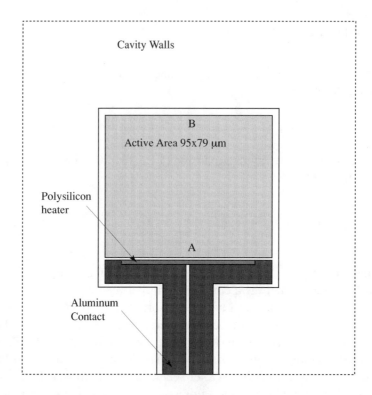

Figure 3.16 The layout of the microhotplate. The polysilicon heater is contacted left and right by aluminum contacts. The active area is heated to radiate, so forming a thermal pixel.

electrothermal coupling, and a reduced model for the radiation losses in the third dimension, the thermoelectric equations were discretized using the control volume method and implemented as a SPICE cell[42] [see Fig. 3.17]. Using effective electrical and thermal material properties through the thickness of the plate, and discretizing the planar geometry with the thermoelectric cells of Fig. 3.17, a complete SPICE model of the device was built and connected to the external driving circuitry. Using SPICE, the design was then optimized to arrive at a uniformly heated thermal pixel of about 110°C for a heater bias of 1.5 V. The steady-state thermal response is shown in Fig. 3.18 for the surface of the hotplate. The transient thermal response of the active area at two locations [A and B indicated in Fig. 3.16] is shown in Fig. 3.19.

3.5.2 OPTOMECHANICAL PRESSURE SENSOR

We consider a micromechanical pressure sensor with optical readout.[18,43] The pressure sensor consists of a multilayer rectangular diaphragm, 3 by 15 mm in

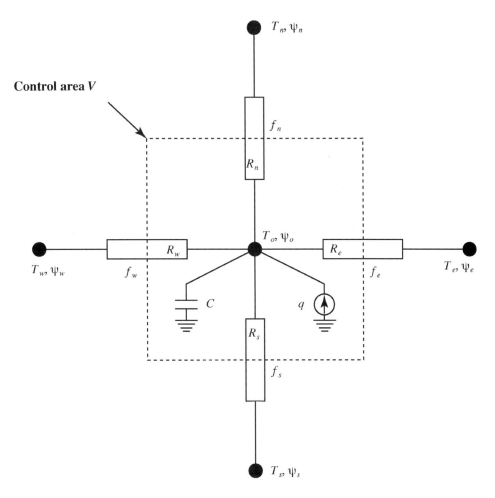

Figure 3.17 The equivalent circuit SPICE model used to implement the finite-volume discretization (corresponding to the control area) of the governing thermal and the electrical conduction equations for the microhotplate.

lateral dimensions and 60 μm thick, manufactured with silicon IC technology [see Fig. 3.20]. A pressure difference about the membrane causes the membrane to deflect from its planar position. Employing ARROW (antiresonant reflecting optical waveguide) technology, a low-loss single-mode planar waveguide was formed from the SiO_2 and Si_3N_4 layers, as shown schematically in Fig. 3.21 (top). Using the rib spectral index method, or SIM, the field profile for the geometry was obtained [Fig. 3.21, bottom]. This was used to find practical waveguide dimensions as a function of the oxide patterning etch depth. The ARROW waveguides were used to form a Mach–Zehnder interferometer (MZI) about the pressure membrane. An input Y-junction is used to form the two parallel optical paths of the MZI, one arm running across the membrane where the mechanical strain will be greatest, the other arm along a path where the mechanical strain is minimal. An output

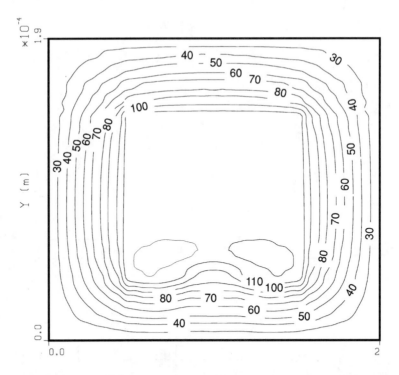

Figure 3.18 SPICE-simulated surface temperature contour for the microhotplate when subjected to a 1.5-V heater bias. For a steady state, the hotplate has good uniformity.

Y-junction joins the two paths again. Coherent light passes along the waveguide, is split among the two optical paths, and is joined again after traversing the pressure membrane. A perturbation to the sensing arm causes a phase difference, $\Delta\Phi$, between the optical signals along the two paths, which is noticeable as a variation in the output intensity of the MZI of $I = 0.5I_0(1 + \cos\Delta\Phi)$. The phase difference is caused by a change in the geometrical length of the optical path ΔL as well as variation in the propagation constant $\beta = 2\pi n_{\text{eff}}/\lambda$ along the waveguide, with an effective refractive index n_{eff}, due to the anisotropic stress distribution in the pressure membrane:

$$\Delta\Phi = \beta\Delta L + \int_L \delta\beta dL. \tag{3.112}$$

Numerical computations were used to determine each of these two contributions separately. Using thin-plate theory, both the plate deflection and the anisotropic stress distribution were obtained numerically, using a trigonometric series ansatz for the plate deflection and a multilayer representation to account for the layered nature of the diaphragm. The computed deflections and stress are summa-

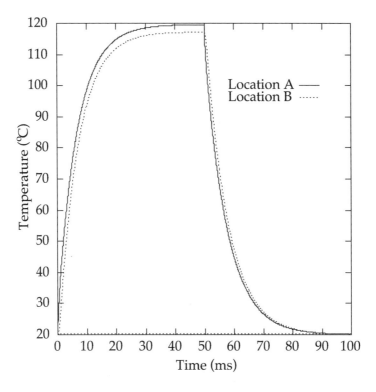

Figure 3.19 SPICE-simulated transient response of the microhotplate at two locations on the active area [see Fig. 3.16] for a pulsed bias of 1.5 V to the microhotplate polysilicon heater. Good pixel uniformity was also obtained for a transient response.

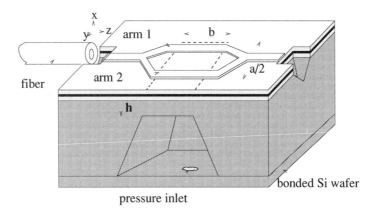

Figure 3.20 Schematic view of the membrane pressure sensor based on a Mach–Zehnder interferometer.[18,43] The membrane dimensions are $a = 3$ mm, $b = 15$ mm, and $h = 60$ μm. The membrane sandwich contains one of the two arms of the Mach–Zender interferometer.

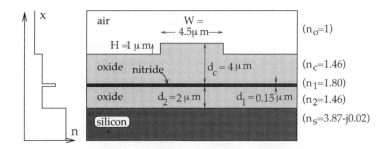

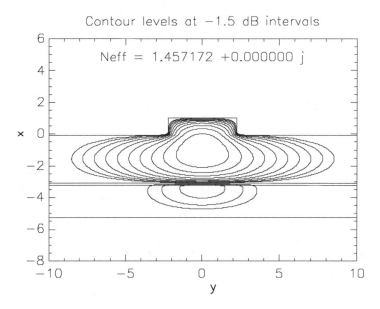

Figure 3.21 The ARROW structure cross-section (top) and its field distribution (bottom) corresponding to the fundamental TE0 mode, obtained using SIM.[18]

rized in Fig. 3.22. The photoelastic effect [see Section 3.2.5.3], written in revised notation as

$$\begin{bmatrix} \Delta n_{xx} \\ \Delta n_{yy} \\ \Delta n_{zz} \\ \Delta n_{xy} \\ \Delta n_{xz} \\ \Delta n_{xz} \end{bmatrix} = \begin{bmatrix} C_1 & C_2 & C_2 & & & \\ C_2 & C_1 & C_2 & & & \\ C_2 & C_2 & C_1 & & & \\ & & & C_4 & & \\ & & & & C_4 & \\ & & & & & C_4 \end{bmatrix} \begin{bmatrix} \sigma_{xx} \\ \sigma_{yy} \\ \sigma_{zz} \\ \sigma_{xy} \\ \sigma_{xz} \\ \sigma_{xz} \end{bmatrix}, \quad (3.113)$$

relates the stress σ to a change in refractive index Δn through the stress-optic constitutive parameters C_P. For SiO_2 two unique parameters are specified:

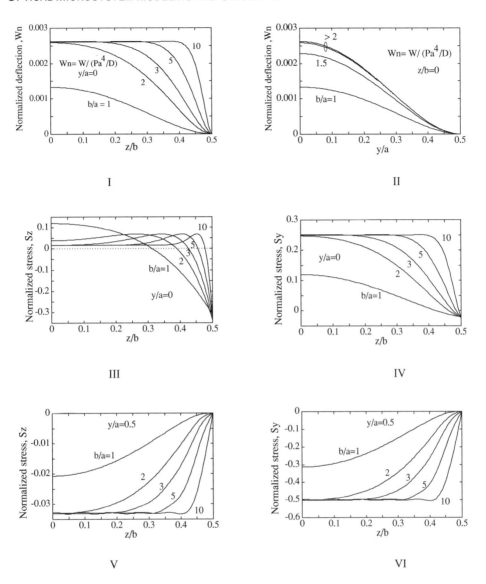

Figure 3.22 Computed normalized deflections along the long and short symmetry lines of the pressure membrane (I and II). Normalized stresses along the active waveguide (III-IV) and along the membrane edge (V and VI) for the multilayer membrane structure. All calculations are for a set of different membrane aspect ratios.

$C_1 = -0.65 \times 10^{-12}$/Pa and $C_2 = -4.22 \times 10^{-12}$/Pa, so that $C_4 = C_1 - C_2 = 3.57 \times 10^{-12}$/Pa. For the TE modes, the following equations are thus relevant: $\Delta n_{yy} = C_1 \sigma_{yy} + C_2 \sigma_{zz}$ and $\Delta n_{xx} = C_2(\sigma_{yy} + \sigma_{zz})$, which are now computed from the results of Fig. 3.22. Using these intermediate results, simulations of the pressure sensor response in light intensity were compared with measurements [see Fig. 3.23].

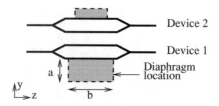

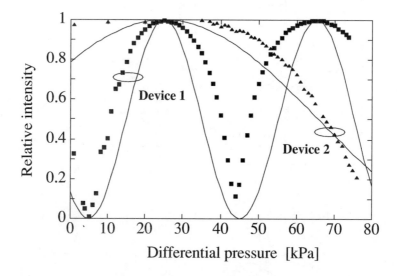

Figure 3.23 Comparison of measurement and simulation for two Mach–Zehnder interferometer pressure sensor devices. The top figure shows the geometry of the sensors. The lower figure compares light intensity measurements (squares and triangles) and simulations (solid lines) for a range of differential pressures.

3.5.3 THE SHAPE OF MICROLENSES

It is possible to form small lenses from patterned photoresist by melting the resist so that in its liquid state the surface tension causes the resist to form a rounded droplet.[44] The shape of the droplet is determined by the forces acting in and on the liquid–environment surfaces, and include all interface surface tensions and the liquid's own weight. The incompressible liquid's volume is a constraint for the process.

It is a straightforward matter to describe each of these effects as an energy contribution to the liquid–gas surface of the droplet. Dividing the surface into a triangular mesh (much as for a BEM mesh), the energy can be expressed in discrete terms as an equation for each mesh node. The nodes of the surface are now constrained to move in a direction that both minimizes the surface energy and conforms to any constraints (such as liquid incompressibility and wetting edges).

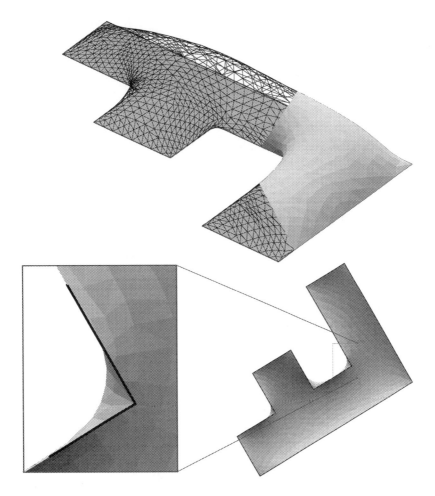

Figure 3.24 The shape of a liquid droplet computed using an explicit numerical method that minimizes the surface energy of the liquid interface while maintaining volume constraints and edge pinning. Note that the test-case droplet forms both convex and concave angles. Such droplets may be used to produce microlenses.[44]

We used an explicit numerical method[45] to evolve the shape of the surface. We are able to verify the plausibility of the results by invoking static equilibrium of the surface tension forces at a drop-let's edge, which places a requirement on the liquid/solid/gas interface angle, for which we found perfect agreement. We illustrate the method for a test case involving a wetting surface with both convex and concave corners [see Fig. 3.24].

The computed microlens shape may be easily extracted from the surface evolver to be passed on to suitable software so as to characterize the lens from an optical viewpoint. This was not done here. Clearly, careful use of a combination of surface evolution and optical characterization tools will lead to microlens designs with the required characteristics.

3.6 Discussion

Much work remains before developers of MOEMS will have access to a (commercial) TCAD environment for the complete design and testing of MOEMS prototypes within a modeling environment. While process simulation tools can profit from the advances made in electronic semiconductor process modeling, specialized processing steps, such as ultrathin film deposition, are not yet adequately considered. Individual physical effects, however, as used in sensor devices, are already covered by a variety of specialized modeling tools. Here MOEMS can readily profit from current advances in the simulation of MEMS and microelectronics. Compact modeling seems, to date, not to have been considered for optical microsystems, hampering the development of complete systems that necessarily include microelectronic interface circuitry. For geometric-based optical devices, many specialized ray-tracing tools already exist in the marketplace. The combination of existing MOEMS simulation tools and an integrated TCAD simulation environment will necessitate the development of the "missing links." Following the lead by the microelectronic industry, we predict that such an environment will greatly enhance the development of MOEMS-based products by industrial companies and research groups alike.

Acknowledgments

We express our indebtedness to our past and present co-workers who have contributed to this article through numerous discussions and collaboration, and by extensive proof-reading of the manuscript. In particular, we sincerely thank Dr. Martin Bächtold of Coyote Systems in San Francisco, California; Dr. Kamel Benaissa of Texas Instruments in Dallas, Texas; Mr. Markus Emmenegger and Mr. Stefano Taschini at the Physical Electronics Laboratory, ETH Zurich, Switzerland; Prof. W. P. Huang of McMaster University in Hamilton, Ontario, Canada; Dr. Andreas Greiner, Mr. Jens Müller, and Mr. Ricardo Osorio of IMTEK-Institute for Microsystem Technology, The Albert Ludwig University, Freiburg, Germany; and Dr. Nicholas Swart of Analog Devices, Cambridge, Massachusetts.

References

1. H. Baltes, "Is Technology Transfer Possible?" *Sensors Materials* 10(6), 313–323 (1999).
2. J. Doscher, *Innovations in Acceleration Sensing Using Surface Micromachining*. Analog Devices, Inc., Cambridge, MA (1995).
3. G. Binning and H. Rohrer, "Scanning Tunneling Microscopy," *Physics* 127B, 37–45 (1984).

4. L. J. Hornbeck and W. E. Nelson, "Bistable Deformable Mirror Device," in *OSA Technical Digest Series*, Vol. 8, *Spatial Light Modulators and Applications*, p. 107 (1988).
5. J. G. Korvink, "SOLIDIS Reference Manual 1.0," Internal Report No. 95/01, Physical Electronics Laboratory, ETH Zurich (1995); ISE Integrated Systems Engineering AG, Technopark Zurich, Technoparkstrasse 1, CH-8005 Zurich, Switzerland.
6. Microcosm Technologies, 5511 Capital Center Dr., Suite 104, Raleigh, NC 27606, USA.
7. IntelliSense Corporation, 16 Upton Dr., Wilmington, MA 01887, USA.
8. Coyote Systems, 2740 Van Ness Ave. #210, San Francisco, CA 94109, USA.
9. J. G. Korvink, ed., *Proc. CAD for MEMS '97*. Technopark Zurich, Switzerland (1997).
10. *Tech. Proc. MSM '98* (1998).
11. A. Nathan, ed., "Special Issue on Microsensor Modeling," *Sensors Materials* 6(2,3,4) (1994).
12. J. G. Korvink, ed., "Special Issue—CAD 4 MEMS," *Sensors Materials* 10(7) (1998).
13. A. Nathan and H. Baltes, *Microtransducer CAD, Physical and Computational Aspects*. Springer-Verlag, Berlin (1999).
14. J. G. Korvink and H. Baltes, "Microsystem Modeling," in *Sensors Update*, Ch. 6. VCH Publisher, Weinheim (1996).
15. H. Baltes, O. Brand and O. Paul, "CMOS MEMS Technology and CAD: The Case of Thermal Microtransducers," *Proc. SPIE* 3328, 2–12 (1998).
16. M. Bächtold, "Efficient 3D Computation of Electrostatic Fields and Forces in Microsystems," Ph.D. dissertation No. 12165, ETH Zurich, Switzerland (1997).
17. J. Bühler, J. Funk, J. G. Korvink and H. Baltes, "Electrostatic Aluminum Micromirrors Using Double Pass Metallization," *IEEE Micro-Electro-Mechanical Systems* (1997).
18. K. Benaissa, "Integrated Silicon Opto-Mechanical Sensors," Ph.D. dissertation, Electrical and Computer Engineering, University of Waterloo, Waterloo, Ontario N2L 3G1, Canada (1996).
19. J. H. Williams, *Fundamentals of Applied Dynamics*. Wiley, New York (1997).
20. D. F. Nelson, *Electric, Optic and Acoustic Interactions in Dielectrics*. Wiley, New York (1979).
21. IEEE Standard on Piezoelectricity, ANSI/IEEE Std. 176-1987, IEEE (1988).
22. F. Scheck, *Mechanics: From Newton's Laws to Deterministic Chaos*, 2nd edn. Springer-Verlag, Berlin (1994).
23. B. H. Lavenda, *Thermodynamics of Irreversible Processes*. Dover, New York (1993).
24. O. Madelung, *Introduction to Solid State Theory*. Springer-Verlag, Berlin (1978).

25. N. W. Ashcroft and N. D. Mermin, *Solid State Physics*. W. B. Saunders, Philadelphia, PA (1988).
26. A. Zangwill, *Physics at Surfaces*. Cambridge University Press, Cambridge (1988).
27. J. H. Visser, "Surface Acoustic Wave Filters in ZnO-SiO2-Si Layered Structures," Ph.D. dissertation, Delft University of Technology, Delft, the Netherlands (1989).
28. J. D. Jackson, *Classical Electrodynamics*, 2nd edn. Wiley, New York (1975).
29. J. Funk, "Modeling and Simulation of IMEMS," Ph.D. dissertation No. 11378, ETH Zurich, Switzerland (1995).
30. H. Pham, "Numerical Capacitance Extraction for Large-Area Systems," Ph.D. dissertation, University of Waterloo, Waterloo, Canada (1998).
31. W. G. Strang and G. J. Fix, *An Analysis of the Finite Element Method* (1973).
32. C. A. Brebbia, J. C. F. Telles and L. C. Wrobel, *Boundary Element Techniques, Theory and Applications in Engineering*. Springer-Verlag, Berlin (1994).
33. A. Iserles, *A First Course in the Numerical Analysis of Differential Equations*. Cambridge University Press, Cambridge (1996).
34. S. Taschini, Electrostatics, private communication.
35. W. H. Press, S. A. Teukolsky, W. T. Vetterling and B. P. Flannery, *Numerical Recipes in C—The Art of Scientific Computing*, 2nd edn. Cambridge University Press, Cambridge (1992).
36. T. Belytschko and T. R. J. Hughes, eds., *Computational Methods for Transient Analysis*, Elsevier Science, Amsterdam (1983).
37. G. Golub and C. F. van Loan, *Matrix Computations*, 3rd edn. Johns Hopkins University Press, Baltimore, MD (1996).
38. S. Wolfram, *Mathematica*, 4th edn. Cambridge University Press, Cambridge (1999).
39. A. Liegmann, *Efficient Solution of Large Sparse Linear Systems*. Hartung-Gorre Verlag, Konstanz, Germany (1995).
40. R. Barrett, M. Berry, T. Chan, J. Demmel, J. Donato, J. Dongarra, V. Eijkhout, R. Pozo, Charles Romine and Henk van der Vorst, *Templates for the Solution of Linear Systems: Building Blocks for Iterative Methods*. SIAM, Philadelphia, PA (1994).
41. P. Arbenz, "Short Course on Large Scale Eigenvalue Problems," course notes, ETH Zurich (1998).
42. N. R. Swart, "Heat Transport in Thermal-Based Microsensors," Ph.D. dissertation, University of Waterloo, Waterloo, Canada (1994).
43. K. Benaissa and A. Nathan, "IC Compatible Optomechanical Pressure Sensors Using Mach–Zender Interferometry," *IEEE Trans. Electron Devices* 43, 1571–1582 (1996).
44. H. P. Herzig, http://www-optics.unine.ch/research/Microopt/microlens.html
45. K. Brakke, "The Surface Evolver," *Exp. Math.* 1(2), 141–165 (1992).

CHAPTER 4

THE DIGITAL MICROMIRROR DEVICE—A MICRO-OPTICAL ELECTROMECHANICAL DEVICE FOR DISPLAY APPLICATIONS

Michael A. Mignardi
Richard O. Gale
David J. Dawson
Jack C. Smith
Texas Instruments, Inc.

CONTENTS

4.1 Introduction / 170

4.2 CMOS Fabrication / 171

4.3 Superstructure Fabrication / 172

4.4 Die Separation and Pixel Release / 176

4.5 Package Assembly / 179

4.6 Wafer, Package, and Projection Level Testing / 181
    4.6.1 DMD electrical, optical, mechanical test / 182
    4.6.2 DMD image quality / 194
    4.6.3 Projector-based image quality evaluation / 195

4.7 Device Reliability / 203

4.8 Summary / 203

References / 204

Appendix A: List of significant publications on DMD and DLP technology / 204

4.1 INTRODUCTION

Texas Instruments has developed a range of projectors using Digital Light Processing (DLP™) technology. DLP is a reflective and truly digital projection/display technology in full-scale commercial production. DLP has enabled the development of XGA (Extended Graphics Adapter) resolution ultraportable (i.e., <10 lbs.) projectors. There have been dramatic leaps since the first DLP-based projectors entered the market in 1996. The Digital Micromirror Device (DMD™) is the heart of DLP technology. The DMD is a semiconductor-based array of fast, reflective digital light switches that precisely controls a light source using a binary pulse width modulation technique. This device is a unique combination of optical, mechanical, and electrical systems. Built on standard CMOS technology base wafers, the DMD mirror "super structure" provides an all-digital display solution capable of high brightness, high contrast and high reliability.

Several DLP designs allow the technology to be used in very compact and lightweight applications such as portable projectors, as well as in very high-brightness fixed installation applications such as digital cinema and boardroom projectors. This range of design options based on the common DMD "optical element" has allowed DLP technology to compete in all segments of the projection display market against both liquid crystal display (LCD)- and cathode ray tube (CRT)-based systems. The all-digital architecture, small size and weight, and simple optical designs make possible DLP-based systems for high-quality HDTV products of the future.

The invention of the transistor and the integrated circuit began a technological revolution—the miniaturization of electronics. Mechanics is now poised on the verge of a similar revolution with its own miniaturization.[1] The driving forces for the miniaturization of mechanical systems are cost, size, speed, weight, and precision, while providing an effective interface between the macro- and the microdynamic worlds. The term for these small mechanical systems is microelectromechanical systems (MEMS). Unlike most MEMS devices, the DMD is fully integrated and monolithically fabricated on a mature SRAM CMOS address circuitry. Thus, the substrate for the mirrors and the drive circuitry is integrated on silicon. The level of development and production effort to fabricate the DMD is equally divided. The DMD superstructure, package assembly, and electro-optical test all required equal development. Although one of the tangible products produced from Texas Instruments' DLP group is the DMD itself, the end customer is purchasing a projected image. Thus, optical performance, or image quality, is our most critical metric.

A key strategy established during the early days of Digital Imaging was to fabricate this technology using standard semiconductor processes. As a result, the wafer portion of the processing was just recently transferred to a standard Texas Instruments semiconductor (TI SC) fabrication facility.

The DMD is currently in volume production for SVGA (Super Video Graphics Adapter, 848 × 600 or 508,800 pixels) and XGA (1024 × 768 or 786,432 pixels)

The Digital Micromirror Device

resolution display formats. Soon to go into production is the next higher resolution device—SXGA (Super Extended Graphics Adapter, 1280 × 1024 or 1,310,720 pixels). This chapter provides some of the key details of the development and production effort of the DMD manufacturing process and how it compares to the IC industry as well as other MEMS components.

4.2 CMOS Fabrication

The DMD mechanical element (called either the proof mass, pixel or mirror) is processed over CMOS architecture. CMOS is processed in an existing, mature TI wafer fab and consists of conventional SRAM cells. The SRAM underlying each mirror is a six-transistor cross-coupled inverter circuit or storage cell fabricated using a twin-well CMOS, 0.8 μm, double-level metallization process. Figure 4.1 shows the layout of the integrated DMD structure. The SRAM cell provides addressability to each individual mirror. The mirrors for each device are 16 μm square on a pitch of 17 μm.

Once the double-level metallization is complete, a final interlevel dielectric is applied. This dielectric level is the surface on which the DMD pixel architecture is fabricated. It is critical that this level is locally flat so that the pixels will achieve a high degree of uniformity in landed tilt angle. Tilt angle is one of the most sensitive parameters controlling optical efficiency or brightness in typical DMD-based projection systems. Since the human eye is very sensitive to brightness variations

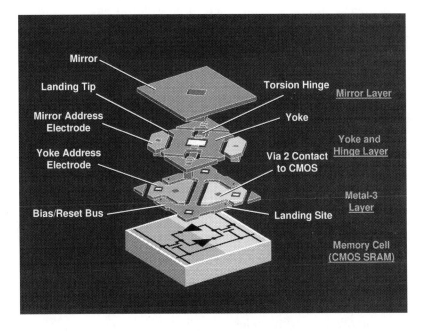

Figure 4.1 Exploded view of the DMD pixel and underlying CMOS.

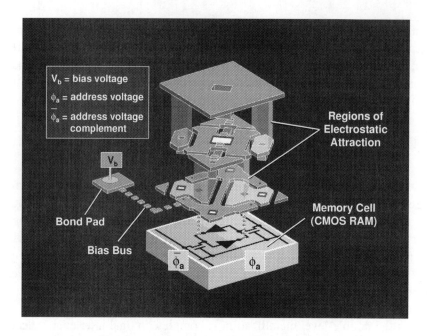

Figure 4.2 Electrical schematic of DMD pixel.

(as little as 0.5% differential brightness can be noticed) and pixel-to-pixel brightness uniformity is a key element of image quality, chemical-mechanical polishing (CMP) is required for adequate planarization. Planarization topographies in the sub-1,000-Å range need to be controlled to mitigate tilt angle variations. Once CMP is complete, vias are formed in the dielectric for contact to the CMOS metal-2 layer. For each pixel, there is a pair of address electrodes, which are connected to the complementary sides of the underlying SRAM cell (see Fig. 4.2). Depending on the SRAM state, a combination of the bias and address voltages electrostatically attracts each pixel element to one of the address electrodes. One address side represents the "off-state" while the other side represents the "on-state" of the pixel.

4.3 Superstructure Fabrication

IC fabrication employs surface micromachining layering techniques. This involves repetitive deposition, pattern, and removal processes.[2] The DMD utilizes surface micromachining layering techniques, as do some MEMS components. Other MEMS components can be fabricated utilizing bulk micromachining, bonding, or high-aspect ratio (or LIGA, Lithografie, Galvanik, Abformung) techniques.[3] An attractive feature of the processing of the DMD superstructure (i.e., the pixels) is that it is an outgrowth of the IC industry and its fabrication sequence. Many of the processes in the superstructure fabrication (or wafer level) are common with the

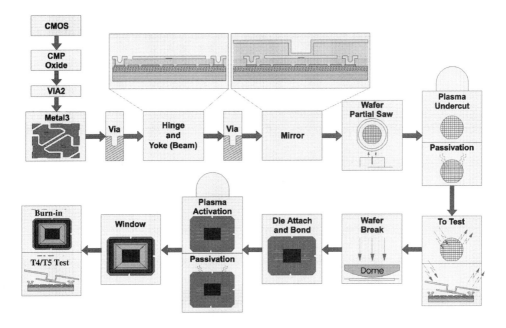

Figure 4.3 Simplified process flow of the DMD.

semiconductor industry. Four additional metal layers define the pixel superstructure. Figure 4.3 provides a simplified process flow of the superstructure, test, and assembly operations.

After contact openings (i.e., the vias), the address circuit electrode is formed on top of the CMOS dielectric layer. An aluminum address electrode film is sputter-deposited onto the dielectric, lithographically patterned, and plasma etched. An organic sacrificial planarizing layer is then spun over the electrode and patterned with vias. These vias will be used subsequently to form the support posts. These support posts provide both mechanical support and electrical contact.

Two proprietary metal alloy layers are employed for the buried hinge process. The hinge alloy sputtered on top of the sacrificial layer is followed by a plasma-deposited masking oxide that is patterned to define the hinges. Deposition of a second alloy metal layer then buries the hinge oxide and is the beam level metal. This, in turn, is followed by still another plasma-deposited masking oxide, which is patterned to form the beam. A single plasma-etching step defines the hinges and beam, stopping on the hinge oxide hard masks. The support posts consist of both the hinge and beam metals. Standard IC technology tightly controls the critical dimensions in the x and y directions. However, the z-axis dimension (i.e., thickness) must also be tightly controlled. Both the sacrificial layer and hinge thicknesses are precisely controlled to achieve reproducible, uniform mirror tilt angles and addressing voltages, respectively.

A second sacrificial planarizing layer is spun over this hinge-beam level and the process is essentially repeated as outlined above. In this second layer, only one

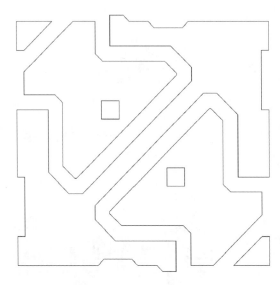

Figure 4.4 Electrode pattern.

metal layer is deposited—a highly reflective aluminum mirror level. The completed superstructure is called the "hidden hinge" since the hinge is literally hidden by the mirror. Figures 4.4 through 4.8 provide an overview of the pixel fabrication process as described above.

This hidden hinge structure is a fourth generation design. The impetus for this new design was the need to improve the optical efficiency of the device. This new pixel also provided for a more robust design in terms of electrostatic operation and reliability. Figure 4.8 depicts a transparent view of two completed pixels.

Particle control throughout the superstructure process is critical. Since the DMD is a three-dimensional component with mechanical movement, particle control involves both intralevel and interlevel monitoring. The smallest spacing at the intralevel is on the order of 1 µm. However, some critical interlevel gaps can approach spacings as small as 0.2 µm during the mechanical operation of the DMD pixel. For this reason, particle control well into the submicron range is critical. Equipment and process engineering teams are heavily involved in the manufacturing of the DMD as they continue to achieve defect density levels that are comparable to the semiconductor field. Similar to IC design, development activities are underway to make the DMD architecture less sensitive to these submicron particles while still maintaining robustness for electrostatic operation and reliability, as well as optical performance.

A primary strategic requirement of the process development for the DMD was to be compatible with standard IC processing and equipment. Today, the DMD process can be transferred to a majority of standard IC wafer fabs with minimal equipment and process modification.

The Digital Micromirror Device

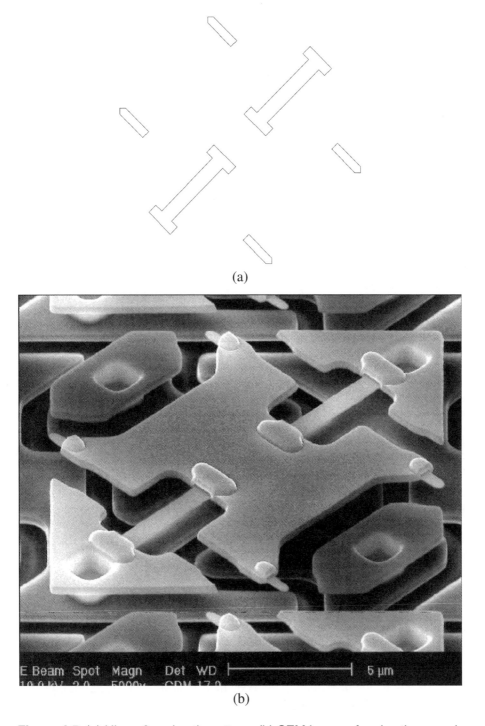

Figure 4.5 (a) Hinge & spring tip pattern, (b) SEM image of spring tips on yoke.

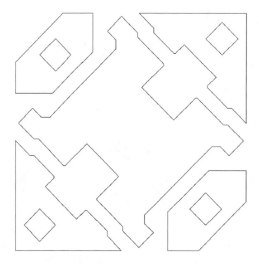

Figure 4.6 Beam (or yoke) pattern.

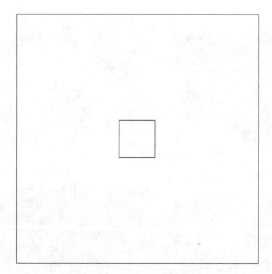

Figure 4.7 Mirror pattern.

4.4 DIE SEPARATION AND PIXEL RELEASE

At this point in the fabrication of the DMD, the superstructure process is complete. This is essentially the demarcation between the front-end and back-end of the process. The individual pixels, however, have not been released, so mechanical motion of the pixels is not possible at this time.

Die separation is the typically the first step where the MEMS industry (and the DMD technology) deviates from that of the IC industry. For silicon-wafer applica-

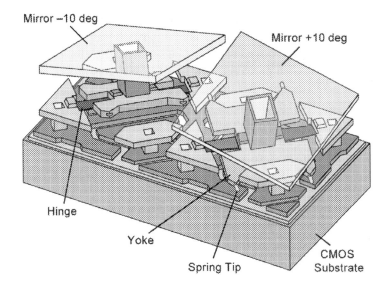

Figure 4.8 Transparent view of two pixels.

tions, the IC industry can employ three die separation techniques: laser scribing, diamond scribing and diamond-wheel sawing.[4] Sawing is the common die separation technique. Due to the expense of most MEMS packages, and the fact that the package is integral to the function of the device, it is imperative that only good functional devices are assembled. Therefore, the MEMS mechanical element must be released for mechanical functionality testing prior to assembly. As with any high-volume process, wafer batch or single wafer processing is the preferred manufacturing strategy. Processing individual dies is typically a time-consuming and labor-intensive process—even with automation tools in place. The separation of DMD dies on a silicon wafer and the subsequent packaging of those dies are technical challenges for the DMD program.[5]

Within the silicon-wafer MEMS industry, there are several approaches to die separation. One is to fully saw the wafer, separate the dies, remove the sacrificial layer, and then assemble the dies into a package. This method is typically reserved for low-volume applications since post-saw processing of a die can be labor intensive. Another approach is to remove the sacrificial layer at the wafer level. This is attractive as it allows for batch level processing; that is, all dies on the wafer have their sacrificial layer removed at the same time. This also allows one to test the full functionality of each die, including mirror movement, before any package commitment is made.

Batch level sacrificial layer removal can employ two saw techniques: a full or partial saw process. Analog Devices Inc. (ADI) employs a full saw process after the wafer's sacrificial layer is removed.[6] To protect the proof mass of their accelerometer, the wafer is mounted facedown onto a saw-tape grid and then sawed from the backside of the wafer. ADI's approach is attractive for MEMS devices with a small proof mass-to-die area ratio. The DMD, however, exhibits a rather large proof mass-to-die area. At TI, the wafer is first partially sawed (i.e., some fraction of the wafer thickness is removed in the scribe streets). Then the organic sacrificial layer is removed in a batch process that utilizes a plasma isotropic ash. Both Analog Devices' and TI's approaches allow for wafer-level electrical and mechanical testing prior to any die separation.

The partial saw approach utilizes a spun-on layer over the surface of the mirrors to protect them from water during saw operation. After the partial saw process the wafer undergoes a series of wet cleans to remove residual silicon debris. The release process involves an isotropic plasma ash that removes the saw protective layer and both sacrificial layers from the DMD pixel.

The TI partially sawn wafer is obviously fragile, and special handling techniques are employed during wafer transportation, wet cleaning, sacrificial layer removal, and wafer test. The wafer-level test provides valuable and quick feedback to the wafer facility and process engineering teams. (Testing of the wafer will be discussed later in this chapter.) The partial saw process is attractive in that fully functional dies (i.e., the proof mass is functional) can be identified (using electro-optical testing) prior to packaging. After wafer test, the wafers are prepared for die separation. The wafer is placed on dicing tape while a domed surface is used to break the wafer along the partially sawn scribe streets from the backside. Ionized air is used throughout the process to reduce the attraction of particles that are liberated during the actual breaking of the wafer. This process must be extremely clean, since any debris generated during the break process can be a killer defect to the DMD. Similar to standard sawing of IC components, the separated dies remain on saw frames with dicing tape so the entire wafer can be presented to and processed on a die attach assembly tool.

Another approach being employed by TI involves a full saw followed by sacrificial layer removal in package form. The full saw and package-level sacrificial layer removal process is a dramatic change from the partial saw process and was developed to reduce the amount of time the active mirror array is exposed during the assembly operation. This method also allows for a wet clean-up step during assembly. Approaches where the sacrificial layer is removed prior to packaging do not allow for a clean-up step during assembly. In this approach, the dies go directly from the wafer saw to die attach. Picked dies are based only on the electrical CMOS test performed after the CMOS process is complete. The packaged dies undergo a wet clean prior to the sacrificial layer removal. Thus, any particles generated during assembly are removed during the clean process, a major advantage over the partial saw process. The package process employs a plasma process chemistry similar to the wafer level sacrificial layer removal. From this point onward, this process and the partial saw process follow similar assembly. Figures 4.9 and 4.10 show a schematic representation of two die separation flows.

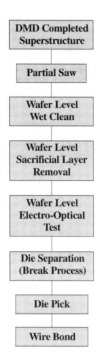

Figure 4.9 Partial saw flow.

Figure 4.10 Full saw flow.

4.5 Package Assembly

The majority of the DMD assembly is discussed in Chapter 9, which is dedicated to packaging and assembly. For most MEMS technologies, packaging is considered

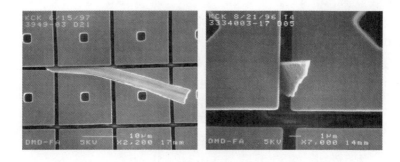

Figure 4.11 Particulate generated defects during assembly.

the primary challenge of the manufacturing team.[2] The package is an integral part of the DMD. It provides not only signal and power lead connections to the die, but also protection from the outside environment. The package is also a critical component in the optics system. The DMD package (as is also true for most MEMS products) is application specific. Again, the manufacturing strategy for DMD assembly is to harmonize with standard SC equipment. The DMD, however, cannot rely on conventional packaging technology. Within the MEMS industry (and unlike the IC industry) there is limited MEMS packaging infrastructure, experience base, and depository of assembly knowledge.[5]

Most IC assembly is performed in non-cleanroom environments. Since the DMD pixels are critically vulnerable to particulate contamination, the entire assembly process is performed in a Class 10 (measured at 0.3 μm particle sizes) clean room. Scrapping a DMD due to particle contamination at this stage in the process flow is costly: Fig. 4.11 shows an example of particulate contamination and its potential effect on mirror functionality. All of the packaging machines are commercially available semiconductor tools. Each, however, has been modified for low particulate operation. Package materials, such as die attach adhesives, are chosen with low outgassing and low-temperature process capabilities so as not to affect key device parametric and yield qualities such as mirror stiction, mirror planarity, and hinge compliance.

As with any mechanical system, friction is a chief concern. Since the DMD is a mechanical component and the DMD pixels make surface contact, a TI proprietary process was developed to mitigate pixel stiction. This process is performed at the package level just prior to placement of the optical lid. During the early stages of DMD development, stiction was a primary focus. Most MEMS products have the same stiction reliability concerns and many employ some type of stiction mitigation chemistry during the sacrificial layer removal process. Introduction of the spring tip design [see Fig. 4.5(b)] was an excellent mechanical design innovation which greatly reduced stiction issues with the DMD.

The final stage of the assembly process is window attachment. The window is seam welded to form a hermetic seal. Protection from the environment is the chief concern and thus requires a hermetic seal. The window must also be of high optical quality for image quality reasons in projection-based systems. Again, more detail on the packaging is provided in Chapter 9.

4.6 WAFER, PACKAGE, AND PROJECTION LEVEL TESTING

Mature parametric test structures are utilized on production IC components to monitor the health of the semiconductor process. Such parametric test structures have yet to be developed, characterized, and implemented for most MEMS components.[2] The DMD does utilize standard CMOS test parametric structures for electrical characterization, e.g., resistance and capacitance. Mechanical parametric test structures are not yet in place to monitor such things as the residual stress of deposited films. Currently, the DMD pixel itself is an excellent parametric monitor.

Since the released pixels are very sensitive to particles, the DMD probe area must remain in a cleanroom environment. Typical IC wafer probe areas are outside of this environment. Like any IC component, there are a number of wafer-probing tools that can be purchased to perform the electrical testing. The DMD, however, requires electrical, mechanical, and optical testing. The SRAM CMOS functionality can certainly be tested on most electrical testers. The read-write function of the SRAM, however, does not guarantee that the pixels are operational. Thus, each pixel must be tested for its optomechanical functionality. Herein lies a major challenge—electrical optomechanical testers are not off-the-shelf tools. This challenge is similar to many MEMS products where the tester needs to be application specific. The DMD test development group built very low cost, high speed, low particle generating test sets which perform the SRAM CMOS test while also determining the functional state of each mirror on a device. For a typical SVGA product, the tester determines the functionality for over 0.5 million pixels per device. The tester output indicates the number of single bit fails in the SRAM circuitry as well as the pixel array. This determines which dies are selected for assembly. This test output also provides process engineering with valuable mechanical parametric information—hinge compliance and mirror stiction in particular.

After package assembly is complete, the device is ready for a mechanical burn-in and then final package test. The intent of burn-in is to mechanically stress the pixel to uncover any process problems related to stiction mitigation and hinge deposition. At package test, the wafer level tests are repeated, but with the addition of measuring pixel reflectivity, contrast ratio, and tilt angle. Again, valuable mechanical parametric information is provided to product and process engineering. A projector test is the final test performed prior to shipment. This test uncovers problems that are mostly related to the optical lid (such as window blemishes) and border problems. The purpose of a projector test is to assess the image quality of the device.

4.6.1 DMD ELECTRICAL, OPTICAL, MECHANICAL TEST

There are many commercially available electrical CMOS test systems; however, the same cannot be said for electrical optical/mechanical testers. These types of testers, like the DMD Mirror Master tester shown in Fig. 4.12, are very application specific and therefore require complex custom design. The tester must be capable

Figure 4.12 Mirror Master test set.

of providing a variety of functions. It must screen out electrical CMOS failures, mechanically dysfunctional mirrors, and optically imperfect mirrors. It must also provide parametric data to enable process control, as well as defective mirror coordinate data to facilitate failure analysis.

Maintaining acceptable test times is another challenge when testing the large number of pixels on the DMD (786,432 on the XGA device, for example). Large die size, optical alignment, image capture, image processing, and stage movement all contribute to considerably longer test times than the semiconductor industry standard. This drives up the number of required test systems needed to maintain equivalent throughput, which in turn drives the need to minimize the test system cost.

Even such routine operations as wafer probe need careful consideration when dealing with micromechanical structures. Exposed DMD mirrors are extremely sensitive to particulate contamination, so if full mechanical and optical testing is to be performed at wafer level, extreme caution must be taken to keep the probe operation, an inherently dirty process, as clean as possible. This necessitates locating the probe test operation inside the wafer fabrication cleanroom (the same Class 10 facility as for the wafer and assembly fabrication), or the use of special filtered probers and sealed wafer pods to transfer the wafers from the cleanroom to the prober. The contact force between the probe needle and the die bond pad must also be minimized to avoid damaging the bond pad aluminum by the probe tips. Aluminum debris could attach to the probes and then fall onto the mirrors as the wafer indexes from die to die during the probe operation. A certain amount of probe "over travel" is required to maintain good electrical contact to the device, so a delicate balance has to the maintained. Frequent probe needle cleaning is required to remove any accumulated debris.

4.6.1.1 Mirror Master 2000 Concepts

Initial prototype DMD testing was performed using manual microscope inspection, a very laborious and time-consuming task. As the DMD matured and market segments emerged, it became clear that an automated tester would be needed to meet the increasing demand. The lack of a commercially available test system, which could perform both CMOS electrical and optical image capture, spawned the development of the Mirror Master 2000 (MM2000). The Mirror Master 2000 is based around a Wentworth MP2000 semi-automatic wafer probe station which provides x, y and theta motion of either the wafer or packaged die under test. Later Mirror Master package test systems utilize faster Aerotech translation stages to provide the necessary motion control. A Sony 3-CCD NTSC camera is mounted to a programmable microscope mount (PMM) which provides the x, y movement needed to align the camera to each tested frame of the DMD. A System PC provides prober control, user interface, and data handling, while a separate dual-processor Vision PC, containing a high-speed data translation board, handles the captured image analysis. A single lamp source and a mechanical shutter provide illumination to

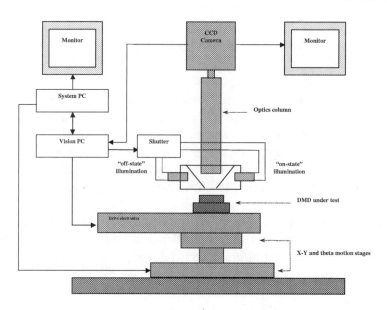

Figure 4.13 MM2000 block diagram.

the DMD via fiber optic bundles to either the "on" or "off" mirror direction. Unlike DMD projector applications, where the device is illuminated from only one direction, bidirectional illumination is required to provide mirror defect classification. It is important to distinguish whether a mirror is not functional in a particular direction, which in turn will determine if the particular mirror will appear as a bright or dark defect in the projector application. A block diagram of the Mirror Master 2000 test system is shown in Fig. 4.13.

The custom-designed TI C50 DSP-based Universal Drive Electronics (UDE) board provides the drive signals and test pattern algorithms needed to control the DMD. A block diagram of the UDE board is shown in Fig. 4.14. C50 program code, which provides all timing and test algorithms to the DMD, is stored on flash EPROM. An ALTERA programmable logic chip provides latching and buffering of I/O signals, as well as control signals for the TI SR16C custom ASIC reset driver chip. A programmable power supply module provides the necessary voltages to the UDE and the DMD under test.

A typical DMD production test sequence is shown in Fig. 4.15. During the initial current test the CMOS Icc and bias-reset currents are measured. This test is performed at a static bias-reset voltage of eight volts, high enough to allow leakage currents to be measured without landing the mirrors. The CMOS tests verify the operation of the DMD shift register and block stepped address circuitry, along with the functionality of each memory cell. True and inverse 1×1 checkerboard patterns are used to ensure that each cell can retain both "1" and "0" data states without any signal interference from neighboring cells. Since no electrical outputs from the DMD are required in the projector application, the device has only one data output pin for electrical test (the XGA and SXGA devices have two data output pins, for

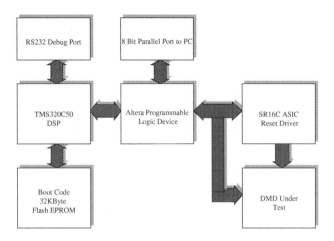

Figure 4.14 UDE/C50 board block diagram.

upper and lower sections of the memory array). This keeps device pin count to a minimum, but restricts data reading to serial only. Data input can be either serial or parallel via the full data input bus.

Upon successful completion of the CMOS tests the device is powered-up to full operating voltages. All mirrors are first addressed to the $+1$ direction and then the -1 direction. The "landed" bias and reset currents are measured at each mirror state. The optical parametric test measures the land and release performance of the mirrors, which in turn is used to determine hinge stiffness and surface adhesion performance. There are two distinct varieties of parametric tests: a Turbo Optimization Voltage (TOV) test which utilizes average light intensity measurements from dynamically operating mirrors, and Bias Adhesion Mapping (BAM) which measures individual mirror performance under static bias conditions. TOV, as the name suggests, is a faster but less comprehensive technique. To reduce test time, the parametric tests are usually performed only on sample areas of the active mirror array.

During the optical defect inspection test the CCD camera is stepped across the DMD array in a series of frames. Individual mirror functionality is evaluated at each frame. Two inspection modes are utilized: the Quickscreen method, which uses a scan technique to detect variation in image intensity, and Cross Pattern Analysis at failing frames to determine the type and exact location of the detected defects. The Sea Of Mirrors (SOM) border of the DMD (to be discussed later in the Image Quality section of this chapter) is also inspected using the Quickscreen method. After defect inspection, all voltages are returned to the initial values. Final current measurements are taken to verify that no shorts occurred during mirror operation. Finally, a power-down sequence is issued to the device to ensure that all mirrors are released prior to complete power-down. Early out options, which terminate program execution, can be enabled throughout the test program flow to minimize test time. A typical test sequence is depicted in Fig. 4.16.

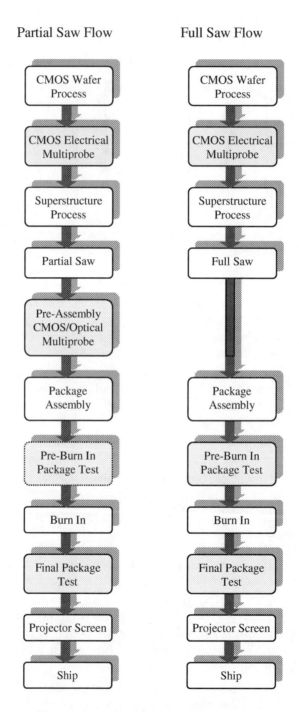

Figure 4.15 Production test sequence.

THE DIGITAL MICROMIRROR DEVICE 187

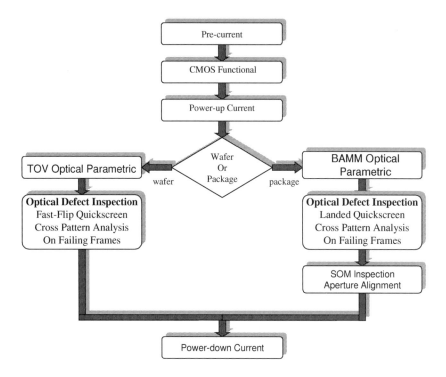

Figure 4.16 DMD test sequence.

4.6.1.2 Optical Defect Inspection Techniques

A microscope objective lens between the DMD under test and the CCD camera ensures a magnification factor of approximately 16 CCD pixels (4 × 4) to one DMD mirror. Each DMD mirror is exactly square, however the CCD pixels are not, so alignment is not exact. The CCD camera is not large enough to capture the whole DMD at one time at this 16:1 magnification, so the camera is stepped around the DMD in a series of frames. Figure 4.17 shows the inspection frames of the XGA (1024 × 768) resolution DMD. An additional number of frames are required to test the SOM region of the device. A new large CCD pixel array camera is under evaluation, which will significantly reduce the number of required frames.

Defect inspection begins with precise microscope movement (PMM) to the first frame location of the DMD where the camera is aligned to the edge of the active mirror array. Next, the Quickscreen inspection test is performed to detect any defective mirrors in the current frame. The mirrors are continuously operated in fast-flip mode while illuminated from the on-state direction. During fast-flip all mirrors are successively addressed to the on-state then the off-state. This produces a reflected light intensity level approximately midway between the on-state and off-state levels. The CCD camera captures this image, and alternate CCD pixel rows are scanned to detect variations above or below this average intensity. An intensity level below the average value indicates a dark defect such as a mirror, or mirrors,

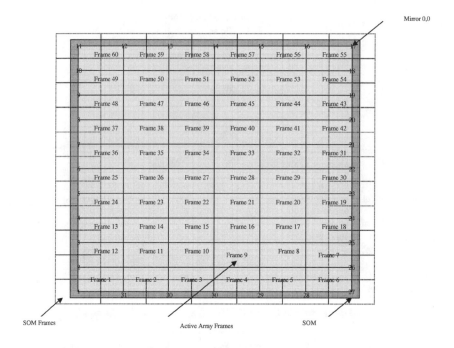

Figure 4.17 Inspection frames for XGA resolution DMD.

stuck in the off-state. Conversely, an intensity level above the average value indicates a bright defect. Threshold levels somewhat below and above the average levels are chosen to prevent false fails. This procedure is repeated for alternate CCD pixel columns to confirm any defect found. Figure 4.18 shows Quickscreen detection of two adjacent dark defects and a single bright defect. The Quickscreen process is repeated at each inspection frame as the PMM steps in a serpentine pattern around the DMD. The frame number of each frame containing a defect is logged. Inspection frames are overlapped to avoid missing a defect at the edge of the inspection area.

It should be noted that a defective mirror, which produces an intensity level close to the average value (such as a stuck, slightly torqued mirror), would not be detected with this fast-flip single illumination method of Quickscreen. A "landed," dual-illumination method would be required to detect this defect mode. With this method all mirrors are first addressed to the $+1$ direction where an image is captured with on-state illumination. The mirrors are then addressed to the -1 direction and a second image captured, this time with off-state illumination. Both of these "bright" images are then scanned for intensities below the bright level threshold. The penalty for this method is, of course, additional test time, since two images have to be captured and scanned instead of one. Typically, the fast-flip method is used at wafer probe and the landed method used at final package test. If defects were detected during the Quickscreen inspection, the PMM steps back to the first frame where an accurate CCD pixel to DMD mirror alignment is performed using

The Digital Micromirror Device

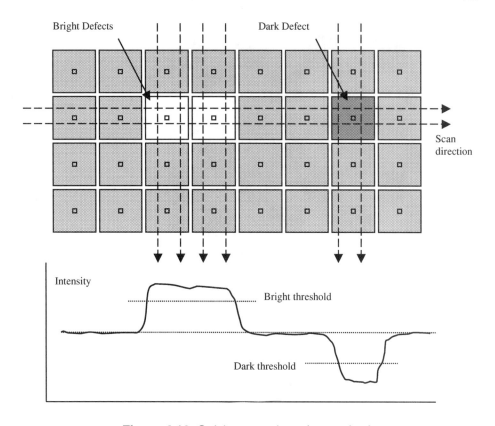

Figure 4.18 Quickscreen detection method.

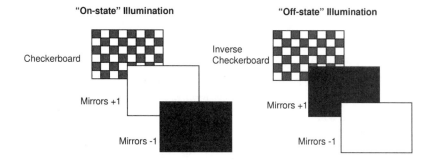

Figure 4.19 Image capture.

a 16-mirror by 16-mirror checkerboard pattern. The camera then steps to each defective frame in turn, where a Cross Pattern Analysis inspection is performed to identify the type and exact location of the defective mirrors.

During Cross Pattern Analysis inspection, six images are captured: a 16×16-mirror checkerboard, an all mirrors $+1$ and an all mirrors -1 image, each with on-state and then off-state illumination (Fig. 4.19). The checkerboard images are

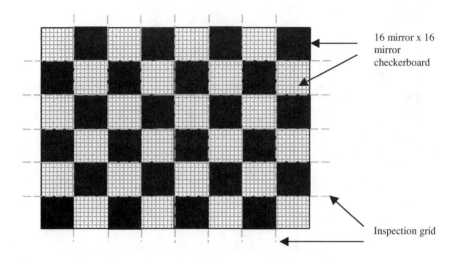

Figure 4.20 Frame inspection grid.

used to establish an inspection grid so the exact location of each mirror can be calculated (Fig. 4.20). The $+1$ and -1 images are then analyzed to determine the location and category of each defect. To avoid false defect detection due to the gaps between the mirrors (and because the CCD pixels don't align exactly to the DMD), only five CCD pixels, in the form of a cross nearest the center of each mirror, are considered during the analysis. If three or more of the five pixels are above (or below) the expected threshold, a defect is logged at that location. (The defect location is obviously critical information for failure analysis.) For example, if the image under analysis is expected to be bright and three of the cross pattern pixels fall below the threshold then a dark defect is logged at that location. Figure 4.21 shows a truth table of defect categories. A "stuck" defect is when the mirror stays landed in one direction. A "not" defect is when the mirror will land in one direction and release flat, but will not land in the opposite direction. A mirror is "stuck flat" if it will not land in either direction. An "invalid" defect is logged if, for example, the mirror appears stuck in both directions (which can be due to a broken or melted mirror).

During the image capture sequence, various memory patterns can also be applied to verify that the mirrors cannot be "upset" to the wrong state. For example, while the mirrors are latched in the $+1$ direction, the memory cells are addressed to the opposite state without applying a reset pulse. All mirrors should remain landed in the $+1$ state. Figure 4.22 shows some SEM photographs of typical pixel failure modes.

Once the main defect inspection is complete, the SOM is inspected in a series of frames around the periphery of the main (active) array. Since SOM mirrors are always landed in the -1 (or "off") direction when bias voltage is applied and accurate pixel alignment is impossible, a landed Quickscreen inspection method is used. With off-state illumination and the main array addressed to the $+1$ state the

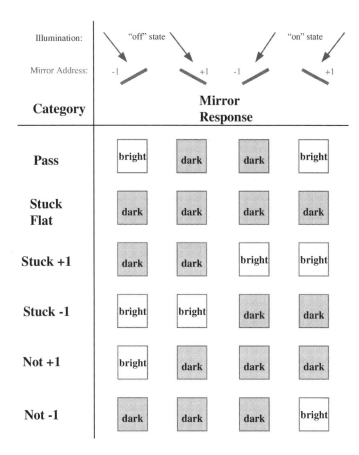

Figure 4.21 Defect categorization.

SOM will appear bright while the window aperture and main array appear dark. SOM mirrors with intensities below the expected threshold are counted as defects. The width of the SOM can also be measured, and alignment and rotation of the aperture relative to the active array calculated (see Fig. 4.23).

4.6.1.3 Optical Parametric Techniques

Optical parametric tests are used to measure the mechanical performance of the hinge, along with the surface adhesion effects at the spring tip landing points of contact. The parametric values, along with their corresponding Statistical Process Control (SPC) charts, provide timely feedback for the DMD superstructure process.

The Turbo Optimization Voltage, or TOV, test is a fast method for measuring the mirror landing response to the applied bias voltage. During this test, all memory cells are first addressed to the $+1$ state and the device illuminated from the on-state direction. The bias voltage is then stepped up in half-volt increments and the aver-

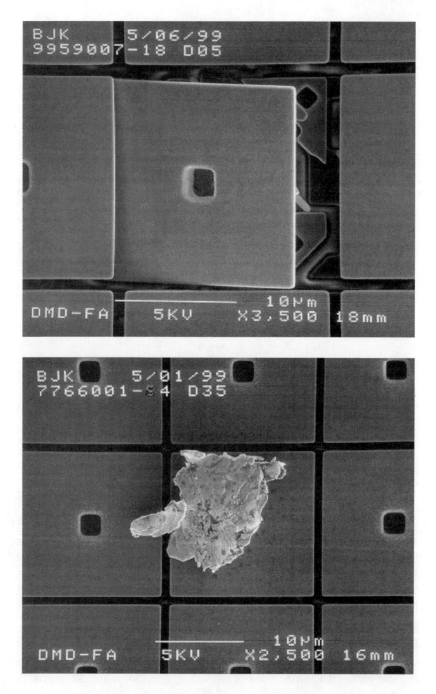

Figure 4.22 Examples of a pixel failure mode.

age frame intensity level captured at each step. As the bias voltage increases, more and more mirrors land until a maximum intensity level is reached with all mirrors landed. The DMD hinge compliance landing response is the main process para-

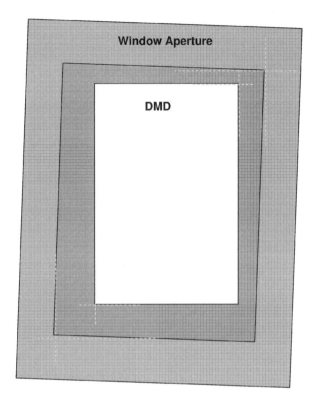

Figure 4.23 SOM inspection and aperture alignment.

metric determined by this bias voltage mapping. The compliance is determined for both the +1 and −1 state of the pixel and provides valuable uniformity information to the process engineering team. The pixel release response is also measured during this bias voltage mapping. Essentially, the bias voltage is stepped down (i.e., voltages lowered) in half-volt increments until the mirror releases from the surface. The release response provides hinge compliance, spring tip compliance, and surface adhesion information.

The pixel tilt angle as well as the device reflectivity and contrast ratio also constitute valuable optical parametric information. Again, the parametric information is collected on TI-built tools. Tilt angle is measured using a light source at a specified illumination angle to the DMD and the measured tilt angle is collected with all mirrors in a landed state. The reflected light from the DMD is measured on a position detector which can determine tilt angle within 0.1 degrees. Reflectivity and contrast ratio are measured on a stand-alone tool where light amplitude is measured with mirrors in the on state (reflectivity) and off state (contrast ratio). It is important to note that test set or projection system has a significant influence on the DMD contrast ratio: the DMD by itself has no contrast ratio. For parametric information, the tool utilized for this measurement is used mainly as a monitor of the DMD fabrication process.

4.6.2 DMD IMAGE QUALITY

A significant effort involves continued improvement for the DMD in projection-based systems. Many new pixel designs are developed to improve the overall optical performance of the system. The first pixel developed from the research laboratories had poor optical performance from a fill factor standpoint. This pixel design had hinges and mechanical support posts (i.e., vias) within the optical path. This spawned the development of the hidden hinge pixel architecture where over 88% of the optical area is the mirror (Fig. 4.24). Further improvements to the hidden hinge mirror have been made for optical efficiency enhancements. Two of these are the small rotated via (SRV) within the mirror and smaller mirror gaps (SMG)—Figs. 4.25 and 4.26, respectively. Improvements in these areas not only allowed for more lumens on the screen but greatly enhanced the black levels by mitigating the amount of light getting under the mirrors or being scattered/diffracted from various geometries within or under the pixel. To further improve the black levels of

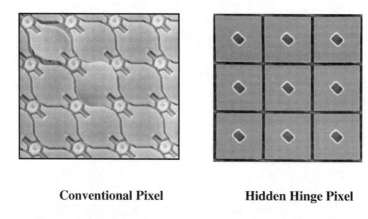

Figure 4.24 Conventional DMD pixel compared to hidden hinge pixel design.

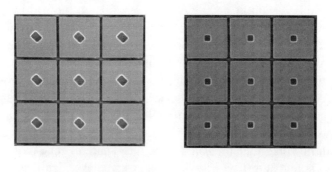

Figure 4.25 Small rotated via (SRV).

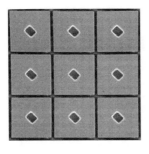 

Hidden Hinge Pixel **Small Mirror Gap + Small Rotated Via**

Figure 4.26 Small mirror gap (SMG).

the device within a system, we are investigating utilizing an absorptive film on the metal 3 layer of the device. This would further reduce light reflections from under the mirror and would greatly improve the system contrast ratio.

Since the DMD is placed within an optical system, special attention is focused on ensuring that light from outside the active array does not interfere with the main image, nor add unattractive bright features to the border of the image. An array of border pixels surrounds the active array to mitigate any border artifacts. We call these pixels the Sea of Mirrors (SOM). These pixels are globally addressed and only tilt toward the off direction. Thus, light is steered to an optical dump within the projection system. The pixels also hide features within the CMOS that provide signal to the array (Fig. 4.27). In addition to the SOM, the window of the DMD has an aperture to prevent light from illuminating the bond wires and other features outside of the SOM region. The SOM design for each type of DMD resolution unfortunately adds an additional number of mirrors to the device. As with any random defect distribution, the mirrors within the SOM region can be defective. The development of the absorptive film on the metal 3 layer as mentioned above is attractive in that it will allow for the elimination of the SOM region. A flat absorptive film within the border of the device will hide underlying CMOS metal layers while providing an attractive dark border within the projected image.

4.6.3 PROJECTOR-BASED IMAGE QUALITY EVALUATION

The DMD electro-opto-mechanical testing section has a key function in determining the pixel functionality. Image quality testing is not fully realized until a device is placed within a projection-based test system. This test is useful in identifying artifacts which can be subjective and are related to human perceptive factors. This projection test is our final outgoing test and looks for such details and defects as shown in Table 4.1.

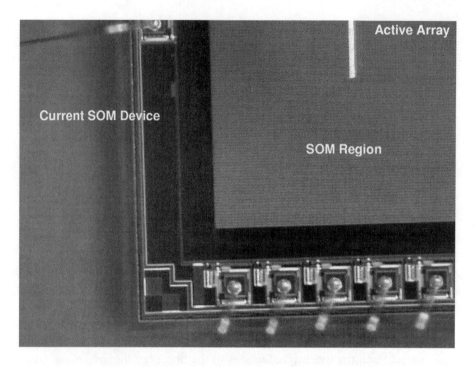

Figure 4.27 Sea of Mirrors region on the DMD.

Table 4.1 Image quality defects.

Pixel-related defects	Packaging-related defects	Soft errors
• Non-functional elements • Local (high spatial frequency) nonuniformities • Global (low spatial frequency) nonuniformities • System sensitivities/design considerations	• Active area borders • System aperturing, window and external • Particles and window material defects • System sensitivities/design considerations	• Signal timing • Fringe field

4.6.3.1 Pixel-related Defects

Nonfunctional element defects are mirrors that for one reason or another don't operate properly. The mirrors can be saturated dark (always off) pixels, saturated light (always on) pixels, or nonsaturated defects showing a gray level somewhere between full off and full on. The most objectionable defects are full or near-full on intensity pixels that show up as small, intense spots on otherwise dark screens. Acceptable levels for these defects are low, and judged primarily on count and location. Saturated bright defects anywhere in the field of view, and saturated dark defects in the center 50% of the screen are generally not acceptable in any venue. Adjacent saturated dark defects are shown in Fig. 4.28.

Figure above shows cluster of 2 saturated dark pixel defects. Inset to right shows higher magnification view of same defect.

Figure 4.28 Adjacent saturated dark defect.

Local (high-spatial-frequency) nonuniformities are relatively small variations in landed tilt angle, and can result in noticeable variations in pixel brightness. These are most objectionable at Nyquist spatial frequency, i.e., for adjacent pixels. The range of noticeability is variable in the population and the use venue, with limiting values of approximately 0.5% for very sensitive individuals in dark environments looking at low- to medium-brightness material, to approximately 2% for average individuals in low to medium ambients looking at average-brightness material. Especially objectionable are high-spatial-frequency variations showing some level of systematicity, with alternating rows or columns being the worst case. As above, the acceptable level is highly dependent on the target audience and use venue. An example is shown in Fig. 4.29.

Global (low-spatial-frequency) nonuniformities are much harder to detect due to the more gradual changes. Device-related variations can easily be washed out by other system effects, such as poor homogeneity in the illumination path, or vignetting in the projection path. Uniformities (also called rolloff) of 1.2:1 to 1.5:1 from center to corner are commonly acceptable, and device-related variables make this degree of uniformity easily achievable. The situation is slightly more compli-

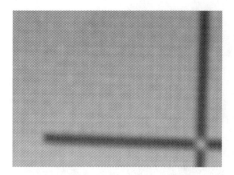

High-magnification, contrast-stretched view of high-spatial-frequency nonuniformities. Defects of this type give rise to a "dirty window" appearance, and are the result of small variations in tilt angle and/or pixel flatness (planarity).

Figure 4.29 High-spatial-frequency nonuniformity.

cated in multichip configurations where the chips can come from different wafers and lots, but to date we have been able to meet device tolerance allocations.

System sensitivities/design considerations are incorporated into the design of the DMD. One of the reasons we have been able to meet device tolerance allocations is that it is straightforward to design optical systems that are relatively insensitive to many MEMS device parameters. Another is that mature semiconductor processes and materials are very uniform on the order of the device diagonal. It is important that some optical deadband be incorporated into the optical system. This means that the cone of light originating from the lamp and reflected from the device should not completely fill the input pupil of the projection lens. If this is the case, small variations in landed tilt angle and/or pixel planarity will not immediately cause a dramatic change in the brightness of the pixel.

4.6.3.2 Package-related Defects

Packaging-related artifacts are more difficult to control in DMD-based systems. This is because in most cost-effective reflective modulator configurations, the area on the chip immediately surrounding the active area (the fully addressable array) will be visible. This is also true for the parts of the package immediately adjacent to the device.

Since the package is visible in almost all configurations for a reflective spatial light modulator, the package is an integral part of the MEMS functionality for DMD-based projection displays. The window aperture and shadow have image quality challenges like those of the active array. Figure 4.30 shows an example of border appearance quality. Border appearance is properly comprehended as a direct function of system design and package optical performance. The single remaining artifact in the top right corner of the top image is probably acceptable in most conference room venues. The border appearance of the bottom image would not be acceptable in any darkened ambient use. Pinholes in the opaque deposited-film aperture on the window generally create high-spatial-frequency bright spots which are unacceptable, and the edge definition can be challenging as well. We initially

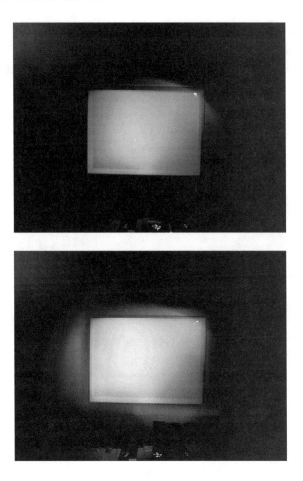

Figure 4.30 Examples of border appearance artifacts.

designed a reflective window aperture, with thermal management concerns playing a big role in this choice. However, as system throughput has become more of a customer concern, the illumination and projection lens cones have been opened up to the extent that the reflective window aperture is no longer acceptable, since it creates a bright "picture frame" effect in the display. Hence, we have developed absorptive window apertures and are challenged to develop thermal management strategies that continue to meet higher and higher illumination requirements. It should be added that both the aperture and the thermal management that goes with it are significant contributors to the overall package cost, and that package cost is a significant part of the total modulator cost. A second aperture in the system, if properly designed and placed, can play a pivotal role in reducing the cost of the packaged device, and make a significant difference in overall manufacturability.

Since the device must be sealed, and the mirrors optically accessible to the system, a high-quality transparent window must cover the active area. Further, system efficiency or throughput (brightness, for a given lamp size) and contrast (ratio

Figure 4.31 Window-related defects.

between on and off brightness levels) can be limited at some point by the transmissivity of the window, so typically a multilayer antireflection coating is required. The processing of the window to achieve a reliable mounting to the header and alignment to the active area involves a number of steps. These requirements demand processing that can result in particles and other defects present in the optical path. Examples of bright- and dark-window-related defects are shown in Fig. 4.31. In the top picture, a light-scattering particle on the inside surface of the window is depicted. The photographic exposure of a dark screen has been extended to 16 seconds to enhance the visibility of the defect. In the bottom picture the doubled effect of an opaque window particle is seen in the top right quadrant. Neither of these glass-related defects is acceptable.

A simple model of this optical path and the sensitivity of the system to several parameters is instructive. The figure of merit used for evaluating the artifacts resulting from these defects is the differential brightness of the defects compared to the undisturbed areas immediately adjacent to the manifestation of the defect. This is shown in Fig. 4.32. The figure is based on a model that assumes each point on the surface of the device is illuminated by a cone coming from the lamp. This is approximately correct in most useful configurations, and exact in the case of critical illumination, in which the lamp arc is imaged on the device surface. Then the differential brightness of each point in the defect area will depend on the ra-

THE DIGITAL MICROMIRROR DEVICE

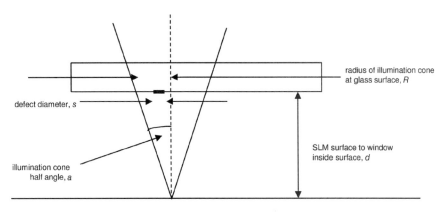

cone half angle *a*: $\tan a = 1/(2 * f/\#)$

radius of illumination: $R = d * \tan a = d * [1 / (2 * f/\#)]$

area of illumination: $A = pi * R^2 = pi * d^2 * [1 / (2 * f/\#)]^2$

area of blemish: $B = pi * (s / 2)^2$

differential brightness of defective area relative to clear area:

d.b. = area of defect * opacity / illuminated area = $(s / d)^2 * (f / \#)^2$

Figure 4.32 Optical path and sensitivity model.

tio of the area of illumination for that point compared to the area blocked by the blemish. An interesting feature of this model is that within large bounds, and certainly within the 0.5–2.0% differential brightness acceptable in most venues, the size of the shadow or image defect is fixed by the illumination cone. The size of the defect determines the contrast or differential brightness of the defect, but not its real extent. This feature, and the general numerical validity of the model, has been confirmed in many systems and configurations.

It should be realized that any defect on the window as shown will create two spots, one resulting from the shadow cast on the device in the illumination cone, and one resulting from the blockage of the projection cone by the particle. If it is assumed that the cone characteristics ($f/\#$) are determined by other system design considerations, which is usually the case, and we limit ourselves to opaque (worst case) particles, the only free parameter left to the package designer is the spacing between the window surface and device active area. This leads to the set of curves illustrated in Fig. 4.33 showing the differential brightness versus particle size, with window spacing as the parameter.

It is immediately apparent from the above analysis that the entire system must be comprehended when designing or making changes in the modulator/package configuration, and when designing the system to utilize a given device configuration. Although it would seem from the window defect analysis that moving the window further away from the device is always a good idea, it is fairly easy to show that this broadens the area of shadow cast by the window aperture, and can

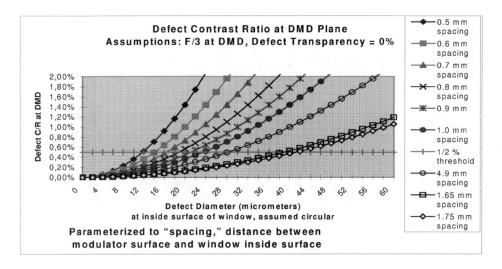

Figure 4.33 Differential brightness as a function of particle size and window spacing.

make alignment of the window to the die more difficult, which has its own yield and cost impact. Another example is the possible conclusion that faster optical systems (smaller f/numbers) are desirable. This is true if window defect sensitivity is the paramount consideration, and has the additional benefit that it will generally provide better coupling to the lamp and deliver higher system optical efficiency. However, at $f/\#$s larger than about $f/2.8$, the available contrast in the system begins to be limited by the nominal device tilt angle of 10 degrees. A full treatment of the available design space for DMD-based projection displays is clearly beyond the scope of the present work.

4.6.3.3 Soft Errors

Similarly, a comprehensive treatment of the electromechanical and optomechanical performance of the MEMS structure is the subject of an extended article in itself. Indeed, a major thrust of the MEMS technology development effort should be the creation of accurate and precise modeling capabilities. When the design and processing of the structure involves nonrecurring engineering costs and development timelines similar to semiconductor devices, there is considerable value to an integrated modeling effort that combines these effects. Some of the considerations central to DMD modeling include pixel dynamics and signal timing, the effects of data-dependent fringing electrical fields, and how device dynamics change over the extended array and through use conditions seen by fielded systems. The mission profile or expected use environment is a key input to the range of variables that should be comprehended.

Close working relationships are maintained with our customers to establish pass/fail criteria for each system within specific market segments. The image quality results are continually fed back to our suppliers and device and system design teams to ensure that the product performance is not limited by the DMD, to optimize a fit to system requirements/capabilities, and to comprehend trade-offs and system cost/benefit issues.

4.7 DEVICE RELIABILITY

Eliminating defective pixels is the main quality objective. Reliability of the DMD means that all pixels remain functional over time, and the device works over a wide range of operating conditions. The DMD product could not depend solely on Military Specification reliability standards as used in the IC industry. A number of reliability tests had to be developed to identify failure causes and validate the effectiveness of corrective actions. As with most MEMS devices, reliability tests that are developed are application specific. The DMD undergoes a number of reliability tests both with and without projection illumination. Key factors investigated involve the following:

- Stiction and accelerators;
- Fatigue;
- Metal creep;
- Compromised packaging; and
- Photochemistry and other environmental factors.

To date, we have identified and eliminated all life-limiting failure modes. Future failure rate improvements will result from continued particle reduction efforts as well as design improvements.[7]

4.8 SUMMARY

Having the core competencies of CMOS, DMD, window and projector engine fabrication in close proximity has allowed very rapid product development and maturation. The DMD is currently in volume production for both SVGA and XGA display formats. It is highly unlikely that specialized (i.e., one-of-a-kind) DMD processing equipment is going to influence any semiconductor equipment manufacturers. Therefore, it is our responsibility to continue to develop processes that are within the mainstream of the IC industry. Compared to other high-volume MEMS companies, we are producing a much more complex device.

The manufacturing team has continually demonstrated increased shipping rates each month in order to meet the business needs for digital imaging. A very significant percentage of SVGA and XGA devices we ship today are defect-free. The demonstrated ability to transfer this technology to a mature wafer fab clearly meets the business and technology objectives as we become more in line with the semiconductor industry. This transfer further assists the business strategy by not being capacity-limited. The technological and manufacturing capability to fabricate devices configured for other applications, including higher-resolution (SXGA—1280 × 1024) displays has also been demonstrated. The DMD is meeting customer performance and reliability expectations as well as the business yield and cost goals.

Recently, a comprehensive review of this technology was published in the "Texas Instruments Technical Journal."[8] This journal discusses topics from the history of electronic displays to marketing a new technology. For additional reading and information on this technology, a list of significant publications is shown in Appendix A.

References

1. K. Gabriel, et al., Small Machines, Large Opportunities: A Report on the Emerging Field of Microdynamics, National Science Foundation Workshop, 1987–1988.
2. K. W. Markus, The Challenges of Infrastructure—Supporting the Growth of MEMS into *Production, Commercialization of Microsystems '96*, Sponsored by EF and SEMI (October) (1996).
3. M. Mehregany, Introduction to Microelectromechanical Systems and the Multiuser MEMS Processes, Case Western Reserve University—MCNC Short Course Handbook (August) (1993).
4. P. S. Burggraaf, "Sawing Systems Update," *Semiconductor International* (December), 47 (1982).
5. M. A. Mignardi and T. R. Howell, "Fabrication of the Digital Micromirror Device (DMD)," in *Commercialization of Microsystems '96*, Sponsored by EF and SEMI (October) (1996).
6. Discussion with Bob Sulouff, Analog Devices, Inc., Director of Business Development and Marketing, Micromachined Products Division (October) (1996).
7. M. R. Douglass, "Lifetime Estimates and Unique Failure Mechanisms of the Digital Micromirror Device (DMD)," in *IEEE International Reliability Physics Proceedings*, pp. 9–16 (1998).
8. J. L. Bender and R. L. Knipe, (ed.), "Digital Light Processing," *Texas Instruments Technical Journal* 15(3), July–September (1998).

Appendix A: List of Significant Publications on DMD and DLP Technology

1. L. J. Hornbeck and W. E. Nelson, "Bistable Deformable Mirror Device," *OSA Technical Digest Series Vol. 8, Spatial Light Modulators and Applications*, p. 107 (1988).

2. W. E. Nelson and L. J. Hornbeck, Micromechanical Spatial Light Modulator for Electrophotographic Printers, in *SPSE Fourth International Congress on Advances in Non-Impact Printing Technologies*, p. 427 (1988).
3. L. J. Hornbeck, "Deformable-Mirror Spatial Light Modulators (Invited Paper)," Spatial Light Modulators and Applications III, *SPIE Critical Reviews*, Vol. 1150, pp. 86–102 (1989).
4. T. J. Sheerer, W. E. Nelson and L. J. Hornbeck, "FEM analysis of a DMD," in *Seventeenth Nastram User Colloquium, NASA Conf. Pub. 3029*, p. 290 (1989).
5. J. B. Sampsell, "An Overview of Texas Instruments Digital Micromirror Device (DMD) and Its Application to Projection Displays," in *Society for Information Display Internatl. Symposium Digest of Tech. Papers*, Vol. 24, pp. 1012–1015 (1993).
6. L. J. Hornbeck, "Current Status of the Digital Micromirror Device (DMD) for Projection Television Applications (Invited Paper)," in *International Electron Devices Technical Digest*, pp. 381–384 (1993).
7. J. M. Younse and D. W. Monk, "The Digital Micromirror Device (DMD) and Its Transition to HDTV," in *Proc. of 13th International Display Research Conf. (Late News Papers)*, pp. 613–616 (1993).
8. J. B. Sampsell, "The Digital Micromirror Device," in *7th ICSS&A*, Yokohama, Japan (1993).
9. J. M. Younse, "Mirrors on a Chip," *IEEE Spectrum* (November), 27–31 (1993).
10. J. M. Younse, "DMD Projection Display Technology (Invited Talk)," *IEEE Multimedial Symposium*, Richardson, TX (1994).
11. R. J. Gove, "DMD Display Systems: The Impact of an All-Digital Display," Society for Information Display International Symposium (June 1994).
12. M. A. Mignardi, "Digital Micromirror Array for Projection TV," *Solid State Technology* 37 (July), 63–66 (1994).
13. K. Ohara and R. J. Gove, "DMD Display and its Video Signal Processing," The Institute of Television Engineers Japan 1994 Annual Convention, ITEC'94, Saga, Japan (1994).
14. V. Markandey, T. Clatanoff, R. Gove and K. Ohara, "Motion Adaptive Deinterlacer for DMD (Digital Micromirror Device) Based Digital Television," *IEEE Trans. on Consumer Electronics* 40(3), 735–742 (1994).
15. J. M. Younse, "Commercialization of Digital Micromirror Display Technology," in *Conference for Commercialization of Microsystems*, Banff, Alberta, Canada (1994).
16. V. Markandey and R. Gove, "Digital Display Systems Based on the Digital Micromirror Device," in *SMPTE 136th Technical Conference and World Media Expo* (1994).
17. D. W. Monk, "Digital Micromirror Device Technology for Projection Displays," in *EID Exhibition & Conference*, Sandown, U.K. (1994).
18. J. B. Sampsell, "An Overview of the Performance Envelope of Digital Micromirror Device Based Projection Display Systems," in *SID 94 Digest*, pp. 669–672 (1994).

19. C. Tew, L. Hornbeck, J. Lin, E. Chiu, K. Kornher, J. Conner, K. Komatsuzaki and P. Urbanus, "Electronic Control of a Digital Micromirror Device for Projection Displays (Invited Paper)," in *IEEE International Solid-State Circuits Digest of Technical Papers* 37, 130–131 (1994).
20. G. Deffner, M. Yuasa, M. Mckeon and D. Arndt, "Evaluation of Display Image Quality: Experts vs. Non Experts," *1994 SID International Symposium, Digest of Technical Papers*, pp. 475–478 (1994).
21. G. Deffner and M. Yuasa, "Understanding Perceived Image Quality: New Applications of Verbal Protocol Methodology," in *Proceedings of the Human Factors and Ergonomics Society 38th Annual Meeting*, Santa Monica, CA (1994).
22. G. A. Feather and D. W. Monk, "The Digital Micromirror Device for Projection Display," in *IEEE International Conference on Wafer Scale Integration*, San Francisco, CA (1995).
23. W. E. Nelson and R. L. Bhuva, "Digital Micromirror Device Imaging Bar for Hardcopy," *Color Hardcopy and Graphic Arts IV, SPIE*, Vol. 2413, San Jose, CA (1995).
24. M. R. Douglass and D. M. Kozuch, "DMD Reliability Assessment for Large-Area Displays," *Society for Information Display International Symposium Digest of Technical Papers*, Vol. 26 (Applications Session A3), pp. 49–52 (1995).
25. E. Chiu, C. Tran, T. Honzawa, and S. Namaga, "Design and Implementation of a 525 mm^2 CMOS Digital Micromirror Device (DMD) Chip," in *Proc. 1995 International Symposium of VLSI Technology Conf.*, p. 127, Taipei, Taiwan (1995).
26. B. R. Critchley, P. W. Blaxtan, B. Eckersley, R. O. Gale and M. Burton, "Picture Quality in Large Screen Projectors Using the Digital Micromirror Device," in *SID 95 Digest,* pp. 524–527 (1995).
27. G. Sextro, T. Ballew and J. Iwai, "High-Definition Projection System Using DMD Display Technology," in *SID 95 Digest*, pp. 70–73 (1995).
28. G. Feather, "Digital Light Processing^{TM}: Projection Display Advantages of the Digital Micromirror Device^{TM}," in *19th Montreux Television Symposium* (1995).
29. J. M. Younse, "DMD Microelectromechanical (MEMS) Technology (Keynote Address)," in *Society of Experimental Mechanics (SEM) UACEM Conference*, Worcester, MA (1995).
30. J. M. Younse, "DMD Display Technology (Invited Talk)," Naval Research Center, Washington, D.C. (1995).
31. G. Hewlett and W. Werner, "Analysis of Electronic Cinema Projection with the Texas Instruments Digital Micromirror Device^{TM} Display System," in *137th SMPTE Technical Conference*, Los Angeles, California (1995).
32. J. M. Younse, "DMD Display Technology (Invited Talk)," in *U.S. Government's Large Display Working Group Meeting*, NUWC, Newport, RI (1995).
33. L. Yoder, "The Fundamentals of Using the Digital Micromirror Device (DMD) for Projection Displays," in *Proc. International Conference on Integrated Micro/Nanotechnology for Space Applications*, sponsored by NASA and Aerospace Corp., Houston, Texas (1995).

34. T. Hogan and J. Hortaleza "Low Temperature Gold Ball Bonding for Micromechanical Device Interconnections," Ultrasonic Industry Association, Columbus, Ohio (1995).
35. J. M. Younse, "Projection Display Systems Based on the Digital Micromirror Device (DMD)," SPIE Conference on Microelectronic Structures and Micrelectromechanical Devices for Optical Processing and Multimedia Applications, Austin, Texas, *SPIE Proceedings*, Vol. 2641, pp. 64–75 (1995).
36. L. J. Hornbeck, "Digital Light Processing and MEMS: Timely Convergence for a Bright Future (Invited Plenary Paper)," *SPIE Micromachining and Microfabrication '95*, Austin, Texas (1995).
37. W. E. Nelson, "Tutorial on Optical Printheads," in *IS&T Eleventh International Congress on Advances in Non-Impact Printing Technologies*, Hilton Head, South Carolina (1995).
38. J. Iwai and G. Sextro, "High Definition Projection System Using DMD Technology," in *IEE 1995 International Workshop on HDTV and the Evolution of Television*, Taipei, Taiwan (1995).
39. G. Deffner, "Eye Movement Recordings to Study Determinants of Image Quality in New Display Technology," in *Eye Movement Research*, J. M. Findlay, R. Walker and R. W. Kentridge (eds.), pp. 479–490. Elsevier, Amsterdam (1995).
40. R. M. Wallace, P. J. Chen, S. A. Henck and D. A. Webb, "Adsorption of Perfluorinated n-Alkanoic Acids on Native Aluminum Oxide Surfaces," *Journal of Vacuum Science and Technology* A 13, 1345 (1995).
41. S. Heimbuch, V. Liu, V. Markandey and D. Fritsche, "Integration of Today's Digital State With Tomorrow's Visual Environment," *SPIE Photonics West '96*, San Jose, California (1996).
42. V. Markandey, T. Clatanoff and G. Pettitt, "Video Processing for DLP Display Systems," *SPIE Photonics West '96*, San Jose, California (1996).
43. L. Yoder and J. Florence, "Display System Architectures for Digital Micromirror Device (DMD) Based Projectors," *Proc. SPIE*, Vol. 2650, Projection Displays II, pp. 193–208 (1996).
44. V. Markandey, D. Doherty and G. Pettitt, Investigation of DMD Based Display for Integrated Computer and Television Application, in *SMPTE Advanced Motion Imaging Conference* (1996).
45. R. Howell and M. Mignardi, "Fabrication of the Digital Micromirror Device (Invited Paper), in *SID Displaywork 1996, Manufacturing Technology Conference* (1996).
46. V. Markandey, G. Hewlett and G. Pettitt, "Digital Light Processing: The Convergence of Television and Computer Display," in *30th SMPTE Advanced Technical Conference 1996 Proceedings*, pp. 150–160 (1996).
47. D. Monk, "Digital Light Processing With the Digital Micromirror Device," *Electronic Displays 96*, Chemnitz, Germany (1996).
48. Kunzman and A. Wetzel, "1394 High Performance Serial Bus: The Digital Interface for ATV," *Trans. on Consumer Electronics* 41(3), 893 (year?).

49. G. Pettitt, A. DeLong and A. Harriman, "Colorimetric Performance Analysis for a Sequential Color DLP Projection System," in *SID '96*, San Diego, California (1996).
50. R.G. Fielding, M. Burton and T. Bartlett, "Colorimetry Performance of a High Brightness DMD Based Optical System," in *SID '96*, San Diego, California (1996).
51. M. R. Douglass and C. Malemes, "Reliability of Displays using Digital Light Processing," in *SID '96*, San Diego, California (1996).
52. G. Deffner, "Determinants of Perceived Image Quality in Digital Printing," in *Print & Image Quality Conference*, sponsored by Information Management Institute, San Diego, California (1996).
53. L. J. Hornbeck, "Digital Light Processing: A New MEMS-Based Display Technology (Keynote Address)," in *IEEJ 14th Sensor Symposium*, Kawasaki, Japan (1996).
54. L. J. Hornbeck, "Digital Light Processing and MEMS: Reflecting the Digital Display Needs of the Networked Society (Invited Plenary Paper)," in *SPIE/EOS European Symposium on Lasers, Optics, and Vision for Productivity in Manufacturing I, Conference on Micro-Optical Technologies for Measurement, Sensors & Microsystems*, Besancon, France (1996).
55. R. L. Knipe, "Challenges of a Digital Micromirror Device: Modeling and Design," in *SPIE/EOS European Symposium on Lasers, Optics, and Vision for Productivity in Manufacturing I, Conference on Micro-Optical Technologies for Measurement, Sensors & Microsystems*, Besancon, France (1996).
56. V. Markandey, G. Wolverton, W. Werner and J. Younse, "Digital Light Processing Systems for Standard and High Definition Displays," in *IEEE International Conference on Consumer Electronics, ICCE '96*, Chicago, Illinois (1996).
57. L. J. Hornbeck, "Digital Light Processing and MEMS: An Overview (Invited Paper)," in *IEEE/LEOS/OSA 1996 Summer Topical Meetings, Optical MEMS and Their Applications*, Keystone, CO (1996).
58. W. E. Nelson, "Digital Light Processing for Color Printing (Invited Paper)," in *IEEE/LEOS/OSA 1996 Summer Topical Meetings, Optical MEMS and Their Applications*, Keystone, CO (1996).
59. F. Skaggs, "Automatic Testing of the Digital Micromirror Device, DMD," in *IEEE/LEOS/OSA1996 Summer Topical Meetings, Optical MEMS and Their Applications*, Keystone, CO (1996).
60. K. Ohara, A. Takeda and G. Sextro, An Optimal Video Signal Enhancement on SVP2, in *IEEE International Conference on Image Processing, ICIP '96*, Lausanne, Switzerland (1996).
61. L. J. Hornbeck, "Digital Light Processing for Projection Displays: A Progress Report (Invited Paper)," in *SID EuroDisplay '96, The 16th International Display Research Conference*, Birmingham, England (1996).

CHAPTER 5

OPTICAL WAVEGUIDES AND SILICON-BASED MICROMACHINED ARCHITECTURES

Christophe Gorecki
Institut des Microtechniques de Franche-Comté—LOPMD

CONTENTS

5.1 Introduction / 210

5.2 Status of Optical Waveguide Technologies on Silicon / 211
    5.2.1 Advantages of optical waveguides on silicon / 211
    5.2.2 Basic optical waveguide geometries / 212
    5.2.3 Fabricating the index contrast / 214
    5.2.4 Antiresonant reflecting optical waveguide / 217
    5.2.5 Deposition methods / 218
    5.2.6 Properties of deposited films / 230
    5.2.7 Fabrication of channel waveguides / 236
    5.2.8 Basic guided-wave optical components / 249
    5.2.9 Propagation loss and fiber-to-waveguide coupling loss / 251

5.3 Silicon Micromechanics / 254
    5.3.1 Materials and micromachining / 254
    5.3.2 Microactuators / 255
    5.3.3 Simplest microstructures combined with optical waveguides / 259

5.4 Examples of Micromachined Guided-Wave MOEMS / 260
    5.4.1 Micromachined integrated optic switches / 261
    5.4.2 Micromachined optical modulators / 267
    5.4.3 Micromachined optical sensors / 274
    5.4.4 Micromachined nanoprobes for near-field optical microscopy / 292

5.5 New Trends and Conclusions / 295

References / 297

5.1 Introduction

In the past decade, rapid advances in silicon-based microelectromechanical (MEMS) technology have enabled the realization of a variety of novel functions in the micromechanical world similar to the functions performed by microelectronics. One motivation for developing MEMS is their small scale and the ease of integrating them with electronic circuits and sensors, resulting in miniaturized and smart microsystems with moving parts. The dimensional scale of MEMS devices is immediately compatible with the size of integrated optics (IO) and micro-optical devices such as optical fibers, laser diodes, channel waveguides, and diffractive or refractive microlenses. This scale (1 to 500 µm) is compatible with the size of optical beams, and is appropriate for controlling or manipulating optical radiation.

MEMS technology exploits photolithographic techniques to fabricate microstructures with extremely high dimensional tolerances, precise positioning on the chip, and well-controlled microactuation. This potential is suitable for fabricating precision-defined optical components and offers relatively easy alignment procedures for optical parts.

For all these reasons, MEMS technology has benefited from the integration of microstructures with free-space micro-optics and guided wave optics. This category of photonic devices constitutes a new extension of MEMS, called micro-optoelectromechanical systems (MOEMS). Many novel MOEMS architectures, including micromirrors, microlenses, modulators and switches, microchoppers, microplatforms for integrated optics, fiber optics, scanners, optically driven actuators, integrated optical sensors, and other devices with optical functions have been developed. In this chapter we focus our attention on guided-wave MOEMS, combining micromachined structures and optical waveguides on silicon substrates. Since guided-wave MOEMS are fabricated by batch process, these devices are potentially low cost, and their monolithic structure improves their reliability. This approach of integrating optics with micromechanics has applications for integrated sensors, microactuators, and optical communication.

A wide variety of fabrication technologies have been developed for optical waveguides on silicon. These include chemical vapor deposition (CVD), flame hydrolysis deposition (FHD), and spin-coating deposition of polymers. Waveguide structures are based on depositing silica layers for the cladding and doped silica, silicon nitride or silicon oxinitride, and polymers for the core layer, for example. The core layer is commonly structured by reactive ion etching (RIE). The waveguide-related structures include directional couplers, Y junctions, multiplexers, and possibly insertion of standard fiber optics via V-groove alignment procedures. Optical waveguides may be integrated with micromechanical deformable structures such as diaphragms, beams, bridges, and membranes. The micro-optomechanic element can serve as a sensor to detect a physical parameter of the mechanical deformation, as well as a microactuator, and also as an optical modulator or switch.

Micromechanical structures may be fabricated using bulk or surface micromachining where a number of different thin films are deposited on top of the silicon wafer, and the desired microstructure is obtained by successive deposition, etching, and patterning operations on the surface of the wafer. The working principle may be based on the physical or mechanical effect that changes the effective refractive index of guided waves induced by motion or deformation of a micromechanical structure. Electrostatic, piezoelectric, shape memory alloy (SMA), elasto-optic, thermal, or mechanical forces or pressures can actuate the micromechanical structure. The conversion of electrical signals results in the generation of functions through phase or intensity modulators, switches, directional couplers, and tunable frequency filters or Fabry–Pérot resonators.

Alternatively, if the guided-wave MOEMS acts as an optical readout of movable micromechanical elements, converting displacements or strains in an optical signal via interferometric or intensity modulation techniques, the function is the sensing. Thus, the light propagating in a deformed waveguide is used to determine the amount of deformation induced by external physical forces (pressure, displacement, force, or acceleration).

In the following sections, we will discuss the above developments in detail and provide a map of advances in the field of guided-wave MOEMS. Section 5.2 reviews the processes involved in the fabrication of silicon-based optical waveguides and describes and analyzes how this photonic technology may be integrated within the micromachined MEMS environment. Section 5.3 briefly reviews the MEMS materials and microfabrication technologies by which guided-wave MOEMS are made and describes the principles of microactuation involved. Section 5.4 provides some examples of optical functions that can benefit by the integration of optical waveguides with MEMS. It discusses fabrication aspects and specific experimental realizations, including integrated optic switches, integrated optical modulators, micromachined optical sensors, and some promising applications such as microprobes for optical near-field microscopy. Finally, we introduce some new trends and draw some conclusions.

5.2 STATUS OF OPTICAL WAVEGUIDE TECHNOLOGIES ON SILICON

5.2.1 ADVANTAGES OF OPTICAL WAVEGUIDES ON SILICON

The combination of integrated optics, micro-optic technologies, and micromechanical structures on silicon created a new class of MOEM devices and offers new integration potentials for MEMS applications. In particular, the integration of planar optical waveguides on a silicon substrate permits a wide variety of sensor applications as well as communication devices. This technological approach of integrated optics associated with silicon micromachining technology is very attractive because it offers some major advantages over previously developed and competing technologies such as ion-exchanged glass waveguides, LiNbO$_3$, or III-V technologies using GaAs and InP components:

- The fabrication process is compatible with monolithic integration of detectors and microelectronic circuits.
- Hybrid integration may be used to add optical sources.
- It is amenable to dry and wet etching, yielding the potential for integration of micromechanical components.
- It provides easy optical fiber connections with a good match through V-grooves etched into the silicon substrate.
- There are well-developed deposition technologies for waveguide fabrication, yielding the potential for cheap mass production.
- The methods are lower cost and less toxic than III-V compounds.

Thus, silicon is not only an excellent substrate for passive optical waveguides, but also could incorporate all optoelectronic and micromechanical components on the same wafer.[1] In the integration of waveguides both microelectronics and micromechanical technologies of MEMS are viewed as a common technology platform with which the addition of photonic functions must be compatible. Integration of waveguide components within the MEMS environment is expected to open a wide range of new applications, leading to a richer set of optical capabilities when embedded in the MEMS technology.

5.2.2 BASIC OPTICAL WAVEGUIDE GEOMETRIES

The most widely used optical waveguide architecture fabricated on silicon is the channel waveguide, shown schematically in Fig. 5.1(a). This 3D waveguide with a rectangular section is usually formed by a combination of dielectric layer depositions and etch operations performed on a platform that is a crystal silicon substrate of refractive index n_s. The waveguide structure consists of a core region with a refractive index n_1 higher than that of its surrounding confining medium n_2. The channel waveguide guidance is based on the same principle as that of optical fibers, confining all the light rays that suffer total internal reflection. If the core size is large and the refractive index difference $\Delta n = n_1 - n_2$ is large, there is multimode propagation. A waveguide with a very small Δn and small core size allows single-mode propagation. In the particular case of a rib waveguide, illustrated in Fig. 5.1(b), the light is confined in the vertical direction by a step in the refractive index Δn, and in the lateral direction by a step in the core thickness between the center and outer regions. In practice, a buffer and a cladding surround the core medium, as shown in Fig. 5.2(a). The buffer layer is of considerable thickness (>4 µm), which is necessary to isolate the waveguide from the high-index silicon substrate ($n \sim 3.5$). The cladding may be air, but often the cladding materials are of higher refractive index in order to make the modal fields symmetric. The cladding layer is used to bury the core region and to isolate it from the environment (air). The refractive index profile

OPTICAL WAVEGUIDES AND SILICON-BASED MICROMACHINED ARCHITECTURES 213

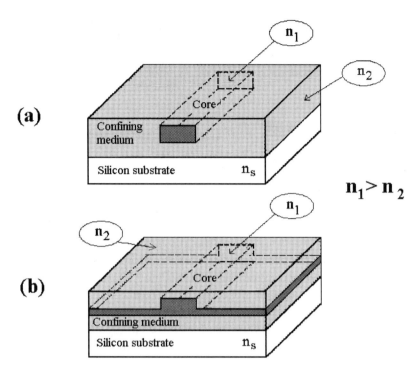

Figure 5.1 Types of optical waveguides: (a) general channel waveguide, (b) rib waveguide.

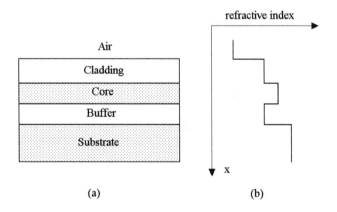

Figure 5.2 Schematic of a planar optical waveguide: (a) multilayer structure, (b) refractive index profile.

of a typical multilayer planar waveguide is illustrated in Fig. 5.2(b). The refractive index of the confining medium is not necessarily uniform on all sides. Different materials with indices less than n_1 may be used to surround the core region. In that case, the waveguiding modes are asymmetric.

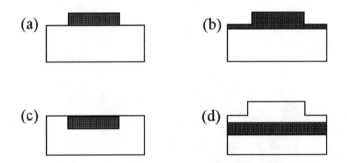

Figure 5.3 Examples of channel optical waveguides: (a) raised waveguide, (b) rib waveguide, (c) embedded-strip waveguide, (d) strip-loaded waveguide.

Figure 5.3 shows the cross-sections of four different types of channel waveguides.[2] The raised waveguide, shown in Fig. 5.3(a), may be fabricated by starting a planar slab waveguide, masking the strip, and removing the lateral regions of cladding by etching techniques. If the removal of the surrounding layer is incomplete, the rib waveguide is fabricated, as shown in Fig. 5.3(b). By combining depositions and etching, it is also possible to fabricate an embedded-strip waveguide and strip-loaded waveguide, shown respectively in Figs. 5.3(c) and 5.3(d).

5.2.3 Fabricating the Index Contrast

It is desirable to fabricate optical waveguides with low propagation loss and a well-controlled refractive index profile. The main category of silicon-based waveguides uses pure silica (SiO_2) as the basic material. Deposition and etching of multiple material layers or selective modifications of regions within a single material (e.g., dopant implantation) mostly fabricate the resulting waveguide 3D structure. Single-mode waveguides may be fabricated using as cladding material the silica and as core materials silicon nitride (Si_3N_4), silicon oxinitride (SiO_xN_y), and doped silicon oxide.[3-5] In principle, all these materials are compatible with MEMS technology, including both the deposition of distinct waveguide layers and the processing to pattern the waveguide structure with a refractive index difference, Δn, between core and cladding, called also index contrast. Several deposition techniques are used to fabricate these materials; they are characterized by different chemical reactions and operate with different reactors at different temperatures. An index contrast may be obtained by the addition of dopants in SiO_2, transformation of SiO_2 chemical content, or juxtaposition of different thin films obtained by a chemical vapor deposition process with a well-controlled refractive index profile. For example, pure silica may be deposited at low temperatures, lower that 500°C, by reacting silane and oxygen:

$$SiH_4 + O_2 \rightarrow SiO_2 + 2H_2. \tag{5.1}$$

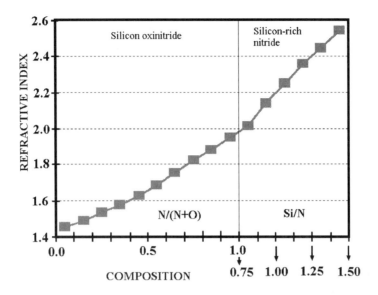

Figure 5.4 Variation of refractive index of SiO_xN_y and Si_3N_4 films with composition (deposition at temperatures from 700 to 900°C).

Pure silica can be deposited with or without dopants. Undoped silicon dioxide is used as buffer and cladding layers, and doped silicon dioxide plays the role of a core layer. For example, phosphorus-doped oxide can be produced in the following way:

$$SiH_4 + O_2 \rightarrow SiO_2 + 2H_2 \quad \text{and} \quad 4PH_3 + 5O_2 \rightarrow 2P_2O_5 + 6H_2. \tag{5.2}$$

Silicon nitride can be formed by reacting silane and ammonia at temperatures between 700 and 900°C or by reacting dichlorosilane and ammonia:

$$3SiH_4 + 4NH_3 \rightarrow Si_3N_4 + 12H_2, \tag{5.3a}$$
$$3SiCl_2 + 4NH_3 \rightarrow Si_3N_4 + 6HCl + 6H_2. \tag{5.3b}$$

Figure 5.4 shows how the refractive index of silicon nitride films changes with composition, where the lines are least-square fits of published data.[6] Silicon nitride layers that contain oxygen are silicon oxinitride. The composition of SiO_xN_y can vary from silicon dioxide with a refractive index of about 1.46 to silicon nitride with an index of 2.01. Films that have silicon-to-nitrogen ratios greater than 0.75 are free of oxygen and are called silicon-rich silicon nitrides. The corresponding refractive index is in the range from 1.8 to 2.5.

Most published work on silica-based waveguides refers to the following waveguide structures: silicon substrate/SiO_2/Si_3N_4/SiO_2, silicon substrate/SiO_2/SiO_xN_y/SiO_2, and silicon substrate/SiO_2/doped SiO_2/SiO_2 multilayer waveguides. Since Si_3N_4 and SiO_xN_y are used as core materials, requiring low attenuation of light

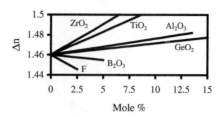

Figure 5.5 Modification of refractive index of SiO_2 for different dopant concentrations measured at 0.633-μm wavelength.

and adjustable refractive index, the optimal deposition conditions differ from those used in integrated circuit (IC) technology. There are two different approaches to generating the difference in refractive index Δn. One is based on low Δn, the other is based on high Δn. Each of them offers advantages and disadvantages in terms of fiber coupling efficiency and propagation loss.[7] Thus, the waveguides with Si_3N_4 as core layer have a large index contrast, greater than 0.55. This produces a tightly confined modal field inside the core medium and is well suited for matching to semiconductor lasers. The waveguides with SiO_xN_y as core layer have a lower index contrast, in the range from 0.05 to 0.55, improving the fiber-to-waveguide coupling efficiency. The silica-doped waveguides have a lower index contrast than that of multilayer waveguides because Δn is in the range from 5×10^{-3} to 0.05.

Generally, the dopants are similar to those used in optical fiber technology. The most common dopants producing an increase in refractive index are GeO_2, P_2O_5, N_2, As_2O_3, Al_2O_3, TiO_2, and ZrO_2. Dopants causing a decrease in refractive index are B_2O_3 and F. Figure 5.5 illustrates the refractive index change as a function of dopant concentration for some of the above materials.[8] The lines are least-square fits of published data. The refractive index of pure silica is 1.46 for an operating wavelength of 0.633 μm. Waveguides with doped SiO_2 as a core medium do not produce a confined modal field because of low index contrast. This is suitable for matching to single-mode fibers owing to the possibility of tailoring the mode-field profile of such waveguides to that of silica-based optical fibers. All Si_3N_4, SiO_xN_y, undoped and CVD techniques, or flame hydrolysis deposition may fabricate doped SiO_2 layers.

All waveguides are based on depositing silica layers for the cladding (thermally oxidized silicon is often used) and depositing a core layer commonly structured by reactive ion etching. During the past decade, a number of polymer optical waveguides on silicon have been fabricated. Low optical loss, high temperature stability, and adjustable refractive index have been demonstrated.[9] Because of their ease and low-temperature processing (~150°C) and compatibility with integrated circuit (IC) fabrication techniques, polymer waveguides should be considered as a potential optical material for integration with micromechanical elements. Polymer technology is relatively nonintrusive for MEMS micromachining because the polymer waveguide appears as an "add-on" component whose deposition and patterning may present in the future interesting fabrication compatibility issues. Waveguide fabrication based on each of the described methods, and suitable deposition techniques are described in Sections 5.2.5 and 5.2.7, respectively.

5.2.4 ANTIRESONANT REFLECTING OPTICAL WAVEGUIDE

As we have seen, the common approach to constructing optical waveguides on silicon is to grow a layer of pure silica and then to deposit on top of this layer another dielectric layer such as Si_3N_4 or SiO_xN_y having a substantially higher refraction index than that of SiO_2. Another approach is based on deposition of SiO_2 doped with elements that increase the index of refraction. One potential problem with this approach is that the deposition of a substantial thickness of silica buffer layer is required to reduce mode leakage losses into the silicon substrate, requiring a long deposition time. This can be overcome in the configuration of an antiresonant reflecting optical waveguide (ARROW), shown in Fig. 5.6.[10,11]

ARROW depends partially on antiresonant reflection rather than total internal reflection for wave guidance. The bottom silica layer ($d_2 \sim 2$ μm) is the second cladding, whereas the top silica layer ($d_1 \sim 4$ μm) acts as the core of the waveguide. The intermediate high-index layer, which is about 0.1-μm thick between both the silica regions is the first cladding. Light propagates through the SiO_2 core by repeating total internal reflection at the air/SiO_2 interface, whereas the high-index first cladding layer sandwiched between the two silica layers acts as an ultrahigh reflectivity interface. Thus, ARROW uses the silica as a core like that in optical fibers, providing a good match for fiber coupling. In the original configuration, ARROW involved polycrystalline silicon (poly-Si) as the thin first cladding. The thickness of this layer is chosen to be small to play the role of a Fabry–Pérot resonator, and high reflection (99.96%) occurs because of the antiresonant nature of this Fabry–Pérot cavity. The fundamental mode propagates in the thick core ($d_1 \gg \lambda/2n_1$) with a glancing angle incident from the SiO_2 core into the silicon layer, so the large index

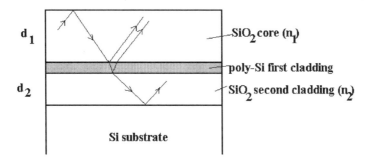

Figure 5.6 Structure of antiresonant reflecting optical waveguide.

discontinuity at the interface of core SiO$_2$/poly-Si generates a large reflectivity. Antiresonance of the thin poly-Si layer is obtained with a thickness t:[10]

$$t \approx \frac{\lambda}{4n_2}(2N+1)\left(1 - \frac{n_1^2}{n_2^2} + \frac{\lambda^2}{4n_2^2 d_1^2}\right)^{-1/2}, \quad N = 0, 1, 2, \ldots, \quad (5.4)$$

where λ is the wavelength and n_1 and n_2 are the refractive indices of the top and bottom silica layers, respectively.

Another feature of ARROW is its high polarization selectivity. TM modes have much higher optical attenuation than TE modes since TM reflections are always lower because of the same phenomenon that gives rise to the Brewster angle.

5.2.5 DEPOSITION METHODS

Silicon-based waveguides with multilayer structures or that have doped silica as a core material are widely used for the fabrication of optoelectronic devices. In fabricating these waveguides, the silicon substrate may first be thermally oxidized in order to grow the buffer layer of silica before the fabrication of the waveguide core. The core layer is obtained by deposition, for example, of silicon nitride, silicon oxinitride, or phosphorus-doped silica or undoped silica films. Suitable processes are low-pressure chemical vapor deposition (LPCVD), plasma-enhanced chemical vapor deposition (PECVD), germanium- or titanium-doped silica growth by flame hydrolysis deposition, as well as spin coating of polymer films on a silicon substrate. Some of the major deposition technologies used for fabrication of optical waveguides are described in the following sections.

5.2.5.1 Thermal Oxidation

High-quality, thick SiO$_2$ films may be grown on a silicon substrate by thermal oxidation. Since the silicon surface has a high affinity to an oxidizing ambient, thermal oxidation may be obtained in an oxidation furnace in a temperature range of 600 to 1250°C. When dry or wet oxygen is introduced into the chamber at a constant rate, the oxidation is obtained following two reactions:

$$Si + O_2 \rightarrow SiO_2 \text{ [dry process, furnace in Fig. 5.7(a)]}, \quad (5.5a)$$

$$Si + H_2O \rightarrow SiO_2 + 2H_2 \text{ [wet process, furnace in Fig. 5.7(b)]}. \quad (5.5b)$$

In this process, the silicon at the surface of the wafer combines with oxygen gas to form SiO$_2$. The Si/SiO$_2$ interface is moved into silicon, so the silicon slowly recedes from its original location.[6] The volume of silica, however, is larger than the volume of the silicon consumed, and the consequence is an increase in thickness. The ratio of these volumes is equal to 0.44, meaning that to grow a 1-μm SiO$_2$ layer, 0.44 μm of silicon is "consumed." Initially, the oxide growth rate is constant,

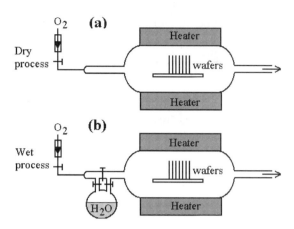

Figure 5.7 Tube furnace for thermal oxidation: (a) dry process, (b) wet process.

but as oxide grows, the diffusion of oxygen through the existing oxide is the rate-limiting step, and the thickness becomes proportional to the square root of the oxidation time.

Deal et al. defined a model that permits one to calculate the rate of oxide growth by the following equation:[12]

$$d_0^2 + Ad_0 = B(t + \tau), \quad (5.6)$$

where d_0 is the oxide thickness, A and B are constants, t is deposition time, and τ represents the time to account owing to the presence of an initial oxide film.

For short oxidation times, when $(t + \tau) \ll A^2/4B$, we have a linear law:

$$d_0 = \frac{B}{A}(t + \tau), \quad (5.7)$$

where B/A is the linear rate constant.

For long oxidation times, when $t \gg \tau$ and $t \gg A^2/4B$, we have a parabolic law:

$$d_0^2 = Bt. \quad (5.8)$$

Table 5.1 shows the temperature dependence of rate constants for wet and dry oxidation, respectively.[6]

These data were obtained for <111> oriented silicon. The optimal quality of silicon dioxide for optical waveguide applications is obtained by dry oxidation in a temperature range of 900 to 1150°C in pure oxygen. Since the dry oxidation time for 1 μm thickness of SiO_2 at 1000°C is 85 hr, growing thermal oxides by dry oxidation to a thickness greater than a few microns is difficult. Wet oxidation works 5–10 times faster, but generates a lesser quality oxide that is subject to diffusions

Table 5.1 Comparison of rate constants for wet and dry oxidation of silicon.

Oxidation temperature (°C)	Wet oxidation		Dry oxidation	
	A (µm)	B (µm^2/hr)	A (µm)	B (µm^2/hr)
1200	0.05	0.720	0.04	0.045
1100	0.11	0.510	0.09	0.027
1000	0.226	0.287	0.165	0.0117
920	0.50	0.203	0.235	0.0049
800			0.37	0.0011

Table 5.2 Deposition rates for silicon wet oxidation (in H_2O) at 640 torr.

Oxidation temperature (°C)	Orientation	Linear rate constant B/A (µm/hr)	B/A ratio <111>/<100>
900	<100>	0.150	
	<111>	0.252	1.68
1000	<100>	0.664	
	<111>	1.163	1.68
1100	<100>	2.977	
	<111>	4.926	1.65
			Average 1.675

of impurities. It was also demonstrated that the oxidation kinetics are strongly dependent on the position of the silicon surface. Table 5.2 shows the linear deposition rates for <100>, and <111> oriented silicon, for silicon oxidation in water at a pressure of 640 torr.[6]

These data present linear rates for <111> silicon that are an average of 1.675 times those of <100> oriented silicon at a corresponding temperature. This means that <100> silicon oxidizes 1.675 more slowly than <111> silicon. In anisotropic wet etching, the sequence of etch rate is inverted because a <100> silicon etches up 100 times faster than <111> silicon.

In conventional MEMS technology, the silicon dioxide obtained by thermal oxidation of silicon is commonly used as an insulating layer. In optical waveguide applications, the major disadvantage of thermal oxidation is the slow deposition rate and the fact that this process is not appropriate for the formation of doped oxides with an easily controlled refractive index. Even if silicon oxinitride core layers are fabricated by thermal nitration of SiO_2, the thermal oxide is mainly used for the formation of waveguide buffer layers.

5.2.5.2 Plasma-Enhanced Chemical Vapor Deposition

Chemical vapor deposition CVD processes originated in very large-scale integration (VLSI) microelectronics, and the principle of deposition is based on a surface

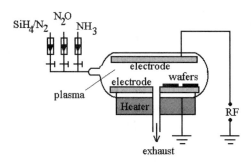

Figure 5.8 Parallel plate PECVD reactor.

chemical reaction of one or more reactant gases, where the thermal energy source is solely based on high-temperature reactors such as atmospheric pressure CVD (APCVD), LPCVD, or an additional energy source such as electrically excited plasma used by PECVD (plasma enhanced).

In PECVD, radio frequency (RF)-induced plasma activation provides thin-film decomposition at low operational temperatures because the transfer energy needed for deposition is transferred from a plasma to reactant gases. The discharge ionizes the gases, creating radicals that react at the wafer surface. Typically, pure silica, silicon nitride, and silicon oxinitride thin films can be fabricated using a conventional parallel-plate PECVD reactor, shown in Fig. 5.8. Substrates can be placed horizontally on a heated susceptor, which acts as one pair of RF electrodes.[13] A mixture of four processing gases, namely, silane, nitrogen, ammonia, and nitrous oxide, are supplied from a flow control system into the region between the electrodes generating RF-induced plasma in this region. PECVD pure silica films may be deposited by reacting silane (SiH_4) and nitrous oxide (N_2O) in argon plasma. N_2O produces atomic oxygen that reacts with SiH_4 to form SiO_2.[14] The PECVD process exhibits a disadvantage based on the presence of a large hydrogen content owing to the dissociation of silane in the plasma. Hydrogen is bonded to the silicon as Si—H bonds and the PECVD silicon nitride is in general an amorphous hydrogenated material (a-Si_3N_4:H).[15] This is usually deposited from a mixture of SiH_4 and NH_3, often diluted with a carrier gas such as He, Ar, or N_2.[14] It was demonstrated that the deposition of PECVD silicon nitride fabricated from a mixture of 2% SiH_4 in N_2, where N_2 instead of NH_3 is used as the source of nitrogen, permits the H content of deposited films to be reduced.[16] Even with low H content, PECVD silicon nitride films have a high mechanical stress, which may lead to cracking. Heat treatments after deposition could lead to a decrease in cracking resistance in the nitride films.[17] By introducing oxygen, the silicon nitride may be converted to silicon oxinitride. In the presence of oxygen, the mechanical stress can be lowered and the oxygen permits adjustment of the refractive index.

Usually, silicon oxinitride is formed by reacting silane and a mixture of several reactants (among them nitrous oxide, oxygen, nitrogen, and ammonia) in argon or nitrogen plasma. PECVD-grown oxinitride layers with excellent optical quality were obtained from an SiH_4-NH_3-N_2-N_2O gas mixture.[18] In PECVD-grown silicon oxinitride, the hydrogen is bonded to the silicon as Si-H and to the nitrogen as N-H. This is a stochiometric amorphous hydrogenated material (a-SiN_xO_y:H) with a hydrogen content from 10 to 35%, depending on the different reactors used.[19]

Si—H and N—H bonds cause absorption in the IR spectrum, and their second-order harmonics produce absorption peaks in the visible and near IR domain. This may create high waveguide losses. To obtain a low-attenuation waveguide, the hydrogen content needs to be reduced by using the appropriate reactant gas mixture and suitable adjustment of the processing parameters. It has been demonstrated that the deposition of PECVD oxinitride fabricated from a mixture of 2% SiH_4 in N_2 with N_2O as the source of oxygen reduces the H content of the film.[16] In this case, the H content is about half that of films deposited with ammonia as the source of nitrogen. The refractive index of these films is adjusted by varying the deposition temperature and N_2O flow. In terms of uniformity, N_2O was found also to be the better oxygen source for PECVD deposition of silicon oxinitride.

Finally, the stress of a-SiN_xO_y:H films is less than that of a-Si_3N_4:H. This improves the thermal stability and cracking resistance of deposited films. Particular attention is necessary to control the PECVD parameters influencing thin-film

Table 5.3 Deposition data of PECVD silica, silicon nitride, and silicon oxinitride films.

Material	Processing conditions	Refractive index/ uniformity	Deposition rate (nm/min)/ thickness uniformity	Optical loss (dB/cm)	References
SiO_2	Gas mixture: SiH_4-N_2-NO_2 $T = 300°C$ $P = 85$ Pa RF parameters: 187.5 kHz, 60 W	1.48	28	?	20
Si_3N_4	Gas mixture: SiH_4-NH_3-N_2 $T = 300°C$ $P = 33$ Pa RF parameters: 57 kHz, 0.17 W/cm^2	2.02	38 (4% on full wafer)	?	14
SiO_xN_y	Gas mixture: SiH_4-Ar-N_2-O_2 $T = 300°C$ $P = 5$ Pa RF parameters: 13.56 MHz, 200 W	1.46–1.8	5–7 (2% on full wafer)	2	21
SiO_xN_y	Gas mixture: SiH_4-Ar-NH_3-N_2O $T = 570°C$ RF power=60 W	1.52	?	<0.5	22
SiO_xN_y	Gas mixture: SiH_4-N_2-NH_3-N_2O $T = 300°C$ $P = 85$ Pa RF parameters: 187.5 kHz, 60 W	1.48–1.75	28–45	<0.5	20

properties and evaluating intrinsic stresses in those deposited films. Mechanically speaking, high stress can cause cracking of the film. Optically speaking, the strict control of stress may be useful in optical applications using the elasto-optic properties of waveguides obtained via stress-induced birefringence. In PECVD-grown silicon oxinitride films, the mechanical stress may decrease with increasing oxygen content.[16] High-temperature anneals up to 700°C may reverse the compressive stress character into a tensile type. Table 5.3 summarizes the processing data and optical properties of PECVD silica, silicon nitride, and silicon oxinitride films fabricated by several different gas mixtures.

Since the film quality depends strongly on processing conditions such as RF excitation, total pressure, reactant partial pressure, temperature, and gas mixture, each of these has a predictable effect on deposition results. The influence of these parameters on the properties of PECVD-grown films is investigated in Section 5.2.6.1.

5.2.5.3 Low Pressure Chemical Vapor Deposition

The LPCVD process is carried out in a high-temperature furnace where the surface reaction is obtained with a thermal energy source. LPCVD silicon nitride and silicon oxinitride films are grown at low pressure, below 130 Pa, and high temperatures between 700 and 950°C.[23] Under these processing conditions, nitrides and oxinitrides are stochiometric and mostly amorphous. These materials are also hydrogenated materials with low hydrogen content (up to 5%).[19] The low-pressure deposition is chosen to allow surface-catalyzed reaction of an otherwise highly reactive gas mixture. Figure 5.9 illustrates a horizontal "hot wall" LPCVD reactor where substrates are vertically placed on a quartz support. Processing gases are introduced from one extremity of the tube reactor, heated prior to arrival at the substrate by contact with the walls, and then absorbed on the surface when reacting to form the deposited thin film. Usually, two different processing gases, one the silicon-containing precursor and the other the nitrogen-containing source gas, are used in deposition of silicon nitride layers.[24] The usual silicon-containing gases are silane and dichlorosilane (SiH_2Cl_2). The usual nitrogen-containing sources are ammonia and pure nitrogen. Silicon oxinitride may be considered as a mixture of Si—O and Si—N bonds, where the refractive index can vary from that of silica (1.46) and those of silicon nitride (about 2). Reacting silane or dichlorosilane and

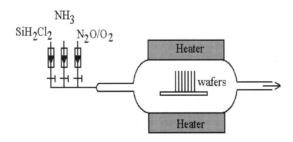

Figure 5.9 Horizontal "hot wall" LPCVD reactor.

Table 5.4 Deposition data of LPCVD silicon nitride and silicon oxinitride films.

Material	Processing conditions	Refractive index/ uniformity	Deposition rate (nm/min)/ thickness uniformity	Optical loss (dB/cm)	References
Si_3N_4	Gas mixture: SiH_2Cl_2-NH_3 $T = 800°C$ $P = 26$ Pa	2.01 (0.2% on 50 × 50 mm)	4.85 (2.3% on 50 × 50 mm)	0.2	23
SiO_xN_y	Gas mixture: SiH_2Cl_2-NH_3-N_2O $T = 850°C$ $P = 26$ Pa	1.46–2.0 (0.2–2.5% on 50 × 50 mm)	0.26–9.3 (2.3–7% on 50 × 50 mm)	0.2	23
SiO_xN_y	SiH_2Cl_2-NH_3-O_2 $T = 800$–$950°C$ $P = 3$–50 Pa	1.43–2 (<0.5% on full wafer)	4–40 (<3% on full wafer)	<0.5	26
SiO_xN_y	SiH_4-NH_3-NO_2 $T = 850°C$ $P = 65$ Pa	1.5–2	3.1–3.7	3	46

several mixtures of gaseous reactants such as ammonia or nitrous oxide can produce LPCVD-grown silicon oxinitride.[25] LPCVD silicon dioxide may be deposited by reacting silane or dichlorosilane with nitrous oxide. Table 5.4 summarizes the processing data and optical properties of LPCVD silicon nitrides and silicon oxinitrides fabricated using different reaction mixtures.

Successful deposition of silicon nitride films was reported for both SiH_4-NH_3 and SiH_2Cl_2-NH_3 gas mixtures.[23,26] In the deposition of silicon nitride, better results in terms of film uniformity and optical loss were obtained by reacting dichlorosilane and ammonia. For example, Gleine et al. obtained a refractive index and thickness nonuniformity across a wafer at 0.5% and 3%, respectively, under such conditions.[4,26] The typical value of optical loss is below 0.5 dB/cm at a 0.633-µm wavelength and a deposition rate larger than 25 nm/min. Silicon oxinitride layers have been deposited successfully from different gas mixtures, such as SiH_4-NH_3-O_2, SiH_2Cl_2-NH_3-N_2O, and SiH_2Cl_2-NH_3-O_2. The uniformity of the deposited film with respect to refractive index, layer thickness, and optical losses is similar to that of silicon nitride. Throughout the entire refractive index range, the deposition rates depend strongly on processing conditions (temperature, pressure) and vary with the gas input ratio of NH_3/N_2O that defines the refractive index value of the deposited layer. Table 5.5 summarizes the properties of LPCVD- and PECVD-deposited films.

The high temperatures and low deposition rates of the LPCVD process may help the formation of single-crystalline materials, improving the micromechanical properties of deposited thin films, but increasing the optical losses. Thus, Gleine et al. fabricated LPCVD silicon nitride films by reacting SiH_4 and O_2-NH_3 mixtures beyond 800°C. These films presented polycrystalline behavior exhibiting important optical losses as high as 30 dB/cm.[26] For lower temperatures (750°C), such

Table 5.5 Comparison of LPCVD and PECVD methods.

	LPCVD	PECVD
Materials	SiO_2, Si_3N_4, SiON, P-glass, BP-glass, poly-Si	SiO_2, SiON
Temperature/ pressure	550–900°C / below 130 Pa	300–400°C / below 650 Pa
Uses	Core of waveguides, insulation, gate metal, doped and undoped films, structural layers	Core/buffer/cover of waveguides, insulation, passivation (nitride)
Advantages	High-purity and high-uniformity films, high reproducibility, uniform step coverage, low optical loss	High-uniformity films, moderate reproducibility, good coverage step, low optical loss, low temperature, good adhesion, low mechanical stress, fast deposition
Disadvantages	High-temperature, low deposition rate, high mechanical stress	Moderate-purity films (hydrogen content), moderate reproducibility

films are amorphous and present an optical attenuation below 2 dB/cm. A major advantage of LPCVD over PECVD is the higher uniformity and better reproducibility of deposited films. This is partially due to the low content of atomic hydrogen in the deposited thin films. Two main disadvantages of LPCVD are the low deposition rate and relatively high operating temperatures. Two major attributes of PECVD are the relatively high average deposition rate (30–40 nm/min) and the low-temperature processing in the range of 300 to 400°C. Such a range of processing temperatures is compatible with well-established microelectronic processing and permits the deposition of thin films over the final device metallizations. PECVD is also suitable for deposition on the optical chips of components, such as GaAs or InP lasers.

5.2.5.4 Deposition of Phosphosilicate Glass by LPCVD and PECVD

An extension of CVD techniques would be the preparation of phosphorus-doped silica films obtained by introduction of few percent of phosphine (PH_3) into the LPCVD or PECVD process. Adjusting the amounts of phosphine incorporated during CVD can control the refractive index of silica film. Since there is a linear relation between phosphine concentration and increase in refractive index, very small index differences can be achieved. When the phosphorus concentration in SiO_2 increases, the refractive index also increases. For a typical process, a 1% increase in the weight of phosphorus increases the index by roughly 0.003.[27] In addition, since optical fibers are made from similar materials, phosphorus-doped waveguides on silicon offer high-power fiber-to-waveguide coupling efficiency. Finally, this doping technique offers a serious reduction in complexity over more conventional doping techniques, which are usually based on the postdeposition implantation of the dopant, or an infusion of dopants from a dopant-rich film applied to the substrate surface.[28]

Phosphorus-doped silica may be deposited at temperatures lower than 500°C, reacting silane, phosphine, and oxygen.[6] This reaction can be carried out at reduced pressure in an LPCVD reactor. The advantage of this deposition is the low operating temperature and fabrication of phosphorus-doped silica films with mostly compressive stress. The disadvantage is the poor step coverage due to the silane-oxygen reactions. Adding phosphine to the silane LPCVD process results in severe reduction of film deposition rates, accompanied by an altering of the thickness uniformity of the deposited films. This can be corrected by heating the sample. By this process, called P-glass flow (or PSG-flow), the oxide softens and flows, providing a smoother topography and reducing the surface roughness that causes scattering losses. P-glass flow improves the stability of deposited films because of the tendency of the glass to absorb water vapor from the environment.[29,30] P-glass flow requires a temperature range from 950 to 1100°C, depending on the phosphorus concentration, which is usually from 4 to 8 wt.%. Thermal annealing subsequent to the deposition of P-glass has a marked effect on decreasing the optical loss of waveguides. Usually, after PSG films are deposited on a thermally oxidized silicon substrate, the films are heat treated in various ambients; annealing with nitrogen and annealing in oxygen containing water are the most effective in causing the P-glass to flow.

Meyerson et al. deposited LPCVD phosphorus-doped silica layers by adding 2.5% phosphine to the silane gas source with argon as the inert purge gas.[28] Deposition was carried out at 623°C and a pressure of 0.1 torr. Thin films prepared in this manner are shown to be polycrystalline.

Pure silica can be also deposited at 650 to 750°C in an LPCVD reactor by decomposing tetraethoxysilane [$Si(OC_2H_5)_4$], also called TEOS, instead the traditional silane-based reaction.[6] TEOS-based depositions lead to improved film quality in terms of step coverage and reflow properties. The disadvantage is the high operating temperature. With a PECVD reactor, the deposition temperature for TEOS is as high as 300 to 400°C. This PECVD process offers the same good step coverage compared with low-pressure processes, presents higher film density, and exhibits relatively neutral stress. Phosphosilicate TEOS glass, called P-TEOS glass, can be formed from LPCVD by adding small amounts of phosphine (2 to 5 wt.%) in tetraethoxylane and oxygen.[31]

The fabrication of optical waveguides on P-glass and P-TEOS glass has two natural advantages. First, there is a linear relation between phosphine concentration and an increase in refractive index. Consequently, very small index differences can be achieved by precisely controlling dopant amounts. Second, the resulting phosphorus-doped waveguides on silicon offer high-power fiber-to-waveguide coupling efficiency.

5.2.5.5 Flame Hydrolysis Deposition

Nippon Telephone and Telegraph (NTT) developed "high silica technology" in which thick layers of silica are deposited on a silicon substrate by flame hydrolysis, a technique originally used for fiber preform fabrication.[7,32] By this technology, silica is fabricated by a hydrolysis reaction of $SiCl_2$ in an H_2/O_2 glass burner:

$$SiCl_2 + 2H_2O \rightarrow SiO_2 + 4HCl. \qquad (5.9)$$

In practice, a mixture of silicon tetrachloride ($SiCl_4$), titanium chloride ($TiCl_4$), or $SiCl_4$ and germanium tetrachloride ($GeCl_4$) are fed into an oxygen–hydrogen torch, as shown in Fig. 5.10.[7] The torch distributes over the substrate small silica particles, called "soots," which are deposited on a silicon wafer placed on a rotating turntable that permits homogeneous surface coverage. A turntable of 100-cm diameter can hold a large number of 3-inch silicon wafers (about 30). A variety of dopants such as BCl_3 and PCl_3 (in a mixture of $SiCl_4$-$TiCl_4$) may be incorporated in the gas mixture to lower the melting point of the synthesized glass particles. The refractive index of synthesized glass particles can be controlled by adjustment of the $TiCl_2$-$GeCl_4$ flow rate. Thus, the composition of each deposited layer may be varied through variations of gas flow, and graded-index waveguides can be fabricated. The deposited material is porous. After deposition, heating in an electric furnace at 1200 to 1300°C consolidates the glass layer. The consolidation produces glasses of optical quality with a resulting film shrinkage of as much as 90%. The FHD technique is capable of depositing thicknesses as high as 100 μm. The process requires high-temperature heating for consolidation, which can be incompatible with other thin-film operations. In this case, the FHD process should be used with the standard MEMS micromachining only as a postprocessing of the wafer.

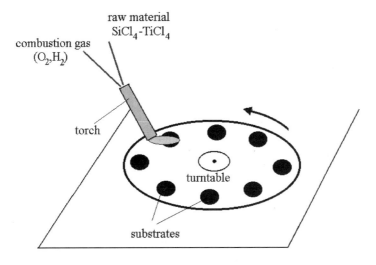

Figure 5.10 Arrangement for FHD process.

5.2.5.6 Polymer Deposition by Spin-Coating Technique

Polymer materials can be considered as promising candidates for constructing waveguide devices with adjustable refractive indices, low optical propagation loss, and good thermal and environmental stability. Several polymeric films have been investigated, including polymethyl methacrylate (PMMA), polystyrene, polyimide, polycarbonate, and polysiloxane.[9,33] Many of them can be fabricated very simply by a spin-coating method on a silicon substrate. To illustrate this deposition technique, the fabrication sequence of PMMA-based film will be described. This polymer is a widely used polymeric material in optical applications. The processing starts by dissolving PMMA using inorganic solvents with a high boiling point, such as cyclohexanol or xylene.[34] The appropriate choice of solvent is crucial because the polymer is allowed to dry and the solvent evaporation must be relatively slow to avoid an orange-peel surface effect, introducing scattering losses. A viscous solution of the polymer is dispensed onto a wafer lying on a wafer platen, as shown in Fig. 5.11. A vacuum chuck holds the wafer substrate in place. The wafer is then spun at high speed (number of rotations/min in the range of 1500–8000 rpm) to make a spin-coated, perfectly uniform film. Centrifugal forces cause the solution to flow to the edges, where it builds up until it is expelled when the surface tension is exceeded. The resulting film thickness is proportional to the spin-coater speed, the concentration of the solution, and the viscosity of the film:[13]

$$T = \frac{K C^B \chi^\gamma}{\omega^\alpha}, \tag{5.10}$$

where K, α, β, γ are calibration constants; C is the polymer concentration; χ is the viscosity; and ω is the number of revolutions per minute.

After spin coating, the film contains a certain proportion of solvent that must be removed from the polymer by evaporation. This can be done by baking (also drying in laminar nitrogen flow) the wafer at 80–85°C in a vacuum atmosphere for 30 to 60 min. For pure PMMA films fabricated by this technique, the refractive index is constant at 1.489, and optical waveguides operating at a 1.3-µm wavelength have

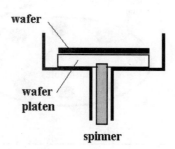

Figure 5.11 Arrangement for spin-coating deposition.

roughly a 0.2-dB/cm loss. Ulrich et al. demonstrated the fabrication of PMMA-based films with an adjustable refractive index by blending the PMMA in solution with a styrene-acrylonitrile copolymer (SAN).[35] The mixing of these polymers is compatible on a molecular scale because the composite system has a single glass transition temperature between that of PMMA (80°C) and that of SAN (105°C). The resulting film is homogeneous, exhibiting low propagation loss (0.2 dB/cm) and a linear dependence of the refractive index on the mixture composition (1.489–1.563). The index of refraction measured as a function of the compositions of PMMA and SAN is shown in Fig. 5.12.[34] This curve was obtained with samples baked 24 hr at 80°C temperature. More recently, NTT workers demonstrated interest in using modified PMMA polymer as waveguide core material in devices for optical fiber communication systems. Deuterated and fluorinated PMMA was used to form waveguides, and optical losses below 0.1 dB/cm at a 1.3-μm wavelength were obtained.[36] In fluorinated PMMA, a highly controllable refractive index is obtained by adjusting the fluorine content during the copolymerization, because the refractive index decreases with increasing fluorine content.

Nevertheless, there are two major limitations of PMMA-based thin films. One is that the propagation loss at a 1.55-μm wavelength is higher than 1.5 dB/cm, owing to carbon-hydrogen (C−H) bond vibrational absorption. High performance in this infrared region is particularly crucial in optical fiber telecommunication, which is actually the main application field of polymer waveguides. Another problem is the limited thermal stability of PMMA (<80°C), which is much less than that of other materials such as silica. This is due to the extremely low glass transition temperature of PMMA. To solve these problems, several new polymeric materials were recently synthesized, such as deuterated polysiloxane with improved environment stability and low optical losses in the entire 1.3–1.55-μm wavelength range.[37]

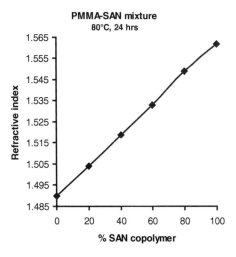

Figure 5.12 Measured index of refraction as function of compositions of PMMA and SAN.

Deuterated polysiloxane is obtained by hydrolysis of deuterated phenylsilyl chloride ($C_6D_5SiCl_3$) monomers, followed by condensation. The chemical structure of the new polymer presents two major improvements. One is the substitution of deuterium for hydrogen in the phenyl group (C_6D_5); another is the presence of a polysiloxane backbone $-Si-O-$. Since the phenyl group has a large molecular refraction, the refractive index of the polymer can be controlled by adjusting the content of the phenyl group. Thus, the refractive index can be controlled in the 1.48–1.6 range with a precision of 0.001. For example, an 8 mol.% increase in the content of the phenyl group increases the refractive index of 0.0046. On the other hand, the substitution of deuterium for hydrogen in the phenyl group generates lower absorption than the normal nondeuterated phenylsilyl chloride ($C_6H_5SiCl_3$) monomer. Because of the presence of $C-H$ bonds, the vibrational absorption peaks of this new polymer are shifted to higher wavelengths, with a maximum absorption at a 2.3-µm wavelength. The propagation loss is less than 0.2 dB/cm at 1.3 µm and below 0.5 dB/cm at a 1.55-µm wavelength. Compared with carbon-carbon ($C-C$) bonds, the adoption of a polysiloxane backbone constitutes an improvement of environmental stability. A heat resistance test demonstrated that the refractive index of deuterated polysiloxane film deposited by spin coating on a silicon wafer changed less than 0.1% after being heated at 120°C for 1000 hr. The propagation loss at a 1.3–1.55-µm wavelength range of this film remained unchanged after heating at 75°C for 1000 hr under 90% relative humidity (RH) (humidity resistance test), demonstrating a long-term stability in a humid atmosphere.

5.2.6 PROPERTIES OF DEPOSITED FILMS

In MEMS technology, a wide variety of materials are being used, including semiconductors, insulators, metals and alloys, and polymers as well as special functional materials. Materials properties are in general dependent on the sample size and the microstructure. For example, the mechanical deformation of a thin layer of material used as a membrane is based on different deformation mechanisms than a macroscopic membrane. Because of the miniaturization and the use of various thin films, the micromechanical properties of MEMS are different than their bulk mechanical counterparts and are more strongly dependent on the fabrication processes. Interface effects in MEMS also tend to be different than their bulk counterparts. In particular, the nucleation and movement of dislocations in ductile materials is strongly depressed in microstructures, which can lead to catastrophic failure after relatively short times in comparison with macroscopic systems. Additional internal stresses and thermal stresses induced by a thermal mismatch at the interface between different materials strongly influence the mechanical stability and long-term performance of such microparts. However, the mechanical properties governing elasticity, strength, creep, fatigue, and fracture resistance are controlled by defects such as grain or interface boundaries, and pores typically in the submicrometer range.

To summarize, the micromechanical characteristics of structures fabricated by MEMS micromachining critically depend on the residual stress in thin films as well as on the stress variation in the direction of deposition. The integration of optical waveguides on MEMS microstructure films must satisfy a large set of rigorous chemical, structural, mechanical, and optical characteristics. In Sections 5.2.6.1 and 5.2.6.2 it is demonstrated that all material properties of thin films depend strongly on the deposition process and growth conditions. Experimental results on specific properties of deposited films are presented, with special emphasis to explain the interdependencies between thin film composition, film mechanical stress, refractive index characteristics, and optical birefringence.

5.2.6.1 Influence of Processing Conditions on Thin Film Quality

In the following paragraphs, a description of dependencies of thin films on processing conditions is given. In the particular case of a PECVD process, the quality of deposited thin films depends strongly on reactor parameters such as RF excitation, total pressure, reactant partial pressure, temperature, and gas mixture, each of them having a predicable effect on deposition results. This dependence on deposition conditions makes it difficult to compare films fabricated by different reactors. All deposition conditions must be carefully specified when discussing the properties of PECVD films. Some of different deposition variables have the following influence on PECVD-deposited thin films:[13]

Deposition temperature: The deposition temperature strongly influences the properties of thin films. At low temperatures and high deposition rates, the diffusion of molecules on the substrate surface is slow, generating amorphous films. At higher temperatures and low deposition rates, the process helps the formation of single-crystalline materials. This last situation improves the micromechanical properties of deposited thin films, but may be not optimal in terms of optical losses. How to build a compromise between the micromechanical properties of microstructures and deposition of low-loss optical waveguides will be explained in Section 5.2.6.2.

RF excitation conditions: RF excitation frequency combined with appropriate RF power permits the quality of deposited films to be improved. Low RF excitation frequency promotes the fabrication of mostly compressive films. An increase in RF power produces more intense ion bombardment. The deposition rate increases linearly with RF power. High film density is obtained with low RF power.

Total reactor pressure: Deposition at low reactor pressure and a low deposition rate improve the quality of thin films in terms of high film density and low compressive stress. Experiments demonstrated that when the reactor pressure is decreased, the stress goes from a tensile situation to a compressive one. Too high pressures tend to increase the density of defects of deposited materials. On the other hand, too low pressures also have negative effects on the thin film quality because of low cracking resistance due to columnar films with defects.

To summarize, low deposition rates obtained at high temperatures, low RF frequencies, and low reactor pressures improve deposited film quality, as evidenced by stress evaluation of different deposition parameters. Claassen et al. investigated how thin film quality is influenced by mechanical stress. They demonstrated that the stress of PECVD-grown silicon nitride films at a temperature of 300°C can be lowered by increasing total pressure and/or RF frequency.[17] At frequencies above 4 MHz, the stress even becomes tensile. However, increasing the total pressure and increasing RF frequencies lead to silicon nitrides with low densities, high hydrogen content, and low cracking resistance.

5.2.6.2 Influence of Mechanical Stress on Thin Film Quality

Mechanical stresses in thin films are mainly caused by the difference in thermal expansion of the substrate and the grown layer. The stress may be compressive or tensile. Compressive stress produces convex bending, and tensile stress causes concave bending of a thin substrate (Fig. 5.13). The stress developing in a film during the initial step of a CVD process may be compressive, tending to expand parallel to the substrate. This may cause buckling of the substrate. Alternatively, the stress may be tensile, tending to contract the film parallel to the substrate. This may cause the film to crack. A subsequent rearrangement of atoms, owing to the remainder of the deposition or addition of thermal processing such as annealing, can modify tensile or compressive stresses, respectively. The stress of the film can be determined by measuring the bending of the silicon wafer. This bending needs to be measured before and after processing. The stress of the film and the resulting film curvature are given by the Stoney formula:[13]

$$\sigma = \pm \left| \frac{E}{6(1-\nu)} \frac{d^2}{t} \left(\frac{1}{R_a} - \frac{1}{R_b} \right) \right|, \qquad (5.11)$$

where E and ν are the Young's modulus and the Poisson ratio of the substrate, R_a and R_b represent curvature radii after and before deposition, d is the film thickness, t is the wafer thickness, and $+$ and $-$ denote the tensile and compressive nature

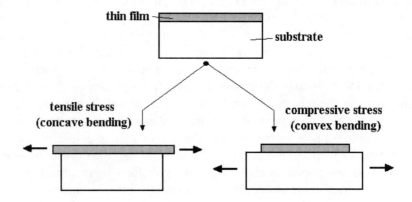

Figure 5.13 Schematic effects of compressive and tensile stresses.

of the stress. For most films we assume that $d \ll t$; for example, d/t measures 200–1000 for thin films on silicon. The term $E/(1-\nu)$ is the biaxial modulus of the substrate.

Stress in silicon dioxide depends on deposition parameters, but also on dopant concentration, porosity, and water content. Stress in oxides is strongly influenced by impurities. Oxides often are porous enough to absorb water. The stress is more compressive when phosphorus is added, but it is more tensile if the film contains water. Undoped PECVD silicon dioxide has been reported with a compressive stress from 0.1 to 4×10^9 dyn/cm^2 (10^7 dyn/cm^2 = 1 MPa), and thermally grown silicon dioxide exhibits a compressive stress from 3 to 4×10^9 dyn/cm^2.[38]

Stress in PECVD silicon oxinitride is mostly compressive and rapidly decreases with increasing oxygen content. Figure 5.14 shows the stress of the SiN_xO_y film, plotted as a function of N_2O flow in a PECVD reactor, operating at a constant flow of 2% SiH_4 in N_2 with N_2O as the source of oxygen.[16] Reactor pressure is 0.36 torr and RF power is 0.7 kW. The films were deposited at different temperatures of 250, 300, and 350°C. The films are all compressed in the absence of oxygen, become less compressed, and start to be tensile as the N_2O flow increases. A decrease in the refractive index and a decrease in the hydrogen content of the deposited film accompany the increase in N_2O flow. With less hydrogen content (or decreasing refractive index), SiN_xO_y films have lower stress.

Figure 5.15 illustrates this dependence between the stress and H content for films deposited with refractive indices fixed at 1.85 and 1.75, respectively.[16] With less hydrogen content, the stress becomes less compressive. The fabrication of PECVD silicon oxinitrides with refractive indices in the range 1.65–1.85 was reported; all films were deposited with compressive stresses from -1 to -2×10^9 dyn/cm^2.[39] For instance, high-temperature annealing leads to hydrogen reduction and shifts the nature of the stress toward a more tensile form. Physically this increases the density of the deposited film, driving the stress from a compressive situation toward a more tensile one. This is illustrated in Fig. 5.16, in which the variation in stress is plotted as function of annealing temperatures (400, 500, and 600°C).[18]

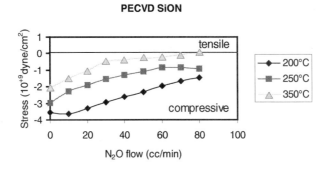

Figure 5.14 Stress of the SiN_xO_y film plotted as function of N_2O flow in the PECVD reactor (SiH_4-N_2-N_2O mixture) at 200, 250, and 350°C.

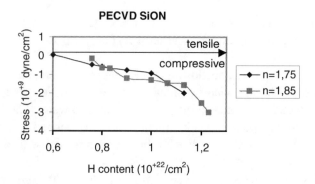

Figure 5.15 Dependence between the stress and hydrogen content for SiN_xO_y films deposited with refractive index values at 1.85 and 1.75.

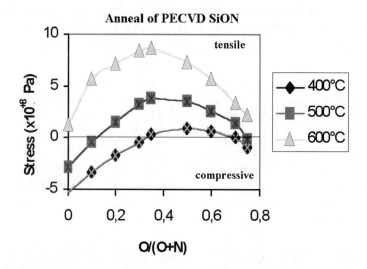

Figure 5.16 Effects of anneal process on stress in SiN_xO_y films for annealing temperatures at 400, 500, and 600°C.

Stress in silicon nitride is mostly tensile. LPCVD silicon nitride films are tensile, with a stress about 10^9 dyn/cm^2. The values for the stress in PECVD nitrides are between 2×10^9 dyn/cm^2 (compressive nature) and 5×10^9 dyn/cm^2 (tensile nature).[6] By increasing the silicon content in LPCVD silicon nitrides, the tensile stress is reduced (even to compressive). This may be obtained by increasing the SiH_2Cl_2/NH_3 flow ratio.

All crystal silicon substrates exhibit a moderate degree of anisotropy with regard to their physical properties, including biaxial stress and refractive index. For <100> oriented silicon substrates, the biaxial modulus of Eq. (5.11) is $E/(1-\nu) = 1.8 \times 10^{11}$ Pa and 2.3×10^{11} Pa for <111> oriented silicon, respectively.[18] The presence of biaxial stress causes optical birefringence. This is an anisotropism of the refractive index that varies as a function of polarization as

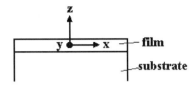

Figure 5.17 Placement of E-field coordinate systems.

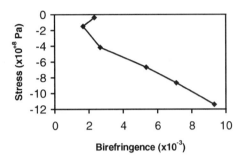

Figure 5.18 Dependence of birefringence from mechanical stress in SiN$_x$O$_y$ films (LPCVD processing with N$_2$O and O$_2$).

well as orientation with respect to the incident light. In the geometry of Fig. 5.17, the change in the refractive index seen by fields polarized in the $x-y$ and z directions is:

$$\Delta n_{x-y,z} = n_{x-y} - n_z, \qquad (5.12)$$

where n_{x-y} and n_z represent the refractive indices corresponding to electrical fields parallel (TE mode) and perpendicular (TM mode) to the substrate, respectively.

The stress, for example, in doped silica waveguides on silicon is compressive because the thermal expansion coefficient of silicon is larger than that of doped silica.[7] In this category of waveguides, a stress-induced birefringence $\Delta n_{x-y,z}$ of about 5×10^{-4} with an effective index for TM polarization larger than that for TE polarization is observed. Such magnitude of birefringence is comparable to those of polarization-maintaining optical fibers. In silicon oxinitride and nitride films grown by LPCVD in the presence of nitrous oxide and oxygen, the stress is tensile for refractive indices above 1.6.[23] For these processing conditions, the dependence of birefringence on mechanical stress is shown in Fig. 5.18, demonstrating that silicon oxinitride is an elasto-optic material.

In conclusion, the precise control of stress-induced birefringence is sometimes useful for constructing waveguides with polarization-sensitive or insensitive operations. Another application of stress-induced birefringence is as a measuring principle in integrated optical sensors to evaluate external parameters such as force, pressure, or displacement. A portion of the optical waveguide is deformed and this deformation is used to determine the amount of deformation. The deformation of structure leads to an elasto-optic effect in the waveguide and influences the propagation constant of the guided mode. Both approaches to birefringence control are discussed in Sections 5.2.7.3, 5.2.7.4, 5.4.2.2, and 5.4.3.3.

5.2.7 FABRICATION OF CHANNEL WAVEGUIDES

5.2.7.1 Waveguide Categories

Channel waveguides on silicon substrates may be formed by a variety of methods. In topographic waveguides, photolithographic pattern definition processes followed by anisotropic etching (e.g., reactive ion etching) may form cores of rectangular cross-section. Depending on the material used, the core may be reflowed to decrease the surface roughness and reduce optical losses by scattering in the final waveguide. The reflow process typically requires high-temperature processing (1100°C for phosphosilicate glasses). In some cases the incompatibility of reflow with metallizations and other thin-film processes precludes the addition of this high-temperature operation to the MEMS micromachining process; it should has carried out only in a postprocessing of the wafer. The fabrication of channel waveguides using the CVD process, silica-doping techniques, FHD, and ARROW technology will be given in Sections 5.2.7.2–5.2.7.5.

No topographic waveguides may be formed by local densification by irradiation (laser irradiation or electron-beam irradiation), ion exchange, or selective diffusion of dopant (polymer waveguides), modifying locally the refraction index profile of the core layer. Fabrication of buried waveguides using electron-beam irradiation of silica was reported by Barbier et al., leading to a slab waveguide geometry with the lowest loss of about 0.3 dB/cm at a 0.633-μm wavelength.[40] Densification is caused by a combination of heating and bond breaking and is obtained by irradiation of samples with an electron accelerating voltage of 25 keV. Ion exchange is a local thermal diffusion process with dopant ions such as thallium and silver. The substrate is an optical glass containing alkaline oxides immersed in the dopant salt bath at temperatures of 300 to 400°C. Channel waveguides have been formed by both CVD and ion-exchange processes, with losses as low as 0.05 dB/cm.[41] Irradiation techniques and ion-exchange methods are somewhat restrictive in regard to compatibility with MEMS micromachining, and only polymeric waveguide fabrication will be described in Section 5.2.7.6.

5.2.7.2 PECVD and LPCVD Silicon Oxinitride Channel Waveguides

In this chapter we are mainly interested in optical waveguides that have a good match with optical fibers. Silica waveguides with silicon nitride as core layer have a large refractive index contrast (∼0.5) and are not suitable for matching with optical fibers. Hence waveguides with silicon oxinitride as a guiding layer have a smaller index contrast that can be varied between 0.005 and 0.5, these waveguides are suitable for matching with optical fibers.

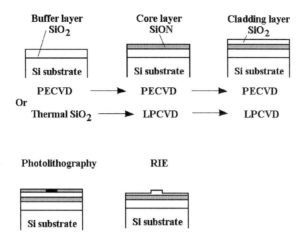

Figure 5.19 Fabrication sequences of a typical oxinitride strip-loaded waveguide.

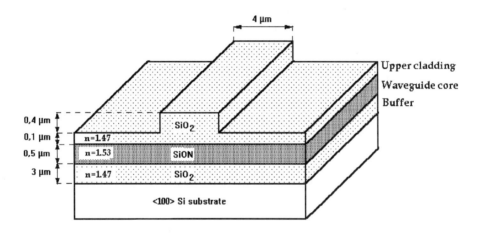

Figure 5.20 Basic structure of a single-mode silicon oxinitride strip-loaded waveguide operating at 0.633-μm wavelength.

Two types of deposition processes have been developed and optimized for fabrication of SiO_xN_y channel waveguides with regard to low optical losses: PECVD in parallel-plate reactors and LPCVD in "hot wall" reactors. Both methods are comparable in terms of optical performance. The fabrication sequences of a typical oxinitride strip-loaded waveguide are shown in Fig. 5.19. This high refractive index contrast waveguide is used in sensor applications (i.e., temperature, pressure, and distance measurements). In this case, the dependence of the refractive index variations on temperature effects or externally induced mechanical stresses (i.e., bending of a waveguide that is crossing a membrane) are used as the principle of optical interrogation obtained via interference fringe analysis.

Figure 5.20 depicts the basic structure of a single-mode silicon oxinitride strip-loaded waveguide, operating at a 0.633-μm wavelength. Bonnotte et al. fabricated

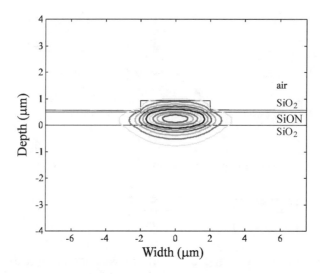

Figure 5.21 Optical field distribution of the strip-loaded waveguide of Fig. 5.20 calculated using 3D beam propagation method.

this structure by a PECVD technique.[20] The core of this high-index contrast waveguide ($\Delta n = 3.4\%$) is a PECVD SiO_xN_y layer with refractive index $n = 1.53$ and a thickness of 0.5 µm. The core layer is sandwiched between two PECVD SiO_2 cladding layers with refractive index $n = 1.48$. A 4-µm wide rib is etched in the SiO_2 upper cladding layer. This overlayer structure laterally confines the optical field. For the strip-loaded waveguide of Fig. 5.20, the calculated optical field distribution is shown in Fig. 5.21. This calculation was performed for TE polarization using a 3D beam propagation method (BPM). This confirms an excellent confinement of the optical field for the previously described structure and permits visualization of the lateral size of the guided TE_{00} mode. The waveguide is single mode up to a 6-µm width. Using the configuration of a horizontal parallel-plate PECVD reactor (Fig. 5.8), the operating RF frequency was at 187.5 kHz, the plasma power at 60 W, the chamber pressure at 85 Pa, and the substrate temperature at 300°C. Refractive indices of the layers to be deposited are controlled by changing the ratio between the processing gases, 2% SiH_4 diluted in N_2, N_2O, and NH_3.

Figure 5.22 shows the refractive index and deposition rate of silicon oxinitride as a function of the silane/nitrogen to nitrous oxide flow ratio. This technique allows fabrication of silicon oxinitride films with an index range from 1.48 to 1.75 and with deposition rates of 28–45 nm/min. SiO_2 layers have been deposited with 200 sccm of 2% SiH_4/N_2 and 710 sccm N_2O, resulting in a thin film with a refractive index of 1.48. A silicon oxinitride layer having a refractive index of 1.53 has been grown by applying 1000 sccm 2% SiH_4/N_2, 900 sccm N_2O, and 6 sccm NH_3. In the upper SiO_2 cladding, a 4-µm-wide rib is structured with a resist mask in a reactive ion etching process using CHF_3 plasma. Typically, the optical losses are less than 0.5 dB/cm. A single-mode inverted-rib oxinitride waveguide with the con-

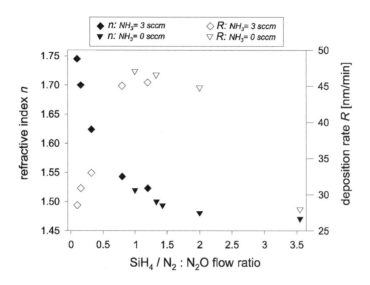

Figure 5.22 Refractive index and deposition rate of PECVD SiN_xO_y film as function of SiH_4/N_2 to N_2O flow ratio.

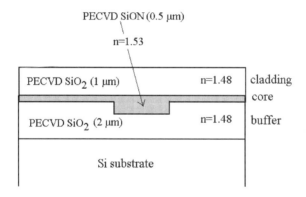

Figure 5.23 Structure of single-mode inverted rib oxinitride waveguide.

figuration of Fig. 5.23 was fabricated using the same PECVD-based technology.[42] The waveguide core is a PECVD SiO_xN_y thin film 0.5 µm thick. This core is sandwiched between a 2-µm-thick buffer layer and a 1-µm-thick cladding layer of PECVD SiO_2. In this configuration, a metal mask (Cr) is used as a stop etch for RIE etching of a 4-µm-wide groove in the SiO_2 buffer layer. The measured optical losses are less than 0.75 dB/cm.

A PECVD-based technology was also used to form buried oxinitride waveguides with high refractive index contrast ($\Delta n = 3.5\%$) and small core size, as shown in Fig. 5.24.[43] This waveguide is well adapted to be mode matched to standard single-mode fibers for optical communication applications (1×8 power splitters, thermo-optical switches, Wavelength Division Multiplex (WDM)). In this

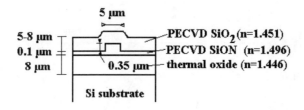

Figure 5.24 PECVD oxinitride buried waveguide with refractive index contrast $\Delta n = 3.5\%$.

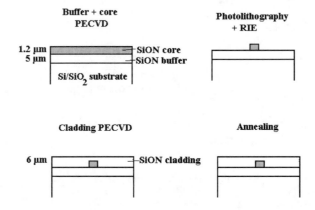

Figure 5.25 Fabrication by PECVD of a single-mode SiO_xN_y buried waveguide.

case, the silicon wafer is covered with 8.5–11 μm thermal oxide or PECVD oxide. A 0.35-μm-thick $Si_xO_yN_z$ core is deposited using SiH_4-NH_3-N_2O or SiH_4-N_2O mixtures operating in Ar plasma in a 13.56-MHz PECVD parallel-plate reactor. The core layer is patterned with a resist mask in an RIE etch using a CHF_3-O_2 gas mixture. A layer of 5–8 μm PECVD SiO_2 is used as a cladding. The measured waveguide attenuation is less than 0.2 dB/cm at a 1.3-μm wavelength and the fiber-waveguide coupling loss is about 0.5 dB per facet.

Imoto et al. reported the fabrication of a single-mode buried waveguide that consists of the buffer, core, and cladding layers, all formed with PECVD SiO_xN_y film.[44] The refractive index contrast Δn is obtained by adjusting the nitrogen concentration. The fabrication sequence is illustrated in Fig. 5.25. The buffer, core, and cladding layers are deposited using a PECVD process operating at a temperature of 270°C. The refractive index of these layers is modified by adjusting the gas flow of SiH_4-N_2O-N_2 mixture under plasma. A high refractive index contrast of 2% is obtained with core thickness of 1.2 μm and a refractive index of 1.5605. The thicknesses of the buffer and cladding layers are 5 and 6 μm, respectively. The core ridge is defined and patterned through conventional photolithography and RIE patterning. To stabilize this structure and decrease the optical losses, an annealing process at 500°C under nitrogen atmosphere was performed. The measured propagation loss is 0.12 dB/cm with a fiber-to-waveguide coupling loss of 3.1 dB/connection

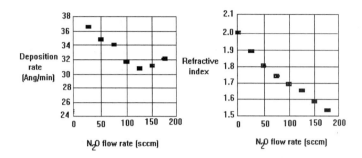

Figure 5.26 Refractive index and deposition rate of LPCVD SiN_xO_y film as function of SiH_4/N_2 to N_2O flow ratio.

at 1.3-μm wavelength. The main advantage of this waveguide is that this SiO_xN_y triple layer presents a very low intrinsic stress because of low differences of thermal expansion between the core, cladding, and substrate.

To summarize, by using PECVD techniques, buffer/core/cladding triple layers may be deposited in one continuous run. In the case of an $SiO_2/SiO_xN_y/SiO_2$ sandwich, structures with low stress can be fabricated. An $SiO_2/SiO_xN_y/SiO_2$ sandwich formed by LPCVD generates higher stress, increasing the optical losses. This effect can be reduced when SiO_xN_y LPCVD starts from a thermally grown SiO_2 buffer layer because the stress is compensated when a thermal SiO_2 is sandwiched between an SiO_xN_y layer and the silicon substrate. Hamburg-Harburg University workers formed low-loss (0.2–0.3 dB/cm at a 0.633-μm wavelength) single-mode strip-loaded oxinitride waveguides on a silicon substrate using the PECVD process or thermal oxidation followed by an LPCVD process, respectively.[45] They used a mixture of O_2 and 2% SiH_4 diluted in N_2 and a deposition temperature of 350°C for PECVD. LPCVD was performed in a mixture of $SiCl_2H_2$, NH_3, and O_2 at 930°C. The refractive index was controlled by adjusting the O_2 gas flow to the total gas flow.

Using a horizontal "hot wall" LPCVD reactor, Gorecki et al. fabricated a single-mode strip-loaded silicon oxinitride waveguide similar to that of Fig. 5.20.[46] The waveguide core is an SiO_xN_y thin film of 0.5 μm sandwiched between two SiO_2 layers deposited by LPCVD on a silicon substrate. The SiO_xN_y layer is fabricated with a refractive index $n = 1.51$ at a temperature of 850°C by adjusting the flow of 50 sccm SiH_4/He, 375 sccm N_2O, and 25 sccm NH_3 reactant gases in a pressure range of 65 Pa. This technique permits fabrication of silicon oxinitride films with an index range from 1.5 to 2, at deposition rates of 3 nm/min. Figure 5.26 shows the refractive index and deposition rate of silicon oxinitride as function of silane/nitrogen to nitrous oxide flow ratio. SiO_2 layers with a refractive index of $n = 1.454$ are deposited at a temperature of 600°C by adjusting the SiH_4-O_2 mixture in a pressure range of 48 Pa. The deposition rate is 30 nm/min. By means of RIE in CHF_3 plasma, a 4-μm-wide rib is etched in the top SiO_2 layer. The obtained etch rate is 30 nm/min. Optical losses lower than 3 dB/cm were measured at a

0.633-μm wavelength. By substituting silane with dichlorosilane and nitrous oxide with oxygen, Gleine et al. improved the propagation loss of LPCVD waveguides.[4] A $Si_xO_yN_z$ core waveguide 0.25 μm thick was fabricated by LPCVD at pressures of 3–19 Pa from a mixture of SiH_2Cl_2, NH_3, and O_2 on top of a 4-μm-thick SiO_2 layer. The core layer was covered with a silica cladding. Optical losses less than 0.5 dB/cm were obtained at a 0.633-μm wavelength.

To improve the optical quality of silicon oxinitride and silicon nitride films formed by PECVD and LPCVD processes, annealing techniques are used. Thus, annealing with a CO_2-pulsed laser permits reduction of optical losses to 0.1–0.2 dB/cm by reduction of stresses in the deposited films. Thermal annealing is also used to reduce the waveguide losses, but operating at relatively high temperatures (1000°C).

5.2.7.3 LPCVD and PECVD Phosphosilicate Glass Waveguides

Fiber-to-waveguide coupling efficiency is one of main motivating factors for fabrication of phosphosilicate glass channel waveguides, as reported by Verbeek et al. at ATT.[47] The fabrication sequences of this waveguide are shown in Fig. 5.27. The process starts by deposition of a buffer silica layer of 10 μm, which is grown through oxidation of the silicon substrate in high-pressure steam. A 2-μm-thick phosphosilicate core layer is fabricated by incorporating 8 wt.% of phosphine in silica through LPCVD using silane and oxygen, operating at a temperature of 450°C. The LPCVD process involves a chemical reaction from a mixture of tetraethylorthosilane, ammonia, and phosphine. In this core, 5-μm-wide mesas were etched by RIE after a photolithography step. The mesa is flowed by annealing at 1000°C in nitrogen, so a bell shape is obtained. A 3-μm-thick cover layer of silica with 4 wt.% of phosphorus was deposited by LPCVD at 730°C using TEOS, oxygen, and phos-

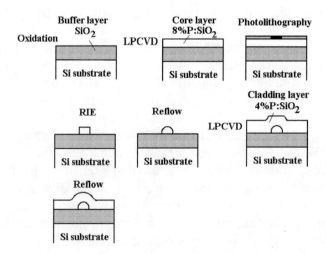

Figure 5.27 Fabrication by PECVD of a P-glass channel waveguide.

phine and then annealed at 800°C. This last annealing is made to relieve strains and to densify the film. The resulting refractive index difference between core and top layer is 7.3×10^{-3} at a 1.5-μm-wavelength. The measured fiber-to-fiber coupling loss of straight waveguides at 1.5 μm is 0.5 dB/cm and 0.6 dB/cm at a 1.3-μm wavelength. Waveguide losses as low as 0.05 dB/cm at 1.5-μm wavelength and 0.12 dB/cm at 1.3-μm are obtained. Since this waveguide has essentially the same refractive index as an optical fiber, it is possible to tailor the mode-field of such waveguides to perfectly match the fiber-mode profile. This technology was used successfully in the fabrication of optical WDM devices such as Mach–Zehnder interferometer multiplexers.

The category of previously described waveguides has a moderate birefringence-preserving polarization. Mechanical strain is released by thermal annealing at temperatures of 1000°C. In cooling to room temperature, this strain reappears because the silicon substrate contracts more rapidly than the SiO_2 film. The resulting birefringence is about 5×10^{-4} for TM polarization.[48]

LETI in Grenoble developed an approach to low-loss phosphosilicate channel waveguides on silicon.[49] The structure of these waveguides is Si substrate/SiO_2/P-doped SiO_2/SiO_2, as shown in Fig. 5.28. All silica layers are deposited by a PECVD process operating at 800°C, where the phosphorus-doped core layer exhibits a higher refractive index than the surrounding media. These waveguides exhibited optical losses in the range 0.2 to 0.3 dB/cm at a 0.633 μm wavelength. If the deposition by PECVD of silica layers requires a relatively low temperature (about 400°C), the fabrication of these waveguides at 1.5 μm wavelength with low optical loss needs a 3-hr thermal annealing process at 1000°C. The change in the refractive index of the core layer is obtained by doping the silica with phosphine. Figure 5.29 shows the change in refractive index as a function of phosphorus content in weight percent.[50] The core layer is doped with 5–10 wt.% of phosphine, whereas the buffer and cover layers are fabricated by incorporating 3 wt.% of phosphine. An accuracy of refractive index control of about 10^{-3} is obtained. RIE etching of the core layer in doped silica followed by the deposition of a silica overlayer and a second dry etching are performed to fabricate the channel waveguide structure. The coupling efficiency between these channel guides and single-mode fibers is greater than 90% for channel widths between 3.6 and 5.4 μm (when the waveguide structure is optimized to the fiber mode profile). This technology has been used to

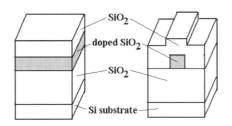

Figure 5.28 Structure of P-doped SiO_2/SiO_2 channel waveguide.

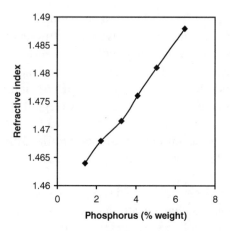

Figure 5.29 Change in refractive index as function of phosphorus content in P-doped SiO_2.

fabricate integrated four-channel multiplexers with a micromachined Fresnel mirror.

5.2.7.4 FHD Channel Waveguides

The combination of FHD and RIE is an alternative to fabricating several guided-wave components on silicon with a modal field well matched to optical fibers. This combination readily produces single-mode buried waveguides with a core made of titanium-doped silica. The fabrication sequences are shown in Fig. 5.30.[7] After the FHD process ($SiCl_4$-$TiCl_4$-$SiCl_4$-$GeCl_4$ mixture into an oxygen/hydrogen torch), the silicon wafer with deposited porous glass is consolidated by heating at temperatures of 1200 to 1300°C. The buffer and core layers are 20 μm and 8 μm, thick, respectively. To form a core ridge of a channel waveguide, a 2-μm-thick overlayer of amorphous silicon (a-Si) is deposited by magnetron sputtering. The combination of this layer with a resist mask acts as an etch stop for RIE etching of core ridge. The ridge pattern is defined by photolithography, followed by RIE of the overlayer using $CBrF_3$ gas. RIE using a gas mixture of C_2F_6-C_2H_4 is employed to remove the core layer, except for the photolithographically defined ridge structure, until the buffer layer is exposed. The core ridge is finally covered with a 30-μm-thick cladding of silica deposited by FHD. The resulting waveguide is a buried waveguide with a core size of about 8×8 μm^2 and a refractive index contrast Δn between the core and cladding of about 0.25%. The propagation loss in these waveguides is about 0.1 dB/cm. To obtain a low bending loss for a small-curvature radius, waveguides with a core size of 6×6 μm^2 and a refractive index contrast $\Delta n = 0.75\%$ were also fabricated. But in this case the propagation loss was increased to 0.3 dB/cm. This can be overcome by using SiO_2-GeO_2 glass with a lower melting temperature than SiO_2-TiO_2 glass. GeO_2-doped silica single-mode waveguides with a refractive index contrast $\Delta n = 0.75\%$, 40 cm long and

OPTICAL WAVEGUIDES AND SILICON-BASED MICROMACHINED ARCHITECTURES

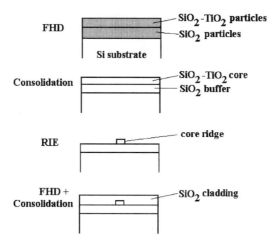

Figure 5.30 Fabrication by FHD of a single-mode buried waveguide with core made of titanium doped silica.

constructed from fourteen 90° bends with a bending radius of 5 mm have been fabricated. They exhibit a very low propagation loss of 0.04 dB/cm at a 1.55-μm wavelength. To obtain this high performance, special attention is necessary to prevent the volatilization of GeO_2 during the FHD process. These waveguides have been used in directional couplers, ring resonators, and multichannel optical frequency switches.

Two different techniques have been proposed by NTT to control the birefringence between the TM and TE modes of the previously described waveguides. In one method a pair of grooves is etched symmetrically by RIE around the ridge of the waveguide and along its length in the cladding, as shown in Fig. 5.31.[51] These grooves help to release the stress on the ridge core and induce a variation of birefringence within the waveguide. The resulting birefringence reaches zero below the ridge width of 50 μm and increases as the ridge width increases from 50 to 500 μm. This stress-releasing method was used in construction of TM/TE mode converters.

Alternatively, birefringence may be controlled by deposition of a magnetron-sputtered 6-μm-thick film of amorphous silicon on the top of the cladding layer, as shown in Fig. 5.32.[7] The presence of this a-Si strip locally changes the birefringence of the waveguide. The magnitude of the resulting birefringence is a function of the width of the deposited a-Si strip. As the strip width increases, the birefringence starts from a value of 4×10^{-4}, passes a peak value at 5.5×10^{-4} for a strip width of 50 μm, and then decreases to 2.5×10^{-4} for 200-μm wide a-Si film.

5.2.7.5 Fabrication of Antiresonant Reflecting Optical Waveguide

A combination of silicon with silicon oxide is used to form an ARROW waveguide. This multilayer waveguide, which is based on antiresonant reflection, rather than

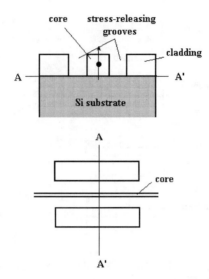

Figure 5.31 Control of birefringence of silica waveguides by stress-released grooves.

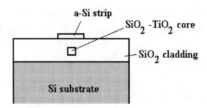

Figure 5.32 Control of birefringence of silica waveguides by deposition of a magnetron-sputtered a-Si strip.

internal reflection, can be designed to carry leaky modes with relatively low losses. Its functional principle was discussed in Section 5.2.4. Duguay et al. reported one implementation of ARROW starting by deposition of a 2-μm-thick SiO_2 second cladding grown on the silicon substrate by thermal oxidation (Fig. 5.6).[10] A thin layer (0.1 μm) of high refractive index polycrystalline silicon is deposited by CVD. The waveguide core is a 4-μm-thick undoped silica. In this waveguide, an optical loss as low as 0.4 dB/cm was obtained at a 1.3-μm wavelength and for TE modes. The choice of material for the thin first cladding depends on the operational wavelength. In near IR and visible wavelengths shorter than 0.9 μm, TiO_2 is preferred ($n \sim 2.1$) because poly-Si suffers from high absorption in these wavelengths. In IR wavelengths longer than 1.1 μm, poly-Si is used ($n \sim 4.5$). CVD silicon nitride can be used to form the thin first cladding, working in both the IR and visible wavelength ranges.[52]

The fabrication of the ARROW waveguide is relatively simple, as illustrated in Fig. 5.33.[53] The second cladding material is 1-μm-thick thermally grown sil-

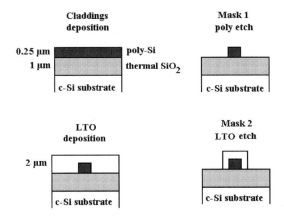

Figure 5.33 Fabrication of ARROW waveguides.

icon oxide at 1100°C deposited on a <100> oriented silicon substrate (c-Si). Subsequently, 0.25-μm-thick undoped polysilicon (first cladding) is deposited by LPCVD at 585°C from silane. The first and second layers form the antiresonant reflector necessary for low radiation losses into the high index substrate. Subsequent patterning and wet etching with KOH of the poly-Si layer define the first cladding. To form the core, a layer of low-temperature oxide (LTO) is deposited by CVD to a thickness of 2.1 μm. Subsequent patterning to form the waveguide ridge is obtained by etching in buffered HF. This creates a channel waveguide formed by ARROW confinement vertically and effective index confinement horizontally.

5.2.7.6 Fabrication of Polymer Waveguides

Waveguides based on CVD deposition and patterning of inorganic thin films (nitride, oxinitride, doped silica) typically require high temperature deposition and in certain cases an annealing processing. Such processes may be viable but need a specific redefinition of individual sequences of fabrication steps. Polymers such as polyimide, siloxane, or PMMA can be easily spun on the surface of a silicon wafer and processed at low temperatures. Optical polymers have advanced significantly, driven by application requirements such as low optical loss and environmental stability. A number of single or multimode waveguide devices have recently been developed, including splitters, couplers, and switches for wavelength division multiplexing. There are several different routes to fabricating polymeric channel waveguides, including polymer removal by etching of the core layer, externally induced dopant diffusion by selective polymerization, modification of molecular orientation of dopants by an externally applied electric field, and local polymerization-driven diffusion reactions without the presence of dopants.[9] Externally induced dopant diffusion uses the fact that the photopolymerization reaction alters the diffusion rates. Consequently, when the material is heated after removal of the photomask, unexposed regions are thermally outdiffused, lowering the index of refraction. In

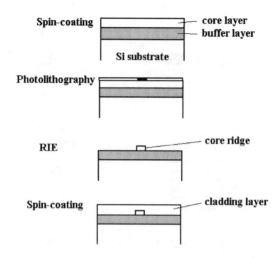

Figure 5.34 Fabrication by spin coating of polymeric waveguides.

molecular orientation of dopants by an externally applied electric field, the induced local orientation of dopant molecules generates high index regions (core waveguide). In polymers such as PMMA, the dopant molecules are electro-optically active.

Our goal is to demonstrate performances and application capabilities of polymeric waveguide fabrication that constitute a relatively nonintrusive processing step of the basic MEMS micromachining. Fabrication techniques require only low-temperature processing because the polymeric materials may be spin coated in multiple layers to form waveguides; patterned; and then consolidated by solder reflow at temperatures not exceeding 300°C. In this section, two different approaches to polymer waveguide fabrication are illustrated. The first one uses conventional photolithography and dry etching and was developed by NTT.[37] Both the buffer/cladding and core regions are made using nondeuterated and deuterated polysiloxane, respectively. The process starts by spin coating deposition on a silicon substrate of buffer and core layers dissolved in organic solvent, as shown in Fig. 5.34. After each successive spin coating, the deposited polymer layer is baked at 150°C to remove the remaining solvent by evaporation. The channel waveguide is structured by photolithography followed by etching of the core ridge via RIE using a gas mixture of CF_4 and O_2, operating at a pressure from 1 to 5 Pa. After the etch mask is removed, the cladding layer is spin coated, covering the core ridge structure. A buried ridge waveguide is formed with a core dimension of 8×8 μm^2. In order to achieve the small refractive index contrast of 0.3%, the refractive index of the core layer is 1.5365 at a 1.5-μm wavelength. The propagation loss of this waveguide was measured at less than 0.2 dB/cm at a 1.31-μm wavelength and less than 0.45 dB/cm at 1.55 μm.

Du Pont de Nemours developed a different technology for polymer waveguide fabrication based on internal and local polymerization-driven diffusion reactions,

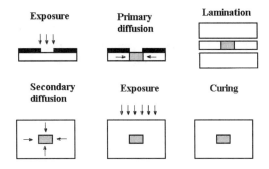

Figure 5.35 Fabrication by Polyguide technology of a single-mode polymeric waveguide.

generating very low refractive index contrast.[9,54] In this process, called Polyguide, the waveguide formation is based on internal diffusion of acrylate photopolymerizable monomers. This technique uses the polymeric property that high-mobility, low-molecular weight monomers can rapidly diffuse within a polymeric binder matrix. The process is dry, with no etching or molding, and the waveguide formation is completely light induced at low (15 to 45°C) temperatures followed by photo and thermal fixing. The fabrication of a single-mode waveguide starts by diffusion into the exposed waveguide region. The waveguide layer is laminated to a mask followed by a buffer and cladding layer diffusion between all layers, as shown in Fig. 5.35. Usually by large area of photoexposure locks up, remaining a monomer after diffusion has occurred, thus fixing the system. Finally, the baking and curing steps stabilize the structures. Baking is usually at 135°C for 2 hr. Waveguide refractive indices relative to the uniform lower index surrounding the cladding region can be controlled by adjusting the exposure parameters of power, energy, and temperature with the index contrast in the range from 0.003 to 0.01. A recently fabricated single-mode polycarbonate 1×8 splitter produces an optical loss of less than 0.5 dB/cm at a 1.3-µm wavelength.

To conclude, polymers exhibit interesting physical properties such as electro-optical or piezoelectric effects. These materials are attractive for sensor applications, because it is possible to incorporate functional organic groups that react with the surrounding media.

5.2.8 BASIC GUIDED-WAVE OPTICAL COMPONENTS

Several components known from bulk optics can be miniaturized and integrated on a single optical chip. Some of them are shown schematically in Fig. 5.36. Various kinds of optical components have been realized on silicon-based waveguides, including Y-junctions, X-crosses, couplers, polarizers, beam splitters, TE/TM polarization converters and splitters, modulators, and interferometers. All these components may be fabricated by standard IC fabrication methods. Two more important

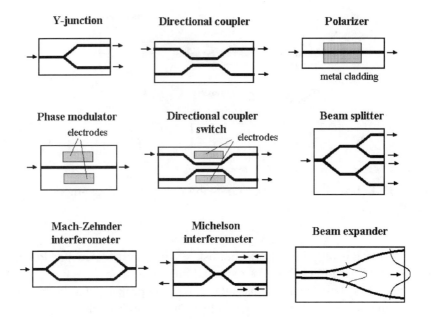

Figure 5.36 Schematic of some IO components.

elements in Fig. 5.36 are the Y-junction and directional coupler because most of the other components consist of combinations of both these elements. Thus, cascading of Y-shaped branches or a combination of directional couplers may form a beam splitter. Combining two single-mode Y-junctions may form a Mach–Zehnder interferometer, where the incoming light is split into two parts and guided in the two arms. Each Y-junction acts as a half mirror with a 50:50 splitting ratio. An alternative to a Mach–Zehnder interferometer configuration consists of two directional couplers with a 50% coupling ratio. A Michelson interferometer, in which the light in the two arms is reflected at the polished substrate end face, may consist of a directional coupler or a combination of two Y-junctions. In the case of Y-junctions, a variable splitting ratio can be obtained by asymmetric branching. A Y-junction is relatively sensitive to fabrication defects, especially in the junction region. A directional coupler is less sensitive to fabrication tolerances. The coupling is based on the energy exchange that occurs between closely spaced optical waveguides. Light from the input waveguide couples into the second because of the overlap of evanescent fields of two guided waves. NTT has fabricated both Y-junctions and directional couplers, designed by photolithography and using the technology of doped-silica FHD single-mode buried waveguides on silicon.[7] A directional coupler is formed by two waveguides separated by a few micrometers and the intermediate region between the two cores is filled with cladding glass during the consolidation step of the FHD. The excess loss due to the coupling from fiber-to-waveguide-to-fiber in a Y-junction is about 1 dB in the 1.2–1.6 μm wavelength range. For directional couplers, the corresponding excess loss is less than 1 dB.

5.2.9 PROPAGATION LOSS AND FIBER-TO-WAVEGUIDE COUPLING LOSS

In silicon-based waveguiding devices, the loss is generally attributable to three different factors: scattering, radiation, and fiber-to-waveguide coupling. Scattering loss usually predominates in glass and dielectric waveguides and has two principal sources. One is based on volume scattering, where the scattering is caused by imperfections such as crystalline defects or contamination atoms within the volume of the waveguide. The second category of scattering is based on waveguide surface irregularities and roughness. A surface scattering loss can be created, for example, by scattering from the rib side walls formed by RIE. RIE can faithfully transfer the etch mask with any irregularities without smoothing. Thus, the roughness results in a scattering loss that can be exacerbated by a high index difference between the rib and cladding layer. Scattering loss generally contributes from 0.5 to 2 dB to the losses of channel multilayer waveguides with a silicon nitride or oxinitride core on silica. Radiation losses become significant when waveguides are bent through a curve. The radiation loss is low compared with the scattering loss, but it is not negligible in the case of curved channel waveguides. Because of distortions of the optical field that occur when guided waves travel through a bend in these waveguides, the magnitude of radiation loss depends on the radius of curvature and the lateral mode confinement.

Besides permitting the construction of low-loss channel waveguides, silicon-based waveguides offer efficient fiber-to-waveguide connections. Efficient fiber-to-waveguide coupling assumes perfect alignment between the fiber and the waveguide. The two main sources of coupling loss are reflection (called Fresnel loss) and the loss caused by mismatching between the fiber and waveguide modes. The Fresnel loss is mainly due to the difference between the effective refractive indices of the fiber and waveguide modes. The contribution of Fresnel loss is rather small and may be greatly reduced by use of an index-matching fluid between the fiber and channel waveguide. The loss resulting from mode mismatching is then the main source of fiber-to-waveguide coupling loss. To reduce the mode mismatching, it is necessary to fabricate waveguides with a mode size and shape well matched to that of the fiber. Usually the overlapping between the fiber and the channel mode is small. The waveguide mode shape is typically asymmetric, while that of a single-mode fiber is circularly symmetric, with greater dimensions. For any optical fiber $F(x,y)$ and channel waveguide $W(x,y)$ mode amplitudes, the coupling efficiency is given by the normalized field overlap integral:[55]

$$\eta = \frac{|\iint F(x,y)W(x,y)dxdy|^2}{\iint F^2 dxdy \iint W^2 dxdy}. \quad (5.13)$$

As discussed in Section 5.2.7.4, Kawachi et al. demonstrated a coupling efficiency of $\eta > 98\%$ (for an optical loss of 0.04 dB/cm), when connecting a standard 1.55-µm single-mode fiber with 8-µm mode-field diameter to GeO$_2$-doped silica single-mode buried waveguides with a refractive index contrast $\Delta n = 0.75\%$, and 8×8 μm^2 core size.[7] In this configuration of a low-index contrast waveguide, the coupling loss is negligible because of an excellent mode overlapping between the fiber and waveguide. The high-index contrast waveguides typically have a coupling loss in the range of 0.5 dB/cm. Consequently, the low-index contrast waveguides are preferred to high-index waveguides in terms of loss coupling. This advantage is also verified in terms of propagation losses, as shown in Section 5.2.7.4. To achieve reasonable bend angles and radii of curvatures with a minimum of loss, the lateral confinement of the waveguide must be carefully controlled. The lateral confinement, for example, in rib waveguides with low index contrast is not high enough for applications with bent waveguides presenting a small radius of curvature. These waveguides cause excessive propagation loss. High index contrast rib waveguides, however, have better lateral confinement and bends, and Y-branches have been made using this type of waveguide.

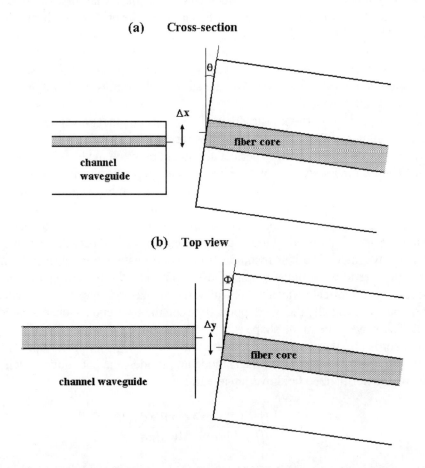

Figure 5.37 Sources of fiber-to-waveguide coupling loss: (a) cross-section with vertical misalignment Δx and tilt angle θ, (b) top view with lateral misalignment Δy and tilt angle Φ.

The sensitivity of fiber-to-waveguide coupling to the precision of mechanical positioning includes the vertical misalignment Δx, lateral misalignment Δy, and tilt angles θ and Φ. All these sources of misalignment schematically illustrated in Fig. 5.37(a–b) represent an important issue in considering manufacturable fiber attachment techniques. If the channel mode size is relatively large to closely match single-mode fibers, the coupling efficiency is not very sensitive to translation misalignments. For both lateral and vertical misalignments, an offset of ±2.5 µm increases the coupling loss in the range from 0.1 to 0.25 dB relative to the best alignment. Both the angular tilts must be maintained below 0.5 deg to produce a loss below 0.25 dB. When using silicon-based channel waveguides, these alignment restrictions can be accomplished via photolithographically defined V-grooves, wet etched in <100> oriented silicon substrate. A schematic representation of a fiber to channel waveguide-coupling silicon V-groove is shown in Fig. 5.38(a–c). Commonly used anisotropic etchants are solutions of KOH or EDP that permit the V to be made with an angle of 54.74 deg between the <100> and <111> crystallographic directions of silicon. The use of a silicon V-groove permits one to precisely control the desired groove shape and to ensure that the vertical positions of the cores of fiber placed in the V-groove are aligned. A typical angular precision of ±0.2 deg and a lateral alignment accuracy less than ±0.5 µm were demonstrated.[56] Alternatively, Yamada et al. proposed to fabricate by RIE etching a V-groove on silicon for single-mode optical fiber alignment.[57] A coupling efficiency in the range of 75% can be obtained with the silicon V-groove technique.[58]

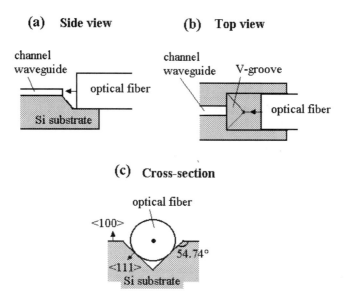

Figure 5.38 Schematic representation of fiber-to-waveguide alignment system using a silicon V-groove: (a) side view, (b) top view, (c) cross-section.

5.3 SILICON MICROMECHANICS

Optical waveguides may be integrated with a micromechanical deformable structure and the microstructure can serve as a sensor to detect physical parameters of deformation, as well as a passive microactuator to move optical parts. This chapter focuses on the integration of silicon guided-wave architectures within the MEMS environment rather than on the fabrication technologies involved in forming micromechanical platforms for optics. There are numerous articles in the literature, that treat this in detail. In the context of this chapter, it is appropriate to briefly examine the global characteristics of microstructures driving the optical applications in order to relate micromechanical characteristics to needs in optical waveguide technology. The main interest is then the types of structures that may be fabricated and the types of proposed MOEM function, rather than the processing by which they are made. To help our readers, however, we start with a very short review of materials and fabrication technologies of silicon micromachining.

5.3.1 MATERIALS AND MICROMACHINING

The simplest micron-sized movable micromechanical structures are membranes, bridges, beams, and cantilevers. The basic material used in fabrication of these microstructures is silicon because (1) the process is well established, (2) it has good mechanical properties, and (3) integration with electronics is possible. The attributes of silicon are well known, and were described eloquently by Petersen.[1] In addition to single-crystalline silicon and previously described silicon-derived materials such as silicon oxide or silicon nitride, other materials are also used for specific tasks: metallic thin films (chromium, gold, etc.) as etch masks or thin structures, shape memory alloy metals (such as TiNi), CVD, or sputtered piezoelectric materials (ZnO, PZT) in actuation, polyimides, etc. Table 5.6 summarizes these materials and their useful characteristics.[59]

Microstructures fabricated by micromachining are planar in nature and have a thickness of 10 μm in most cases. Some applications use 3D integrated cir-

Table 5.6 Materials in micromachining.

Material	Use	Process	Characteristics
Polyimides	Structure	Thin film	Soft and flexible, easy coating
Tungsten	Structure	Thin film	Not attacked by HF, thick structure
Ni, Cu, Au	Structure	Electroplating	Thick structure
Quartz	Actuation	Anisotropic etching	Piezoelectricity, insulator
ZnO	Actuation	Thin film	Piezoelectricity
PZT	Actuation	Thick film	Large piezoelectricity
TiNi	Actuation	Thin film	Shape memory alloy
GaAs	Optics	Thin film	Semiconductor laser, LED, detector

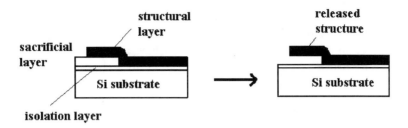

Figure 5.39 Surface micromachining of a structure.

cuits in which structures deeper than 10 µm are sculpted into the substrate using a process compatible with standard VLSI batch processing. Today, all required micromechanical structures can be routinely fabricated using bulk or surface micromachining.[60,61] Bulk micromachining is used to sculpt the bulk of the silicon substrate from all sides to fabricate the desired 3D structures in the same material. It uses wet and dry etching techniques in association with etch masks and etch stops to sculpt MEM devices. Isotropic etching of silicon (using, for example, nitric acids) is insensitive to the crystal orientation of particular silicon planes being etched, while anisotropic etchants (such as KOH or EDP) selectively attack the different crystallographic directions (<100> and <111>) at different rates. Etch mask and etch stop techniques are used in conjunction with anisotropic etching to selectively prevent regions of silicon from being etched (for example, formation of thin diaphragms or beams).[61] Surface micromachining permits the formation of 3D structures without sculpting the substrate itself, providing a significant gain in time in etching operations. The process is achieved by successive deposition and etching of structural and sacrificial layers on the surface of the silicon wafer, using polysilicon as the structural material and silicon dioxide as the sacrificial layer. This simple process is illustrated in Fig. 5.39. A sacrificial layer is deposited on a silicon substrate, which is coated first with an isolation layer. Windows are opened in the sacrificial layer, followed by deposition and etching of the structural layer. Selective etching of the sacrificial layer leaves a free-standing microstructure such as a beam or cantilever.[62]

5.3.2 MICROACTUATORS

There are many mechanisms of actuation such as electrostatic, magnetostrive, piezoelectric, thermal, or shape memory alloy, to name a few. Each actuation technique has its specific advantages and disadvantages. The force developed, the displacement, or the available power depend strongly on the type of actuator. Piezoelectric actuation is often considered the most promising technique in certain applications. One of the functions of a piezoelectric actuator is to deliver a large force, but through very limited displacement. Conversely, electrostatic-based actuators can move through large displacements, but are only capable of delivering small forces. In practice, the electrostatic actuator is the most common and well-developed method in MEMS applications; it is easily fabricated by batch techniques of IC-compatible micromachining technology. This is suitable for performing positioning tasks, which can be completed within a chip. Other types of actuators may be more robust, may offer larger forces or larger displacements, but they require more specialized processing and are not easily integrated within the silicon MEMS environment because of material compatibility. Magnetic actuation, for example, generally requires relatively high currents and high energy. Thermal actuators also require relatively large amounts of electrical energy, and the heat generated has to be dissipated. Here we focus our attention on electrostatic and piezoelectric microactuators, considering that the use of electrostatic microactuators is most suitable for positioning applications while piezoelectric microactuation is appropriate for both actuator and sensor applications.

5.3.2.1 Electrostatic Microactuators

An actuator is defined as an element that applies a force to an object through a distance. It performs work as it moves from point a to point b:

$$W = \int_a^b F\,dx, \tag{5.14}$$

where F is the force vector and x is the displacement vector.

The power is defined by

$$P = Fv \quad \text{or} \quad P = T\omega, \tag{5.15}$$

where v is the velocity at the point of application of F, T is the torque, and ω is the angular velocity.

A parallel-plate electrostatic capacitor with two plates in which the lower plate is fixed and the upper plate is movable is shown in Fig. 5.40(a). It presents a developed force such as[13]

$$F = \frac{\varepsilon_0 A}{2d^2} V^2, \tag{5.16}$$

where each plate has an area A, d is the separation gap, V represents the constant potential applied to the upper plate, and ε_0 is the dielectric constant. For actuators with plates 50×50 μm^2 and the gap $d = 2$ μm, the force obtained is 2.5 μN at potential $V = 30$ volts.[63]

To generate a larger force, a large change of capacitance with distance is required. This has led to the development of a comb-drive electrostatic actuator.[64] It consists of a pair of comblike electrodes (one is suspended and other is fixed) with a total of $2n$ interdigitated fingers, as shown in Fig. 5.40(b). When a voltage V is applied, an attractive force developed between the fingers moves the suspended

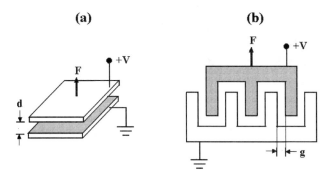

Figure 5.40 Two electrostatic microactuators: (a) parallel plate capacitor, (b) comb-drive actuator.

part. The increase in capacitance is proportional to the number of fingers (or capacitors). The developed force is:

$$F = n\varepsilon_0 \frac{t}{g} V^2, \qquad (5.17)$$

where t is the thickness of the finger.

The force of a parallel plate capacitor varies as $1/d^2$ while the force of a comb drive is constant for any value of the gap. The thickness t is typically limited to about 2 to 3 μm in surface micromachining technology, and typical gap widths are about 1 μm.

An example of a highly efficient electrostatic microactuator is the scratch drive actuator (SDA) developed by Akyama et al.[65] The SDA performs a linear displacement Δx, owing to the motion, of a polysilicon plate with a bushing side on the silicon substrate. This is shown in Fig. 5.41, where the applied driving signal is plotted. Successive bending and relaxation of the plate produce the motion. The bending is induced by electrostatic force and during the relaxation steps, the effect of asymmetrical torsion produces a serious motion. For SDAs with a plate size of 80×80 μm^2, the force developed is typically from 50 to 60 μN and the incremental steps are in the range of 0.1 μm. Multiple SDAs can be grouped to increase force.

5.3.2.2 Piezoelectric Actuators

The small strains and high stresses generated by piezoelectric actuators have permitted the development of several micromotors and micropumps. These applications were based on bulk piezoelectric ceramics such as lead zirconate titanate (PZT), but thin-film piezoelectric materials are relatively new arrivals. Besides PZT, ZnO and AlN are the most popular thin-film piezoelectric materials. MEM devices using the piezoelectric effect are usually operated in one of two modes: sensor or actuator.[66] When used in sensor mode, changes in the internal dipole moment of a piezoelectric crystal under externally applied strain or change of temperature are detected. The alteration of dipole moment detected is converted in terms of the external physical perturbation (pressure, temperature). When used as an actuator, a voltage is applied across the electrodes of a piezoelectric thin-film transducer, inducing a strain or deformation and generating a force. Both controlled deformation and controlled force are used to do work. The contribution of silicon micromachining is to fabricate mechanical structures capable of controlled motion when the microstructure is under activated deformation (actuator) or under externally subjected physical force (sensor). How this may be used in actuator mode is shown in Fig. 5.42, where two microstructures currently used are represented: a cantilever microbeam [Fig. 5.42(a)] and a thin membrane [Fig. 5.42(b)].

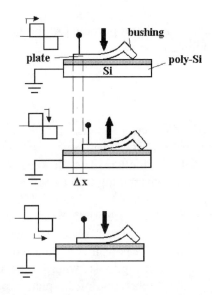

Figure 5.41 Schematic view of SDA actuator.

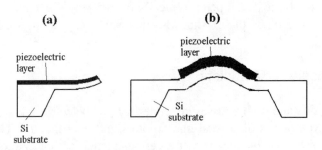

Effects of an applied voltage

Figure 5.42 Schematic of piezoelectric microactuators using: (a) a cantilever microbeam, (b) a thin membrane.

Microstructures may be defined by using a photolithographic process and patterned in silicon using a carefully chosen anisotropic etching technique. The piezoelectric layer is deposited on the top of the microstructure, and transducer electrodes are patterned. When a voltage is applied, the stress generated causes the beam to bend or the membrane to deform, respectively. The use of piezoelectric thin films (PZT, ZnO) for microactuation has certain advantages, including the ability to use surface and bulk micromachining to build complex mechanical structures that provide stable motion control at moderate voltage levels (in general above 100 V for displacements of a few micrometers). Additional advantages are features such as large force generation, high-frequency responses, and compatibility with silicon electronics.

5.3.3 SIMPLEST MICROSTRUCTURES COMBINED WITH OPTICAL WAVEGUIDES

The simplest microstructure that may be combined with an optical waveguide is the cantilever beam, as shown in Fig. 5.43. Knowing the elastic properties of the cantilever, the deflection along the x-axis resulting from the application of a force F can be approximated by[67]

$$X(z) = \frac{F}{6EI_Y}(3b - z)z^2, \quad (5.18)$$

where E is Young's modulus, I_Y is the moment of inertia of the cross-section of the beam, and b is the beam length. At the coordinate $z = b$, Eq. (5.18) helps to direct the force to the deflection:

$$F = kX(b), \quad (5.19)$$

where $k = 3EI_Y/b^3$.

By interfacing the cantilever beam with optical waveguides, an integrated optical read out of the micromechanical element is obtained. The deformation of such cantilever-shaped waveguide structure leads to an elasto-optic effect in the waveguide and influences the propagation constant of the guided mode. Figure 5.44 illustrates two examples of the utilization of such structures in sensing applications. As

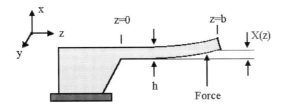

Figure 5.43 Deflection of a simple cantilever beam.

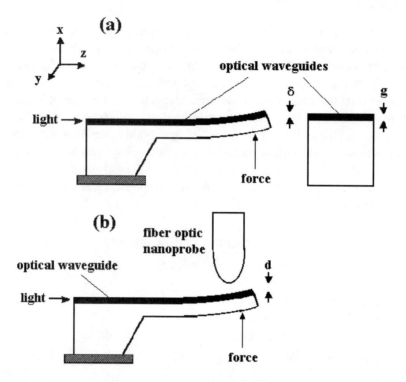

Figure 5.44 Utilization of a cantilever beam interfaced with an optical waveguide as an optical interrogation: (a) amplitude detection, (b) near-field detection.

the deflection of the cantilever beam increases the propagation loss of the guided wave, the amplitude of the guided light is directly affected. The device can be used as an amplitude-sensitive technique for detection of mechanical oscillations of such resonant structures [Fig. 5.44(a)]. It can also be operated as an intensity or phase modulator. Finally, it is also possible to construct near-field detection systems based on the evanescent interaction between an optical fiber optic nanoprobe and the guided-wave light structure [Fig. 5.44(b)].

Any interactions affecting the wave-guided light can be used as sensing principles. Exposure of the microstructure to environmental factors can determine the variation of the mechanical properties due to such factors. Physical parameters such as temperature or pressure can therefore be measured as a consequence of the deformation produced when they act as resonators, thereby inducing changes in its resonant frequency. Finally, a cantilever with an incorporated waveguide can also act passively as an optomechanical switch.

5.4 Examples of Micromachined Guided-Wave MOEMS

In the most general form, guided-wave MOEMS would consist of mechanical microstructures, integrated optic elements, microsensors, microactuators, and electronics integrated in the same environment. The combination of micromachine technology and optoelectronics leads to advantages of small scale. This permits the development of new photonic systems and applications that perform optical functions previously deemed impossible. This section reviews the field of silicon guided-wave MOEMS, including significant applications in communication systems as well as microsensors.

5.4.1 MICROMACHINED INTEGRATED OPTIC SWITCHES

Some integrated optical communication applications involve switches and modulators. To construct these devices, it is possible to use electro-optic materials such as $LiNbO_3$, acousto-optic materials, and liquid crystals. However, it would be also interesting to use, for example, silicon-based passive materials because of their compatibility with fiber optics and the ability to be fabricated using standard IC technologies. The disadvantage of passive materials is that the required change in refractive index must be externally induced. This may be obtained by using some mechanical or other perturbations of waveguide material, inducing optical modulation of light. In addition, silicon-based materials permit the batch fabrication of mechanically movable microstructures and the inclusion of electronics to modulate the guided-wave light as well as to adjust the alignment of optical elements. One optical element that has seen considerable interest with regard to the need for microactuation is the optical switch. The main field of application for optical switches is high-speed optical data communication networks. There are several methods of accomplishing the switch function, which have been attempted via optical waveguides combined with micromechanical structures. The microstructure can play a passive or active role in the switching operation. Several integrated optical switches that simply move and align an initial optical waveguide to one or more alternative waveguides have been proposed. In these mechanically passive switches, the use of MEMS microactuators is particularly suitable because their motion can be controlled accurately, available output forces are adequate, and total travel is also acceptable. Mechanically active switches use a movable microstructure that interacts with a guided wave through evanescent coupling or by spoiling the waveguide transmission characteristics. In any of these methods, the interaction of microstructure and waveguide induces an optical phase or intensity modulation and thus a switching function may be performed. All these approaches offer some advantages over conventional fiber optic switches and modulators. Since the entire device is made by a photolithographic process, the optical elements can be pre-aligned during the layout of the photomasks. The assembly and packaging cost can also be reduced. Low insertion loss and small crosstalk can be achieved. MOEM switches can be made compact and lightweight, and are potentially integrable with optical sources and detectors and controlling electronics.

5.4.1.1 Optical Switches with Passive Microactuation

An electrostatically driven micromachined 2×2 optical bypass switch was proposed by Hogari et al., as shown in Fig. 5.45.[68] Two optical waveguides form an X-cross configuration in which a small air gap is created in the cross region, separating the X-cross into two V branches. Waveguides are integrated on an insulation layer deposited on a metallic substrate. A metallic cantilever beam is mounted above the cross region, with the side wall inserted into the gap. The switch operates in two different states. When the cantilever is electrostatically undeflected, the membrane side wall slides into the gap and the light signal is reflected to the same waveguide. When a voltage is applied, the cantilever beam is bent electrostatically and the side wall is moved away from the gap. The light signal can then be transmitted through the gap and the waveguides along the same diagonal directions are allowed to communicate with each other. The insertion loss of this switch was less than 3.1 dB; its crosstalk is less than -40 dB at a 1.55-µm wavelength. The operating voltage is 100 V and the switching time about 2 ms for displacement of 15 µm of the membrane tip.

Ollier et al. proposed an integrated optical 1×2 switch based on a micromachined cantilever beam on the path of an input optical waveguide, as shown in Fig. 5.46.[69] Electrostatic deflection of the cantilever beam moves the input waveguide laterally between two fixed output waveguides. By adjusting the beam deflection, the input waveguide can be aligned on one of two output waveguides. The waveguide structure is of the buried channel type, fabricated by waveguide technology developed at LETI.[70] It is a sandwich of three PECVD phosphorus-doped layers, obtained with different doping levels. The buffer layer is doped with a phosphorus content of 3 wt.%, the ridge core with 6 wt.%, and the cladding with 3 wt.%, respectively. The core ridge of the waveguide is patterned by conventional photolithography followed by an anisotropic RIE etching. The microcantilever beam

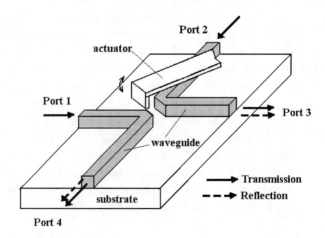

Figure 5.45 Electrostatically driven micromachined 2×2 optical bypass switch.

is fabricated with three layers of silica (25 μm thick). The beam is structured by an anisotropic RIE etch followed by an isotropic etching of silicon via microwave RIE. The last etch permits the release of the mechanical structure, leaving an air gap under the beam. A typical cantilever beam is 2 mm long, 26 μm wide, and 2 μm thick. A subsequent deposition of metallic layers for driving electrodes is performed by evaporation of aluminum on both side walls of the beam. The release of the beam structure generates the residual stress in the silica that deflects the beam tip. To avoid undesirable vertical misalignments of the input waveguide with output waveguides, this residual stress is compensated by a thermal treatment. The insertion loss of this switch was −3.5 dB at a 1.3-μm wavelength; its optical isolation is −35 dB. This switch has a switching speed of 0.6 ms and its driving voltage is 300 V.

To reduce the high driving voltage, a second generation of the device with a different concept of electrostatic microactuator has been achieved. Some improvements are obtained by use of a comb-drive actuator with 15 μm of gap, requiring a switching voltage less than 30 V. Figure 5.47 shows a scanning electron micrograph (SEM) of this switch configuration.

More recently, NTT proposed the architecture for the 2 × 4 optical switch that is shown in Fig. 5.48.[71] The switch consists of a micromachined silica planar lightwave circuit (PLC) and a bulk magnetic actuator to perform the switching operations of the micromachined part. The micromachined movable structure is composed of a pair of beam actuators, each of them in the path of an input optical waveguide, and a head shared by the two beams. Each beam is used to laterally align the input waveguide with a pair of output waveguides. The actuation is per-

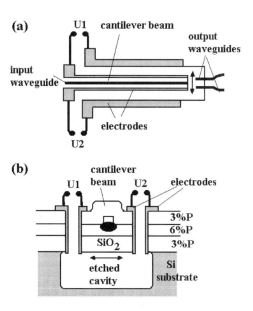

Figure 5.46 Electrostatically driven micromachined 1 × 2 optical switch.

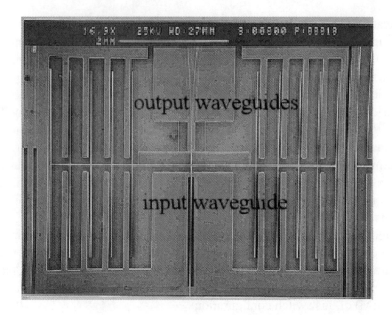

Figure 5.47 SEM photograph of the optical switch using a comb-drive actuator.

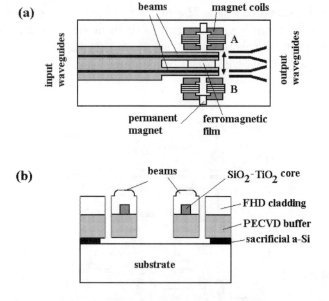

Figure 5.48 Micromachined 2 × 4 optical switch driven by bulk magnetic actuator: (a) top view, (b) cross-section view.

formed by a laterally placed magnetic actuator that consists of a ferromagnetic plate attached on the head and magnetic circuits containing magnet coils and a permanent magnet. When a current is applied to the coils on one of two sides A or B, the head is driven by a switching current and attracted to that side. This motion permits the switching operations. The bulk architecture of the magnetic actuator is mounted on the micromachined PLC chip. The fabrication sequence of this switch starts by deposition of a sputtered sacrificial layer of 1-μm-thick amorphous silicon (a-Si) on a silicon substrate. The waveguide buffer layer is a 15-μm-thick film of PECVD silica. An 8-μm-thick film of TiO_2-SiO_2 deposited by electron-beam evaporation is used as the core layer. The core ridge of the waveguide is patterned by RIE operating with CHF_3 as an etchant gas. A 27-μm-thick cladding layer is formed by a conventional FHD process followed by a second deep RIE using a WSi mask as an etch stop mask. The sacrificial layer of a-Si is removed with TMAH solution, releasing the movable structure of microbeams [Fig. 5.48(b)]. The insertion loss of this switch is 4.4 dB; its crosstalk was −40 dB and its switching time around 40 ms.

The main disadvantage of this device is the use of the bulk magnetic actuator. The difficulty of magnetic microactuation is that the three-dimensional coils made in ferromagnetic materials are not available using micromachining technology. The available two-dimensional coil presents relatively small actuation forces, reducing interest in integrating such microactuators. In this sense, electrostatic microactuators that involve only a pair of planar electrodes separated by one insulator are typically stronger than magnetic devices for equivalent volumes, and are easily integrated within the MEMS environment.

5.4.1.2 Optical Switches with Active Microactuation

Various integrated switches and modulators that take advantage of the interaction between the guided wave and a movable microstructure placed above the waveguide are reported. The switching principle is to simply deform and touch the waveguide by the active microactuator in order to spoil the waveguiding action.

The first example of this approach is an optical crossbar switch developed by Texas Instruments.[72] This is currently based on the technology of digital micromirrors (DMDs), where the electrostatically driven membranes are monolithically integrated with a phosphosilicate glass (PSG) waveguide. The waveguide is located under the membrane, separated from the membrane by an air gap, as shown in Fig. 5.49. The switch is in the "on" state when the membrane is undeflected, and the guided wave is propagated down the waveguide without perturbation. When a voltage is applied between the membrane and the substrate, the membrane makes contact with the waveguide surface. The waveguiding action is spoiled because of high loss due to the absorption in the metal (30 dB/mm with an aluminum membrane). In this situation, the switch is in the "off" state. The membrane fabrication is similar to those of DMDs used in projection displays. After the deposition of a 15-μm-thick buffer of silica and depositing, patterning, and reflowing the waveguide core (∼5 μm), the polymer spacer is deposited by spin coating. An aluminum film with a thickness of 0.5 μm is deposited on the spacer. When the membrane is defined, an array of access holes is etched through the metal. The central part of the

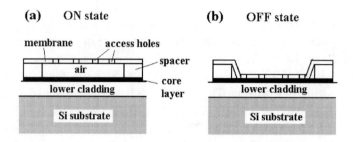

Figure 5.49 Micromachined optical cross-bar switch developed by Texas Instruments.

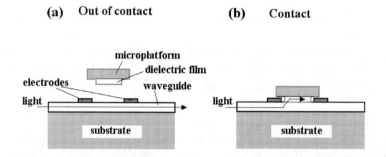

Figure 5.50 Micromachined optical switch with a polyimide microplatform.

spacer is removed by RIE etch from the region under the holes, creating an air gap under the membrane. This switch has an actuation speed of 20 μs and a scaleable actuation voltage of 4 to 45 V for membrane sizes varying from 300 × 300 to 5000 × 5000 μm^2.

A second example of optoelectromechanical switching with active microactuation is the switch proposed by Kim et al.[73] This device, which is fabricated by surface micromachining, consists of a polyimide microplatform suspended few micrometers above the waveguide surface, as shown in Fig. 5.50.[73] The movable microplatform may be electrostatically actuated when a voltage is applied between the microplatform and the substrate. The back side of the microplatform is coated with a thick dielectric film that makes contact with the waveguide surface. Upon contact, the guided wave is coupled into the dielectric film. The coupled guided wave experiences high loss due to absorption in the dielectric. To permit the coupling, the refractive index of the dielectric film must be greater than the effective refractive index of the waveguide.

A borosilicate glass (BK-7) is used as the substrate where a channel waveguide is made by ion-exchange process. After definition of the waveguide by photolithography, the substrate is immersed in a solution of silver nitrate melt buffered with sodium nitrate at 325°C for 10 min. The process is stabilized by an annealing step at 500°C for 30 min. The resulting waveguide propagation loss is less than 0.2 dB/cm at a 0.633-μm wavelength. The fabrication sequence of the polyimide

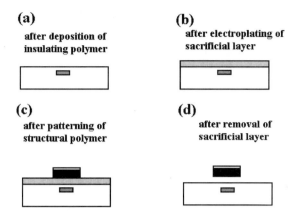

Figure 5.51 Fabrication sequence of the polyimide microplatform.

microplatform is shown in Fig. 5.51. The microplatform is fabricated on an ion-exchanged BK-7 glass substrate. The metallic layer is deposited by evaporation (chromium and gold) and patterned for use as electrodes of an electrostatic actuator. The surface wafer is then spin coated with an insulating layer of polyimide DuPont PI-2555, followed by baking at 130°C for 20 min and a final cure at 300°C for 1 hr [Fig. 5.51(a)]. A 3-μm-thick copper layer (seed layer for electroplating) is then evaporated and patterned to form an anchor for the arms of the movable plate. An additional layer of copper (the sacrificial layer) is electroplated to a final thickness of 15 μm [Fig. 5.51(b)]. A 7.5-μm-thick layer of structural polyimide is then deposited. To pattern this structural layer, a gold mask is evaporated and patterned. The exposed polymer is etched by RIE in CF_4-O_2 plasma. The resulting structure is a gold-covered microplatform over the sacrificial layer [Fig. 5.51(c)]. The final fabrication step permits removal of the copper sacrificial layer in a solution of ferric chloride operating at room temperature [Fig. 5.51(d)]. This switch has a scalable actuation voltage of 50 V to deflect a 100×100-μm^2 microplatform 15 μm from the surface of the wafer.

5.4.2 Micromachined Optical Modulators

The phase of a guided wave can be affected by local modification of the effective refractive index of the waveguide through changes in its boundary caused by the evanescent interactions between the optical waveguide and a small microstructure coated with a dielectric film. Alternatively, phase modulation can be obtained by altering the refractive index of the guiding region using electro-optic, thermo-optic, and acousto-optic effects, respectively. These modulators consist of a dielectric waveguide and an appropriate electrode structure. The device structure can be classified into two generic categories: (1) the substrate on which the optical waveguide is constructed is a piezoelectric material or (2) the substrate is a nonpiezoelectric material. The linear electro-optic effect, which allows operation in the first category, offers efficient phase modulation up to a 10-GHz frequency range. It is restricted to piezoelectric substrates such as LiNbO$_3$, LiTaO$_3$, or GaAs with large Pockels coefficients. In terms of the physical characteristics required to generate highly efficient phase modulation, one concludes that LiNbO$_3$ is the best material. However, the choice of a substrate material is not only based on its physical features. Since MOEMS are mainly constructed on substrates with silicon-based layers, it is attractive to fabricate modulators on nonpiezoelectric substrates. Thermo-optics and acousto-optics in principle allow operation in this category. In any of these approaches, the resulting phase shift may be converted to a change in the dielectric constant of the waveguide material or more simply in terms of its amplitude by use of interferometric or noninterferometric architectures.

5.4.2.1 Phase Modulation by Evanescent Interaction

One approach to constructing optical phase modulators is the use of a movable microstructure such as membrane or cantilever that interacts with a guided beam through evanescent effects. The principle of evanescent interaction is shown in Fig. 5.52.[74] When an optically guided wave is confined within a dielectric guide, the field is not strictly confined to the core layer of the waveguide, but shows exponentially decaying tails in the surrounding materials via evanescent fields. Its amplitude decreases with increasing distance from the waveguide core. The decay length of the evanescent field depends on the waveguide geometry and the refractive index characteristics of the buffer and core materials. On typical single-mode waveguides, the decay length is less than the emission wavelength or on the order of this wavelength. If a movable microstructure with a dielectric film is placed within the reach of the evanescent field, creating a small air gap d (with $d \leq \lambda$, where λ is the wavelength of light), new boundary conditions are imposed on the wave propagating in the waveguide. The interaction causes changes in the effective refractive index n_{eff} of the waveguide that depends on d. A displacement Δd of the

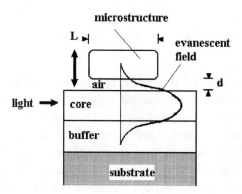

Figure 5.52 Switching principle based on an evanescent interaction.

microstructure induces a variation Δn_{eff} of the effective refraction index over the length L of the microstructure. The resulting phase shift is

$$\Delta \Phi = \frac{2\pi L}{\lambda} \Delta n_{\text{eff}}. \qquad (5.20)$$

The consequence of this interaction may be a phase modulation or intensity modulation. Lukosz et al. proposed a switch based on a Mach–Zehnder interferometer schematically shown in Fig. 5.53. It uses an electrostatically actuated microbridge on one arm of the interferometer.[74] The interferometer transforms the phase modulation $\Delta \Phi$ into an intensity modulation with the output power proportional to $1 + \cos(\Delta \Phi)$. The modulation is obtained by varying the air gap between the microbridge and the reference arm of the interferometer. Electrostatic actuation deflects the bridge and alters the propagation constant of the guided mode. Since the TE and TM modes are affected differently, the polarization also is modified. The waveguide is an Si_3N_4 rib waveguide with a thickness of 0.162 to 0.164 μm and a refractive index of about 2.0. It is grown by an LPCVD process on a thermally oxidized silicon substrate. A 2.02-μm-thick silica buffer with a refractive index of 1.46 is grown by wet oxidation. Etching the silicon nitride layer in buffered hydrofluoric acid (BHF) patterns the Mach–Zehnder interferometer. The resulting interferometer is 35 mm long, with an arm separation of 500 μm. The microbridge is made of oxidized 390-μm-thick silica and is fabricated by two-stage etching in the BHF solution. The resulting air gap is $d = 0.18$ μm. The response time of the microbridge is about 100 μs.

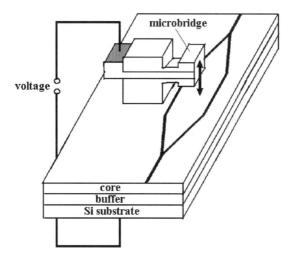

Figure 5.53 Schematic diagram of a Mach–Zehnder optical switch based on an evanescent interaction.

5.4.2.2 Phase Modulation by Elasto-optic Effect

To obtain a device under active phase modulation, Bonnotte et al. proposed to generate acoustic surface waves (SAW) by means of a thin-film piezoelectric transducer deposited near the optical channel waveguide.[46] The acousto-optic interaction mechanism is then based on the change in the index of refraction caused by a mechanical strain, which is introduced by the passage of an acoustic wave. An optical spectrum analyzer using a similar modulation technique has been reported, working up to a 1-GHz frequency range.[75]

The proposed Mach–Zehnder interferometer architecture is shown in Fig. 5.54. Phase modulation is obtained by passing the guided reference beam through a SAW generated on a ZnO thin-film transducer driven by an interdigital electrode structure. To avoid perturbations on the measuring arm of the interferometer, acoustic waves have to be confined to the region of the reference arm by means of an isolation trench. An SiO_xN_y strip-loaded waveguide operating at a 0.633-μm wavelength serves as the waveguiding structure, such as that shown in Fig. 5.20. When a sinusoidal SAW propagates along the x-axis, the refractive index in the region of the acousto-optic interaction is

$$n(x,t) = n + \Delta n \sin(\Omega t + Kx), \tag{5.21}$$

where $K = 2\pi/\Lambda$ is the wave number, n is the average refractive index of the medium, and Ω is the frequency of the SAW.

Assuming the medium is isotropic, the change in the index of refraction n is

$$\Delta n = -\frac{1}{2}n^3 pS, \tag{5.22}$$

where p is the strain-elasto-optic coefficient and S is the strain amplitude.

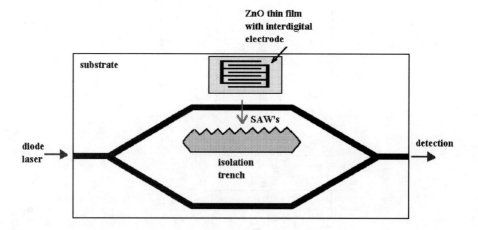

Figure 5.54 Architecture of Mach–Zehnder interferometer with phase modulation by elasto-optic effect.

Figure 5.55 illustrates main fabrication sequences of a Mach–Zehnder integrated interferometer. A strip-loaded waveguide consisting of an SiO_2-SiO_xN_y-SiO_2 sandwich has been deposited by PECVD with the deposition parameters described in Section 5.2.7.2.[20] The strip-loaded waveguide propagation loss was 0.5 dB/cm for TE0 polarization. To pattern the isolation trench, a deep RIE etching using a mixture of $He/SF_6/O_2$ gases was performed. To pattern the interdigital electrode of the piezoelectric transducer, a thin film of Cr was deposited by vacuum evaporation on the front surface of the wafer. The metal layer was subsequently wet etched to pattern the transducer electrode. The deposition condition to generate SAWs is obtained when the c-axis of the ZnO layer is perpendicular to the substrate. ZnO thin film is deposited by sputtering directly on the top of the SiO_2 upper cladding near the reference arm of the interferometer. A 3-μm-

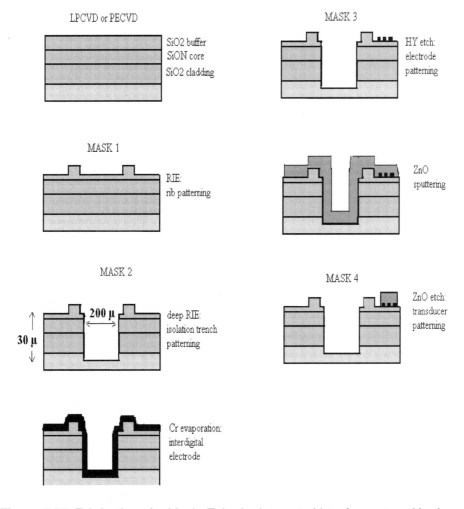

Figure 5.55 Fabrication of a Mach–Zehnder integrated interferometer with phase modulation.

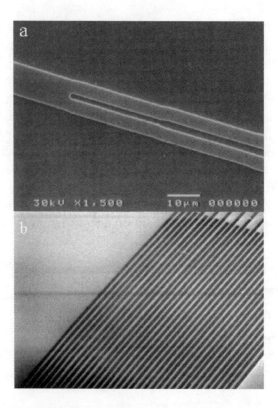

Figure 5.56 SEM photographs: (a) one of Y-junctions just after RIE patterning of the rib, (b) structure of interdigital electrodes.

thick ZnO film is fabricated by sputtering, operating with targets of pure Zn in pure oxygen atmosphere at temperature of 150°C. In the last fabrication sequence, the ZnO layer is removed outside the piezoelectric transducer area. Figure 5.56(a) is an SEM photograph of one of the Y-junctions just after the patterning of the SiO_2 rib by RIE. Figure 5.56(b) shows the structure of the transducer interdigital electrodes. Phase modulation is demonstrated by detecting the spectrum of the interference signal via a spectral analyzer, as shown in Fig. 5.57. There was a component corresponding to the SAW frequency $f_0 = 48$ MHz (first harmonic). The second and third harmonics also appeared as frequency components spaced by f_0. From the detected ratio of the first to third harmonics, it was possible to estimate the phase shift. The phase shift was determined to be $\delta\varphi = 1.3$ rad. The application of the present device is microsensing.

5.4.2.3 Amplitude Modulation by Thermo-optic Effect

Cocorullo et al. reported the constructions of an optical modulator using the thermo-optic effect in amorphous silicon based on a Fabry–Pérot cavity architecture.[76] Figure 5.58 shows the cross-section of the proposed planar waveguide. The

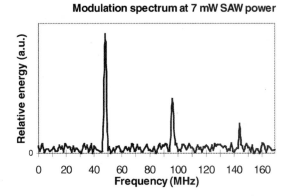

Figure 5.57 Spectrum of the interference signal.

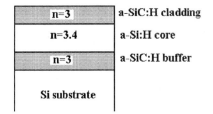

Figure 5.58 Cross-section of amorphous silicon-based planar waveguide.

fabrication process is based on PECVD deposition of core and cladding layers. An Sb-doped <100> silicon wafer is used as substrate material. The structure of the waveguide is a sandwich of an undoped 2.2-μm-thick core layer of hydrogenated amorphous silicon (a-Si:H) between two 0.4-μm-thick cladding layers of hydrogenated carbon silicon (a-SiC:H). The sandwich fabrication starts by PECVD deposition of the first a-SiC:H cladding layer. For deposition of both the cladding layers, the PECVD reactor is operating at a temperature of 180°C, using a gas mixture of SiH_4 and CH_4 at a pressure of 0.7 torr. The refractive index of the resulting cladding layer is 3.0. The deposition of an a-Si:H core layer requires the PECVD process in an SiH_4 atmosphere operating at temperature of 220°C and a pressure of 0.7 torr. The resulting refractive index is 3.4. The propagation loss of this waveguide is 1.8 dB/cm at a 1.3-μm wavelength. A Fabry–Pérot cavity is fabricated by cleaving the output facets of a 2.52-mm long waveguide. The reflectance of the resulting Fabry–Pérot resonator is

$$R = \frac{(n_{Si} - n_{Air})^2}{(n_{Si} + n_{Air})^2} \approx 0.3. \tag{5.23}$$

The Fabry–Pérot interferometer is mounted on a Peltier thermoelectric cooler. When tuned in the temperature range from 30 to 40°C, a periodic ampli-

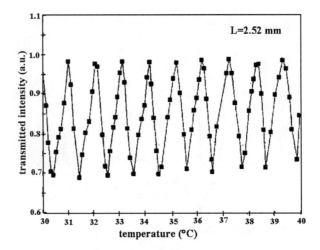

Figure 5.59 Transmitted intensity as temperature function of Fabry–Pérot resonant cavity.

tude modulation is observed, as shown in Fig. 5.59. The Fabry–Pérot cavity transmission is

$$I \approx \frac{I_0}{1 + 4\pi R(1-R)^2 \sin^2 \phi}, \qquad (5.24)$$

with $\phi = \dfrac{2\pi n_{\text{Si}} l}{\lambda}$,

where I_0 is the intensity of incident light, l is the cavity length, and λ is the optical wavelength. The effect of temperature tunability of the Fabry–Pérot interferometer is due to the generation of a strong thermo-optic effect in amorphous silicon materials.

5.4.3 MICROMACHINED OPTICAL SENSORS

For many optomechanical sensor applications, a single-crystal silicon is the substrate of choice because of its excellent micromechanical and electronic properties, and the facility of integrating sensing and electronics on the same chip. Even if the silicon presents a relatively high cost per unit area, this material demonstrates greater flexibility in design and machinability than other substrates. In optomechanical sensors, the active structural element converts a mechanical external input signal (force, pressure, acceleration, etc.) into an electrical signal output via an optical waveguide readout of the micromechanical sensing element. The optical interrogation of micromechanical structures can be obtained by interferometric or

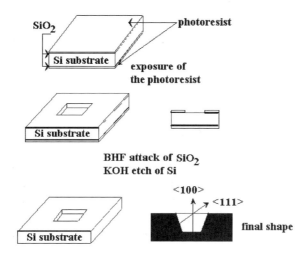

Figure 5.60 Fabrication sequence of the stress-induced membrane.

intensity modulation techniques, for example. Structurally active elements are typically high aspect ratio components, such as suspended beams or membranes. Certainly the most successful application of micromachined integrated optical sensors is in the area of pressure transducers. The measuring architecture typically requires optical waveguides crossing the SiO_2/Si membrane near the area where the strains are largest. A pressure differential across the membrane causes deflections that induce strains in the membrane, thereby modulating the optical signal.

SiO_2 membranes may be currently fabricated using the KOH anisotropic etching. An example of fabrication of a stress-induced membrane is shown in Fig. 5.60.[77] The membrane was fabricated from a <100> silicon wafer where two films of SiO_2 (0.52 μm thick) were formed on both the sides of a silicon substrate. A part of the top side of the SiO_2 film was patterned by lithography and then the SiO_2 layer was removed by wet etching with BHF. This serves as an etch mask during the diaphragm patterning; the silicon substrate was etched by a KOH solution from the top side. The residual stress in a 4×4 mm^2 membrane has been measured experimentally in the absence of pressurization by an interferometric fringe projection technique.

Figure 5.61 shows the residual stress at the Si/SiO_2 interface. The stress distribution of this square membrane presents one symmetrical oscillation. The rectangular membranes present a stress distribution with a number of oscillations equal to the membrane length-to-width ratio. The stress magnitude is strongly dependent on the fabrication process because the residual stress varies with CVD deposition parameters (temperature, pressure, doping) and thermal annealing conditions. Consequently, if the principle of optical interrogation requires elasto-optical properties of the membrane, the optical readout sensitivity will depend on the membrane design and the location of the optical waveguide crossing the membrane, as demonstrated in Section 5.4.3.3.

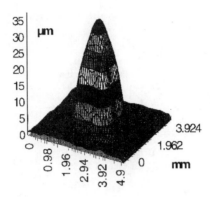

Figure 5.61 Residual stress distribution at the Si/SiO$_2$ interface (experimental data).

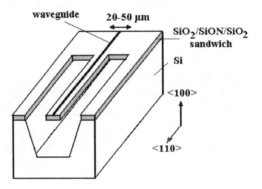

Figure 5.62 Microstructure with an integrated optical readout (silicon oxinitride strip-loaded waveguide).

The use of stress-induced membranes increases the sensitivity of optical readout due to increasing the stress-induced birefringence of the optical waveguide. In other applications, it is suitable to compensate for the residual stress. Thus, Bezzaoui et al. observed that thermal oxide membranes have a tendency to crack that is caused by the difference in thermal expansion of Si-SiO$_2$ (during oxidation).[22] When etching SiO$_2$-SiO$_x$N$_y$-SiO$_2$, membrane microcracks are not observed because a low stress in this sandwich structure is obtained. An excellent strip-loaded waveguide can be fabricated with this three-layer structure. An example of such a microstructure carrying an integrated optical readout is shown in Fig. 5.62. This is a vibration sensor using an SiO$_x$N$_y$ strip-loaded waveguide deposited on a <100> oriented silicon substrate as an optical interrogation. Figure 5.63 shows an SEM image of a microresonator whose mechanical resonance frequency changes with thermal activation; the cavity, the waveguide, and a thin film heater are visible. Similar microstructures have been used successfully for light coupling between

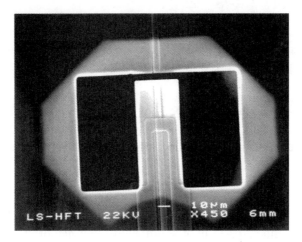

Figure 5.63 SEM image of a microresonator with a mechanical resonance frequency that changes with thermal activation.

two waveguides located on cantilevers and separated by an air gap (coupling efficiency of 30–50% through a 4-μm gap).

Guided-wave micromachined sensors combine a number of different technologies and new sensing materials that are much superior to other MOEM devices. A variety of wet and dry etching and patterning techniques are used to produce microstructures that either provide mechanical support for the sensing materials, or are themselves part of the sensing structures. Thin films of varied materials are also used to convert a number of physical and chemical parameters into an electric signal, which is obtained via a waveguide interrogation. In the following sections different working principles of micromachined optical sensors are discussed and a number of examples of recently developed microsensors are given.

5.4.3.1 Micromachined Amplitude-Modulated Optical Sensors

Wu et al. demonstrated an amplitude-modulation sensor based on a microcantilever beam and a microbridge that are interfaced with optical waveguides, as shown in Fig. 5.64.[78] These guided-wave microstructures have been applied in acoustic signal detection. A micromechanical cantilever beam and bridge deflects when an acoustic pressure is applied. In the bridge structure, this deflection increases the propagation loss of the guided wave propagating through the bridge. Thus, a mechanical vibration is directly converted into an intensity-modulated signal, measured in the output section of the waveguide. In the cantilever structure, apart from the propagation loss due to the deflection, a decrease in light coupled across the gap into the output waveguide is measured.

The waveguide consists of a 0.3-μm-thick core layer of Al_2O_3 sandwiched between a 1.7-μm-thick silica buffer and a 0.5-μm-thick silica cladding. The propagation loss of this planar waveguide is about 1 dB/cm at a 0.633-μm wavelength.

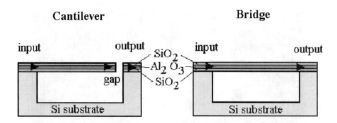

Figure 5.64 Micromachined amplitude-modulation sensors for acoustic signal detection.

The typical microbridge is 100 μm in length, and 5 μm in width. Microcantilevers are 100 μm in length and 30 μm in width, with an air gap of 25 μm. The coupling efficiency of the microcantilever may be obtained by calculating the overlap integral of the Gaussian profile emerging from the input section to the output section through the gap d:

$$\eta = \frac{2w_0 w_1}{w_0^2 + w_1^2} e^{2\delta^2/(w_0^2 + w_1^2)}, \qquad (5.25)$$

with $w_1^2 = w_0^2 \left[1 + \left(\frac{\lambda d}{\pi w_0^2} \right)^2 \right]$,

where δ is the deflection offset between the center of the input and output waveguides, and w_1 determines the beam waist of the Gaussian laser beam.

According to Eq. (5.19), the deflection of the cantilever subjected to a static force F is $\delta = Fb^3/3EI_Y$. Figure 5.65 illustrates the response of the microcantilever and microbridge structures to an acoustic signal frequency modulated at 1 kHz. Both microstructures are capable of measuring deflections in the range of 0.01 nm and forces on the order of 10^{-10} N.

Burcham et al. developed a micromachined cantilever beam accelerometer with optical readout based on an integrated optical waveguide located in the path of the cantilever beam, as shown in Fig. 5.66.[79] Light coupled across the air gap (4, 6, or 8 μm) is related to the amount of deflection of the beam tip. Displacements on the order of a few hundred angstroms could be measured. A foot mass incorporated onto the end of the beam permits the measurement of acceleration. Standard photolithography and anisotropic etching of the silicon substrate are used to fabricate the cantilever. The cantilever structure is patterned from the wafer back side so as to leave a gap between the tip of the diving board and the opposite fixed edge on the front side of the wafer. Subsequently, a planar waveguide is deposited on the top of the wafer. The buffer layer is a 2-μm-thick layer of thermal SiO_2 ($n = 1.46$) and the core layer is a 0.12-μm-thick LPCVD silicon nitride film ($n = 2.02$).

OPTICAL WAVEGUIDES AND SILICON-BASED MICROMACHINED ARCHITECTURES   279

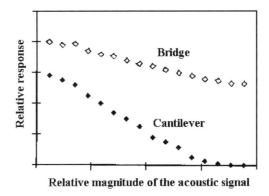

Figure 5.65 Response of microcantilever and microbridge structures to an acoustic signal frequency modulated at 1 kHz.

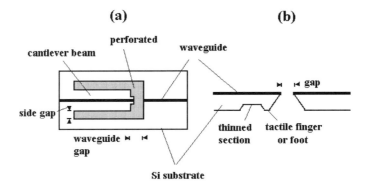

Figure 5.66 Micromachined cantilever beam accelerometer with an optical readout: (a) top view, (b) edge view.

5.4.3.2 Micromachined ARROW Waveguide Optical Sensors

An example of application of ARROW waveguides is the micromachined optical pressure sensor shown in Fig. 5.67.[53,80] This silicon Mach–Zehnder interferometer has its sensing arm suspended over a micromachined cavity. An applied pressure causes deflection of the suspended arm, resulting in a change in the optical path length and a modulation of the output light intensity. Dimensional parameters and fabrication sequences of the ARROW waveguide are given in Section 5.2.7.5 (Fig. 5.33). After the definition of the ARROW by patterning of the top layer of LTO, the field oxide is patterned and etched until the c-Si substrate is exposed for micromachining. The anisotropic etching of the underlying substrate is performed using an EDP solution at 105°C for 1 hr. Thus, the etched cavity below the suspended Mach–Zehnder arm is formed. Prior to the EDP etch, the device is subjected to an annealing procedure to reduce the residual stress in the suspended sensing arm of the interferometer (at temperatures from 1100 to 500°C). The cs-

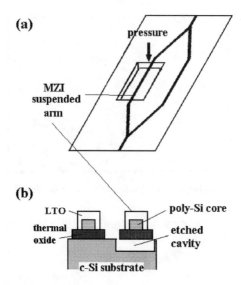

Figure 5.67 Micromachined ARROW Mach–Zehnder interferometer with a suspended sensing arm.

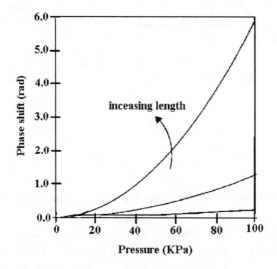

Figure 5.68 Estimated phase shift as a function of applied pressure for three suspension lengths (250, 200, and 150 μm).

timated change in phase as a function of applied pressure for three suspension lengths (250, 200, and 150 μm) is shown in Fig. 5.68. The sensitivity of this interferometer increases if the suspension length is increased. Thus, a pressure of 100 kPa applied to a suspension of length 250 μm produces a phase shift of about 2π. This determines the maximum operating pressure range for a specific sensing arm length.

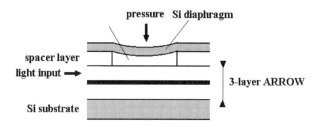

Figure 5.69 ARROW pressure transducer based on evanescent modulation.

Figure 5.69 shows a noninterferometric version of the ARROW pressure transducer, based on the evanescent modulation of the output power.[81] The architecture is based on a deformable silicon diaphragm clamped at a fixed distance from the waveguide surface. When external pressure is applied to the diaphragm, it deflects and goes into closer proximity with the waveguide, producing the light coupling from the waveguide into the diaphragm. To promote the coupling, the transducer operates at a 0.633-μm wavelength, where the silicon diaphragm is absorbing. Pressure can be measured by monitoring the light intensity modulation caused by the deflection of the diaphragm. The advantage of evanescent coupling is the high sensitivity to the gap modification between the interacting structures. The upper half part of the device, including the diaphragm and spacer layer, is formed by deposition of thermal oxide on a <100> oriented silicon substrate for a thickness ranging from 0.1 to 1 μm, followed by plasma etching of the silicon to form the diaphragm. Waveguide and diaphragm wafers are anodically bonded together at 450°C for 30 min. After the bonding, the entire device is thermally annealed at 1000°C for 1 hr.

5.4.3.3 Micromachined Optical Pressure Sensors Using Elasto-optic Effect

Fischer et al. proposed an integrated optical pressure sensor, as shown in Fig. 5.70.[45] It uses two Mach–Zehnder interferometers crossing a silicon membrane with a force plate located at the center. Both arms of each interferometer are situated inside the square membrane, one at its center and other at its outer edge. When a pressure is applied on the force plate, the deflection causes strains in the membrane, thereby modifying the optical path length of the interferometer. An He-Ne source laser is coupled via a fiber to the input Y-junction of the interferometer and the pressure induced phase shift is detected at the output Y-junction via an integrated detector. The waveguide technology is based on an SiO_xN_y single-mode strip-loaded waveguide, similar to that described in Section 5.2.7.2 (Fig. 5.20). It consists of an $SiO_2/SiO_xN_y/SiO_2$ sandwich with the refractive indices of the buffer and cladding layers at 1.46 and that of core layer fixed at 1.50. SiO_2 and SiO_xN_y layers are deposited on a silicon substrate by two different processes: one consists of thermal oxidation and an LPCVD process, the other requires only the PECVD. Thus, the elasto-optic properties of LPCVD- and PECVD-deposited waveguides

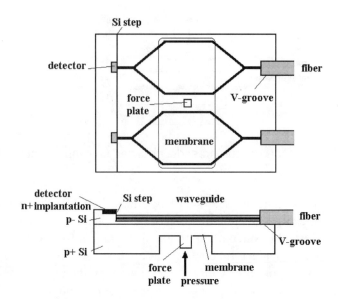

Figure 5.70 Micromachined interferometric pressure sensor using elasto-optic effect.

have been compared and their impact on the measuring performance of the interferometer has been analyzed. A set of membranes with different areas (4×4 and 4.5×4.5 mm^2) and thicknesses (50 to 100 μm) is fabricated by plasma etching in an SF_6-O_2 gas mixture. A p-i-n detector is integrated in the epilayer on the silicon substrate. It is based on the p^+-doped silicon, p^--doped epilayer and n^+-doped phosphorus implanted top contact.

When the waveguide is deformed, the strain-induced optical birefringence changes the refractive index through the elasto-optic effect. The relationship between the principal refractive index n_i and the principal stress σ_1 is given by

$$n_i = n + C_1\sigma_1 + C_2(\sigma_j + \sigma_k), \qquad (5.26)$$

with $C_i = -\dfrac{n^3}{2E}(p_{11} - 2v_{12}p_{12})$ and $C_2 = -\dfrac{n^3}{2E}(p_{12} - vp_{11} - vp_{12})$,

where n is the refractive index of the unloaded material, C_1 and C_2 are elasto-optical coefficients, and v is the Poisson ratio.

When a pressure is applied on the force plate, the strain distribution is not homogeneous over the entire membrane. The magnitude of strains is highest at the edges of the membrane, compressive at the outer, and tensile at the inner rim. There is also a difference between the strain distribution generated by thermal oxidation followed by a LPCVD and the PECVD process only. Since the resulting elasto-optic effect is not uniform over the entire membrane, the measuring sensitivity of the interferometer depends on the location of the sensing arm inside the

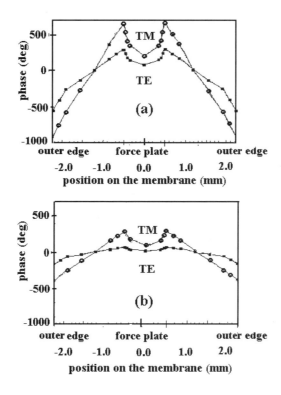

Figure 5.71 Phase shift of the interferometer as function of the waveguide location: (a) PECVD waveguide, (b) LPCVD waveguide.

membrane. The Mach–Zehnder interferometer formed by thermal oxidation and an LPCVD process exhibits lower elasto-optic coefficients C_1 and C_2 than an interferometer made by PECVD deposition. The elasto-optic effect is also stronger in PECVD-deposited waveguides. PECVD waveguides demonstrate higher sensitivity in a TM-polarization mode than in the TE mode, due to the stronger phase shift obtained with this polarization mode. Alternatively, the TM-polarization mode of LPCVD waveguides exhibits very high propagation loss. Figure 5.71 shows the phase shifts for PECVD and LPCVD waveguides plotted as function of the waveguide location around the center of the membrane. This FEM calculation was performed for a 100-μm-thick membrane and both TE and TM modes of polarization. The maximum stress of this membrane type is calculated at 100 MPa.

Ohkawa et al. developed a micromachined integrated optic pressure sensor, as shown in Fig. 5.72. It is based on the Mach–Zehnder interferometer architecture with a thin diaphragm structure used as the sensing elements, while the sensing arm of the interferometer crossing the diaphragm acts as an optical readout of the deformable diaphragm. With the application of the pressure, a mechanical strain is generated and a change in the refractive index of the waveguide is induced via the elasto-optic effect. The interferometer detects the pressure difference between the front and back surfaces of the diaphragm. An n-type <100>-oriented silicon

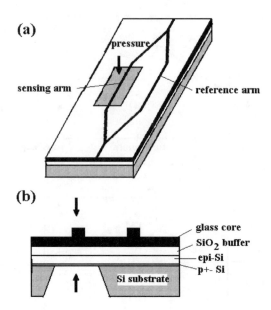

Figure 5.72 Schematic diagram of a micromachined integrated optic pressure sensor with optical glass waveguide: (a) top view, (b) cross-section.

substrate, p^+-doped on the front side with a 5-μm-thick epitaxial layer, is used. The fabrication sequence starts by thermal oxidation of the silicon on both the sides of the substrate. Thus, 0.3-μm-thick layers of SiO_2 are grown. To pattern the diaphragm, the backside SiO_2 layer is etched by the buffered HF solution, while the front side of the wafer is protected by an etch stop made of a 0.1-μm-thick film of Cr deposited by evaporation. Using the patterned film of SiO_2 as an etch mask, the diaphragm is etched from the back side with an EDP solution. Anisotropic etch stops on the p^+-doped layer. The diaphragm size is 1×0.22 mm^2. The waveguide consists of a 1.8-μm-thick buffer layer of silica, and a 1.2-μm-thick core layer of Corning 7059 glass fabricated by sputtering. Photolithography and RIE etching in CHF_3 gas form the rib-waveguide structure. The rib is 0.14 μm deep and 3.6 μm wide. The measured propagation loss of this waveguide is 0.1 dB/cm at a 0.633-μm wavelength. Figure 5.73 shows the output intensity measured as a function of the pressure difference. The maximum applied pressure is 0.6×10^5 Pa.

5.4.3.4 Micromachined Bragg Gratings for Optical Sensing

The combination of micromachined Bragg gratings with an optical waveguide can be used as strain-sensing elements in distributed optical sensors, providing high resolution in acceleration measurement or optical motion control systems. The reflecting wavelength λ_B of the first-order Bragg grating is

$$\lambda_B = 2\Lambda n_{\text{eff}}, \tag{5.27}$$

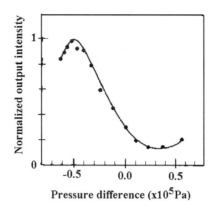

Figure 5.73 Measured output intensity as function of the pressure difference.

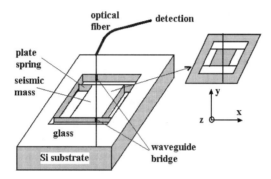

Figure 5.74 Micromachined optical accelerometer with waveguide Bragg gratings.

where Λ is the grating period and n_{eff} represents the effective refractive index of the waveguide.

An integrated accelerometer based on strain sensing by an optical readout using a Bragg grating in an optical waveguide has been proposed by Storgaard-Larsen et al.[83] The architecture of this 2D accelerometer is shown in Fig. 5.74. Four parallel springs and two waveguide bridges suspend a seismic mass. The sensitive y-axis is perpendicular to the spring plates that ensure the movement of the seismic mass only along the x-axis. The symmetry axis of the waveguide bridges is parallel to the y-axis. Waveguides act as dominating springs in this direction, operating at the same time as strain gauges. When the structure is subjected to acceleration along the y-axis, the waveguide bridges come under compression. Optically speaking, the compression produces the strain-induced birefringence in the waveguide.

The fabrication sequence of a waveguide bridge with a Bragg grating is shown in Fig. 5.75. A <110>-oriented silicon is used as a substrate material. The formation of the waveguide starts by PECVD deposition of a 5-μm-thick SiO_xN_y buffer layer with a refractive index of 1.47–1.48. The core layer of 1-μm-thick germanium-doped glass is deposited by PECVD. This structure is stabilized by

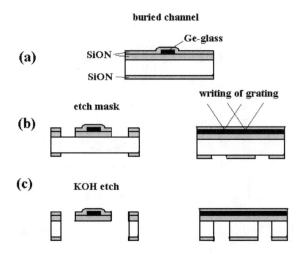

Figure 5.75 Fabrication sequence of the waveguide bridge with Bragg gratings.

thermal annealing at 800°C in an N_2 atmosphere. To define the core ridge of the buried channel waveguide, the core layer is patterned and etched by RIE, followed by PECVD deposition of an SiO_xN_y cladding layer grown to a thickness of 2 μm. Then an etch mask of a 1-μm-thick SiO_xN_y layer is deposited by PECVD on the back side of the substrate [Fig. 5.75(a)]. A second annealing process is repeated and the PECVD layers on both the sides of the wafer are patterned and etched by RIE to create an etch mask for the final step of anisotropic etching of the silicon [Fig. 5.75(b)]. Then Bragg gratings are written in the surface of the waveguide using UV two-beam interferometry. By using a 0.193-μm excimer laser, a change in the refractive index of about 3×10^{-3} can be photoinduced. Finally, wet etching in KOH solution is used to release the bridge structures [Fig. 5.75(c)]. Etching is stopped when the <111>-oriented planes are released in the corners of the plate springs (the top angle of these corners is 70.52°). An accelerometer structure with a waveguide propagation loss less than 1 dB/cm and a Bragg grating reflectance of 80% has been obtained. With a spring dimension of $15 \times 450 \times 1000$ μm^3, a waveguide bridge size of $6.5 \times 30 \times 1000$ μm^3, and a 0.8-mm long Bragg grating operating at a 1.3-μm wavelength, the mechanical sensitivity of the accelerometer is 1.4 nm/g.

Tomiyoshi et al. developed a micromachined optical bending sensor with Bragg grating along an elastomer waveguide on a silicon substrate.[84] When the waveguide is subjected to a deformation, the pitch of the grating is modulated by the strain. Polystyrene thermoplastic elastomer (TPE) and silicon rubber are used as core and cladding materials with refractive indices of 1.51 and 1.41, respectively. Figure 5.76 shows the fabrication sequence of this ridge waveguide with Bragg gratings. This starts by the spin coating of a 5–10-μm thick core layer of TPE diluted at 15-wt.% in a solution of toluene [Fig. 5.76(a)]. To promote a good uniformity of the spin-coated film, a special spin coater was developed with a well-

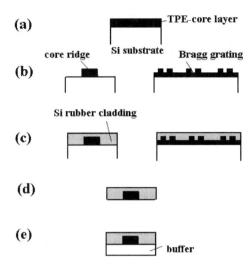

Figure 5.76 Fabrication sequence of elastomer ridge waveguide with Bragg gratings.

controlled rotation speed. Subsequently, the sample is baked at 100°C for 1 hr. An etch mask of a 0.3-μm-thick film of Al is evaporated and patterned for the subsequent RIE processing. To form the ridge core of the waveguide, an RIE process in an oxygen atmosphere is used [Fig. 5.76(b)]. It operates at a low temperature (−5°C) to avoid thermal damage to the sample and at a pressure of 12 mtorr. The Bragg gratings are fabricated by electron beam lithography directly on the surface of the TPE core ridge. After the Bragg gratings are written, the cover silicon rubber cladding layer is spin coated [Fig. 5.76(c)]. The sample is baked in a vacuum atmosphere at 150°C for 30 min. Finally, the silicon is etched out using XeF_2 vapor [Fig. 5.76(d)] and the buffer layer is dip coated [Fig. 5.76(e)].

5.4.3.5 Micromachined Interferometric Distance Sensors

In contrast to optomechanical sensors, the interest in constructing optical distance sensors by micromachining technology is based mainly on the ability to deposit and pattern the waveguide layers and to integrate the processing electronics and photodetectors on the same chip.

LETI developed a micromachined interferometric displacement sensor, shown in Fig. 5.77.[50] It is a Michelson interferometer fabricated on a silicon substrate and using the structure of single-mode strip-loaded waveguides made of a three-layer $SiO_2/Si_3N_4/SiO_2$ sandwich. The first 2-μm-thick SiO_2 buffer layer is grown by thermal oxidation of silicon, whereas the cladding layer of silica is obtained through PECVD. The core layer of silicon nitride is deposited by LPCVD with a thickness between 0.12 and 0.18 μm and a refractive index of 2.014. The change in the effective index of the guided mode is obtained through a controlled RIE etching of a strip in the top silica cladding. Etching of the SiO_2 cladding forms the

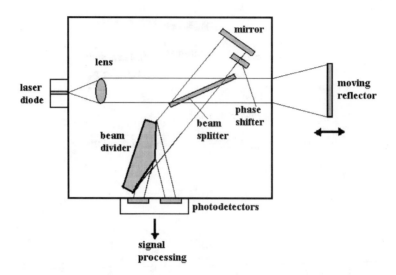

Figure 5.77 Schematic diagram of integrated displacement sensor developed by LETI.

collimating step index integrated lens, the beam splitter, and the phase shifter. The mirror and the beam divider are fabricated by etching the structure up to the silicon substrate. Finally, the input and output facets of the interferometer are obtained by cleaving. The light source is a commercially available Hitachi diode laser coupled to the input of the interferometer, and the detector is a twin photodiode. After collimating by the step lens, the light beam is divided into reference and sensing arms. The light in the sensing arm is guided up to the output plane, where it is then diffracted out. After reflection on a moving reflector, the sensing beam is coupled back into the waveguide in which it interferes with the reference beam. The beam divider permits the observation of two interferograms in quadrature (a phase shift of 90 deg). With a 0.78-μm wavelength, a resolution of 100 nm was obtained at a distance of 200 mm.

An integrated Michelson interferometer for high-resolution displacement measurement of an external reflector has been proposed.[85] Figure 5.78 shows the architecture of this device. It is a Michelson interferometer with two reference arms and one sensing arm, each of them connected to a polarization-maintaining optical fiber. The incident He-Ne light is divided by a Y-junction. The symmetric directional 3-dB couplers play the role of a beam splitter. Thermo-optic phase modulators create a phase shift of 90 deg between both reference arms of the interferometer. A 0.5-μm-thick Al-evaporated mirror on the cleaved waveguide end face terminates each of the reference arms. An AR-coated quarter-pitch GRIN microlens is located at the end face of the sensing waveguide. This permits one to extend the sensing arm of the interferometer, providing displacement measurements up to more than 100 mm of distance. The interference signal is fed to a pigtailed photodiode. The waveguide is a single-mode strip waveguide (strip width: 2 μm) fabricated

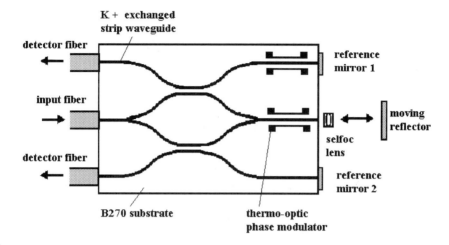

Figure 5.78 Schematic diagram of integrated microdisplacement sensor with thermo-optic phase modulation.

by a K^+ ion-exchange process in B270 glass at an operating temperature of 450°C and under an applied voltage of 50 V/mm. The electrodes of the thermo-optic modulator are delineated from a 0.4-μm-thick Al thin film with a length of 7 mm and a width of 4 μm. This modulator is used for high-resolution optical heterodyning with phase-modulated single sideband detection. At a 0.633-μm wavelength, a resolution of 1 nm at a distance of 35 mm is obtained.

5.4.3.6 Micromachined Optochemical Sensors and Biosensors

Guided-wave optics added to microstructures offer an attractive potential for development of chemical sensors as well as biosensors. In contrast to optomechanical sensors, where the excellent mechanical properties of single-crystalline silicon tend to promote silicon technology, the choice of the optimum manufacturing technology for chemical sensors and biosensors is far less evident. For this category of sensors, silicon often plays no role other than as a passive substrate, and the focus is mainly on new chemistries rather than micromachining. Our interest here is to explore how micromachining combined with guided-wave optics could help this cause. In many cases, micromachining could provide an inexpensive manufacturing platform and solve some of the difficulties involving patterning of relatively thick organic layers such as ion-selective membranes or polymeric active layers. Finally, in optical sensing there is not an absolute need to use reference electrodes, and simultaneous measurement of several chemical species is possible by simply adjusting the optical wavelength.

Chemical sensors and biosensors can be subdivided into three categories according to their optical readout principles: Mach–Zehnder interferometry, surface plasmon resonance, and luminescence quenching.[86,87] As we have seen, the properties of the waveguide's ambient medium affect the propagation mode, owing to

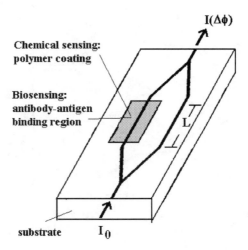

Figure 5.79 Schematic diagram of a Mach–Zehnder interferometer for chemical and biosensing.

the evanescent field illustrated in Fig. 5.52. By using this, the measurement of the refractive index of fluid or gaseous substances on the top of the waveguide is possible. The required detection of the effective refractive index can be done with interferometers of the Mach–Zehnder type. Fabricius et al. report on a gas sensor whose architecture is shown in Fig. 5.79.[88] The working principle is based on the change in the refractive index n_{eff} of a polysiloxane polymer film ($n = 1.4705$) deposited over the sensing arm of a Mach–Zehnder interferometer. The output light intensity $I(\Delta\varphi)$ as a function of the incident light intensity I_0 is

$$\frac{I(\Delta\varphi)}{I_0} = \frac{1}{2}\left[1 + \cos\left(\frac{2\pi}{\lambda}L\Delta n_{\text{eff}}\right)\right] \approx \cos^2\left(\frac{\Delta\varphi}{2}\right), \quad (5.28)$$

with $\Delta\varphi = (2\pi L/\lambda)\Delta n_{\text{eff}}$ where L is the length of the polymer coating.

The polymer coating is 0.5 to 2 μm thick and 13 mm long. A dielectric coating that is thick enough to prevent the penetration of the evanescent field of the modes into the gas protects the reference arm. If organic hydrocarbons are passed over the polysiloxane superstrate, the molecules penetrate this sensitive layer and change its refractive index according to the gas concentration. The phase change with ambient refractive index increases with an increase in refractive index, as the evanescent field increases with the refractive index of the surrounding medium. The measurement range has an upper limit, which is defined by the refractive index of the substrate material. The limit of detection is 100 ppm in the case of gaseous perchloroethylene.

If the sensing architectures described above provide high sensitivity to integrated optics, their specificity as biosensors has to be provided by biochemistry. With a biochemically sensitive coating on their surface, which specifically binds certain analytes from the sample, the IO sensors become biosensors. An example

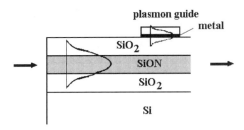

Figure 5.80 An optical waveguide constructed for excitation of plasmon resonance.

of optical immunosensors using a Mach–Zehnder approach is reported by Boiarski et al.[89] It is capable of detecting the protein anti-immunoglobulin-G in blood at concentrations of 1 μg/ml in less than 10 min. The working principle is based on selective binding of antigen–antibody pairs. For this, the sensing arm is coated with a thin film containing an antibody that selectively binds the antigen molecules from the sample to the sensing arm's surface (Fig. 5.79). Interaction with antigen influences the refractive index of the sensitive window. The reference arm is protected from the environment by an SiO_2 buffer layer ($n = 1.468$). The waveguide is formed by KNO_3 ion exchange in a glass substrate with refractive indices of the core and cladding layer of 1.53 and 1.5, respectively. The Y-junctions of the interferometer are made from a 4-μm-wide single-mode channel waveguide operating at a 0.633-μm wavelength. The arm length of this interferometer is 10 mm and the separation gap between the sensing and reference arms is 50 μm. This optical immunosensor can measure refractive index changes of about 0.005.

Another way of detecting the properties of thin films is the excitation of surface plasmon resonance in a metal surface (e.g., 0.5 μm of Ag or Au) deposited on a waveguide surface. A surface plasmon mode travels along the interface between a metal and a dielectric material. These modes are always strongly attenuated. By absorption of species such as gases, or the binding of macromolecules onto the metal film, the effective refractive index of the guided mode is changed and the absorption of species such as gases or the binding of macromolecules onto the metal film can use the modification of the propagation constant as an optical readout. Figure 5.80 shows a schematic diagram of an optical waveguide constructed for excitation of plasmon resonance. The surface plasmon is confined to an area of typically 50 nm around the metal surface. An elegant alternative to implementing surface plasmon resonance by using micromachined planar optical waveguides is shown in Fig. 5.81.[86] It is an array of channel silicon oxinitride waveguides with different cross-sections, each fabricated with a transduction thin film.

A final example of the detection principle is luminescence quenching. In this technique a luminescent agent is optically excited and the intensity of the emitted light gives information about, for example, biochemical reactions that may have taken place in the agent. A decrease in the measured intensity may act as an optical readout of the presence of certain species. Kreuwel et al. proposed using an

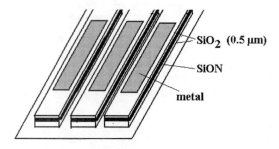

Figure 5.81 Schematic diagram of surface plasmon-waveguide array sensor.

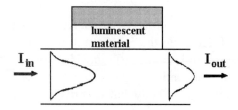

Figure 5.82 Basic principle of luminescence quenching in an optical waveguide.

integrated optical waveguide coated with luminescent films.[90] Through the evanescent field, a strong coupling between the guided wave and the photoluminescence transduction layer is obtained (Fig. 5.82).

5.4.4 MICROMACHINED NANOPROBES FOR NEAR-FIELD OPTICAL MICROSCOPY

Fabrication of nanostructures on silicon is of great interest from both a technological and a fundamental point of view. Technological applications, for example, include the fabrication of nanoprobes for near-field optical microscopy with improved form and nanoscale lateral resolution. Today most near-field nanoprobes consist of specially processed optical fibers. Typically a fiber is processed by chemical etching and then a metal thin film is deposited onto the fiber end, where an aperture is provided, thus forming the metallic aperture apex. These probes are difficult to fabricate with high reproducibility and in large numbers. One solution to overcoming this handicap is to use silicon micromachining for fabrication of such nanoprobes. Thus, the aperture tip may be fabricated in a reliable batch process, which has the potential for implementation of new monolithically integrated architectures for scanning probe microscopy.

Abraham et al. proposed a micromachined aperture probe for combined atomic force/near-field optical microscopy.[91] This consists of a micromachined cantilever with an integrated optical waveguide and a transparent near-field aperture tip, as shown in Fig. 5.83. This figure represents the uncoated version of the tip. Two

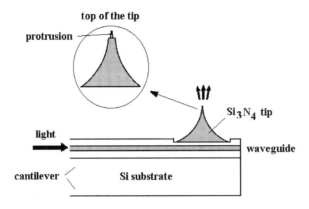

Figure 5.83 Micromachined Si_3N_4 waveguide-cantilever with a near-field aperture tip.

different waveguide technologies are used: an SiO_xN_y strip-loaded waveguide and an SiO_xN_y embedded-rib waveguide. A 3-μm-thick layer of thermal SiO_2 is used as a buffer layer. The core layer consists of a 0.5-μm-thick PECVD SiO_xN_y layer with a refractive index of 1.53. The cladding layer is a 0.7-μm-thick layer of SiO_2 deposited by a PECVD process ($n = 1.49$). The propagation loss measured for the strip-loaded structure is 0.5 dB/cm and for the embedded rib waveguide, 1 dB/cm. The cantilever is formed by isotropic RIE etching followed by anisotropic etching in KOH solution after the protection of the waveguide with a metal mask. The core of the probe tip with a nanometric protrusion is made of silicon nitride with a refractive index of 2.35. The curvature radius of the apex is about 10–20 nm. The fabrication of the tip starts by PECVD deposition of a 4.5-μm-thick layer of Si_3N_4 where the tip shape is defined photolithographically with photoresist 6 × 6 μm² pads. Isotropic dry etching in CF_4 plasma performs the processing of the tip. Subsequently, the tip is coated with a 0.3-μm-thick Al film that is selectively removed at the apex of the tip, resulting in a small aperture. Since the refractive index of the Si_3N_4 tip is higher that than of the waveguide cladding, the light couples from the waveguide into the tip. To promote the coupling, the top silica cladding layer is thinned under the tip base.

As we saw earlier, amorphous silicon (a-Si) is an interesting material for the fabrication of waveguides with high confinement of optical field because of its extremely high refractive index ($n \sim 3.6$). Sasaki et al. proposed an a-Si waveguiding cantilever for optical near-field microscopy.[92] Figure 5.84 illustrates the fabrication sequences of this cantilever. The fabrication starts by thermal oxidation of a <100> oriented c-Si substrate grown to a thickness of 0.44-μm. This SiO_2 layer acts as the buffer layer of the waveguide. To form the core layer, a 0.27-μm-thick a-Si film is deposited by LPCVD. A chromium film is evaporated on the front side of the wafer. To define the tip shape, this metallic layer is used as the mask for the subsequent etch by a focused ion beam (FIB) operating with SF_6 gas. Then the front side of the substrate is coated with a 1-μm-thick PECVD SiO_2 film used

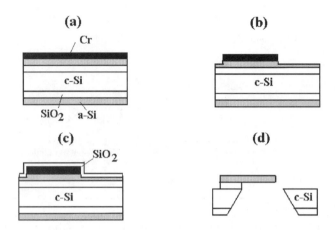

Figure 5.84 Schematic diagram of micromachined a-Si waveguide-cantilever.

Figure 5.85 SEM image of micromachined a-Si waveguide-cantilever.

as the protection from the tetramethylammonium hydroxide (TMAH) etchant that performs the a-Si and c-Si etching. Finally, the whole backside a-Si is removed, and a window is made in the backside SiO_2 layer, followed by etching of c-Si until the etch stops at the silica layer. Thus, the SiO_2 membrane is etched to release the a-Si cantilever. Figure 5.85 shows the SEM image of the fabricated a-Si waveguide-cantilever.

5.5 NEW TRENDS AND CONCLUSIONS

Silicon micromachining, which includes photolithography, deposition, and etching, makes it possible to mass produce high-precision micromechanical structures. Such high-precision components have an evident potential and good matching with integrated optoelectronic systems. Beginning with the development of the first waveguide technology compatible with microfabricated deformable structures, a new category of optomechanical devices was created, the guided-wave MOEMS. Based mainly on silicon surface micromachining or bulk micromachining technology as well as RIE etching, different MOEM architectures have been developed. Basically, this emerging technology has proposed two aspects of planar waveguide technology in which silicon micromachining can be used successfully:

- to manufacture and assemble perfectly aligned precision-defined optical components,
- to construct optical modulation interfaces by using the physical properties of thin-film multilayer structures.

In the past 5 to 8 years, MEMS technology, integrated optoelectronics, and micro-optic technologies have progressed rapidly, enabling the combination of photonic components to activate and power conventional MOEMS structures. The integration of planar optical waveguides with micromachined structures and the inclusion of micro-optic elements within a MEMS environment offer significant promise for achieving the advanced functionality of MOEMS necessary to meet the performance needs of a number of emerging sensor, communications, and near-field scanning microscopy applications. The progress of the state of the art in guided-wave MOEMS is currently impeded by limitations in the enabling technologies. The material base for MOEMS relies on silicon and its related techniques for the most part. Future progress of the field requires a broader materials base, and the integration of photonic components and electronics with MOEMS is a key contributor to the success of most MOEMS devices. Silicon-based micromachining is suitable for fabrication of micromechanical structures integrated with electronic devices, but it cannot monolithically integrate active optical components with micromechanical structures. By placing lasers and photodiodes closer to sensors and actuators, the performance of a device is improved.

Several different approaches have been taken in the integration of photonics with MOEMS. The embedded or monolithic approach, in which the photonics are fabricated on the same substrate as the MOEMS, provides excellent performance and benefits from batch fabrication. The III-V semiconductors in particular offer the potential for performing most of the integrated photonic functions. GaAs-based technology is one of the most common and attractive for fabricating integrated optoelectronic circuits due to its suitability for light emission and efficient electro-optic behavior. Thus GaAs-based micromachining is very appropriate for monolithically integrating micromechanical structures, waveguides, laser diodes, and photodetectors. However, because the photonics and MOEMS are fabricated on the same substrate, the processing is extremely complex and the MOEMS and photonic processes must be compatible. The increased complexity of the process can drive up cost and affect process yield.

A hybrid approach based on flip-chip MOEMS seems to be an excellent alternative to the above methods. This method involves fabricating the photonics and MOEMS as separate parts and then attaching them, face-to-face, with solder bumps. Flip-chip provides good device performance with little or no area penalty, and most important, it allows completely independent processes for the MOEMS and the photonics. This allows the manufacturer to optimize the performance of each component separately. In particular, vertical cavity surface-emitting lasers (VCSELs) have several advantages to be included within the MOEM architectures. Photolithography techniques used for MOEMs production are easily adapted to produce VCSELs, and they can be mounted by flip-chip bonding and integrated with on-chip electronics for feedback and control. Compared with edge-emitted lasers, VCSELs offer surface-normal emission, allowing the MEMS components to operate in-plane, and do not require the erection of standing structures. Flip-chip also allows the electronics and MOEMS to be built on substrate types other than silicon.

Recent advances in thin-film polymer optoelectronics have led to the development of an efficient and versatile material for waveguide applications. Polymer optical waveguides are of particular interest because of their easy processing and the potential to be added to micromachining technology by postprocessing before the micromechanical structure is released.

Another alternative to integrating guided-wave MOEMS is offered by the LIGA technique. LIGA combines lithography, electroplating, and molding. The "photoresist" used in this process is PMMA, which is spin coated to a thickness of up to a millimeter, then exposed through an X-ray mask. In addition to the rigidly fixed structures available in the basic LIGA process, it is possible to make moveable structures by adding an additional (sacrificial) metal layer between the substrate and the plating base. After the structures are plated and the PMMA is removed, the plating base and the sacrificial layer are etched to the surface. Finally, the remaining sacrificial metal is undercut, leaving a free structure. The capacity of LIGA to create a wide variety of shapes from different materials makes it akin to conventional machining, with the added value of unprecedented aspect ratio and high tolerances. Optical waveguides and micro-optical components such as prisms, lenses, and beam splitters can be integrated with very high precision and very smooth vertical walls using the LIGA technique. Thus, the simplest LIGA optical structure based on PMMA is a single-mode waveguide of rectangular cross-section. By varying the composition of the copolymer, a precise refractive index contrast can be obtained.

So far guided-wave MOEMS have found few applications. Although a large number of devices have been developed, only optically integrated optic components for WDM and the interferometric sensors are commercially available. Further developments in the field of photonic circuit integration and packaging are necessary to make guided-wave MOEMS a more attractive alternative to microelectronic counterparts. The penetration of guided-wave MOEMS is relatively modest compared with their potential. Approaches to building MOEM devices that are more accustomed to conventional micro-optics, conventional optoelectronics, or MEMS technology are limited. Even if the understanding of materials problems in MEMS technologies has advanced considerably in recent years, many questions, need to be solved. Realizing a MOEMS requires a multidisciplinary effort, and often more than one microfabrication technology is involved. The choice of the most appropriate one is not evident because of a lack of standardization. Many technical solutions are often only available in unreliable forms at research institutes. In this sense MOEMS manufacturing still presents a low level of industrial maturity. For these reasons, the future advances in guided-wave MOEMS will require the cooperative efforts of all the concerned scientific communities.

REFERENCES

1. K. E. Petersen, *Proc. IEEE* 70, 420 (1982).
2. H. Kogelnik, *Guided-Wave Optoelectronics*, T. Tamir, ed. Springer-Verlag, Berlin (1990).
3. W. Statues and W. Streifer, *Appl. Opt.* 16, 3218 (1977).
4. W. Gleine and J. Müller, *IEEE J. Lightwave Technol.* LT-9, 1629 (1991).
5. A. Neumann and J. T. Boyd, *J. Vac. Sci. Technol.* 17, 529 (1980).
6. S. M. Sze, *VLSI Technology*. McGraw-Hill, New York (1988).
7. M. Kawachi, *Opt. Quant. Electr.* 22, 391 (1990).
8. J. Gowar, *Optical Communication Systems*. Prentice-Hall Englewood Cliffs, NJ (1984).
9. B. L. Booth, *IEEE J. Lightwave Technol.* 7, 1445 (1989).
10. M. A. Duguay, Y. Kokubun, T. L. Koch and L. Pfeiffer, *Appl. Phys. Lett.* 49, 13 (1986).
11. Y. Kokubun, T. Baba, T. Sakaki and K. Iga, *Electron. Lett.* 22, 893 (1986).
12. B. E. Deal and A. S. Grove, *J. Appl. Phys.* 36, 3770 (1965).
13. M. Madou, *Fundamentals of Microfabrication*. CRC Press, Boca Raton, FL (1997).
14. E. van de Ven, *Solid State Technol.* 161(April) (1981).
15. B. Reynes and J. C. Bruyère, *Sensors Actuators* A33, 25 (1992).
16. W. R. Knolle, J. W. Osenbach and A. Elia, *J. Electrochem. Soc.* 135, 1211 (1988).
17. W. A. P. Claassen, W. G. J. N. Valkenburg, W. M. van de Wijgert and M. F. C. Willemsen, *J. Electrochem. Soc.* 132, 893 (1985).
18. W. A. P. Claassen, H. A. J. Th. van den Pol, A. H. Goemans and A. E. T. Kuiper, *J. Electrochem. Soc.* 133, 1458 (1986).
19. W. A. Lanford and M. J. Rand, *J. Appl. Phys.* 49, 2473 (1978).

20. E. Bonnotte, C. Gorecki, H. Toshiyoshi, H. Kawakatsu, H. Fujita, K. Wörhoff and K. Hashimoto, *IEEE J. Lightwave Technol.* 17 (1999).
21. D. Peters and J. Müller, *Proc. Plasma Surface Eng.*, Vol. 2, p. 1296 (1988).
22. H. Bezzaoui and E. Voges, *Sensors Actuators* A 29, 219 (1991).
23. K. Wörhoff, "Optimized LPCVD Si_xON_y Waveguides Covered with Calixarene for Non-critically Phase-matched Second Harmonic Generation," Ph.D thesis, Univ. of Twente, the Netherlands (May 1996).
24. R. S. Rosler, *Solid State Technol.* 63 (April) (1977).
25. K. Watanabe, T. Tanigaki and S. Wakayama, *J. Electrochem. Soc.* 128, 2630 (1981).
26. W. Gleine and J. Müller, *Appl. Opt.* 31, 2036 (1992).
27. H. J. Lee, C. H. Henry, K. J. Orlowsky, R. F. Kazarinov and T. Y. Kometani, *Appl. Opt.* 27, 4104 (1988).
28. B. S. Meyerson and W. Olbricht, *J. Electrochem. Soc.* 131, 2362 (1984).
29. P. Heimala and J. Aarnio, *J. Phys. D: Appl. Phys.* 25, 733 (1992).
30. H. Lee, C. H. Henry, K. J. Orlowsky, R. F. Kazarinov and T. Y. Kometani, *Appl. Opt.* 27, 4104 (1988).
31. K. Law, J. Wrong, C. Leung, J. Olsen and D. Wang, *Solid State Technol.* 80 (April) (1989).
32. T. Kominato, Y. Ohmori, H. Okazaki and M. Yasu, *Electron. Lett.* 26, 327 (1990).
33. S. Imamura, *Proc. IEEE/LEOS Summer Topical Meeting on Organic Optics and Optoelectronics*, 35 (1998).
34. V. Ramaswamy and H. P. Weber, *Appl. Opt.* 12, 1581 (1973).
35. R. Ulrich and H. P. Weber, *Appl. Opt.* 12, 428 (1972).
36. Y. Hida, Y. Inoue and S. Imamura, *Electron. Lett.* 30, 959 (1994).
37. M. Usui, M. Hikita, T. Watanabe, M. Amano, S. Sugawara, S. Hayashida and S. Imamura, *IEEE J. Lightwave Technol.* 14, 2338 (1996).
38. A. C. Adams, F. B. Alexander, C. D. Capio and T. E. Smith, *J. Electrochem. Soc.* 128, 1545 (1981).
39. W. S. Nguyen, W. A. Lanford and A. L. Rieger, *J. Electrochem. Soc.* 133, 970 (1986).
40. D. Barbier, M. Green and S. J. Madden, *IEEE J. Lightwave Technol.* LT-9, 715 (1991).
41. R. V. Ramaswamy and R. Srivastava, *IEEE J. Lightwave Technol.* 6, 984 (1988).
42. C. Gorecki, unpublished data.
43. M. Hoffman and E. Voges, *Proc. 7th Eur. Conf. on Integrated Optics 'ECIO '95*, WeC2, 299 (1995).
44. K. Imoto and A. Hori, *Electron. Lett.* 29, 1123 (1993).
45. K. Fisher, J. Müller, R. Hoffmann, F. Wasse and D. Salle, *IEEE J. Lightwave Technol.* 12, 163 (1994).
46. C. Gorecki, F. Chollet, E. Bonnotte and H. Kawakatsu, *Opt. Lett.* 22, 1784 (1997).

47. B. H. Verbeek, C. H. Henry, N. A. Olson, K. J. Orlowsky, R. F. Kazarinov and B. H. Johnson, *IEEE J. Lightwave Technol.* 6, 1011 (1988).
48. C. H. Henry, G. E. Blonder and R. F. Kazarinov, *IEEE J. Lightwave. Technol.* 7, 1530 (1989).
49. S. Valette, S. Renard, J. P. Jadot, P. Guidon and C. Erbeia, *Sensors Actuators* A21-A23, 1087 (1990).
50. S. Valette, *J. Modern Opt.* 35, 993, 1005 (1988).
51. M. Kawachi, N. Takato, K. Jinguji and M. Yasu, *Proc. C/IOOC '87*, TuQ31, 1998.
52. B. P. Pal, H. Singh, A. K. Ghatak and A. B. Bhattacharya, 23th General Assembly, URSI, Praha, Aug. (1990).
53. A. Vadekar, A. Nathan and W. P. Huang, *IEEE J. Lightwave Technol.* 12, 157 (1994).
54. B. L. Booth, J. E. Marchegiano, C. T. Chang, R. J. Furmanak, D. M. Graham and R. G. Wagner, Du Pont documentation on Polyguide technology (http://www.polymerphotonics.com/).
55. H. Kogelnik, *Proc. Symp. Ouasi-Optics*. Brooklyn Polytechnic Press, Brooklyn, NY (1964).
56. R. G. Heideman and P. V. Lambeck, *Proc. SPIE* 3098 (1997).
57. Y. Yamada and M. Kobayashi, *IEEE J. Lightwave Technol.* 5, 1716 (1987).
58. K. Shem, C. H. Bulmer, R. P. Moeller and W. K. Burns, *Digest of OSA Topical Meeting on Integrated Optics* (Jan.) (1980).
59. H. Fujita, *Jpn. J. Appl. Phys.* 33, 7163 (1994).
60. Y. Uenichi, M. Tsugai and M. Mehregany, *J. Micromech. Microeng.* 5, 305 (1995).
61. K. Najafi, *Proc. SPIE* 1793, 235 (1992).
62. R. T. Howe, *J. Vac. Sci. Technol.* B 6, 1809 (1988).
63. J. J. Sniegowski and E. J. Garcia, *Proc. SPIE* 2383, 46.
64. W. C. Tang, T. H. Nguyen and R. T. Howe, *Proc. IEEE Micro Electro Mechanical Systems Workshop*, 53, 1989.
65. T.Akiyama and H. Fujita, *J. Microelectromech. Syst.* 2, 106 (1993).
66. D. L. Polla, P. J. Schiller and L. F. Francis, *Proc. SPIE* 2291, 108.
67. A. Garcia-Valezuela and M. Tabib-Azar, *Proc. SPIE* 2291, 125.
68. K. Hogari and T. Matsumoto, *Appl. Opt.* 30, 1253 (1991).
69. E. Ollier, P. Labeye and F. Revol, *Digest of Summer Topical Meeting IEEE/LEOS on Optical MEMS and their Applications*, 71 (1996).
70. P. Mottier, *Int. J. Optoelectronics* 9, 125 (1994).
71. D. Kobayashi, H. Okano, M. Horie, H. Otsuki, K. Sato and M. Horino, *Digest of Intern. Conf. IEEE/LEOS on Optical MEMS and their Applications*, 243 (1997).
72. R. M. Boysel, T. G. McDonald, G. A. Magel, G. C. Smith and J. L. Leonard, *Proc. SPIE* 1793, 34 (1992).
73. Y. W. Kim, M. G. Allen and N. F. Hartman, *Proc. SPIE* 1793, 183 (1992).
74. W. Lukosz, *Proc. SPIE* 1793, 214 (1992).

75. S. Valette, J. Lizet, P. Mottier, J. P. Jadot, P. Gidon and S. Renard, *IEE Proc.* 131, Pt. H, 325 (1984).
76. G. Cocorullo, F. G. Delta Corte, I. Rendina, C. Minarini, A. Rubino and E. Terzini, *Opt. Lett.* 21, 2002 (1996).
77. E. Bonnotte, L. Robert, P. Delobelle, L. Bornier, B. Trolard, G. Tribillon and D. Mairey, *Proc. SPIE* 2782, 685 (1996).
78. S. Wu and H. J. Frankena, *Proc. SPIE* 1793, 83 (1992).
79. K. E. Burcham, G. N. De Brabander and J. T. Boyd, *Proc. SPIE* 1793, 12 (1992).
80. A. Nathan, W. P. Huang, Y. K. Bhatnagar, K. Benaisa and A. Vadekar, *Proc. 7th International Conf. on Solid State Sensors and Actuators*, 682 (1993).
81. A. M. Young, C. Xu, W. P. Huang and S. D. Senturia, *Proc. SPIE* 1793, 42 (1992).
82. M. Ohkawa, M. Izutsu and T. Sueta, *Appl. Opt.* 28, 5153 (1989).
83. T. Storgaard-Larsen, S. Bouwstra and O. Leistiko, *Sensors Actuators* A 52, 25 (1996).
84. T. Tomiyoshi, K. Minami and M. Esashi, *Digest of Intern. Conf. IEEE/LEOS on Optical MEMS and their Applications*, 200 (1997).
85. D. Jestel, A. Baus and Voges, *Electron. Lett.* 26, 1146 (1990).
86. P. V. Lambeck, *Sensors Actuators* B 8, 103 (1992).
87. W. Lukosz, *Sensors Actuators* B 29, 37 (1995).
88. N. Fabricius, G. Gauglitz and J. Ingenhoff, *Sensors Actuators* B 7, 672 (1992).
89. A. A. Boiarski, J. R. Bush, B. S. Bhullar, R. W. Ridgway and V. E. Wood, *Proc. SPIE* 1793, 199 (1992).
90. H. J. M. Kreuwel, P. V. Lambeck, J. V. Gent and T. J. A. Popma, *Proc. SPIE* 798, 218 (1987).
91. M. Abraham, W. Ehferld, M. Lacher, O. Marti, K. Mayr, W. Noell, P. Günther and J. Barenz, *Proc. SPIE* 3099, 248 (1997).
92. M. Sasaki, K. Tanaka and K. Hane, *Proc. NFO-5* (1998).

CHAPTER 6

SILICON MICROMACHINES IN OPTICAL COMMUNICATIONS NETWORKS: TINY MACHINES FOR LARGE SYSTEMS

Randy Giles
Vladimir Aksyuk
Chris Bolle
Flavio Pardo
David J. Bishop
Bell Laboratories, Lucent Technologies

CONTENTS

6.1 Introduction / 302

6.2 MEMS Devices / 302

6.3 MEMS in Lightwave Technology / 303
    6.3.1 Variable attenuators: micro tilt mirrors / 304
    6.3.2 Variable attenuator: flag switch type / 305
    6.3.3 Optical switches / 310
    6.3.4 All-photonic switch / 314
    6.3.5 Power limiter / 318
    6.3.6 Add/drop multiplexer / 320
    6.3.7 Active equalizer / 324
    6.3.8 Micro-microscope / 327

6.4 The Future / 327

    References / 329

6.1 INTRODUCTION

Silicon micromechanics is a field with the potential to impact a large cross section of science and technology. In diverse activities such as the automotive industry, aeronautics, cellular communications, chemistry, acoustics, display technologies, and lightwave systems, highly functional microdevices are establishing a large presence despite their diminutive size. In this chapter, we describe interesting new micromachines that enable engineers to build opto-mechanically integrated circuits and subsystems whose capabilities can facilitate the explosive growth in bandwidth and networking features of lightwave systems.

6.2 MEMS DEVICES

Silicon micromachines are the *mechanical* analog of silicon *electronic* integrated circuits and are fabricated by similar methods. Figure 6.1 shows how starting with a silicon wafer, materials such as polysilicon, silicon nitride, silicon dioxide, and gold are deposited and patterned in a sequence of steps to produce a complicated three-dimensional structure. However, unlike an electronic integrated circuit, material is etched away in the final processing step to "release" the device, leaving pieces free to move. Built using IC batch-processing techniques, these devices are inexpensive to produce because large numbers of them can be fabricated simultaneously with high yield. Furthermore, cost reduction is driven by the 200 billion dollar a year VLSI business with its relentless development of new tools, techniques, and processes that apply to the silicon micromachine industry.

VLSI fabrication techniques allow designers to create multifunctional systems by integrating micromechanical, analog, and digital microelectronic devices on the

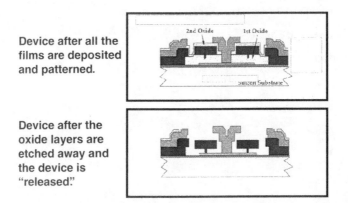

Figure 6.1 Schematic diagram of how a surface micromachined MEMS device is fabricated.

SILICON MICROMACHINES IN OPTICAL COMMUNICATIONS NETWORKS

same chip. Research in MEMS has been extremely active over the last decade, producing microscopic versions of most macro-machines. Contrary to intuition, these mechanical elements have proven to be robust and long-lived, especially ones whose parts flex without microscopic wear points. Examples include rotating gears and pivoting beams being actuated through billions of cycles. The potential for reliable micromachines of a size scale ranging from one to several hundred microns makes MEMS an attractive technology to manufacture opto-mechanical structures for use in waveguide and free-space optical devices in optical communication networks.

6.3 MEMS IN LIGHTWAVE TECHNOLOGY

Figure 6.2 is a schematic representation of several lightwave networks where dark shaded circles indicate potential applications of silicon micromachines. The applications range from data modulators, variable attenuators, active remote nodes, active equalizers, add/drop multiplexers, optical switches, power limiters, choppers for power measurements, and optical crossconnects. The anticipated multibillion-dollar market for micromachines in lightwave systems is attracting the research and development efforts of numerous companies worldwide.

MEMS is a disruptive technology for lightwave networks, poised to abruptly change how things are presently done. One can envision MEMS becoming the

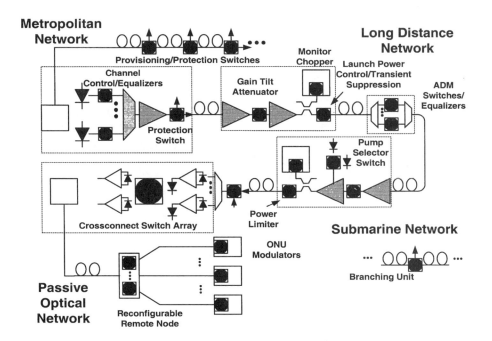

Figure 6.2 Overview of where in lightwave systems we see MEMS being applied.

choice technology for devices ranging from single variable attenuators to large-scale optical cross-connects, and in every case bringing unprecedented advantages in capabilities and cost. Lightwave MEMS may become the first billion-dollar application of micromechanics and the first example in which it revolutionizes an industry.

6.3.1 Variable Attenuators: Micro Tilt Mirrors

Figure 6.3 shows a typical, albeit somewhat simple, silicon micromachine—a tilt mirror for a variable optical attenuator. The device consists of a mirror connected through a pivot point to a suspended capacitor plate. When a voltage is applied to an electrode below the plate, the capacitor plate is pulled toward the substrate, pivoting it and the tilt mirror. In a variable attenuator, the mirror is arranged to reflect light from an input optical fiber to an output fiber with a coupling efficiency determined by the tilt angle.

Figure 6.4 illustrates the power of MEMS technology. Once a single mirror is designed, many hundreds, thousands, or millions more can be easily built. The mirror array of Fig. 6.4 was built for use in an add/drop multiplexer,[1] a device for rout-

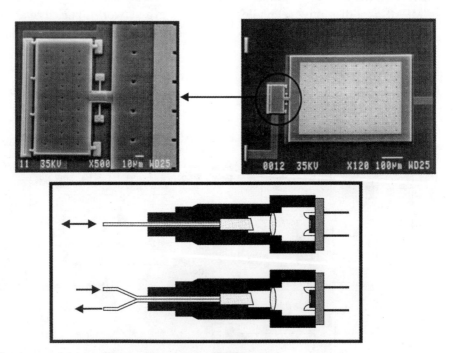

One and two fiber coaxial packages.

Figure 6.3 Chip (upper figures) and device concept (lower figure) for a MEMS tilt-mirror variable attenuator.

ing individual optical channels in a wavelength division multiplexed (WDM) lightwave system. In operation, each WDM channel lands on a particular mirror and is separately routed to either pass through the add/drop multiplexer or be dropped and detected at the node. Without MEMS technology, this capability is achieved with many costly, separately assembled components. A more complicated micromirror capable of two-axis rotation is shown in Fig. 6.5. This device, capable of directing light beams over a wide field, can also be assembled in arrays to facilitate complicated optical signal processing in optical cross-connect and matrix switching.

6.3.2 Variable attenuator: flag switch type

Optical components in lightwave networks operate over a broad range of optical power levels, ranging from greater than 1 W emanating from high-power transmitters and amplifiers, to microwatt-level signals arriving at receivers. The large differences between launched and received optical power levels can result in an unacceptable power incident on a receiver in a system with low span losses or few WDM channels. Fixed optical attenuators are frequently used to increase the loss of short optical fiber spans while variable optical attenuators are proposed for

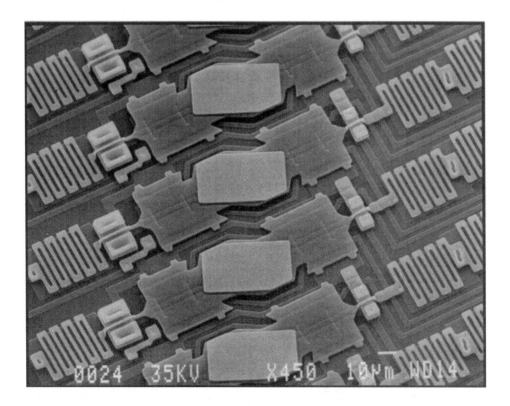

Figure 6.4 Array of tilt mirrors for use in an add/drop multiplexer.

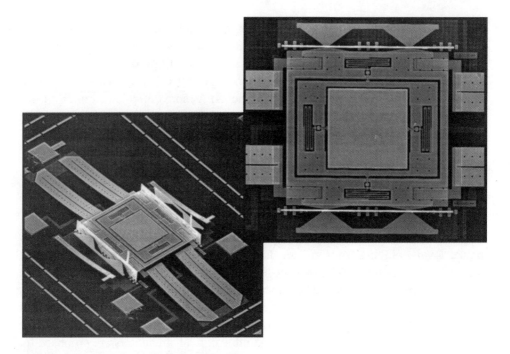

Figure 6.5 Two-axis tilt mirror for use in a more complicated free-space optical switch.

use in amplified WDM networks regulating signal powers as the number of active channels changes. High-speed variable optical attenuators might also be used for transient suppression in these amplified networks.

Several technology choices are available for implementing variable attenuators. Sliding-block mechanical attenuators with > 50-dB dynamic range and 0.1-dB linearity have excellent optical characteristics, but are bulky, costly, and adjust in 0.5–1.0 s. Waveguide thermo-optic attenuators can respond in millisecond time, but cannot achieve the high dynamic range with low insertion loss. Recently, a MARS attenuator has been reported, having 3-dB insertion loss and 30-dB dynamic range.[2] In this section, we describe a MEMS variable optic attenuator that has less than 1-dB insertion loss, > 50-dB dynamic range, and can respond in less than 100 microseconds.

The MEMS variable optical attenuator (VOA) uses our pivoting silicon vane structure[3] with specific placement of the optical fibers to optimize the attenuator's performance. Figure 6.6 is an SEM photomicrograph of the MEMS VOA, showing the 150 × 300-μm capacitor plate spring-suspended over the substrate and levered through a pivot to elevate the shutter platform. The hinged vane is erected on the platform, normal to the substrate, and is raised with a voltage applied between the capacitor plate and the substrate electrode. The vane is placed in an ∼ 20-μm fiber gap and attenuation varies with changes in the applied voltage. The attenuator was packaged in a 16-pin ceramic DIP package with FC-connectorized fiber pig-

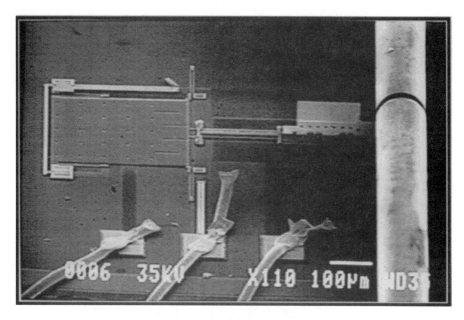

(a)

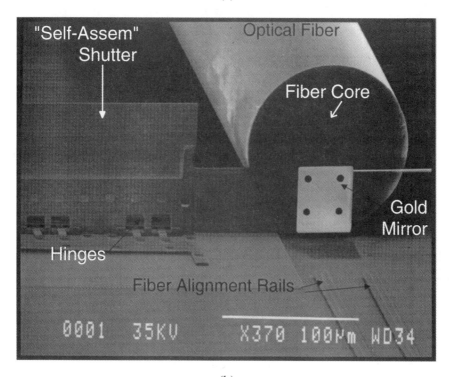

(b)

Figure 6.6 (a) SEM photomicrograph of the MEMS VOA showing capacitor plate, lever arm, and erected vane in front of optical fiber. (b) Shows a closeup of the mirror and fiber.

tails. Minimum insertion loss including two connectors measured in 16 test devices ranges from 0.81 dB to 2.6 dB.

Displacement of the silicon vane of an unmounted MEMS attenuator/switch is directly measured using an optical comparator microscope with 0.5-μm resolution

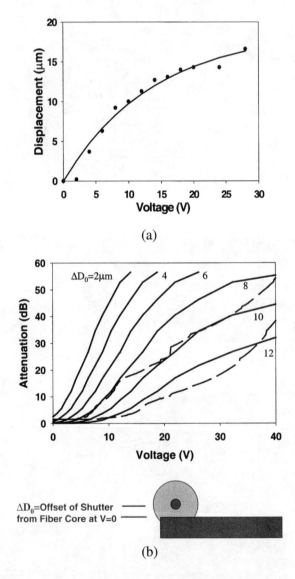

Figure 6.7 (a) Measured relative displacement of shutter as a function of applied voltage. Solid line is an exponential curve fit to the data. (b) Attenuation curves calculated from displacement curve and diffracted-beam calculation of fiber-to-fiber coupling with a knife-edge obstruction and for initial displacement of shutter and fiber axis ranging from 2 to 12 μm. Dashed curves are measured attenuation curves for two MEMS VOAs.

for a voltage range of 0–28 V applied to the capacitor plate. Figure 6.7 shows the results together with an exponential curve-fit, $D = a(1 - e^{-bV})$, where D is the displacement and V is the applied voltage. The best-fit parameters to Fig. 6.7 are $a = 18.8\,\mu\text{m}$ and $b = 0.073\text{ V}^{-1}$. The fiber-to-fiber attenuation as a function of the vane's position is computed using a gaussian beam approximated to the optical fiber mode (mode-field diameter $\omega_0 = 8$-µm at signal wavelength $\lambda = 1550$ nm), a calculation of the free-space diffraction past a knife-edge obstruction based on the Fresnel-Kirkhoff diffraction integral,[4] and the evaluation of the mode-overlap integral between the diffracted beam and fiber mode.[5] This computation is combined with the exponential curve fit $D(V)$ to produce a family of curves, Fig. 6.7(b), of the predicted MEMS VOA attenuation as a function of applied voltage. The varying parameter generating the curves is ΔD_0, the location of the shutter relative to the center of the fiber mode at $V = 0$. This offset value ranges from 2 to 12 µm. The attenuation curves of two MEMS VOAs measured at 1550 nm are superimposed on the predicted responses and show good qualitative agreement with the predictions. Two connectors on each attenuator contributed ~ 1 dB excess loss. Gross differences between the two attenuation curves are explained by < 4-µm differences in ΔD_0 that result from packaging and device variations. As seen in Fig. 6.7(b), a larger offset, of $\Delta D_0 > 8$-µm results in higher operating voltages but yields shallower curves which are better suited for precise attenuator control. Devices having lower offsets and steeper attenuation curves are better suited as optical switches.

Attenuation through the MEMS VOA exhibits a weak dependence on the input signal polarization. Figure 6.8 shows the polarization-dependent loss variation for two VOAs over the attenuation range 0–30 dB. In one device the peak variation is 0.5 dB and only 0.2 dB in the second device. The polarization variation results from vectorial diffraction effects with light grazing exposed silicon on the shutter. The MEMS VOAs also exhibits fine structure in their attenuation curves [Fig. 6.7(b)], produced from microscopic imperfections of the pivot structure.

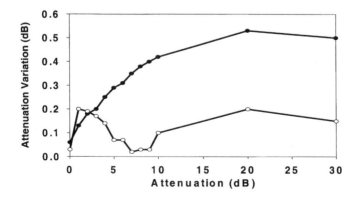

Figure 6.8 Polarization-dependent variation of attenuation for minimum attenuation settings ranging from 0 to 30 dB.

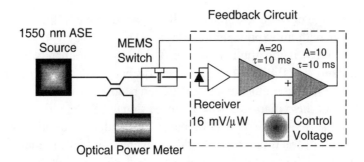

Figure 6.9 MEMS VOA configured with a feedback control circuit to act as an optical power regulator.

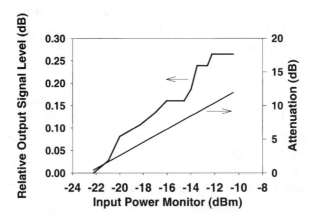

Figure 6.10 Relative change of output signal power versus input power. Corresponding attenuation is also shown.

Figure 6.9 shows the attenuator acting as an automatic optical power regulator with the addition of a simple feedback circuit. Light from a 1550-nm ASE source launched into the MEMS VOA is detected at the output by a DC-coupled InGaAs receiver. The receiver output is amplified and fed back with a control voltage to the VOA's voltage input. Figure 6.10 shows the output power regulation for a 12-dB change of input signal with one control voltage setting. Less than 0.26-dB output power variation is observed and the VOA attenuation varied smoothly with input power. Higher loop gain or use of a proportional-integrator-differential (PID) controller would improve the output power stability.

6.3.3 OPTICAL SWITCHES

Optical switches are vital in many optical systems, particularly in lightwave networks that require rapid and flexible reconfiguration in restoration and protection.

Examples include protection switching of optical transmission links and the reconfiguration of optical cross-connects and WDM add/drop multiplexers. There is a large variety of optical switches, including bulk opto-mechanical switches, liquid crystal switches, lithium niobate switches, thermo-optic switches in silica waveguides, integrated semiconductor switches, and several MEMS optical switches.[6,7] Despite this wide variety of switches, only the silicon MEMS optical switch promises the combination high optical performance, ultracompact size, and ultralow power consumption with the potential of easy fabrication and very low cost. With these attributes a MEMS optical switch may be suitable for widescale deployment in demanding gigabit optical networks and even cost-sensitive applications such as fiber-to-the-home.

Silicon MEMS optical switches can be constructed from the same pivoting structure described for the variable optical attenuator of the previous section. However, as the silicon shutter has a thin gold coating, the shutter placed inside a narrow fiber gap results in a two-port reflective switch. The normal-cleaved fiber ends of the fiber gap require antireflection coating to reduce parasitic Fabry–Perot resonances. The entire switch can be assembled inside a standard 16-pin ceramic dual inline IC package.

Figure 6.11 shows the transmission and reflection at 1550 nm versus applied voltage for eight MEMS switches; four are designed for nominally 20 V actuation, four for 40 V; the low-voltage devices have 575-µm-long lever arms while the high-voltage devices' arms are 500-µm. Figure 6.12 shows the performance of one of the better switches on an expanded scale. Reflection is measured using an optical circulator to redirect reflected light to a separate third port. Switch insertion loss in the transmit state, including loss of two FC-PC connectors, ranged from 0.81 to 3.26 dB, and return loss in the reflection state varied from 2.15 to 4.77 dB, *including* the loss of two connectors and 1.0-dB insertion loss of the circulator. These switch losses are comparable to those achieved with conventional optomechanical switches. Transmission extinction ranged from 38 dB to greater than 80 dB while reflection extinction varied from 15 to 25 dB. The low extinction in reflection resulted from weak Fabry–Perot reflections in the fiber gap, and might be improved using angle-cleaved fiber ends rather than antireflection coatings. Less than 0.05-dB polarization-state induced variation of transmission state insertion loss or reflection-state return loss was observed. Finally, switching speeds with 10 V square-wave drive voltage were approximately 65 µs to close (while in the reflection state) the switch and 95 µs to open.

Differences in the switch characteristics originated from microscopic variation in feature sizes, particularly in the pivot area, edge diffraction effects on the shutter and weak frictional forces in the mechanical motion. Significant responses were observed in some devices with less than 5 V applied to the capacitor plate. Switch hysteresis, resulting from the nonlinear force characteristics between the capacitor plate and substrate, is also observed. Since the electrostatic force on the capacitor plate depends on the *absolute* value of the applied voltage, switch hysteresis is deduced by measuring the switching curves with DC voltages swept from –20 V

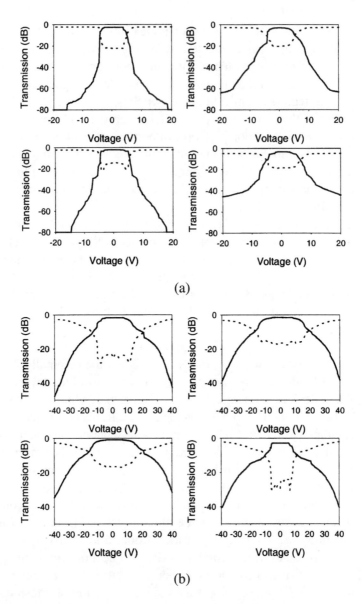

Figure 6.11 Transmission loss (solid line) and return loss with optical circulator (dashed line) versus applied voltage for (a) four low-voltage (20 V) switches and (b) four high-voltage (40 V) switches. Voltage is swept from $-V_{max}$ to $+V_{max}$ to observe hysteresis as the switches change from their reflective state ($V_{switch} = -V_{max}$) to transmissive state ($V_{switch} = 0$) then back to reflective state ($V_{switch} = V_{max}$).

(–40 V) to 20 V (40 V), and comparing measurements at the same absolute voltage. This hysteresis may result in the switches needing feedback control when used as linear devices—i.e., variable attenuators—but would not affect their use as binary-state switches.

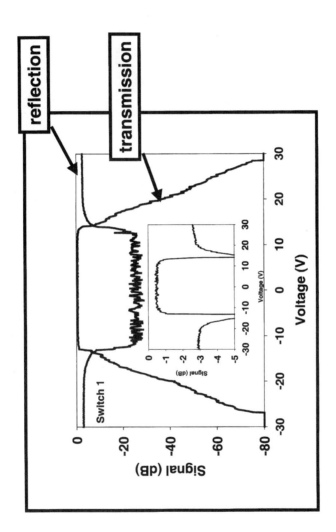

Figure 6.12 Performance of a best-case optical switch on an expanded scale for both transmission and reflection.

Figure 6.13 MEMS 1 × 2 optical switch.

Other switch configurations can be assembled from the basic flag structure. Figure 6.13 shows a 3-port optical switch where light entering from the fiber on the left exits either through the fiber on the top of the figure or the fiber on the right. Clearly this device design has much power and flexibility as it retains the speed and loss properties of the reflective switch with the convenience of a third optical port.

6.3.4 ALL-PHOTONIC SWITCH

Active lightwave circuits are typically powered by external electrical sources, requiring copper wire power delivery to terminal, repeater, and remote node sites. In some networks, providing electrical service may be impractical or too costly, and alternative methods of obtaining requisite system functions are desirable. Additionally, the use of field-deployed optoelectronic circuits is discouraged by the penalties of reduced network reliability and higher cost, resulting in solutions such as passive optical networks (PONs) for high-speed optical access. Constrained to a passive outside plant, these networks incur added complexity at terminal sites and may have unresolved operations and maintenance issues.

Optically powered circuits could enhance the functionality of an otherwise passive outside plant. Telephony circuits powered by Si photodetectors have been described,[8] and a micropower stepper motor switch powered by an InGaAs photogenerator has been remotely actuated through 100 km of transmission fiber.[9] These earlier results needed relatively high transmitted optical power to supply the load current and had low bandwidth and long switching times. Combining a low-power MEMS optical switch with a high-efficiency integrated InGaAs photogenerator enables one to construct an optically powered optical switch. A light-actuated micromechanical photonic switch (LAMPS) made this way requires zero average quiescent current and was actuated in 5 µs with 2.7 µW at 1555 nm illuminating the photogenerator.

Figure 6.14 shows SEM photomicrographs of the InGaAs photogenerator and the reflective micromechanical switch. The 165-µm diameter photogenerator, grown by low pressure MOVPE on semi-insulating InP, is composed of eight series-connected InGaAs PIN photovoltaic diodes; the back surface is antireflection coated to facilitate rear illumination. The photogenerator is attached to a silicon submount and sealed in a fiber-connectorized 8-pin ceramic package. Illumination at 1555 nm of the output from a conventional single-mode optical fiber results in a photocurrent responsivity of 0.057 µA/µW and a greater than 3 V open-circuit voltage at ∼ 100 µW illumination. Significant photocurrent response is obtained for wavelengths from 950 nm to 1650 nm. The silicon micromechani-

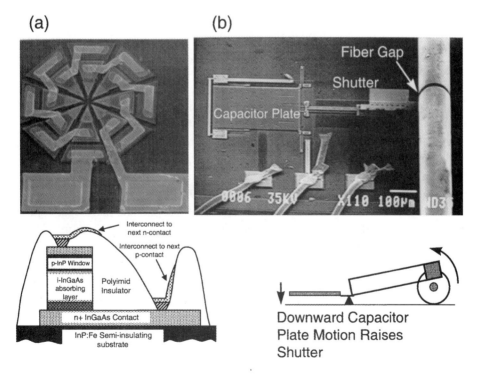

Figure 6.14 (a) 8-sector 165-µm diameter InGaAs photogenerator. (b) Silicon micromechanical reflective optical switch (foreshortened view compresses SEM photomicrograph's y direction).

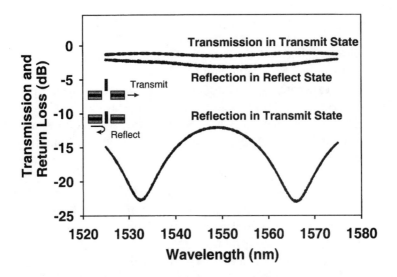

Figure 6.15 Transmission and return loss spectra of the silicon micromechanical switch measured over a 50-nm BW. Fabry–Perot resonances in reflection result from fiber end-face reflections.

cal switch is again made as the MEMS variable attenuator. Switch insertion loss is 1.8 dB, minimum drive voltage of the normally-off device for >57 dB extinction is 1.24 V, and minimum reflective-state return loss is 2.0 dB. As seen in the transmission and reflection spectra of Fig. 6.15, the switch had greater than 50 nm optical bandwidth—any bandwidth limit is expected to arise from the guiding properties of the pigtail fiber. Noticeable Fabry–Perot reflections between the uncoated fiber end-faces could be reduced with antireflection coatings.

Upon connecting the photogenerator to power the optical switch, the minimum optical power needed to actuate the LAMPS was 2.7 µW (−25.7 dBm); higher optical powers increased the switching speed. Figure 6.16 shows the turn-on delay and switching times ranging from 3.7 ms and 1.9 ms, respectively, with 2.7-µW peak power to 700 µs and 100 µs with 95-µW peak power. The inset figure shows typical photogenerator voltage and switch transmission waveforms. Reduced turn-on delay could be achieved by optimizing the initial shutter placement or by operating the LAMPS with a DC optical bias to slightly raise the shutter. With incident powers below 2.7-µW, the shutter partially obstructed the fiber gap and the LAMPS operated as a high-dynamic range, polarization-independent (Δ Loss < 0.5 dB) variable attenuator.

LAMPS can be placed at remote nodes of optical access networks, as illustrated in Fig. 6.17. In this case the LAMPS is a provisioning/interdiction device regulating the customer connections at an unpowered remote node. Individual LAMPS could be addressed with separate optical channels routed through a WDM demultiplexer, but more likely a single photogenerator would power a microcontroller controlling several micromechanical switches. While typical access networks have

Silicon Micromachines in Optical Communications Networks 317

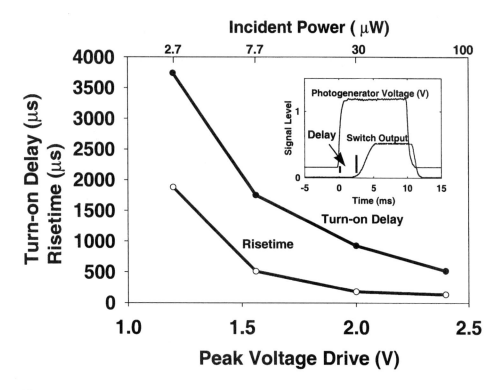

Figure 6.16 Turn-on delay and risetime of optical switch versus peak voltage from the photogenerator. The upper x-axis shows corresponding peak incident power. The inset shows the voltage and switching waveforms with 2.7-µW incident power.

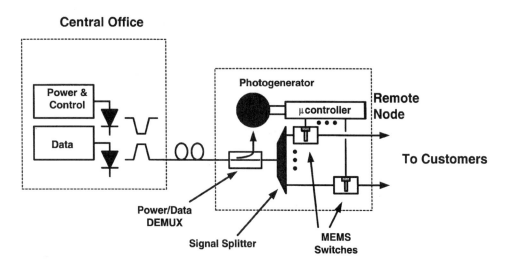

Figure 6.17 Representative application of a LAMPS at a remote node to control data delivery to customers in a passive optical network.

modest reaches, less than 20 km, a LAMPS device might be actuated over very long distances. With a 100-mW (20-dBm) 1550-nm light source and 0.25-dB/km fiber loss, a LAMPS node could be located $(20 + 25.7)/0.25 = 182$ km away from a terminal site, making it attractive for use in branching units of long-distance spans.

6.3.5 POWER LIMITER

In optical communication systems and networks, it is often necessary to regulate or limit optical power to improve system performance. A variety of devices can provide the necessary variable attenuation for power regulation. However, instead of employing an attenuator in conjunction with an electrically powered optical detector and feedback amplifier, it would be beneficial to have a compact limiter which works by utilizing the energy of the incident light at 1550 nm wavelength and requires no additional power source. Since the incident light power is typically very small, less than 1 mW, building such a device is challenging.

A MEMS variable attenuator[10] and a 30-segment InGaAs photogenerator can be combined with an optical tap as shown in Fig. 6.18 to create such a power limiter.[12] The MEMS device is powered by the photogenerator in a feedback configuration with the 2% ($T = -17$ dB) fiber coupler tap diverting a small amount of the output light from the attenuator onto the photogenerator. The photogenerator voltage increases as the input power increases, thereby closing the switch and regulating the output through increased attenuation.

Although providing only miniscule optical power, the unloaded photogenerator produces sufficient voltage to actuate the MEMS variable attenuator, even at incident light intensities below 5 µW. The photogenerator's open-circuit voltage, shown in Fig. 6.19, increases logarithmically with incident power—a maximum voltage of 10.3 V is obtained at 500 µW (-3 dBm). Over the range of incident power (measured in dBm) P_{PGEN} from -30 to -3 dBm, the open-circuit voltage is

$$V_{oc} = V_0 + g P_{PGEN}, \tag{6.1}$$

where $V_0 = 11.02$ V and $g = 0.272$ V/dBm.

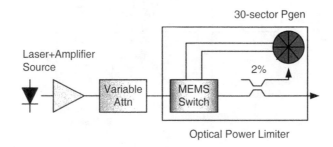

Figure 6.18 Experimental setup schematic for the limiter.

Figure 6.20 shows the output power P_{OUT} versus input power P_{IN} of the optical power limiter. After the threshold output power is reached, limiting starts and the output changes from –7 to –6 dBm with the input power varying from – 4 to +8 dBm. Because the MEMS device does not load the photogenerator substantially, Eq. (6.1) can be used to calculate the limiter response curve along with the following equations:

$$P_{OUT} = P_{IN} - A(V_{oc}), \qquad (6.2)$$

$$P_{PGEN} = P_{OUT} + T, \qquad (6.3)$$

where $A(V_{oc})$ is the attenuation characteristic of the MEMS device.

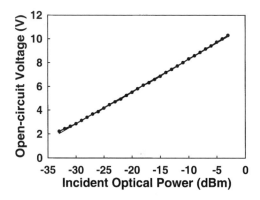

Figure 6.19 Photogenerator open-circuit voltage for range of incident optical power –30 to –3 dBm.

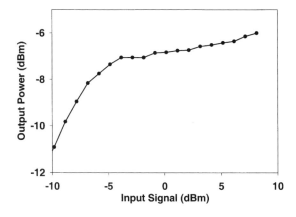

Figure 6.20 Output power versus input power for a photogenerator/MEMS-switch optical power limiter. Solid line is a calculation based on the switch and photogenerator characteristics.

If V_S is a threshold voltage above which the attenuation begins to increase, then the regulation of the output power for $V > V_S$ is given by $R = \Delta P_{OUT}/\Delta P_{IN} = 1/(1 + ag)$ where $a = dA/dV$ for the working voltage region. Good agreement was obtained between $R \sim 0.09$ which is estimated from the switch and photogenerator parameters and the measured $R = 0.083$. The threshold output power $P_{\text{threshold}} = (V_S - V_0)/g - T = -7.7$ dBm is also in good agreement with the experiment. Both R and $P_{\text{threshold}}$ can be tuned for a particular application by choosing the correct $A(V)$ and T.

Using the low-power MEMS optical variable attenuator enables novel optical power limiters that require no external energy source and are capable of operating at less than 10 µW incident optical power levels. The optical limiter offers new capabilities in lightwave networks, particularly in power regulation of wavelength-division-multiplexed channels that minimize degradations caused by interchannel cross-talk and optical nonlinearities.

6.3.6 ADD/DROP MULTIPLEXER

Advanced lightwave systems using WDM channels are capable of supporting functions performing in the optical layer to enhance provisioning and protection of the network. Optical wavelength add/drop multiplexers (ADM) selectively remove one or more WDM channels and replace them with new channels at the same wavelengths. Residual leakage of the dropped channels must be very small to minimize their interference to the added channels. This requires low channel cross-talk through the wavelength multiplexers, and demultiplexers and high-contrast optical switches in a reconfigurable ADM.

Integrated reconfigurable ADMs have been demonstrated using silica on silicon[13] and InP,[14] but their utility is compromised by marginal optical performance relative to that needed in real applications. In this section we describe a reconfigurable drop module (RDM) implemented as a hybrid optical circuit comprising two 16-channel arrayed waveguide grating routers (AWGR),[15] sixteen MEMS optical switches, and ancillary optical components. The RDM is designed with drop-and-transmit (DT) capability for eight channels such that when dropped, they remained combined in a single optical fiber, suitable for WDM transport away from the RDM node. Eight other channels are configured for drop-and-detect (DD) where dropped channels exit on separate fibers, suitable for local reception. Channel-add for full ADM functionality may be trivially obtained using a final-stage coupler.

Figure 6.21 shows the layout of the 100-GHz, 16-channel spaced RDM configured with DT capability for half of the channels and DD capability for the remaining channels. All components have fiber connectors. An input optical circulator (0.6-dB port-to-port insertion loss) redirected DT channels to a transmission fiber where the first AWGR demultiplexed the input channels and recombined any remaining DT channels. Drop-and-transmit is controlled by reflective MEMS optical

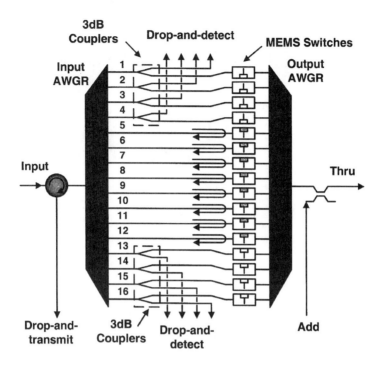

Figure 6.21 Reconfigurable drop module with channels 1–4 and 13–16 arranged for drop-and-detect and channels 5–12 arranged for drop-and-transmit. Arbitrary reconfiguration of all sixteen channels is obtained using voltage-actuated silicon MEMs optical switches.

switches;[3] channels reached the throughport when switches are in the transmission state and dropped when the switches are activated into their reflection state. Channels configured for drop-and-detect are divided after the first silica-waveguide AWGR using 3-dB passive couplers, resulting in a fixed drop port and a second port connected to the output AWGR through a MEMS switch. A channel reaches the throughport when the corresponding MEMS switch is in the transmission state otherwise it is blocked.

As seen in the AWGR transmission spectra of Fig. 6.22, the first AWGR has 40-GHz BW gaussian passbands with a transmission loss varying from 7.1 dB to 12.4 dB, *including* the 3-dB coupler loss of channels 1–4 and 13–16, whereas the second AWGR with flattened 50-GHz passbands has losses ranging from 7.45 dB to 10.61 dB. This combination of two router types is chosen to simplify their channel registration while achieving good cross-talk performance and minimal bandwidth narrowing. Insertion loss of the drop-and-detect channels from the input of the RDM to their respective drop port ranged from 10.8 to 13.8 dB. The MEMS optical switch losses in the transmit state ranges from 0.8 dB to 3.3 dB and the return loss in the reflective state ranges from 2.2 dB to 4.8 dB. As seen in Fig. 6.23, an outstanding feature of these switches is their high extinction ratio, >40 dB, achiev-

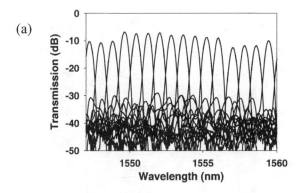

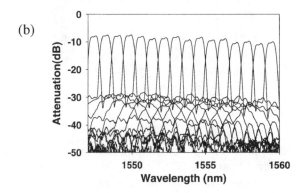

Figure 6.22 Transmission spectra of (a) gaussian passband input AWGR and (b) flattened passband output AWGR.

able with a moderate control voltage (20–40 V). Switching time with a step drive voltage was approximately 95 µs.

Optical characteristics of the assembled reconfigurable drop module are summarized in the spectra shown in Fig. 6.24. Through channel loss of the RDM ranges from 17 to 24 dB, including loss from the 3-dB splitters of the drop-and-detect channels, and cross-talk through it was below –40 dB (noise-limited measurement). High extinction at the throughport is achieved for all dropped channels; the worst-channel extinction is 23 dB, limited by the blocking switch attenuation, while all others exceed 40 dB. Loss of the drop-and-transmit channel ranges from 14 dB to 19 dB and the extinction ratio from transmit to drop state, 11 to 18 dB, is limited by weak reflections from the fiber gaps in the switches. Low-power reflections of the drop-and-detect channels also appear in the drop-and-continue port, but have little impact as they are rejected by a demultiplexer before reaching the drop-and-transmit channel receivers. These values of extinction, loss, and cross-talk through the RDM are sufficient for most network applications where the RDM is used in conjunction with optical amplifiers to compensate for the loss. Full-

SILICON MICROMACHINES IN OPTICAL COMMUNICATIONS NETWORKS

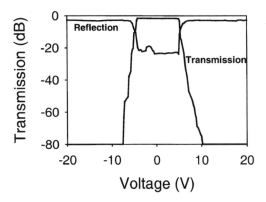

Figure 6.23 Transmission (solid line) and reflection (dotted line) characteristics of a typical MEMS optical switch.

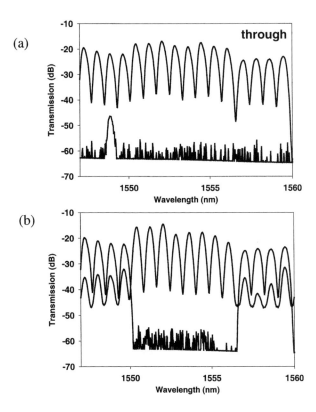

Figure 6.24 (a) Optical spectra at the throughport with all channels transmitted and all channels blocked by the MEMS switches. (b) Optical spectra at the throughport (solid line) and the drop-and-transmit port (dotted line) with all DT channels (throughport 5–12) in the drop state.

channel add/drop capability could be implemented by the inclusion of a passive combiner after the output port of the RDM.

6.3.7 ACTIVE EQUALIZER

WDM lightwave systems are evolving from point-to-point transmission systems to complex networks relying on wavelength and optical path management using optical wavelength add/drop multiplexers and cross-connect switch fibers. Optical signals traversing these networks may be amplified, attenuated, and filtered, resulting in power deviations among channels in WDM systems that may degrade overall performance. In some instances this degradation may be mitigated through pre-emphasis techniques at optical transmitters or by using fixed spectral filters at appropriate places in the system. However, these static equalization methods are not suited for dynamic or reconfigurable networks where adaptive spectral equalization is required. Here we describe a WDM spectral equalizer using athermal grating multiplexer/demultiplexers and micro-electromechanical system (MEMS) variable attenuators. A unique feature of the MEMS optical equalizer is the capability of autonomous power regulation (i.e., *self-powered and self-regulated*) using optical limiters incorporating high-voltage InGaAs photogenerators.[12]

Figure 6.25 shows the channelized WDM spectral equalizer comprising back-to-back athermal grating demultiplexers coupled through two-port MEMS variable attenuators.[10] Maximum insertion loss of the individual 16-channel, 100-GHz-

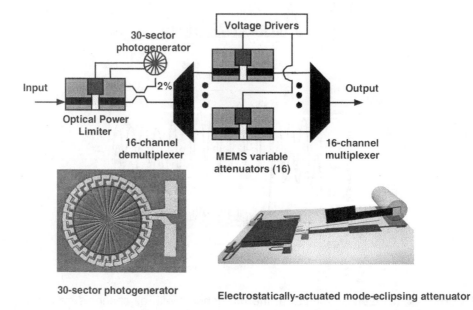

Figure 6.25 Channelized WDM spectral equalizer with autonomous input power limiter. Channel attenuators can be replaced with optical limiters.

spaced gaussian passband demultiplexers is 3.6 dB and cross-talk is below 43 dB. The packaged MEMS variable attenuators are electrostatically actuated flag shutters; the excess loss is 0.9–2.3 dB, transmission extinction > 60 dB and actuation voltages 4 V < $V_{actuation}$ < 30 V. Excess insertion loss of the WDM equalizer ranges from 6.8 dB to 9.1 dB, including four FC/PC connectors. Autonomous power limiters are designed using high-voltage InGaAs photogenerators to power the MEMS attenuators with feedback through a 2% tap at the attenuator output. Figure 6.26 summarizes properties of the optical power limiter.

Representative output spectra of the equalizer are shown in Fig. 6.27, including cases of unattenuated response, maximum attenuation, and arbitrary equalization spectra. Figure 6.29(a) shows minimum excess loss of 6.8-dB on channel 4 and maximum excess loss of 9.1-dB on channel 13. Figure 6.27(b) shows high extinction with > 50 dB dynamic range, and Fig. 6.27(c) illustrates the capability to synthesize arbitrary spectra with smooth (channels 1–7) or discontinous equalization (channels 9–16).

Autonomous power limiting and spectral equalization is tested using MEMS attenuator/InGaAs photogenerator optical limiters placed either at the input to the spectral equalizer or replacing the regular MEMS attenuators inside the equalizer. An input limiter responds to the integrated power of all the channels, augmenting the spectral tailoring of the WDM equalizer and provids an autonomous power control function. Placing optical limiters inside the WDM equalizer yields autonomous spectral equalization, a new function in lightwave networks. Input power limiters and channel limiters working in tandem increase the equalizing dynamic range.

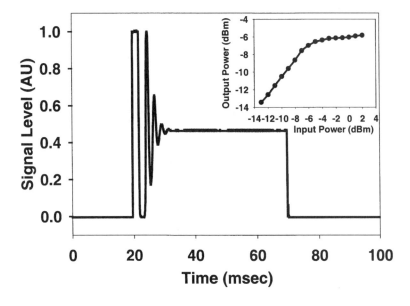

Figure 6.26 Optical power limiter transient response to a −3-dBm peak input pulse and DC transfer function (inset).

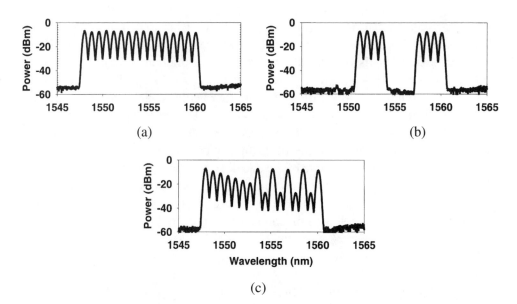

Figure 6.27 (a) Output spectrum of 16-channel WDM spectral equalizer—not active. (b) Channels 1–4 and 9–12 at maximum attenuation. (c) Channels 1–7 showing 2-dB/channel equalization slope and channels 8–16 showing 20-dB adjacent channel equalization.

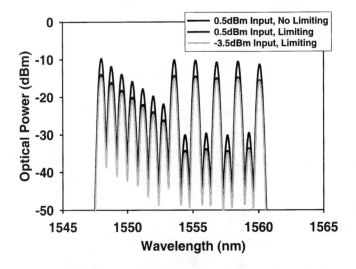

Figure 6.28 Output spectra with input limiter and 4-dB variation in total input power (upper trace has no limiting). Relative channel powers remain fixed.

Figure 6.28 shows three equalization spectra where the input power limiter is inactive (a), active with 0.5-dBm total input power (b), and active with −3.5-dBm total input power (c). From condition (b) to condition (c), a 4-dB change of input power,

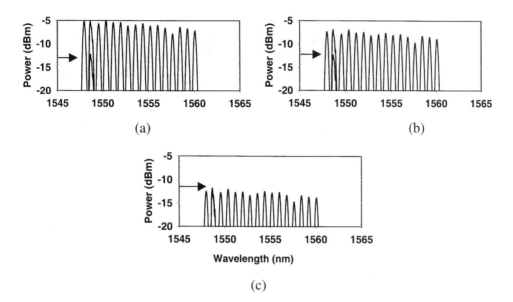

Figure 6.29 Output spectra with channel 2 having autonomous equalizer function ON (lower trace) or OFF (upper trace), showing < 0.3 dB output power variation for 7 dB input power variation.

the output power changed 0.75 dB. Figure 6.29 compares spectra measured with an optical power limiter placed in channel 2 and the input limiter inactive. Three different input power conditions are shown with the channel equalizer in both active and inactive states. The results show that channel 2 is regulated to within 0.3 dB for 7-dB variation in its input power even while all other channels are freely varying in output power.

6.3.8 MICRO-MICROSCOPE

Finally, shown in Fig. 6.30 is a device without short-term practical application, but one that, while somewhat whimsical, demonstrates some of the unlimited power of the MEMS technology. The object is a fully functional micro-microscope. The device is an near-field scanning optical microscope (NSOM) with a tapered optical fiber serving as both the aperture and the micromachined stage for scanning the sample relative to the aperture. The device is fully functional and is capable of a spatial resolution of less than half a micron. Clearly, the design of MEMS devices for optical systems is limited only by the imagination.

6.4 THE FUTURE

We have shown a variety of lightwave network applications that benefit from the small size, scalability, low power consumption, and low cost of MEMS optical

Figure 6.30 Micro-microscope using a MEMS three-axis stage for moving the sample relative to the tapered optical fiber that acts as the NSOM aperture.

circuits. Rapid development in optical MEMS technology is being driven by the urgency of challenging optical network requirements. For example, now that rudimentary *non*reconfigurable WDM optical add/drop multiplexers are deployed in the field, the industry is seeking to upgrade to those that are fully reconfigurable. MEMS optical switches can meet the functional requirements for these complex optical subsystems and may provide the only practical solution when large numbers of add/drop channels are required. Although no MEMS device has yet been deployed in an active lightwave network, the wealth of new capabilities presented by MEMS optical devices places them as certain candidates for imminent commercial success. The optical MEMS industry is estimated to become a multibillion-dollar business in the next five years. The optical communications industry has identified MEMS as a strategic technology that cannot be neglected, with exciting challenges and opportunities for many years to come.

REFERENCES

1. J. E. Ford, V. Aksyuk, D. J. Bishop and J. A. Walker, "Wavelength Add-Drop Switching Using Tilting Micromirrors," *J. of Lightwave Technology* 17, 904–911 (1999).
2. J. E. Ford, J. A. Walker and K. W. Goossen, "Fiber-Coupled Variable Attenuator Using a MARS Modulator," in *Microstructures and MEMS for Optical Processing III, SPIE Proc.*, Vol. 3226, pp. 86–93 (1997).
3. V. Askyuk, C. R. Giles, B. Barber, R. Ruel, L. Stulz and D. Bishop, "Low Insertion Loss Packaged and Fiber-Connectorized MEMS Reflective Optical Switch," submitted to *Electronics Lett.*
4. M. Born and E. Wolf, *Principles of Optics*, pp. 380–381, fifth ed. (1975).
5. H. Kogelnik, "Coupling and Conversion Coefficients for Optical Modes," in *Proc. Sympos. Quasi-Opt.*, Vol. 14, pp. 333–347 (1964).
6. K. W. Goosen, J. A. Walker and S. C. Arney, "Silicon Modulator Based on Mechanical-Active Anti-Reflecton Layer with 1-Mb/s capability for Fiber-in-the-Loop Applications," *IEEE Photon. Tech. Lett.*, 1119–1121 (1994).
7. E. Hashimoto, Y. Uenishi, K. Honma and S. Nagoaka, "Micro-Optical Gate for Fiber Optic Communication," in *Transducers '97, IEEE Int. Conf. on Solid-State Sensors and Actuators*, pp. 331–334 (1997).
8. B. C. DeLoach, R. C. Miller and S. Kaufman, "Sound Alterer Powered Over an Optical Fiber," *Bell Syst. Tech. J.* 57(22), 3309–3316 (1978).
9. A. G. Dentai, E. C. Burrows, C. R. Giles, C. A. Burrus and J. C. Centanni, "High-Voltage (2.1V) Integrated InGaAs Photogenerator," 33(8), 718–719 (1997).
10. B. Barber, C. R. Giles, V. Aksyuk, R. Ruel, L. Stulz and D. Bishop, "A Fiber Connectorized MEMS Variable Optical Attenuator", *IEEE Phot. Tech. Lett.* 10(9), 1262–1264 (1998).
11. M. C. Wu, "Micromachining for Optical and Optoelectronic Systems," *Proc. IEEE* 85(11), 1833–1855 (1997).
12. A. G. Dentai, C. R. Giles, E. Burrows, C. A. Burrus, L. Stulz, J. Centanni, J. Hoffman and B. Moyer, "A Long-Wavelength 10V Optical-to-Electrical InGaAs Photogenerator," *Photonics Tech. Lett.*, 114–116 (1999).
13. K. Okamoto, K. Takiguchi and Y. Ohmori, "16-Channel Optical Add/Drop Multiplexer Using Silica Based Arrayed-Waveguide Gratings," *Electron. Lett.* 31(9), 723–724 (1995).
14. C. G. M. Vreeburg, T. Uitterdijk, Y. S. Oei, M. K. Smit, F. H. Groen, E. G. Metaal, P. Demeester and H. J. Frankena, "First InP-Based Reconfigurable Integrated Add-Drop Multiplexer," *IEEE Photon. Tech. Lett.* 9(2), 188–190 (1997).
15. C. Dragone, "An NxN Optical Multiplexer Using a Planar Arrangement of Two Star Couplers," *IEEE Photon. Tech. Lett.* 3, 812–815 (1991).

CHAPTER 7

ASSEMBLY AND TEST FOR MEMS AND OPTICAL MEMS

Christopher Bang
BFGoodrich

Victor Bright
University of Colorado

Michael A. Mignardi, Thomas Kocian
Texas Instruments, Inc.

David J. Monk
Motorola

CONTENTS

7.1 Introduction / 333

7.2 Packaging Processes / 335
    7.2.1 Wafer Bonding/Wafer-Scale Packaging / 336
    7.2.2 Anti-Stiction Techniques / 344
    7.2.3 Dicing and die separation / 346
    7.2.4 Pick and Place / 348
    7.2.5 Leadframe, Header, and Substrate Materials / 349
    7.2.6 Die Attach / 351
    7.2.7 Interconnection / 353
    7.2.8 Integration with Electronics / 355
    7.2.9 Encapsulation and Overmolding / 369
    7.2.10 Package Housing Materials and Molding / 370
    7.2.11 Trimming and Calibration / 372

7.3 Packaging of Physical Sensors / 373
    7.3.1 Pressure Sensors / 374
    7.3.2 Inertial Sensor Packaging / 400
    7.3.3 Optical MEMS Packaging—The Digital Mirror Display (DMD) Example / 405
    7.3.4 Optical MEMS Packaging—Flip-Up Microstructures Using Microhinges / 413
    7.3.5 Microfluidics Packaging / 422
    7.3.6 Actuators / 425

7.4 MEMS and MOEMS Testing / 427
    7.4.1 Materials Testing, Modeling, and Model Verification / 427
    7.4.2 Testing & Electrical Characterization / 429

7.5 Reliability of MEMS / 432
    7.5.1 Pressure Sensors and Media Compatibility / 432
    7.5.2 Accelerometers and Vibration and Shock Testing / 447
    7.5.3 Microfluidics / 448
    7.5.4 Optical MEMS / 448

7.6 Cost Model and Summary / 449

References / 455

7.1 INTRODUCTION

Micro-electro-mechanical systems (or microsystems) are devices that include mechanical and electrical elements. Transducers, or devices that transform one form of energy into another form of energy, are often the electro-mechanical elements within these systems. The purpose of most MEMS is to act as sensors or actuators: that is, devices that interact directly with the physical environment. Sensors can convert such physical signals as pressure, temperature, acceleration, displacement, light, etc. into electrical signals to be processed by electronics. Moreover, electrical signals can be converted in actuators to perform work on the environment, like pumping and valve operation in microfluidics, and switching in optics and/or electronics. In many cases, physical signals induced by the package can be misinterpreted as signals from the environment.[1] For example, in pressure sensors, packaging stress can affect the stress of the sensor (or transducer), thus affecting the output of the device. Vibrational modes in the package can influence the operation of an accelerometer. In summary, packaging affects the MEMS. This was observed very early in the development of sensors and actuators. Senturia and Smith noted in 1988, "... it is necessary to design the microsensor and the package AT THE SAME TIME.[2]" They made two very important observations in this work: (1) "Packaging people" and "sensor people" are usually not the same people, and they do not always work well together; and (2) consideration of the package can result in elimination of possible sensor designs because they would not be feasible in a particular package.

In addition, the fact that packaging can affect the assembly, testing, and output of a microsystem represents a considerable amount of the cost of a product. Swift[3] and Schuster[4] both showed that the package and test can represent over 50% of the product cost for a microsystem. Discussion during each of the last two Commercialization of Microsystems conferences has confirmed this point. In fact, industrial MEMS engineers and managers often suggest that a typical cost breakdown is roughly 33% silicon content, 33% package, and 33% test. It has been reported, in some cases, as greater than 70% packaging cost vs. silicon content.[5-10] Furthermore, of this silicon content, often less than half of it is the microsensor or actuators. The rest of the silicon content is the control electronics cost.

The package can play a vital functional role in the productization of a microsystem. For example, the package can isolate the induced stress from the mounting of the device,[9,11] it can make the electronics compatible with a harsh environment,[1,9] and it can even be an integral part of the microsystem (e.g., mechatronics packaging). While MEMS engineers often discuss integration as a technique for placing transducers and electronics on the same IC chip, integration can also refer to the combination of typical package functions onto the silicon.[1,12] Wafer bonding is an example of this. In pressure sensors, an absolute vacuum reference can be integrated into the silicon device. Furthermore, many microsystems will not be integrated monolithically; therefore, multichip packaging will be required.[12]

Finally, efforts of packaging and test for MEMS have been significantly underrepresented in academic and trade publications. Considering the cost breakdown, little has been written about the assembly and test for MEMS. Much of the grant money, academic research, and publications focus on the micromachined transducer itself and not the rest of the microsystem. Recently, that point has been highlighted in several conferences.[13–15] Publications that have summarized the slow breakthrough of microsystems in commercial application have pointed to the following limitations of the technology:[6,14–16]

- CAD,
- standards,
- assembly techniques,[12]
- process tools (both front-end and back-end),
- test equipment, processes and standards,
- harsh media compatibility,[1,5,16]
- reliability standards and results, and
- packaging techniques and processes.

Assembly and test for MEMS and Optical MEMS are important topics in the commercialization of microsystems. However, the next question may be, "is the commercialization of microsystems important?" Several marketing surveys have shown the total available market (TAM) for microsystems to be very large. For example, a recent Nexus report shows the TAM to be greater than $34 billion by 2002.[17] It also lists the following reasons for using microsystems over more common "macro-systems:"[17]

- suitability for low cost, high volume production,
- reduced size, weight, and energy consumption,
- high functionality,
- improved reliability and robustness, and
- biocompatibility.

System Planning Corporation issued a report during 1998 that also suggests a large total available market.[16] Figure 7.1 shows their findings and several other groups' findings on the microsystem total available market. The bottom line is that it is, indeed, a large market, and one worth pursuing provided profitability can be attained.

The following chapter summarizes many of the practical issues in MEMS and optical MEMS assembly and test. We start by describing many general process topics for MEMS packaging. This is followed by examples of MEMS and optical MEMS packaging for several common devices, including pressure sensors, accelerometers, optical MEMS, and microfluidics and bioMEMS. The second half of the chapter focuses on test and reliability issues for these types of devices. We conclude the chapter with an overview of several practical issues for commercialization of microsystems.

ASSEMBLY AND TEST FOR MEMS AND OPTICAL MEMS 335

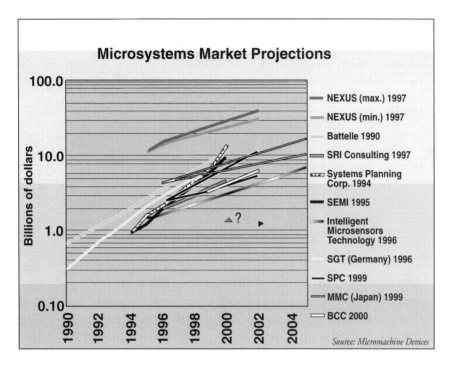

Figure 7.1 Summary of total available market from several sources (from [16]).

7.2 PACKAGING PROCESSES

In general, packaging of MEMS and optical MEMS devices is an outgrowth of the IC packaging industry and the hybrid packaging industry.[1,18] Packages for MEMS and optical MEMS devices typically have the following requirements:[1,5,19–22]

- Interaction with the environment (e.g., media compatibility, or hermetic, vacuum sealing to protect accelerometers, resonators, etc.),
- Low cost,
- Small size,
- High reliability and quality,
- Standardization (although custom packaging for MEMS and optical MEMS is more the rule than the exception),
- Acceptable electrical interconnection (e.g., minimum power supply voltage drop, self-inductance, cross-talk, capacitive loading, adequate signal redistribution, and, perhaps, electrical feedthroughs),
- Thermal management (i.e., power dissipation and matching CTEs with substrates to minimize package-induced stress),
- Acceptable mechanical interconnection (e.g., porting for pressure or flow sensors and mounting techniques that do not apply undue stress to the device, yet support the device appropriately),

- Protection from electromagnetic interference,
- Testability and trimmability (e.g., internal test/trim nodes),
- Precision alignment (especially for optical MEMS),
- Accommodation for optical interconnection either with optical fiber or free-space, and
- Mechanical protection and stress isolation.

To create these MEMS packages, the following common processes are used: bonding, wafer sawing/scribing, pick and place, die attach, wirebonding/interconnection, encapsulation, overmolding, trimming, and final testing. We will discuss each of these topics, except final test, in the following section. Final test is discussed in Section 7.4.

7.2.1 WAFER BONDING/WAFER-SCALE PACKAGING

Wafer bonding or cavity sealing is present on many MEMS products. Wafer-level packaging is used in pressure sensors (Fig. 7.2) to create an absolute vacuum reference for absolute pressure sensors and in accelerometers or micromachined resonators to create a controlled atmosphere for the device (Fig. 7.3). Complete reviews of wafer-to-wafer bonding for MEMS are given by Schmidt[23] and Ko et al.[24]

Figure 7.2 shows an example of an absolute pressure sensor that is created by an adhesive layer bonding, using low temperature (450 to 500°C) glass frit. An optional port through the silicon wafer constraint can be formed with bulk

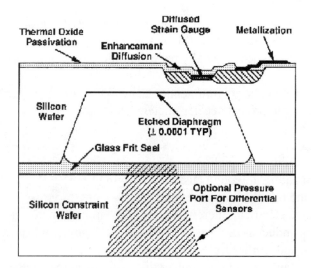

Figure 7.2 A constraint wafer for a presure sensor. In the case of an absolute pressure sensor, the constraint wafer acts to seal in a vacuum pressure on the backside of the diaphragm. In the case of a differential pressure sensor, the constraint wafer acts to help isolate the package stress from the sensor.[23,25]

micromachining to create a constraint wafer for a differential or gauge pressure sensor.

Another function of the wafer-scale package for the accelerometer device, shown in Fig. 7.3, is to provide a hermetic seal to the surface micromachined structure during saw and assembly, processes that could damage or destroy the delicate microstructure (e.g., through stiction).

Examples abound of MEMS that use wafer bonding. More of these devices will be presented as we review the types of wafer bonding processes. In general, wafer bonding can be categorized based on the following processes: (1) direct wafer bonding, (2) anodic bonding, and (3) intermediate layer bonding.

7.2.1.1 Direct Wafer Bonding

Direct wafer bonding dates back to the mid-1960s.[25,26] A complete review of the history, mechanism, processing considerations, and applications for direct wafer bonding to micromachined devices is given in Ristic,[25] and a overview of the applications for MEMS is given by Desmond et al.[27] Two process steps are required for a bond to occur: 1) the polished wafers come into intimate contact with one another, and 2) Si atoms or SiO_4 moieties must form a stable bond. Therefore, the key requirements for this process are to have a specularly smooth surface, to maintain contact between the two wafers throughout the process, and to elevate the temper-

Figure 7.3 This figure shows the Motorola two-chip accelerometer. The micromachined device is located within the wafer-bonded die (right side of the photo). A glass-frit bonded cap wafer is used to eliminate stiction and to protect the sensor from mechanical damage during assembly and to create a constant pressure environment for the operation of the device during the lifetime of the application.[25,297]

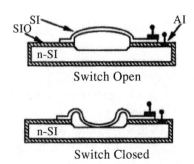

Figure 7.4 Pressure switch fabricated using direct wafer bonding.[25,29]

ature so that the chemical bonding process can form. Typically, bond energies an order of magnitude higher than those obtained at room temperature are observed when bonding at temperatures in the 800 to 1200°C range.[23,28]

Microstructures created with direct wafer bonding include a pressure switch (Fig. 7.4),[25,29] a pressure sensor (Fig. 7.5),[23,25,26,30] a peristaltic pump (Fig. 7.6),[25,31] an accelerometer (Fig 7.7),[23,26] and several other devices.

Provided that one can design around the high temperature requirements for direct wafer bonding, it is a versatile technique that can be used for wafer-level packaging for a variety of MEMS. This works well when the bonding process is the first (or nearly the first) process step. However, the bonding process is often the last step, and high temperature processing is not possible at that point in the process. Alternatively, surface activation bombardment with Ar energetic particle beams,[36–38] chemical cleaning[39] in ultra high vacuum, surface activation for bonding using an ammonium fluoride etch mixture,[40] low vacuum (approx. 700 Pa) followed by a 150°C air anneal,[41] or the storage of wafers at slightly elevated temperatures (<150°C) for a long period of time[42] have shown promising low temperature direct wafer bonding results.

Another concern with direct wafer bonding is the result of diaphragm deflection following bonding of a cavity. Cavities sealed by direct wafer bonding (or anodic bonding) have been observed to contain residual gas pressure (i.e., hydrogen, water, nitrogen, and oxygen) higher than the bonding pressure.[43] Observation of the residual gas in a cavity can be deduced by watching the deflection of a diaphragm Bosch[35,44] or by FTIR measurement of gases within the cavity.[45] Work by Secco d'Aragona et al. and Schmidt et al. suggests that cavities must be sealed in a controlled oxygen environment (e.g., 20% O_2) to eliminate plastic deformation from the diaphragms.[23,46,47] Sealing of the cavities in nitrogen resulted in observed plastic deformation of the single crystal silicon diaphragms.[46]

An interesting alternative use for a wafer bond has been presented as "sacrificial wafer bonding" (Fig. 7.8).[48] In this process, the wafer bond is used only over the areas in which cavities have been defined. Therefore, it is a microcavity-like process (Fig. 7.5). Hypothetically, with the advent of deep RIE, this process could be used with any type of wafer bonding to create a variety of 3D structures.

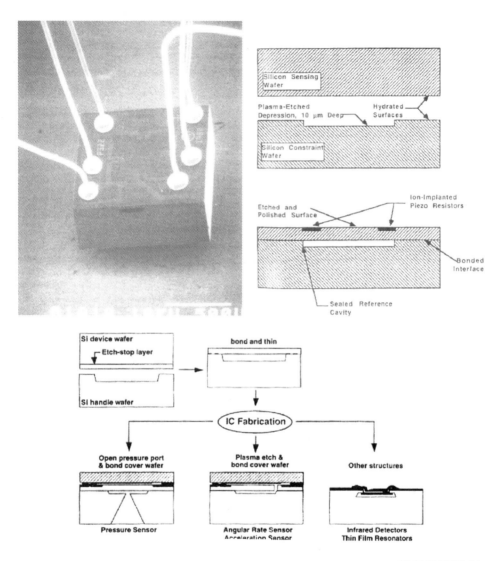

Figure 7.5 Pressure sensor fabricated using direct wafer bonding.[23,25,26,30,32–34]

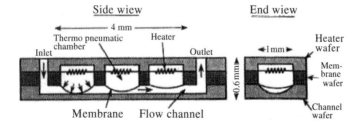

Figure 7.6 Peristaltic pump fabricated using direct wafer bonding.[25,31]

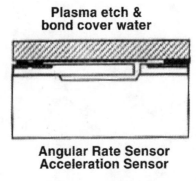

Figure 7.7 Accelerometer fabricated using direct wafer bonding.[23,26] Bosch has also used direct wafer bonding to create an accelerometer.[35]

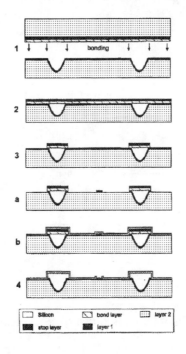

Figure 7.8 The sacrificial wafer bonding process.[48]

7.2.1.2 Anodic Bonding

Anodic, or electrostatic, bonding is accomplished by placing a polished silicon wafer against a glass wafer, heating to between 180 and 500°C,[49] and applying 200 to 1000 volts (Fig. 7.9). Originally, this process was used for bonding metal to glass,[50] but the metal was replaced by silicon for application in the MEMS industry.[23] The glass wafers used are often Pyrex for MEMS applications.[51] These wafers have sodium ions that are mobile during the bonding process and aid in

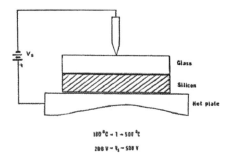

Figure 7.9 A simple mechanism describing the anodic bonding process.[24]

completion of the bond. With the application of high electrical fields, the mobile sodium ions migrate away from the bonded interface. This electrically-driven diffusion creates a fixed charge in the glass, thus inducing a high electric field across the interface. Although the bonding mechanism is not completely understood, it is assumed that a chemical bond forms between the silicon and the glass when they are subject to the combination of high electric field and high temperature.[23] More generally, glasses with positive alkali mobile ions can be used with conductors or semiconductors. With applied temperature, the alkali ions become more mobile and migrate away from the bonding interface with the application of high electrical bias. This leaves oxygen ions at the bonding surface available for chemical bonding.[52]

The advantage of anodic bonding over direct wafer bonding is that the temperatures used are much lower. This allows anodic bonding to be used after the first step in a fabrication process. Anodic bonding is generally considered more reliable than the low temperature wafer bonding processes that are being investigated currently. In fact, several companies have used anodic bonding in producing MEMS devices, including pressure sensors for automotive applications from Motorola,[53] Honeywell, and Bosch[35] (Fig. 7.10).

However, anodic bonding is not without its own limitations:[35]

- it requires non-standard IC equipment;
- alkali metal ions are needed in the glass material which can be a problem with on-chip electronics;
- high voltages are required for the bonding which also can affect on-chip electronics;
- glass wafer structuring can be complicated; and
- the CTE of Pyrex and other glasses is not the same as that of silicon over a wide range of temperature, although it is close.[24]

The following requirements are essential for this process:[24]

- the glass must be slightly conductive;

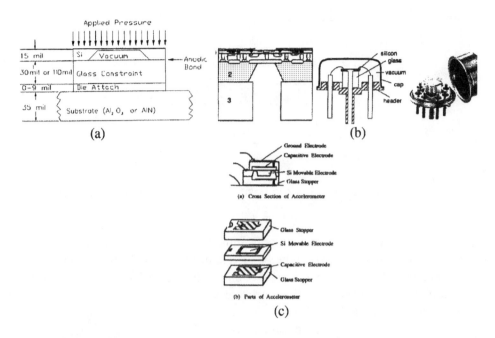

Figure 7.10 Examples of anodic bonding being used in products. (a) An absolute pressure sensor used for automotive applications from Motorola.[53] Bosch has also presented a similar device.[35] (b) A backside exposure "wet-dry" (i.e., harsh media from the backside of the device, benign media on the frontside of the device) differential pressure sensor from Bosch.[35] Honeywell has produced a similar device. (c) A semiconductor accelerometer that has been investigated by NEC.[54] A similar device has been produced by Hitachi.[55]

- the metal used as the anode cannot inject mobile ions into the glass;
- the surface roughness must be small (less than 1 micron) for both the glass and the metal. Experiments have shown that a 20-nm groove under the bond will still allow adequate sealing while a 58-nm groove will not;[35]
- the surfaces must be free of contamination;
- pressure is not required, but has been observed to improve the anodic bonding;[56]
- the CTE of the two materials being bonded should be closely matched (see Table 7.2);
- voltage requirements are dependent upon temperature used, glass type, and thickness; and
- ambient atmosphere can be used for bonding, although nitrogen, forming gas, argon, and helium have also been used.

In addition, these techniques have been used with laser heating[24] and with Pyrex thin film on silicon to create a silicon-to-silicon bond.[24,57] In the latter case, the sputter-coated Pyrex was silicon rich and the film broke down at 20 V causing the bond to fail.[24] Also, Esashi and his group have used nonevaporable getters with anodic bonding to eliminate the problem of residual gas in the bonded cavity.[44]

Bond quality has been investigated with visual inspection, TEM, blade techniques (as described in Section 7.2.1.1), and/or various pull tests. Tatic-Lucic et al. used visual inspection, a design of experiments approach (i.e., a Box-Behnken design), and a bond quality rating scale to optimize a silicon-glass anodic bonding scheme with phosphosilicate glass as an intermediate layer.[58]

7.2.1.3 Alternative Bonding Techniques

Several alternative bonding techniques have been developed for the MEMS industry. Probably the most prevalent today is the use of an intermediate material for bonding two substrates. Ko et al. list such intermediate materials as epoxy and polyimide.[24] Kang et al. have discussed the possibility of using anisotropic conductive films for wafer bonding.[59]

Motorola has been producing pressure sensors since 1979 (Fig. 7.2), and accelerometers since 1994 (Fig. 7.3) using a glass frit intermediate bonding material to form wafer bonds.[60] In this process, glass paste is screen printed onto a "constraint" substrate. Glass pastes are powdered glass with a solvent and a bonder. These are usually proprietary compositions.[51] This material is then glazed at 400°C prior to placing the constraint wafer in contact with the "active" wafer. Once these wafers are joined, they are fired at approximately 550°C to form the bond. The advantage of glass frit bonding over direct wafer bonding and anodic bonding is that it can be performed at low temperature without any large electrical fields. Therefore, with non-CMOS-integrated or bipolar-integrated devices, the glass frit bonding process can be performed at the end of the fabrication and just before wafer sawing and assembly begins. This allows flexibility in the fabrication process and protects the device from damage during the assembly process. Also, glass frit bonding processing can occur at different pressures to create absolute pressure sensors (with vacuum reference cavities) and/or specified damping-level accelerometers.

Eutectic bonding is another alternative.[23,24] The eutectic point in a 2-component phase diagram is where the composition is such that the melting temperature is at a minimum. For example, in the Sn/Pb system, this occurs at 61.9 wt% Sn and 38.1 wt% Pb. The melting temperature is 183°C. In the Au/Si system it is 97.1 wt% Au and 2.85 wt% Si, and the melting temperature is 363°C. Several other systems are possible:[24] lead-indium-silver (Pb/In/Ag), lead-tin (Pb/Sn), gold-tin (Au/Sn), gold-germanium (Au/Ge), and silicon-aluminum-gold (Si/Al/Au).[61] This type of bond was investigated by Ko et al. for pressure sensors, but drift was observed with the devices.[23]

Low temperature glass bonding has been investigated as well. In this process, two wafers are placed in contact in a glass deposition reactor with a weight on them to hold them together.[24] Field and Muller have proposed the use of boron-doped thin film glasses for low-temperature (450°C) bonding.[62] This process uses the same technique as described by Ko et al. It was hypothesized that the bonding occurred because of the low temperature melting of the boron-doped glass at 450°C (and reflow of the glass to promote the bond). Moreover, attempts to broaden the available materials on which bonding is feasible are being made with bonding to silicon nitride[63] and low temperature bonding to silicon nitride.[64]

A group at the University of Michigan is performing wafer bonding on localized areas using a polysilicon heater.[65] This localizes the heat source so a material can form a hermetic bond directly above the heater. This technique can be used for a variety of other applications, including adjusting the resonant frequency of a resonator by adjusting the mass, and trimming an overlying nonstoichiometric metal silicide through localized heating of the material.

A group at the University of California, Berkeley has been working on wafer-to-wafer transfer as a means of integrating MEMS and CMOS devices and, more recently, as a means of sealing a structure within a cavity.[66–69] Even though reliability experiments on this process are still ongoing, this technique may provide an alternative for expensive wafer-to-wafer bonding to create hermetic microstructures.

Finally, several groups have attempted monolithic sealing of microstructures.[70] This was first attempted with the sealing of a light source using surface/bulk micromachining.[71,72] Following this effort, McNair et al. used a thick PSG mesa above a microstructure sealed with low stress silicon nitride or polysilicon using surface micromachining processing.[73] Lebouitz et al.[74] are using permeable polysilicon[75–77] as a shell for later sealing. In this process, thin (1500 Å or less) polysilicon is permeable to HF. The sacrificial layer etching, which does not require etch access holes, etches the underlying glass (particularly phosphorus-doped glass), while leaving the thin polysilicon intact. The permeable polysilicon can then be sealed with a subsequent LPCVD deposition.

7.2.2 Anti-Stiction Techniques

Stiction can be described as the phenomenon where microstructures become stuck to their substrates. Several authors have provided reviews of the stiction phenomenon and literature[25,78–81] and comparative studies of various stiction reduction techniques.[82] Typically, stiction occurs at one of two times (particularly for surface micromachined structures): post-sacrificial-layer-etch and during usage. The former is called "release stiction" and the later is called "in-use stiction." Packaging engineers are most often concerned with in-use stiction because of the potential reliability problems that it can cause; however, a phenomenon similar to release stiction can be a problem if devices are exposed to liquids during assembly, particularly during the wafer sawing operation. Therefore, we will describe both types of stiction and techniques for improving yield and reliability that can be limited by this phenomenon in this section.

Release stiction has been recognized as a problem, especially for surface micromachined structures, since the late 1980s. During drying, surface tension from the liquid-vapor interface causes a downward force on the structural layer. If the layer touches the substrate, it is prone to stick onto the surface. It is hypothesized that etch products and/or contaminants in the rinse water can then precipitate out of solution during drying and cause a bond that is stronger (e.g., a chemical bond) than electrostatic bonding between the two semiconductors.[83,84] Mastrangelo and Hsu were among the first to provide an analytical model for this effect.[34,85,86] Others have followed their lead.[87,88] At the University of California, Berkeley, dimples[89–92] or meniscus-shaped microstructures[93] were designed into the thin films constituting the structural layers of the microstructure to minimize the surface area for stiction. This idea has been developed further into a technique for using "bumpers" to minimize both release and in-use stiction.[94] Fedder and Howe used breakable tethers to hold the device in place temporarily during sacrificial layer etching and drying. These tethers could be broken or melted following release, and stiction was never encountered.[95] Alternatively, roughening the surface of the structural layer (often polysilicon) was attempted through modifications to the sacrificial layer etch process.[96,97] However, the repeatability of this type of solution is suspect; therefore, the general consensus is that surface tension forces must be overcome (i.e., avoided or eliminated) to minimize release stiction. Additional modifications to the etch process include the use of gas phase HF to totally eliminate liquid from the sacrificial etch process[98,99] and the use of high temperature sacrificial layer etch processes.[100,101] The simplest approach to this is to use a rinse solution that exhibits a lower surface tension force during drying. Several groups have used IPA drying or other low surface tension solutions, like n-hexane.[102] But it is not possible to eliminate this problem with this approach because the surface tension force induced by even low surface tension solutions, like alcohols, is still high enough to cause release stiction.[77] Consequently, a group at the University of Wisconsin experimented with freeze-drying and sublimation to eliminate the surface tension forces.[84,103–108] Mastrangelo et al. have presented a process for using polymer (parylene) studs.[109,110] This is similar to the process that Howe et al. has patented for the production of accelerometers at Analog Devices[111] which uses photoresist,[112] and a slightly different process which uses divinylbenzene.[113] Finally, and probably most successful on the research scale, is the use of the supercritical carbon dioxide drying technique.[114,115] Of course, all of this assumes that the structural materials exhibit tensile stress so that they do not inherently bend down onto the substrate.[116]

In the case of in-use stiction, it is hypothesized that moisture from the environment (e.g., relative humidity) comes in contact with the MEMS structural surfaces. If, during operation, these structures come in contact, the moisture can cause a temporary bond, which, like release stiction, can then become permanent with time. To reduce in-use stiction, three basic techniques have been attempted. The first is to use a hermetic seal around the microstructure to eliminate the possibility of moisture encountering the structure. An example of this is the Motorola accelerometer, which uses a glass frit wafer bond that hermetically seals the microstructure from the environment to eliminate the possibility of moisture affecting the device (Fig. 7.3). Secondly, techniques to minimize the work of adhesion have been employed. Specifically, Houston et al. have used ammonium fluoride to reduce the work of adhesion on surface micromachined structures.[117,118] Lastly, a variety of coatings and/or surface treatments have been used on the microstructure to eliminate the chance of contact between two surfaces that have the tendency to stick (e.g., polysilicon and silicon—each material with a native oxide). The University of California, Berkeley has pioneered techniques of using self-assembled layer monolayer coatings to minimize in-use stiction (Alley et al.[83] and Maboudian et al.[79,117,119]). Also, other researchers have used fluorocarbon coatings to minimize in-use stiction.[120–122] Analog Devices uses a controlled amount of moisture and an organic coating that is deposited prior to sealing the package to "seal" the microstructure from moisture during operation.[123]

Lastly, it should be noted that techniques have been published that remove the effect of stiction following the event. Gogoi and Mastrangelo have published a technique for using Lorentz forces to lift the microstructure that is stuck.[124]

7.2.3 Dicing and Die Separation

For microelectronics ICs, dicing or die separation is often not a difficult feat (anymore). Concerns about yield are always present in a manufacturing operation, but for standard microelectronics, typically, a wafer is placed on a mylar film with light adhesive that holds the wafer in place during diamond sawing. The adhesive is strong enough to keep the die on the mylar film during sawing and yet mild enough to allow the "pick and place" operation that occurs next in the assembly process. The diamond-lined blade must be "dressed" in preparation for its initial trial. Once preparations are completed, the saw is used to cut partially or completely through a silicon wafer substrate. A continuous stream of water is used to cool the blade during operation. In production, a vision system with programmable spacing and rotation of a stage carrying the wafer is used to adjust the location of the wafer so that the saw cuts through the wafer "streets" (i.e., areas in between die locations). The width of the saw cut, called the saw kerf, can be from 1 to 10 mils (i.e., thousandths of an inch, or 25 to 250 micrometers). Normally, there are design rules accounting for the roughness of the cut that extend beyond the saw kerf.

The traditional dicing techniques used in the semiconductor industry unfortunately are not satisfactory for several classes of MEMS devices. Surface micromachined devices after sacrificial layer release are easily damaged by moisture and particles, and are completely incompatible with the substantial silicon debris, blade cooling water spray, and vibration generated by diamond wheel dicing. An ideal dicing process for MEMS would cut silicon in a dry fashion, without generating debris. Some candidate processes for improved die separation include:

- dicing with wafer masking,
- wafer scribing,
- laser cutting,

- diamond wire cutting, and
- abrasive jet machining.

Conventional dicing can be used if the sensitive portions of the micromachined device can be protected or masked in some manner. Spin-on polymer coatings, such as polyimide, can be used for this purpose, especially if the polymer can be removed in a plasma ash process afterwards. However, a hard and thick enough polymer to protect the device is usually difficult to remove after dicing without leaving a residue. Furthermore, the coating may be damaging to some types of devices.

When protecting the frontside is not practical or sufficient, wafer scribing is sometimes preferred. Dynatex International manufactures scribers that scribe and break the wafer, while only contacting the wafer backside. Their automated systems can accommodate large substrates and high-throughput requirements. The scribing and fracturing process eliminates the debris and moisture of traditional dicing; however, fracturing can still generate particulate that may be unacceptable depending on the sensitivity of the device to particles and manufacturing yield requirements.

Other cutting techniques which merit consideration include laser cutting, diamond wire cutting, and abrasive jet machining. Laser cutting can be fast and precise, but tends to create a significant amount of splatter and debris that damage devices. Diamond wire cutting has been used in a wide variety of sensitive cutting applications, from tissue samples to moonrocks.[125] Diamond wire cutting uses a fine loop of diamond wire on a motor-driven capstan to cut into the wafer surface. This type of cutting produces very little surface damage, but may not have sufficient throughput for high-volume applications. Abrasive jets are commonly used in repair and rework of conventional electronic components. It is plausible that with a suitably small nozzle and dust collector, abrasive jets could be used to safely dice MEMS devices. While currently no cutting technique meets all of the desired characteristics for MEMS dicing, as the market size for MEMS process equipment grows, manufacturers will undoubtedly develop new and innovative dicing methods.

Micromachined devices are more difficult to saw because of induced stress that affects the device performance, delicate microstructures that can be broken during the sawing operation, and/or stiction resulting from the water used to cool the wafer saw during operation. For example, in the case of the Motorola production accelerometer, glass frit wafer bonding is used to seal the microstructure hermetically within the wafer-level package prior to sawing (Fig. 7.3).[1,126] This eliminates the potential problems listed above. In the case of the Analog Devices production accelerometer, a special upside-down sawing process is used (Fig. 7.11). In this case, the wafer is placed on a special sticky-tape that includes a recess to accommodate the microstructure. By flipping the wafer over, the microstructure can be isolated from the water and from any flying debris during the sawing process. In addition, because of the recess in the sticky-tape, the microstructure is protected

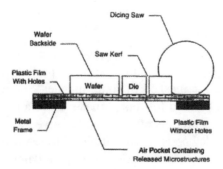

Figure 7.11 Upside-down wafer sawing process used to protect the Analog Devices accelerometer during die separation.[127]

from damage. Finally, a specially designed pick and place tool is then used to remove the dies and turn the water right-side-up for subsequent die attach.

Finally, die separation can be achieved by using micromachining. For example, bulk micromachining can be used to create an opening through a wafer in a pattern such that individual dies can be separated.[5]

7.2.4 PICK AND PLACE

Pick and place is the operation in which dies from a wafer are mounted onto a leadframe or into a package. In a production environment, a wafer is probed as one of the last steps during the fabrication process. During the probing, the wafer is marked to show the "good" die from the "bad" die. This is a process that is used to minimize cost in the assembly area by not building the bad die. Typically, during the probing operation, a bad die is marked with a drop of black ink. This inking shows the operator or the vision system in the assembly area that a particular die is nonfunctional and should not be processed through assembly. Once these dies are determined, dies are picked using a vacuum tool that holds the die during mounting onto a leadframe or into a package.

With micromachined devices, handling of delicate microstructures can be a problem throughout the assembly process. Pick and place is one operation that could cause a potential problem because of mechanical damage or damage to a delicate microstructure caused by the change in pressure (to vacuum). Bulk micromachined pressure sensors are rigid enough structures that this does not represent a significant problem. However, surface micromachined accelerometers can be more difficult. In the case of the Motorola production accelerometer, the wafer-level package provides protection to the microstructure during pick and place, and it provides a large flat area for the pick and place tool. In the case of the Analog Devices production accelerometer, a modified pick and place tool is used to minimize the disturbance of the surface micromachined structure during this operation.[127]

Some research has been performed recently to investigate the possibility of self-assembly. Yeh et al. have observed trapping of semiconductor ICs in micromachined wells.[128,129] Cohn et al. have modified this technique by adding electrostatic alignment of the die prior to settling into the cavities.[130] While these techniques may not be ready for production yet, they represent research efforts to minimize cost of pick and place through batch deposition of dies onto a substrate.

7.2.5 LEADFRAME, HEADER, AND SUBSTRATE MATERIALS

Support materials for micromachined structures come in a variety of styles. MEMS dies are mounted on metal leadframes, pre-molded packages, silicon or other wafer substrates and header package materials. For example, several companies use a leadframe as a mounting substrate for the sensor die. Motorola's two-chip accelerometer is shown in Fig. 7.3. In this case, both the sensor die and the CMOS control die are mounted onto the leadframe prior to molding.

Alternatively, Analog Devices uses a header substrate in many cases, as shown in Fig. 7.12. Many pressure sensor manufacturers (e.g., Lucas NovaSensor and Siemens Semiconductor) also use header package designs.[5] Figure 7.13 shows many of these design variations.

Figure 7.12 Analog Devices BiCMOS integrated surface micromachined accelerometer mounted on a header substrate for packaging.[131]

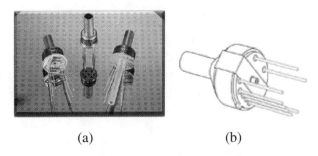

(a)   (b)

Figure 7.13 Pressure sensors mounted to TO-Header packages [(a) http://www.novasensor.com/, ca. 1/31/99, (b) http://www.smi.siemens.com/sensors/, ca. 1/31/99, among other suppliers].

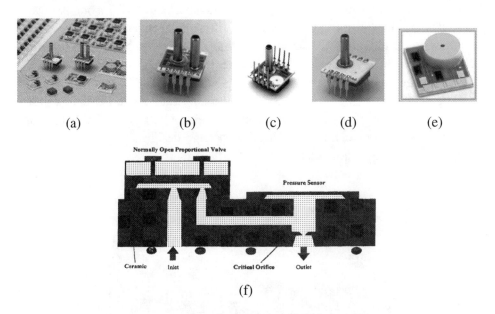

Figure 7.14 Pressure sensors mounted in ceramic packages in various styles [http://www.novasensor.com/, ca. 1/31/99]: (a) a variety of package styles using ceramic supplied by Lucas NovaSensor, (b) a dual topside port package, (c) a backside port version, (d) a single topside port version, and (e) a version used in angioplasty [http://www.novasensor.com/, ca. 1/31/99]. (f) A mass flow controller packaged using ceramic substrates.[132]

Other substrate materials used in MEMS packaging include ceramics (e.g., Analog Devices uses 8-pin CERDIPs for accelerometer package[127]). Also, pressure sensor suppliers use ceramics for packaging (Fig. 7.14).

As mentioned in Section 7.2, Packaging Processes, performance of MEMS and optical MEMS devices can be affected by the package itself. For example, stress in the package can affect the device output for many physical sensors. Thus, thermal management is required when designing the package. In fact, modeling can be performed to show the impact of leadframes, header, or substrate materials on the device performance of a physical sensor (to be discussed in Section 7.4.1, Materials Testing, Modeling and Model Verification). Recently, researchers have begun investigating the possibility of using multi-chip module packaging techniques with MEMS. More will be discussed concerning this packaging technique in the Section 7.2.7, Interconnection.

7.2.6 DIE ATTACH

Once the die is separated from the wafer, it must be attached to the chosen substrate material (e.g., leadframe, header, or alternative substrate). This process precedes wirebonding. Often, this process includes an adhesive bond layer being deposited onto the substrate material prior to placing the die on the substrate. This is then followed by a curing, annealing, or firing step to secure the adhesive bond.

As mentioned earlier, stress induced by the packaging materials is a function of the leadframe, header, or substrate material choice and the die attachment method. Stress is also a function of the material choice for the die attach. Although they have very high CTE values (which are very dissimilar with silicon), silicones are often chosen for stress sensitive MEMS devices, because they are very low modulus materials (often 6 orders of magnitude lower than silicon). In general, stress caused by eutectic (e.g., Au-Si alloy) die attach is higher than epoxy, which is also higher than silicones.[5]

One important challenge in designing with these types of materials is that they are often polymeric. Polymer materials differ from silicon, most fabrication materials (e.g., polysilicon, silicon nitride, silicon dioxide), and metals in that they are viscoelastic. This suggests that they exhibit both a viscous (or lossy term) and an elastic term for modulus. This property makes these materials more difficult to model (to be discussed in Section 7.4.1, Materials Testing, Modeling and Model Verification). Also, material geometry is difficult to determine precisely, because these materials are often drop-dispensed and will flow readily prior to curing.

Location of the sensor die is another concern when die attach methods are considered. For example, often the micromachined die is not the same die as the circuit die. If the output required includes circuit calibration, amplification, or other signal processing, the package may contain two or more dies. Figure 7.3 is an example of the Motorola accelerometer, which contains a micromachined, wafer-bonded accelerometer or "g-cell" die and a CMOS control IC. This side-by-side bonding is one configuration for the die attach. An alternative configuration is "vertical assembly." Kelly et al. (after Lyke[133]) describe this as the "Towers of Hanoi" approach to die attach.[134] In this approach, the sensor die is placed onto the silicon control IC and the two are wirebonded together with chip-to-chip wirebonds. This technique has the advantage of creating a package integrated device instead of doing wafer-level integration, which can affect the efficiency of a "standard" bipolar, CMOS, or other integrated circuit fabrication line.

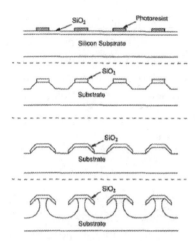

Figure 7.15 Micromachined Velcro fabrication process.[135]

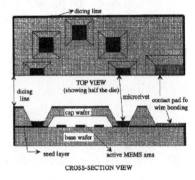

Figure 7.16 Micromachined Rivet fabrication process.[136]

Finally, three micromachining research techniques are being used to attach one die to another. In the first, dubbed "Micromachined Velcro," micromachined structures similar to Velcro are created using a combination of bulk and surface micromachining processes (Fig. 7.15). This "bonding" technique is permanent and has been demonstrated using an areal density of over 200,000 per cm^2 with a bond strength of 1.1 MPa.[135] Another technique applies a riveting process to create a secured bond. Bond strengths have been reported at 7 to 11 MPa. The process uses a base wafer (that could include a micromachined structure) and a cap wafer that has been bulk micromachined with openings for the rivets. Once aligned to the base wafer, the cap wafer is temporarily secured during an electroplating process that fills in the openings in the cap wafer to create the rivet (Fig. 7.16). In a similar process, Cohn et al. have used a metal-metal contact process to transfer a micromachined structure to another die.[66–69] This concept goes beyond die attach because instead of transferring the entire die, the attachment method only transfers the microstructure (Fig. 7.17). Each of these techniques is of interest, although they

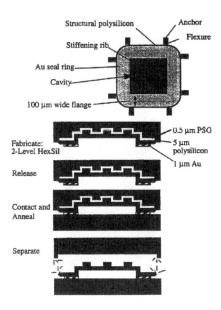

Figure 7.17 Process for transfer of micromachined structures from one wafer to another.

are all probably a long way from being the stable, production-worthy process that drop-dispensed die attach processes are today.

7.2.7 INTERCONNECTION

Typical electrical interconnection for microelectronics includes wirebonding, flip chip or ball bonding, and tape automated bonding (or TAB). Most of the time, MEMS devices utilize wirebonding as the method for electrical interconnection.[25] The wirebonding process is of two types: thermocompression bonding and ultrasonic bonding. Thermocompression bonding uses heat and pressure to create the metal-to-metal bond (usually Au wire). Ultrasonic bonding uses ultrasonic vibration to create the metal-to-metal bond (usually Al wire). Because of the stress isolation that is often essential with MEMS, soft die attach materials are used in many cases. This requires that thermocompression bonding be used. On the other hand, if high temperatures cannot be allowed, ultrasonic bonding is necessary.

Wirebonding is typically a "ball-wedge" bond. The ball is formed initially with the end of the existing wire on the die bondpad. After a stitching process, the wire is placed on the leadframe post with a wedge bond and is cut. Figure 7.18 shows the process.[25] The wirebonding process is fast and automated. Yet, several die and package properties affect this process: the die attach material, the die flatness (or tilt), the allowable wirebond loop height, the level of the bondpad versus the leadframe post, and the cleanliness of the surfaces. In addition, wirebonding properties

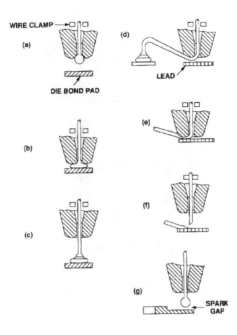

Figure 7.18 Wirebonding process diagram.[25]

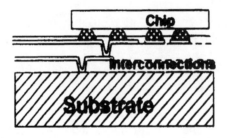

Figure 7.19 An example of flip chip interconnect on an MCM-D substrate.[141]

like the temperature and load with a thermocompression bond or the energy applied with an ultrasonic bond affect the bond quality.

Tape-automated bonding (TAB) can increase the reliability of interconnections, but usually requires gang bonding.[7] Aluminum pads are covered with TiW and Au with a thick photoresist process on the wafer-level. Cantilevered leads are bonded at once to this type of bondpad structure. Although a significant amount of literature exists on TAB processing, little has been done with TAB and MEMS devices.

Flip chip interconnection offers outstanding electrical and density performance because it eliminates leads altogether. It provides a technique for protection of the interconnection from the environment while allowing the sensor area to be exposed to the environment,[137,138] and may be more economically viable than monolithic integration.[139] Furthermore, it provides a method for rapid prototyping of microsystems.[140] The die or chip is "flipped" over and connected to the substrate

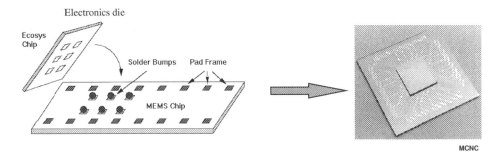

Figure 7.20 SMARTMUMPs flip chip attachment from MCNC.[142]

via solder bumps as shown in Fig. 7.19. The ability to attach (or remove) the die is simply a matter of locally heating the substrate to reflow the solder.

The SMARTMUMPs process available from MCNC is an attempt at multichip packaging of MEMS and electronics using flip chip technology. In this process, a standardized electronics die is flip chip attached onto a MUMPs MEMS die as shown in Fig. 7.20.[142] The MUMPs die is released prior to attachment of the electronics die. A drawback of this integration technique is that the electronics die blocks physical access to the MEMS devices located underneath it.

A group at the Fraunhofer Institute has developed a Microflex Interconnect (MFI) by using multiple layers of metal and polymer insulator (e.g., polyimide or BCB) on the surface of a wafer.[143] Once the interconnect structure is complete, a lift-off process is used to remove the entire structure from the silicon substrate. This MFI structure can then be bonded to a silicon IC with a gold ball stud from a ball wedge bonder.

Finally, the use of micromachining to create interconnections has been presented. Specifically, some groups, especially groups working with chemical sensors, have developed through-wafer interconnections. This will be discussed further during the section on pressure sensor interconnections because it is advantageous to minimize the impact of harsh environments on sensors. Also, research has been pursued to create optical interconnections using micromachining.[144] Lin et al. have shown a micro-optical bench utilizing optical interconnections.[145] Koh and Ahn and Chan et al. have also been working with optical interconnections.[146,147]

7.2.8 INTEGRATION WITH ELECTRONICS

The miniaturization advantages of MEMS are realized only if they can be efficiently integrated with microelectronics. At a first glance, it may seem that the most desirable approach to integration of MEMS and microelectronics would be to create a single or monolithic fabrication process capable of supporting both microelectronics and MEMS. One of the most intense areas of research in the MEMS field is in the development of surface micromachining processes compatible with integrated circuits. This is quite a difficult undertaking. For example, the removal of all oxide layers in the MUMPs process does not allow monolithic integration with CMOS VLSI circuits. Additionally, the high temperature anneal of MEMS devices to relieve internal stress can be harmful to the carefully controlled diffusion budgets of microelectronic circuits. However, some processes, such as Sandia's Modular, Monolithic Micro-Electro-Mechanical Systems (M^3EMS) process, do allow monolithic integration of surface micromachined MEMS and electronics by fabricating the MEMS before the microelectronics.[148]

Although research into monolithic fabrication of MEMS and microelectronics is progressing as demonstrated by the Sandia M^3EMS process, several hurdles must be overcome when integrating MEMS and CMOS. Even though many of the fabrication processes used in the development of MEMS are borrowed from the IC fabrication industry, the necessity of removing sacrificial layers in micromachining presents a set of unique problems. For example, the choice of sacrificial material for the MEMS device may be incompatible with the CMOS devices or microelectronic packaging. In CMOS, silicon dioxide is critical for the transistor gate insulation and for circuit passivation; but many MEMS processes use silicon dioxide as the sacrificial layer. Thus, the selection of structural and sacrificial materials in the MEMS devices impacts the integration with CMOS electronics. Moreover, some etchants for surface or bulk micromachining are not compatible with microelectronic materials or wiring. For example, KOH is often used in bulk micromachining, but it dissolves aluminum, which is commonly used in IC metallizations. Another bulk micromachining etchant, EDP, is not as aggressive in attacking aluminum, but still requires masking the aluminum to preserve the integrated circuits or the package. Other important integration issues include the material properties of the films used (e.g., gate polysilicon may not be the best choice of mechanical polysilicon), the thickness of the thin films, and thermal budget restrictions. As a result, the integration of MEMS with electronics usually requires additional processing steps and materials to protect the CMOS circuits and the package during the final release. Furthermore, electrical and environmental conditions also hamper monolithic integration of CMOS and MEMS. Many MEMS are designed to operate electrostatically with high voltages, which are difficult to implement with digital CMOS technologies. Other MEMS are designed to operate inside living organisms or are exposed to temperatures, radiation, or chemicals that would be destructive to the CMOS integrated circuits.

All of these factors provide challenges for integration of MEMS and CMOS technology. Building monolithically integrated MEMS and electronic circuits in the same process may not always be cost effective or realizable. However, it may be possible to use Multichip Module (MCM) technology to gain the benefits of MEMS and CMOS integration with minimum extra cost or additional technical challenges. MCMs offer an attractive integration approach because of the ability to support a variety of die types in a common substrate without requiring changes or compromises to either the MEMS or electronics fabrication processes. Further-

Table 7.1 Performance benefits of MCM packaging.[149]

MCM Characteristic	Performance Benefit
Hybrid IC Integration	Increased Flexibility & Applications
Higher Packaging Efficiency	More I/O to "outside world"
Reduction in Physical Size of System	Greater Range of Applications
Reduction in Total Interconnect Length	Higher Speed/Lower Power
Reduction in Interconnect Capacitance	Higher Speed/Lower Power
Reduction in Electrical Noise Generated	Higher Speed/Lower Power
Reduction in Connection Levels	Higher Reliability

more, MCMs offer packaging alternatives for applications for which it is cost- or time-prohibitive to develop a monolithic integration solution.

One of the main benefits of MCM packaging for MEMS and IC integration is the ability to combine dies from incompatible processes in a common substrate. Other benefits of MCM technology are the electrical, size, and weight performance improvement over conventional packaging techniques as shown in Table 7.1.

7.2.8.1 Review of MCM Packaging Alternatives

The two common characteristics for MCM classification are the type of substrate used and the means of interconnecting signals between the die.[150] The three dominant MCM substrate technologies available today are MCM-Laminate, MCM-Ceramic, and MCM-Deposited, but other substrate alternatives are emerging.

7.2.8.1.1 MCM-laminate (MCM-L) technology.
Multi-Chip Module Laminate technology is essentially a direct miniaturization of a printed circuit board (PCB) and is often referred to as "chip on a board" technology. In MCM-L packaging, bare die are mounted on a "PCB-like" laminated substrate, wire bonded, and then encapsulated with epoxy to protect the IC and wire bonds from the environment. MCM-L is used in digital watches, calculators, and many other inexpensive consumer electronics. The MCM-L technology is very effective when only a few chips need to be packaged together or the number of interconnects between the die is low. The disadvantages of MCM-L include limited wiring density, relatively thick line widths, and a high coefficient of thermal expansion as compared to silicon.

7.2.8.1.2 MCM-ceramic (MCM-C) technology.
The MCM-C technology represents the class of packages with ceramic substrates. The two main types of MCM-C are High Temperature Cofired Ceramic (HTCC) and Low Temperature Cofired Ceramic (LTCC). HTCC technology is presently the most commonly used. HTCC uses interconnect metals such as tungsten (W), molybdenum (Mo), and manganese (Mn) with high temperature melting points. However, the need for more conductive interconnects has led to the development of LTCC technology so

Table 7.2 Characteristics of MMS MCM-D package.[153]

Property	MMS MCM-D
Substate material	Aluminum
Signal/power wiring layers	3/2
Dielectric material	Polyimide
Conductor metallization	Copper
Die attach adhesive	Ablebond 789-3
Die interconnect method	Wirebond
Die edge-to-edge spacing	>500 µm
Max operating frequency	100–400 MHz

that gold, aluminum, and copper can be used.[150] MCM-C packaging provides the best cost/performance combination of the three predominant substrate choices—it is more costly than MCM-L, but offers a significant increase in wiring density and electrical performance.[151] Moreover, MCM-C has an established infrastructure and ceramic materials are well understood. In addition, LTCC has a limited 3-dimensional capability in that passive components (e.g., inductors, resistors, and capacitors) can be embedded into the substrate.[151]

The primary drawback of MCM-C packaging is the relatively high dielectric constant of the currently used ceramic insulators. Most of the dielectric materials used in MCM-C processing have dielectric constants which are two to three times higher than the materials used in other MCM technologies.[150] The dielectric constant of the insulator impacts the signal speed, power consumption, and wiring density of the packaged system.

7.2.8.1.3 MCM-deposited (MCM-D) technology.

The MCM-D technology borrows heavily from the semiconductor fabrication industry. In fact, the D suffix is associated with the use of IC thin film deposition processes to achieve high density interconnect patterns. The MCM-D systems typically use high conductivity interconnect metals such as Cu, Al, or Au. Deposited dielectric layers with low dielectric constants separate the conductor layers. The metal interconnects can be patterned using the same lithography techniques available to the semiconductor industry; therefore, MCM-D systems are able to achieve very high wiring density and high frequency performance. Special purpose MCM-D prototypes have been designed for clock speeds in excess of 3 GHz.[150]

A foundry MCM-D process was recently studied for MEMS packaging.[152] The Micro Module Systems (MMS) MCM-D represents a traditional packaging approach where the interconnect layers are deposited on the substrate and the dies are mounted above the interconnect layers. The interconnect between the die and the substrate is made through wirebonding. Table 7.2 lists the characteristics of the MMS MCM-D package.

Key considerations for MCM packaging of MEMS include whether to release the micromachined devices before or after packaging and the compatibility of the

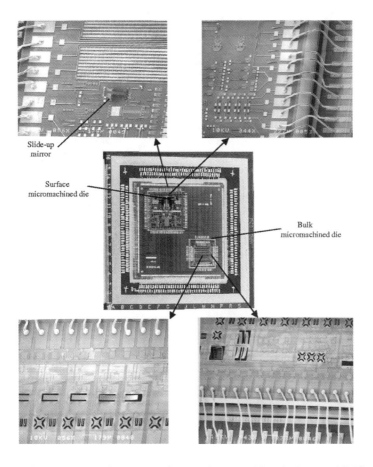

Figure 7.21 Released bulk and surface micromachined die on MMS MCM-D/ MEMS module. The slide up mirror was manually assembled after release.[149]

package materials with the MEMS release procedures. Released MEMS are typically very fragile and require special handling. Consequently, it is desirable to release the devices after packaging. However, many of the release etchants commonly used for MEMS are harmful to microelectronics and microelectronic packaging. The MMS MCM-D process was chosen because its substrate and wiring materials are most compatible with the release procedure for the surface micromachined MEMS.[152] Both bulk and surface micromachined devices have been successfully packaged in the MMS MCM-D process (Fig. 7.21).[152]

7.2.8.1.4 MCM-silicon technology. A derivative of MCM-D is MCM-Silicon (MCM-Si) or "silicon on silicon" technology. In MCM-Si, IC fabrication processes are used to deposit the interconnect layers on a silicon substrate. Aluminum and copper are used as the conductors and silicon dioxide is used as the dielectric (Fig. 7.22). MCM-Si has the highest signal interconnect density of any substrate choice and its coefficient of thermal expansion is an excellent match for any silicon

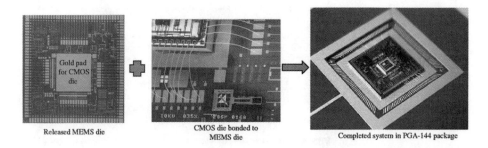

Figure 7.22 CMOS controller and pre-released MEMS die packaged in PGA-144 package. The MEMS die was fabricated through MUMPs.[149]

die.[150] The primary disadvantage of MCM-Si is that silicon is not a good base for the MCM package assembly because it is relatively fragile, and consequently the MCM-Si substrate must be repackaged, resulting in additional cost.[150]

An advantage of MCM-Si technology is that bulk micromachining techniques can be used to pattern the silicon substrate. The substrate can be patterned using silicon bulk etchants and wafer bonding to form useful features such as embedded components and microchannels. Microchannels may be used to align optical fibers or carry fluids to and from MEMS mounted in the module. In addition these microchannels can provide an efficient way to cool the MCM.[154]

7.2.8.2 Flip-Chip Transfer and Assembly of MEMS

Current surface micromachining technologies limit designers with few choices of materials or number of structural layers. Recently, a flip-chip assembly and silicon removal technology has been developed to integrate MEMS with other devices or circuits on substrates other than silicon. Such a technology is particularly beneficial to RF or optical components which require alternate substrates that do not support fabrication of complex MEMS structures. Flip-chip transfer allows silicon-based MEMS to be transferred to another substrate for optimum RF or optical integration. More complex MEMS can be built by flip-chip assembly of surface micro-machined devices onto a variety of material substrates. The flip-chip silicon substrate is then removed to produce a highly advanced micromechanical system where specific material properties are paramount or where more structural layers are required. Major challenges in this technology are the flip-chip bonding with submicron gaps, postassembly release of MEMS and removal of the silicon host substrate. The flip-chip MEMS transfer technology has been demonstrated by transferring surface micromachined resonators and rotary microactuators from the host silicon substrate to a glass substrate.[66,67]

Recently, RF tunable capacitors and micromirrors have been successfully transferred from the silicon to a ceramic substrate.[155,156] Figure 7.23 illustrates major assembly and postassembly release processes:

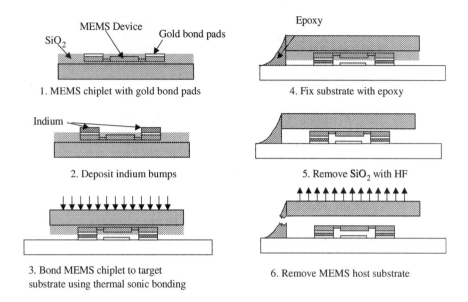

Figure 7.23 Flip-chip assembly of MEMS.

1. Design and fabricate MEMS with a layer of SiO_2 between the MEMS and the host silicon.
2. Deposit indium bumps on the bonding pads.
3. Bond MEMS to a target substrate through thermocompression or thermosonic processes.
4. Glue silicon substrate to the target substrate using epoxy.
5. Remove SiO_2 in HF to separate the host silicon from the MEMS.
6. Remove the host silicon by breaking the epoxy bond.

The bonding pads are only connected to the host silicon substrate with the sacrificial SiO_2 layer. When the SiO_2 is dissolved in HF, the bonding pads and the MEMS device are completely disconnected from the host silicon substrate. The epoxy bond holds the silicon substrate in place during the HF release process to protect the transferred MEMS.

Among the processes illustrated in Fig. 7.23, the bonding and the release are the two critical steps. Submicron height variations are required to assure repeatable MEMS performance that is sensitive to the height, i.e., air gap that affects capacitance for RF applications or applied electrostatic voltage for driving optical mirrors. Indiums bumping and bonding processes have been developed. With a special bonding tool designed for precision force control,[157] the reported height could be controlled within ±0.1 μm.

The second critical step is the postassembly release of the MEMS structure from the host silicon substrate. If the release is incomplete, the MEMS structure can be damaged or destroyed when the host substrate is removed. On the other hand, if the MEMS structure is left too long in the etchant, the structure can be

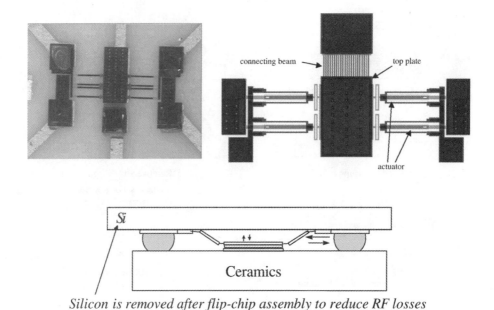

Silicon is removed after flip-chip assembly to reduce RF losses

Figure 7.24 MEMS variable capacitor on a ceramic substrate: photograph, layout, and cross-section of the flip-chip assembly.

over-etched and damaged. A particular challenge for flip-chip MEMS is to remove the SiO_2 when the flow of the etchant is constrained by the narrow gap between the silicon and the target substrates. Therefore, long etching time is needed for effective release, but a long release process can over-etch thin polysilicon structures. A solution to this problem is to include a partial prerelease step before bonding. Essentially, the majority of the silicon dioxide etching is done before bonding. With most of the oxide removed, not only is the overall etching time reduced, but the gap for the etchant flow is also widened by the reduction of oxide.

With the controlled bonding and improved release processes, the flip-chip assembly of MEMS with silicon removal technology has been applied to create novel RF tunable capacitors and optical micromirror arrays on ceramic substrates.

Figure 7.24 illustrates a transferred RF MEMS tunable capacitor on a ceramic substrate.[158] The variable capacitor is driven by vertical electrothermal actuators.[159] The device was fabricated using MCNC's MUMPs tecnology with polysilicon as the structural material and gold film for RF propagation. Vertical electrothermal actuators are used to move the capacitor plate, and thus change the air gap between the center MEMS plate and a corresponding underlying pad on the ceramic substrate. The connecting beams shown in the layout are used to transfer RF signals from the top plate to coplanar waveguide circuits. The demonstrated capacitance change was from 0.1 pF to 1.35 pF.[160] The device can be used for frequencies up to 20 GHz, and demonstrated an excellent Q of 108 at 1 pF and 1 GHz.[161]

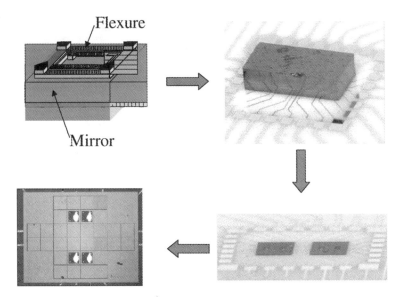

Figure 7.25 Flip-chip assembly of polysilicon micromirrors on a ceramic substrate.[155] Schematic of an individual mirror, a silicon micromirror chip bonded to a ceramic substrate, two micromirror arrays after release with silicon chip removed, and SEM top view of a micromirror array (four mirrors are removed to expose the address electrodes on the ceramic substrate).

The assembly of a piston-type micromirror array on a ceramic substrate is illustrated in Fig. 7.25. A polysilicon mirror is fabricated upside down with the polysilicon flexures on top. Gold bonding pads are connected to the flexure frame. After the flip-chip transfer the flexures are below the mirror surface. This design creates arrays of optically flat mirrors with an array fill factor close to 100%. Such a fill factor could not be accomplished using the MUMPs fabrication processes.

7.2.8.3 Direct Metal Deposition Interconnects

A relatively new technology for making interconnects between dies on an MCM is known as direct metal deposition. Two broad categories of direct metal deposition techniques are patterned overlay and patterned substrate as shown in Fig. 7.26. Patterned overlay occurs when the dies are embedded in the substrate and the metal is deposited above the dies. In a patterned substrate the metal is deposited first then the dies are attached on the surface.

7.2.8.4 High Density Interconnect MCM Technology

An example of direct metal deposition is General Electric's High Density Interconnect (HDI) process. This process uses the "die first" or patterned overlay concept (refer to Fig. 7.27). In the HDI process, holes are milled in the substrate to house

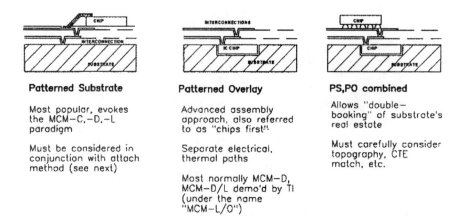

Figure 7.26 Examples of direct metal deposition interconnects.[141]

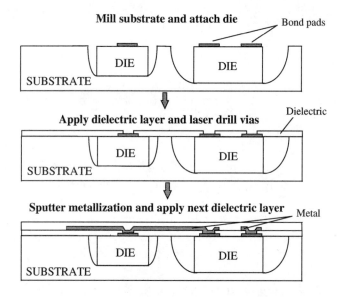

Figure 7.27 The HDI process ([149] after [162]).

the various dies. After the dies are placed and bonded in the substrate, Kapton sheets are glued over the top and via holes are created through a laser drilling process. Metal is then deposited and patterned to form interconnects. The Kapton lamination and metallization process is repeated until all of the interconnect layers are created.[162]

The HDI process is attractive for MEMS and electronics packaging. First, the HDI process can accommodate die from a variety of processes because the die wells can be milled and configured to accommodate each individual die. In addition, the HDI process allows for bond pads to be located virtually anywhere on the chip. This is particularly advantageous for MEMS packaging because many

Table 7.3 Characteristics of HDI and COF packages.[149]

Property	HDI	COF
Substrate material	Alumina	Molded plastic
Signal/power wiring layers	9	3
Overlay dielectric material	Kapton	Kapton/Ultradel
Conductor metallization	Ti/Cu/Ti	Ti/Cu/Ti/TiW/Au/TiW
Die attach adhesive	Ultem	Ultem
Die interconnect method	Direct metallization	Direct metallization
Die edge-to-edge spacing	>375 µm	>800 µm
Max operating frequency	>1 GHz	>1 GHz

MEMS processes have very limited wiring capability. Another desirable feature of HDI packaging is that the use of direct metallization for the die interconnects can result in electrical performance equivalent to monolithic integration.[162] The HDI process can produce systems capable of operating at over 1 GHz, which is a significant improvement over existing available MCM-D technologies. Finally, the HDI overlay can be used as a mask to protect other die in the package from the release procedures of the MEMS die[152] or the environment to which the MEMS die must be exposed. HDI packages can also be designed for repair. Moreover, the Kapton overlay can be removed without destroying the embedded die. The faulty die can be replaced and a new Kapton overlay deposited over top. This repair procedure has been successfully conducted over ten times on an HDI package.[162]

7.2.8.5 Chip-on-Flex MCM Technology

Chip-on-Flex (COF) technology is an extension of HDI technology and was developed only in the last few years.[163] COF was created to provide a low cost, multichip packaging solution which would approach the performance of high-end MCMs.[163] Table 7.3 compares the characteristics of HDI and COF modules.

COF processing uses interconnect overlay technology similar to HDI, but molded plastic is used in place of the ceramic substrate (see Fig. 7.28). Unlike HDI, the interconnect overlay is prefabricated before chip attachment. The chips are attached face down on the COF overlay with an adhesive. After the chips have been bonded to the overlay, a substrate is formed around the components using a plastic mold forming process. Next, the die are electrically connected by laser drilling vias through the overlay to the component bond pads followed by metallization.[163]

7.2.8.6 Packaging of MEMS in HDI and COF Processes

Since the die are embedded in the substrate and covered by an overlay, HDI and COF packages need additional processing in order to provide physical access to the MEMS devices. A modification to the process is to add an additional laser ablation step to allow physical access to the MEMS devices as shown in Fig. 7.29. Addi-

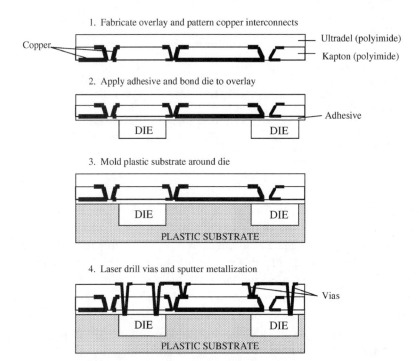

Figure 7.28 Chip-on-Flex process flow ([149] after [163]).

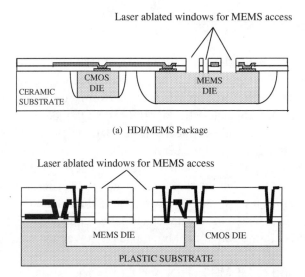

Figure 7.29 Large area ablation for MEMS access in (a) HDI and (b) COF package.[149,152]

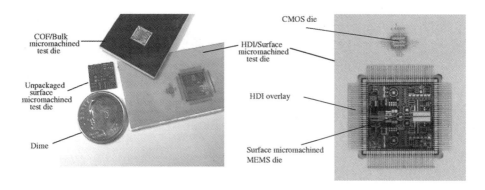

Figure 7.30 HDI and COF/MEMS packages.[152] Each package has a thin profile of approximately 1.5 mm.

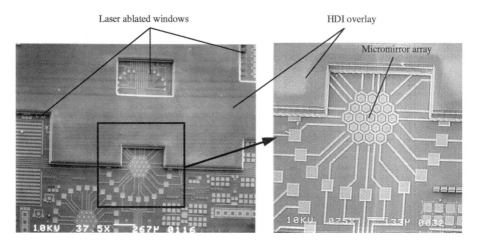

Figure 7.31 Exposed MEMS devices in a HDI package.[152]

tional plasma etching may need to be performed to minimize the ablated dielectric residue which can accumulate in the exposed windows.[152]

Both the surface and bulk micromachined devices were packaged through the General Electric HDI and COF processes.[152] The surface micromachined devices were released in HF after packaging. The bulk micromachined devices were released in XeF_2 after packaging. The packages are shown in Fig. 7.30. Close up views of MEMS devices in HDI and COF packages are shown in Fig. 7.31 and Fig. 7.32, respectively. The bulk micromachined devices in a COF package are shown in Fig. 7.33.

Assessing the performance and reliability of MEMS and advanced packaging concepts may require a more flexible, process-based approach than simply applying existing microelectronic standards.[164] While MEMS devices borrow heavily from microelectronic fabrication, MEMS is not a subset of microelectronics. MEMS devices have failure modes and characteristics not found in microelectron-

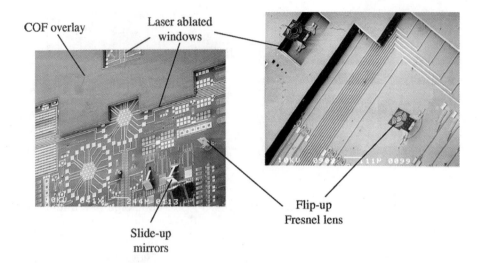

Figure 7.32 Exposed MEMS devices in a COF package.[149] The devices were assembled manually with a microprobe.

Figure 7.33 XeF$_2$ Released bulk micromachined MEMS in a COF package.[152]

ics. For example, the optical reflectivity of a deposited metal layer is typically not a concern for a microelectronics designer, but the same quantity is very important for an optical MEMS designer.

One way to generate a process-based assessment for MEMS packaging is through the development of assembly test chips (ATC). ATCs are special test die which contain sensors and test devices which monitor the health of the die through the packaging process.[165,166] Sandia National Laboratory has used this concept for microelectronics. The same concept can be extended for MEMS packaging by developing MEMS test chips designed to identify and monitor the electrical, thermal, and mechanical properties of critical importance to the MEMS designer.[167]

7.2.9 ENCAPSULATION AND OVERMOLDING

While overmolding of typical IC chips is common,[168] there are more constraints placed on its application when packaging a MEMS device. First and foremost, package stress is often an important consideration for MEMS devices. The difference in coefficient of thermal expansion (CTE) is often the source of the package stress that is observed.[169] Table 7.2 shows typical CTE values for a variety of packaging materials. Package stress can cause catastrophic failure and/or reliability limitations on typical IC devices, but it does not often adversely affect the output of the device. However, stresses in MEMS devices can and do often adversely affect the output of a MEMS device. For example, coatings on a pressure sensor have been shown to affect the device output.[170] Careful consideration must be paid to the package materials that come in contact with the MEMS device. Some solutions to this problem that have been implemented include the use of materials with similar CTE values (this is most often used with die attach materials), use of very thin coatings (e.g., thin film polymer coatings, like polyimide or parylene over a pressure, flow or chemical sensor), and/or the use of materials with very low moduli (e.g., silicone gels on a pressure sensor).[171] In addition, some techniques to isolate the MEMS device from a high modulus material by using an intermediate low modulus material have also been implemented.[172]

Media compatibility is another important issue that constrains the choice of encapsulants or overmolds. For sensors or actuators, where the device must be in contact with an external environment, protection of the device from that environment—that is, media compatibility—is a very important consideration.[25,173] Media protection schemes will be reviewed in more detail in Section 7.3.1.5 on Pressure Sensor Encapsulation and Secondary Diaphragms. In general, there are few techniques for low cost media compatibility.

- Thick, low modulus film coatings (e.g., silicone gel[5,171]).
- Thin, higher modulus film coatings (e.g., polyimide, parylene,[5,174–176] other organics,[177–182] or wafer-level inorganics[183–186]).
- Backside interconnection and typical overmolding processes.[187,191]
- Selective encapsulation to avoid coverage of the sensor.
- Secondary diaphragm and silicone oil.[192]

A third problem with encapsulation and overmolding of MEMS devices is the actual process being used. Because the devices are delicate, problems akin to wiresweep with typical IC packaging are even bigger problems with microstructures. Little has been documented concerning this type of problem; however, generally it is recognized and avoided by using wafer-level bonding (e.g., Fig. 7.3[25]), silicone gel and package "caps" instead of overmolds,[171] and/or TO-style headers (see Figs. 7.13 and 7.14).

Organic encapsulation may be divided into three basic categories:[193] (1) non-elastomeric thermoplastics, (2) nonelastomeric thermosetting polymers, and

(3) elastomers. Several types of polymers can be listed for each of these categories. The common IC encapsulant chemistries include nonelastomeric thermoplastics like poly-*para*-xylylene (parylene) and preimidized silicone-modified polyimides among other chemistries. Thermosetting polymers include silicones, polyimides, epoxies, silicone-modified polyimides, benzocyclobutenes, and silicone-epoxies. Silicone gels and polyurethanes are examples of elastomers. Currently, many pressure sensor manufactureres use various types of silicone gel to encapsulate the device. A drawback of these gels is their limited chemical resistance. Alternative materials that have better chemical resistance are being considered for use in new media-resistant pressure sensor packaging.

7.2.10 Package Housing Materials and Molding

Package materials used to house MEMS are varied. As shown previously, MEMS devices are packaged in metal (Figs. 7.10(b), 7.12, and 7.13), ceramics (Fig. 7.14), thermoplastics (Fig. 7.35), and thermosets (Fig. 7.36). In some cases, package housing materials are used to hermetically seal a device (e.g., the ADI accelerometer, Fig. 7.12) or to seal an atmosphere or fluid within the package (Fig. 7.34).

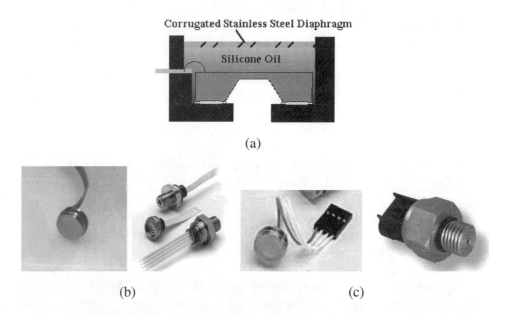

Figure 7.34 A stainless steel package used to house a media compatible pressure sensor. (a) This drawing is a typical cross-section of a stainless steel diaphragm with silicone oil fill used to create media compatible pressure sensor packages. (b) Several examples of stainless steel diaphragm, silicone oil-filled packages from Lucas NovaSensor are presented here [http://www.novasensor.com/, ca. 1/31/99].[192] (c) A high pressure metal package for housing pressure sensors [http://www.mot.com/AECS/General/AIEGSensors/index.html, ca. 2/99].

While this packaging scheme provides the harsh media compatibility function, it is expensive and can affect the device performance over temperature.

Polymer materials are being used more often for packaging MEMS. Motorola has used both thermoplastics (Fig. 7.35) and thermosets (Fig. 7.36) to house MEMS devices. Thermoplastics are materials that are injection moldable. Typically, engineering thermoplastics are used because of their high temperature tolerance properties. For example, polyesters, nylon, poly(phenylene sulfide) (PPS), and polysulfone have all been used to house materials in component-style packages [Fig. 7.35(a)] or in module-style packages [Fig. 7.35(b)]. When molding a material with a metal insert (called "insert molding"), as shown in Fig. 7.35(a), CTE mismatch between the polymer material and the metal can cause leakage of

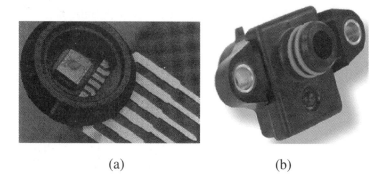

(a) (b)

Figure 7.35 MEMS devices packaged in thermoplastics. (a) The Motorola unibody package using polyester thermoplastic housing material.[25,194] (b) The Motorola manifold absolute pressure (MAP) module [http://www.mot.com/AECS/General/AIEGSensors/index.html, ca. 2/99].

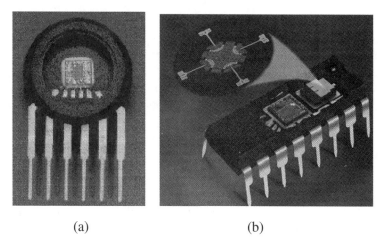

(a) (b)

Figure 7.36 An example of a transfer molded, thermoset (i.e., epoxy) package for housing (a) pressure sensors and (b) accelerometers.

gases or liquids from the inside of the package to the outside of the package (or vice versa). Package designers must consider this as a potential reliability problem when chosing package housing materials.

Thermoset plastics are materials that are transfer moldable. Typically, epoxy materials are used because of their moldability and high temperature tolerance properties. Thermosets are cross-linked plastics. They require very high pressures for molding. Thermosets are generally low-viscosity materials when they enter the mold. Following heating, they cure and chemically react to create the "set" material. Because they enter the mold as a low-viscosity liquid, all metal areas (e.g., leads in Fig. 7.36) must be masked to avoid "flashing" (or coating) of material onto the metal. Therefore, hydraulic-like pins must clamp each metal area. This leaves open areas following molding. In the case of the pressure sensor package in Fig. 7.36, this requires the backside hole around the leads to be filled with another material (called "backfill") to seal the pressure leak. Inevitably, the backfill material will have a different CTE than the housing material, causing another material interface, which becomes a concern for reliability performance.

7.2.11 TRIMMING AND CALIBRATION

In high volume manufacturing, a distribution of electrical parameters from device-to-device is observed. The method for minimizing the effect of that variation on yield loss is called trimming. Manual calibration, laser trimming, zener zap, electronic calibration,[195] polysilicon resistor trimming,[196,197] E/EPROM, and other methods are used for this purpose. In the case of a Motorola bipolar pressure sensor or an Analog Devices accelerometer, laser trimming of the device is used. Laser trimming involves cutting metal silicide (often, CrSi) resistor areas to increase the number of squares, and therefore the resistance value.

Laser trimming (Fig. 7.37) provides a considerable potential for resistance variation. Resistors can be trimmed as much as 300% in some cases. This wide trim range effectively minimizes the effect of device-to-device variation through the manufacturing process. Alternative trimming technology cannot produce this large of a trim range.

However, three limitations exist for laser trimming. First, the equipment is very expensive. Second, the serial nature of this process—testing and trimming each device individually over pressure and temperature—causes this step to be the biggest throughput bottleneck in the process. Finally, laser trimming cannot occur at the end of the assembly process. Because of the heat that is generated at the CrSi surface, laser trimming must occur prior to the final device encapsulation process. Therefore, any variation in the device performance caused by the encapsulation process step must be accounted for during laser trimming.

Advances in calibration techniques for MEMS have been discussed by Schuster and Czarnocki.[198] In this paper, they describe the trend from thick film laser trimming for bipolar, to thin film laser trimming for bipolar and BiCMOS (as shown in Fig. 7.37), to electrical trimming through E/EPROM in CMOS. A key advantage

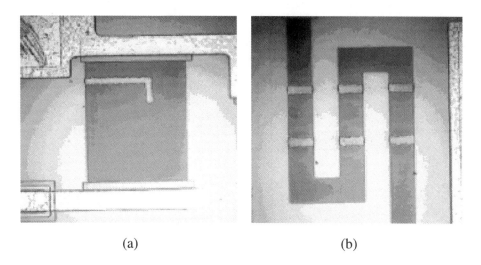

Figure 7.37 Examples of laser trimmed CrSi resistors. In (a), an "L" cut causes the current path distance to increase on the resistor (i.e., the number of resistor squares increases and therefore the resistance, increases). In (b), a resistor is cut (i.e., open circuit) to remove a portion of the circuit.

of using electrical trimming is that it can occur at the end of the assembly process. Laser trimming must occur prior to encapsulation, and one common theme for MEMS packaging applications is that the devices can often be adversely affected by the packaging. Therefore, trim-last can be a differentiator in product yield and specifications.

7.3 PACKAGING OF PHYSICAL SENSORS

Microelectromechanical systems (MEMS), especially physical sensors, can be very sensitive to device packaging and testing techniques. The following is a summary of some of the considerations for assembly processing. In Section 7.5, we will discuss the implications of testing techniques on MEMS output. Each of the following sections describes a process step, arranged chronologically, in the back-end of a typical pressure sensor process flow. Figure 7.38 shows a typical flow diagram for assembly. Specifically, this example is of a pressure sensor.

This process flow includes wafer-level packaging, die attach, wirebonding, trimming, package-level encapsulation, electrical characterization, reliability testing, and media compatibility testing. Some of the assembly flow process steps have been discussed in the previous section (Section 7.2). In this section, we will concentrate on the pieces of the process flow that are specific to pressure sensors, and important problems related to the assembly of pressure sensors. For example, we will discuss some specific wafer bonding issues for pressure sensors, package housing materials and the unique problem of lead leakage, reliability concerns with die

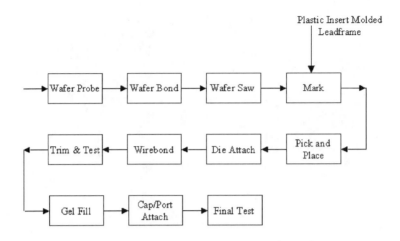

Figure 7.38 An example of a typical pressure sensor assembly process flow.

attach and wirebonding, media compatible packaging, pressure interconnects or ports, second-level packaging or modules, and stress sensitivity of pressure sensor to packaging.

As sensor technology has evolved out of semiconductor processing technology, some of the packaging concepts have also evolved out of the conventional semiconductor or IC packaging technology. The semiconductor package is required to provide: (a) mechanical support, (b) electrical interface, (c) environmental protection for the underlying semiconductor device, and (d) heat dissipation. The same is true for sensor packaging, but there are some additional requirements such as the need for an environmental (pressure or fluid) interface, wafer-level packaging, stress isolation, application specific media testing, and reliability testing. These considerations are essential for advancement of sensor technology from laboratory and small-scale production to large-scale commercialization. Finally, incomplete understanding of the all of the effects and interactions of mechanical, temperature, chemical, and electrical factors on a MEMS device often drives the need for high volume testing of outgoing devices.[199]

7.3.1 Pressure Sensors

Considerable strides have been taken over the last several years in packaging for MEMS devices, yet several obstacles remain to even larger-scale commercialization. Unfortunately, it seems that the MEMS industry may run into packaging limitations earlier in its high-volume manufacturing lifetime than ICs did. To scale up manufacturing of MEMS devices, like pressure sensors, several bottlenecks must be minimized or removed, including:

1) the lack of standards;
2) being a small customer for several materials;

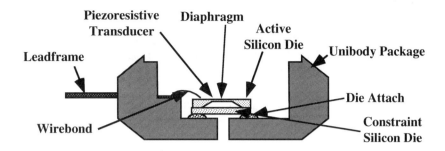

Figure 7.39 A cross-section of the Motorola unibody pressure sensor package with several parts of the package identified. This is an absolute pressure sensor, so glass frit wafer bonding is used to create a stable reference pressure. Following the trimming operation, a silicone gel is dispensed in the opening to cover the die, die attach, leadframe, and wires.

3) the understanding of interactions among the package materials, induced stress, and device performance;
4) the measurement of material properties for all packaging materials;
5) the lot-to-lot variability of some materials;
6) modeling accuracy for the behavior of these materials;
7) modeling of the variability of the manufacturing processes (not just the average);
8) advanced trimming technologies; and
9) accurate simulation of the reliability and/or media compatibility of sensors in customers' environments.

An example of a pressure sensor cross-section is shown in Fig. 7.39.[171] Bulk micromachined pressure sensors have been in production at Motorola and other companies (like Lucas NovaSensor, Sensym, IC Sensors, Bosch, Silicon Microstructures, Honeywell, etc.) for at least fifteen years in small-scale production when compared with other semiconductor devices. Several issues exist with pressure sensors which, to date, have precluded their rapid, high-volume commercialization. Probably first and foremost is the application-specific nature of the pressure sensor business. Presently, only the automotive, medical, and consumer white goods sectors have been able to drive large enough volume business to support customized packaging. The lack of standards—both for the electrical interfacing of a device and the pressure interfacing of a device—has hindered the development of large "across-customer" applications. For example, there are a variety of different pressure ranges in which different devices are built (including differential vs. absolute pressure measurement devices). Motorola and others have, literally, dozens of pressure sensor device types. The level of circuit integration—uncompensated, temperature compensated and calibrated, or integrated—adds to the large number of devices available. Moreover, packaging options are not standardized, so within a given pressure range, a variety of packaging options are available. Both the base package (i.e., the sensor "housing"—the Motorola "unibody" pressure sensor package in Fig. 7.39) and the port options add to the complexity.[171] Several examples of construction problems for sensors (especially pressure sensors) are described in Vaganov.[200] This discussion lists the following problems with sensor construction:

- microstructure connection to a package (discussed in Section 7.3.1.3);
- electrical interconnection (e.g., cables) with the sensor (discussed in Section 7.3.1.4);
- protection from the environment (discussed in Section 7.3.1.5); and
- individual sensor trimming (discussed in Section 7.2.10).

7.3.1.1 Wafer Bonding for Pressure Sensors

In the case of pressure sensors, constraint die can be bonded onto the backside of bulk micromachined pressure sensors to provide device stability, in the differential pressure sensor case, or to provide a constant backside pressure, in the absolute or gage pressure sensor cases (Fig. 7.2). Absolute pressure measurement requires that the reference vacuum be maintained in the cavity under the diaphragm. This is accomplished by bonding the two wafers at very low pressures that approximate an absolute vacuum.

Wafer bonding for pressure sensors has been performed in several different fashions. Figure 7.40 shows examples of bonded pressure sensors using direct wafer bonding [Fig. 7.40(a)], anodic bonding [Fig 7.40(b)], and glass frit bonding [Fig. 7.40(c)].

Finally, wafer-level package represents another means of "integration." Instead of integrating, for example, circuit with sensor, wafer-level package integrates package functionality onto the silicon. In this way, there is the potential for cost reduction, functionality enhancement, or both.

7.3.1.2 Housing Materials, Plastic Packaging, and Lead Leakage

Ceramic and metal housing examples have been used with pressure sensors for some time. Examples are shown in Figs. 7.10(b), 7.13, 7.14, and 7.34. Additional packages are shown in Figs. 7.41 and 7 42.

Packaging of pressure sensors in plastic occurs in either thermoplastics or thermosets. Thermoplastic package examples are shown in Fig. 7.44 (the Motorola "next generation" pressure sensor packages), and thermoset package examples are shown in Fig. 7.22. Two types of plastic molding have been used for pressure sensor housings. In either case, molding occurs around a leadframe where an array of several devices in a "strip" is built at once (Fig. 7.43). Originally, thermoplastics (specifically polyesters) were injection molded for a variety of packaging and porting options. Later, thermosets, like epoxies, were incorporated into pressure sensor housings.

ASSEMBLY AND TEST FOR MEMS AND OPTICAL MEMS 377

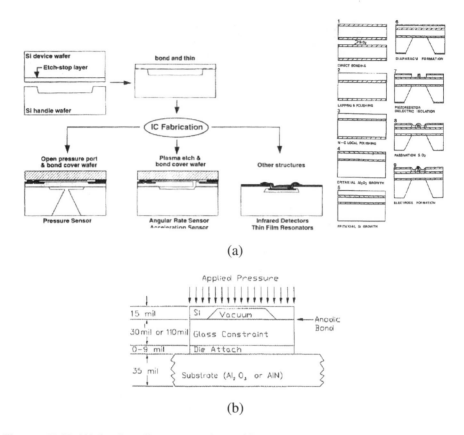

Figure 7.40 Wafer bonding examples with pressure sensors. (a) Direct wafer bonding used for an absolute pressure sensor.[26,30,32–34,201] (b) Anodic bonding used for a differential pressure sensor.[53] (c) Glass frit bonding used for an absolute pressure sensor. (d) Glass frit bonding used to create a backside absolute pressure sensor for media compatibility.[202–206]

Two significant problems plagued the original thermoplastics. First, polyesters are not capable of withstanding high temperature moisture environments.[209] Embrittlement of the package and cracking often occur when these type of packages are exposed to this environment. Therefore, alternative, higher temperature plastics were investigated for these applications. For example, poly(phenylene sulfide) (PPS), polyphthalamide (PPA), and liquid crystal polymers (LCP) can withstand higher temperatures and moisture environments.

The second significant problem with polyesters was that considerable pressure leakage at the leadframe metal/polymer housing interface could be observed, especially after temperature cycling.[208] Most polyesters, even highly filled polyesters, exhibit CTE values (i.e., 50–100 ppm/C) that are very different from those of the metal (i.e., 10–18 ppm/C). "Lead leakage" has been correlated to the mismatch in CTE between the polymer material and the metal leadframe (Fig. 7.45). It is a problem because it causes a systematic error in the pressure reading, and it is a

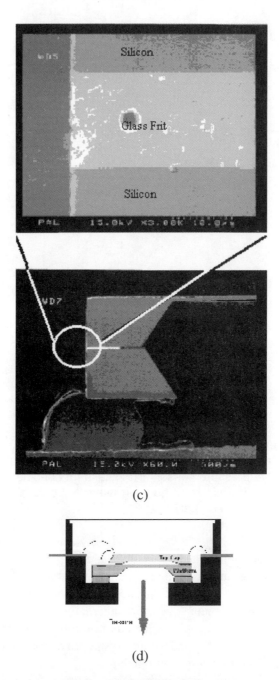

Figure 7.40 (Continued).

leak in the pressure line. The latter becomes a safety problem in the cases of harsh media pressure sensing. However, there are a variety of filled thermoplastics and thermosets that can be used to minimize this CTE mismatch.

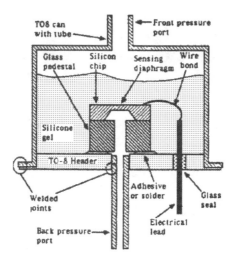

Figure 7.41 Anodic wafer bonding used to place a pressure sensor inside a metal TO header.[207].

Figure 7.42 An automotive pressure sensor on a ceramic substrate with a metal can that isolates the pressure sensor, and harsh media, from the rest of the circuit and pressure sensor module.[25]

Thermosets can be tailored to minimize the CTE mismatch problem with the leadframe, and they (especially epoxies) are resistant to a wide variety of media. Two problems exist with thermoset materials. The tooling cost of the transfer molding operation can be higher than the injection molding operation. Also, the high pressures necessary for transfer molding necessitate the use of pins to hold the leadframe in place during the molding operation. This adds an unwanted cavity to the final package around the metallic leadframe, so a subsequent epoxy backfill operation must be performed. The epoxy backfill/epoxy package is another material interface in which leaks are possible.

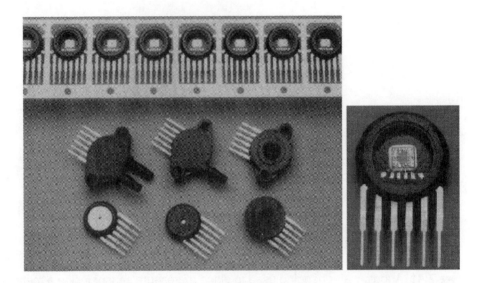

Figure 7.43 Pressure sensor packaging using thermosets. These are examples of Motorola pressure sensor pacakges using epoxy.[25,208]

7.3.1.3 Die Attach Materials

Once the package housing has been built, sawn die are placed on a "square doughnut"-shaped silicone elastomeric die attach material. To provide stable electrical output over a wide temperature range, a soft die attach, like a silicone, must be used. Silicone materials are effective because of their extremely low modulus. Furthermore, the operating range of the device (-40 to $125°C$) is within the rubbery regime of the polymer; therefore, material properties vary approximately linearly within this temperature range. Other die attach methods include the use of glass frit or anodic bonding; however, these will not be discussed here.

One considerable challenge with these die attach materials is the dispense process. Silicone materials have a lot-to-lot variation that can affect the dispense if proper care is not taken.[210] Unfortunately, the material properties prior to curing that cause this variation (e.g., viscosity, molecular weight, etc.) have not been defined; and, furthermore, they are not simple properties to measure in an incoming inspection or quality assurance process step. Optimization of the factors that affect the die attach dispense process is a consistent problem in manufacturing, especially with this lot-to-lot variation. Re-optimization is often required for each lot, which is not conducive to high volume manufacturing. Mahadevan has presented a designed experiment method for optimizing die attach processing for pressure sensors.[210] In the work, he modified the pressure, dispense time, and gap and monitored the height of dispense, thickness, central gap, and dispense width of the resulting die attach material.

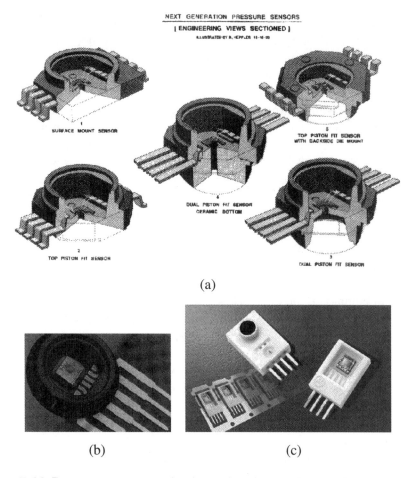

Figure 7.44 Pressure sensor packaging using thermoplastics. These are examples of Motorola pressure sensor pacakges using (a) poly(phenylene sulfide), (b) polyester—this is an example of Motorola's unibody pressure sensor package, and (c) polysulfone—this is an example of Motorola's Chip Pak package.

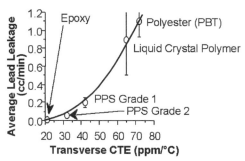

Figure 7.45 A plot showing the correlation between pressure leakage through the package (i.e., "lead leakage") and the CTE of the polymer material. The CTE of the metal leadframe is the same in this comparison, so the CTE of the polymer material represents the mismatch in CTE between the polymer and the metal.

Similarly, the curing of these silicone die attach materials is affected by these lot-to-lot variations. Not enough is known about the effect of silicones and their cured material properties on the electrical output of a device. This is partially because of the difficulty in obtaining material properties for silicone materials. Thus, modeling of these materials requires making significant assumptions, including defining material geometry, modulus (vs. temperature), Poisson's ratio, and CTE. Lot-to-lot and dispense-to-dispense variations make modeling even more uncertain. Often these variations are the cause for rejecting devices in manufacturing final test. Therefore, modeling efforts must consider both the average values and the standard deviations considering the uncertain variable inputs (i.e., a sensitivity analysis must be performed).

In addition, sensors manufacturers buy very small quantities of these materials, and will continue to do so until large-scale commercialization becomes a reality. In some cases, we are the only customers for a given product, and the product generates little revenue for the supplier. Therefore, suppliers (rightfully) are reluctant to spend considerable effort to develop new products. Alternatively, one can buy small quantities of commodity products. However, these may not specified as tightly as required. Again, because of the small quantities that are purchased, suppliers will not respecify products (in many cases) for sensor applications. The same problem is observed in the area of sensor encapsulants, which are often silicone gels.

Once the devices have been assembled, a property that is affected by the die attach material is the backside burst pressure. Burst pressure is affected by adhesion strength, adhesive geometry, and material strength. The type of housing material, the surface preparation of that material, and the chemistry of the die attach material all contribute to adhesion strength "(a typical test method is shown in Fig. 7.46)." The dispense process, the die attach material properties (e.g., the thixotropic nature of the material), and the die attach geometry all influence the adhesive geometry and/or the material strength.

7.3.1.4 Interconnection

Wirebonding can be difficult for three important reasons with these MEMS devices. First, the soft die attach material makes applying a thermosonic wirebond problematic. Heat transfer through the polymer materials in the package is low,

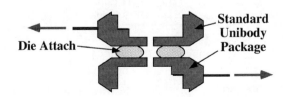

Figure 7.46 A typical test technique for evaluating adhesion strength for die attach materials on pressure sensor.

and applying pressure to the die that is resting on a low modulus, viscoelastic material can cause the die to move, as well as cause a loss of ultrasonic energy. In addition, surface preparation for the Au wire/Al pad bond can cause variation in bond strength. These factors lead to one significant reliability focus in silicon pressure sensors—wirebond integrity. Finally, the high CTE of the silicone gels that are often used to encapsulate pressure sensors can cause reliability problems with wirebonds if the wire is too small and/or if the wire passes through multiple materials with different CTEs. Wire breaking and/or pulling from the bondpad or leadframe post can be a problem unless a package designer uses a large enough bondwire size (e.g., 2 mils is often used) and does not allow the wire to encounter two or more materials with dissimilar CTEs.

Often, pressure sensors are packaged with metal leads that run through the package. In the case of plastic packaging, this requires insert molding. When sensors are packaged in this fashion, they usually employ a leadframe array or strip so that multiple packages can be molded at the same time. Examples of single-stranded leadframes are shown in Figs. 7.43 and 7.47. In Fig. 7.47, the leadframe is designed such that the leads are interdigitated to save area on the metal leadframe and reduce cost. Handling devices in leadframe strips allows for easy handling during high volume automated assembly operations.[211]

7.3.1.5 Encapsulation

Piezoresistive diaphragm pressure sensors present a unique device passivation problem. The requirements that passivation materials have for common integrated circuit devices also pertain to materials used for pressure sensors. Specifically, encapsulation is required to protect the electronic IC device from moisture, mobile

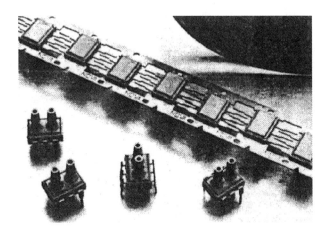

Figure 7.47 An example of a single-stranded, interdigitated leadframe array.[207] Singulated dual-in-line (DIP) packages are shown with ports following completion of the assembly process.

ions (e.g., sodium, potassium, chloride, and fluoride), UV-VIS, alpha particle radiation, and hostile environmental conditions.[212] For pressure sensor devices, the passivation material must form a moisture barrier for corrosion protection, it must provide dielectric isolation, and it must support the wirebonds (or other interconnection). In addition, pressure sensors are affected by the stress that is imparted on the diaphragm; therefore, the applied stress of the passivation material must be controllable, and it must allow the transmission of pressure signals to the transducer. Silicon dioxide and silicon nitride form excellent barriers to moisture and mobile ions,[212] but these materials only cover the silicon die, excluding the bonding pads. Also, the inorganic layers do not provide 100% pinhole-free coatings,[212] so organic encapsulation is essential. Full details on each of these encapsulants can be found in Section 7.2.7.

Encapsulants, whether on the wafer-level or on the package-level, can affect the device performance for a microsystem. Bitko et al.[170,213] have shown that encapsulation stress does alter the output of bulk micromachined pressure sensors. In this work, an empirical result for the output of the pressure sensor device as a function of the stress in the thin film was obtained and used to assist in the trimming of the pressure sensor devices. In the current Motorola pressure sensor assembly process, laser trimming must preceed the package-level encapsulation step. As a result, predictability of the encapsulant effects on device performance is necessary to perform trimming that will compensate for these shifts in device output. The materials used currently are silicone or fluorosilicone gels.[214]

Most of the information known about the effect of silicone gels on properties like sensitivity or offset for a pressure sensor are derived empirically.[215] Material chemistry, mix ratio of two part gels, cure temperature, cure time, cure atmosphere, hardness, and volume dispensed all can contribute to device performance degradation and subsequent yield losses in this process step. Moreover, gels are sticky materials, and some yield loss can result from particulates that are visible in the gel—as an aesthetic failure. In some applications, light transmissivity through the gel (or other encapsulant) can be a problem to the sensor because it can influence the output of the semiconductor device. Therefore, pigments, like titanium dioxide or carbon are often mixed into the silicone gels. However, carbon is a conductor, so it can also be a problem if it comes in contact with a conductive media (e.g., Fig. 7.48). These materials have many of the same complexities that are encountered with the silicone die attach materials: lot-to-lot variation, difficult material properties to measure, uncertainty in modeling these materials, being a small customer, and the general problem that polymer material properties are affected by the processing of these materials. These materials also use long, batch processing bakes which can be a throughput bottleneck.

Finally, and probably most importantly, is the effect of adhesion strength of the material on a variety of surfaces: silicon diaphragms, PECVD silicon nitride passivation, Al/1% Si bondpads, Au wires, Au-plated leadframes, silicone die attach materials, and epoxy (or other polymer) housing materials. This is perhaps the largest unknown when encapsulating a pressure sensor. It is known that sensitivity and offset changes are caused by the encapsulant material properties. However,

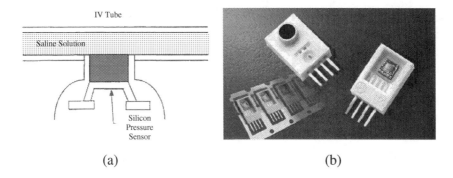

Figure 7.48 (a) The concept of a short term media compatible pressure sensor using the backside of the sensing diaphragm and silicone gel for isolation of the sensor from the harsh environment (i.e., saline solution).[216] (b) An example of this type of sensor: the Motorola Chip Pak.

little is known about the effect of adhesion on electrical properties. In addition, adhesion is a critcally important property for reliability and media compatibility of devices.[175]

7.3.1.6 Media Compatibility

Media compatibility is one of the most pervasive problems affecting the commercialization of pressure sensors.[173–175,207,216–226] Pressure sensors are just one of many type of sensors that encounter the media compatibility problem. Chemical sensors, flow sensors and several types of actuators are also affected by this problem. Several authors have written about or patented packages for media compatibility. A low cost ($3–$5) pressure sensor that is inherently media compatible has yet to be achieved for most applications. However, one example of a successful short-term media compatible pressure sensor that is very low cost is the pressure sensor for intravenous delivery system (Fig. 7.48). These systems use the backside of the sensing diaphragm and silicone gel for media isolation from the saline solution in the IV tube.[216]

The term *media compatibility* has not been defined through a standard for MEMS devices. A device like that shown in Fig. 7.48 is "media compatible" for the disposable blood pressure application but is not "media compatible" for a 10-year automotive application. Therefore, the term *media compatibility* must be defined. It is "the ability of a pressure sensor to perform its specified electromechanical function over an intended lifetime in the chemical, electrical, mechanical, and thermal environments encountered in a customer's application."[173] Several failure mechanisms have been observed and defined that affect MEMS devices during harsh environment exposure, including material swelling and corrosion.

A problem with this type of sensor package for long-term exposure of a sensor to a harsh environment is the effect of the environment on the die attach material. Several groups have studied the impact of organic or aqueous solutions on material properties. Most materials guides will provide the effect of a particular media on percent elongation, tensile strength, or some other mechanical property in a standardized test. However, this is rarely published in an actual application for sensors or MEMS. A group from Lehigh and Motorola have published on the effect of hydrocarbons on the stress induced in the die by using Moiré interferometry.[227] The interferometry technique was used to monitor the curvature of the substrate in situ during hydrocarbon exposure. The results did show that the effect of exposure could be observed in the output of the sensor, establishing this as a viable in situ monitoring technique for screening die attach materials for pressure sensors.

In general, there have been seven techniques that have been investigated for the creation of a media compatible pressure sensor: (1) application of coating materials (either inorganic or organic[225]); (2) a package-level solution, like the use of a secondary diaphragm; (3) configuring the sensor to use the silicon substrate to isolate the device from the harsh environment (e.g., the use of backside exposure for the pressure sensor or bonding techniques[221]); (4) through-wafer interconnects;[221] (5) selective encapsulation techniques; (6) isolating the corrodable areas of the sensor with packaging materials (e.g., die attach material); and (7) module-based solutions.[225] To date, there does not exist a low cost (<$3), high volume (>10 M/yr) media compatible pressure sensor, and media compatible sensors remain one of the most highly investigated areas in MEMS packaging research.

7.3.1.6.1 Coatings.
To provide media protection, barrier coatings were investigated initially. Specifically, fluorosilicone gels and/or parylene coatings, both of which have been identified as possible barrier coatings on pressure sensor packages for some time,[170,173–176,213,217,218,225,228–234] were used to produce these low cost devices. A few groups have published enhancements on the usage of these barrier coatings. Nichols et al. have suggested the utility of adhesion promotion to enhance the lifetime of parylene coated sensors.[182] Dyrbye et al. and Maudie et al. have attempted to use materials with lower permeabilities (i.e., inorganic coatings) to slow the failure mechanism.[183,184,235] And at least two groups have attempted to use a combination of postpackaging barrier coatings to improve lifetime performance.[54,174] Several additional coatings were investigated, but they either did not satisfy the cost or the performance targets. A brief description of the fluorosilicone gels and parylene C coatings is useful to highlight their differences.

Silicone gels
Silicone gels are one or two part siloxanes that are drop dispensed and require a curing step.[215,236,237] Several vendors provide these materials, each with a slightly different formulation, and each with a slightly different curing mechanism. An example are the two-part fluorinated siloxanes that were used for these experiments. Often, these two-part materials consist of one part with a base polymer and a catalyst, and a second part with the same base polymer, a multifunctional moiety, and an inhibitor. When the two parts are mixed and heated, the base polymer reacts (e.g., vinyl addition polymerization) with the multifunction species to form a

gel-like network. These materials are deposited in relatively thick coatings (on the order of millimeters); but because they are extremely low modulus materials, they have little effect on the device electrical performance. Several groups have investigated the possibility of using various coatings to improve the media compatibility for pressure sensors. The most common coating used in production is a silicone or fluorosilicone gel. Examples of silicone gel solutions are shown in some of the patent literature and in production devices from groups like Motorola, Delco, etc. [Fig. 7.50(a)].

Parylene
Parylene C (poly[monochloro-para-xylylene]) coatings are deposited in a chemical vapor deposition chamber.[238–241] Dimer starting material is vaporized and then pyrolyzed to create a monomer gas that is drawn into a vacuum chamber containing the sensors. Because of the vacuum in the deposition chamber (approximately 80 mtorr), the material deposits and polymerizes in a very conformal manner. The nature of the near-room-temperature deposition process allows for deposition on fully assembled devices with reasonably repeatable thicknesses. This has been suggested by several authors for use with pressure sensors.[174–176,225,229,230,232] Thickness control is critical because parylene materials are much higher modulus materials than silicone gels, and they do produce a noticeable effect on electrical output. However, the repeatability of these shifts allows modeling of the effect of parylene on device parameters.[233] Monk et al. have shown that the upper temperature limit for parylene during operation is 105°C as a result of oxidative degradation that occurs to the polymer during heating above that temperature in an oxygen ambient.[231]

Alternative coatings
Figure 7.49 shows the effect of material type on permeability. While permeability is not the only property controlling the corrosion of metals under a coating (adhesion is also very important), it does provide a relative coating effectiveness measure, provided that the coating is uniform and completely covers the substrate.

In general, inorganic films provide better protection against moisture than organic/polymeric films. However, on the package level it is often very difficult to provide a uniform, complete inorganic coating. Nevertheless, several groups have investigated the use of inorganic films to create media compatible pressure sensors. Monk et al., Dyrbye et al., Gardeniers and Laursen, and U. Lipphardt et al. have patented processes for coating pressure sensors with inorganic coatings as a method for creating a media compatible pressure sensor[183–185,235,242] and other sensors.[243,244] Flannery et al., Hegner and Frank, and Kurtz et al. have used silicon carbide with a pressure sensor.[245–247] Although none of the authors have performed media exposure experiments on the silicon carbide coated sensors, Flannery et al. have shown harsh environmental exposure results in H_2SO_4, HF, and KOH. Morever, they have shown the effect of the PECVD silicon carbide on the pressure sensor electrical characterization. Interestingly, they have observed that the CTE of PECVD silicon carbide (deposited as described in their work) is greater than silicon, whereas LPCVD silicon dioxide and PECVD silicon nitride have been

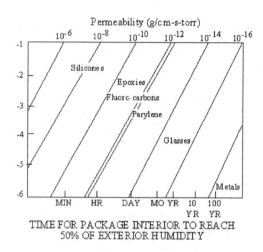

Figure 7.49 Permeability of water through various coatings.

observed to be less than silicon. This provides a possible means for tuning the pressure sensor temperature coefficient average value through mechanical design. Christensen et al.[242] have used tantalum oxide as a protective film over the pressure sensor device on the wafer-level (Fig. 7.50). They have reported less than 0.04 A/h etch rates over a pH range of 2 to 11 with this film as a coating. This is more than an order of magnitude less than silicon nitride and more than two orders of magnitude less than silicon dioxide. Monk et al. have also used this technique with other inorganic films to protect pressure sensors.[235]

7.3.1.6.2 Secondary diaphragms.

Secondary diaphragms are widely used to protect the silicon pressure sensing element from the media environment.[192,249–254] Typically, a 316 stainless steel diaphragm is welded into a stainless steel housing. Examples of stainless steel secondary diaphragms are shown in Fig. 7.27. The silicon pressure sensor is sealed inside a cavity within the housing, and electrical connections to the exterior are made through metal leads shielded from the housing by glass-to-metal seals. External to the cavity, additional electronic components such as a laser-trimmed resistor network, signal conditioners, or calibration constants stored in EPROM are used to set the gain and offset, and improve the sensor output for the application requirements.

The principle of operation for a secondary diaphragm pressure sensor is as follows. First, the pressurized media of interest loads the secondary diaphragm, causing it to deflect. The space between the diaphragm and the pressure sensing element is filled with a fluid, which is pressurized by the deflection of the secondary diaphragm. The fluid pressure is then sensed by a silicon pressure sensor die inside the fluid-filled cavity. If the secondary diaphragm is sufficiently flexible, and the internal cavity is properly filled without air bubbles or an internal pressure offset, the pressure of the external media can accurately be transmitted to the pressure sensor die. The fill fluid must be compatible with the silicon pressure sensor and

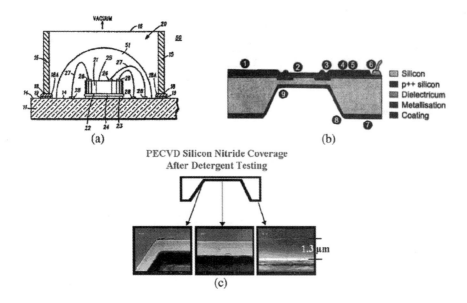

Figure 7.50 Coatings for use in creating "media compatible" pressure sensors. (a) Silicone gels;[214] others include;[248] (b) tantalum oxide;[242] (c) PECVD silicon nitride[235] and others including Lipphardt et al.[243] on a pressure sensor for protection from harsh environments.

die attach materials, and must be electrically insulating. Silicone oil is widely used with secondary diaphragms.

In general, secondary diaphragms provide superior media resistance compared with coatings, but tend to generate more of a disturbance to the true pressure signal. Three disadvantages have been cited for this type of device: (1) the packaged sensor cost is often considered to be too high;[184] (2) the size precludes usage in some systems;[183] and (3) the temperature operating range may be limited by this approach. Secondary diaphragm devices tend to have higher intrinsic temperature sensitivity, since the fill fluid expands and creates a false pressure signal as the temperature rises. This effect can be minimized by increasing the flexibility of the diaphragm, as well as by maximizing the diaphragm surface area to fluid volume ratio. Increasing the diaphragm flexibility also improves accuracy and linearity. For this reason, stainless steel diaphragms are often corrugated. In spite of the technical hurdles, high-performance pressure sensors with high accuracies based on stainless steel diaphragm isolation are widely available.

Advanced MicroMachines Incorporated has developed a unique all-media compatible pressure sensor based on PFA diaphragm isolation. All wetted parts of their sensor are made from PFA or PTFE, a fluoropolymer family which is one of the most chemically resistant materials known and provides superior media resistance even compared with stainless steel. Another advantage of PFA is in purity critical applications where the leaching of metal ions from stainless steel can contaminate critical fluids in medical and semiconductor process applications, for example. Since PFA has a much lower modulus than steel, it is more flexible and can provide better pressure transmission and temperature performance. Furthermore, the use of injection moldable plastic provides cost advanatages compared with the steel sensors.

7.3.1.6.3 Configuring the silicon to isolate the electronics and interconnections.
There are several techniques that allow configuration of the silicon to minimize exposure of the electronics and interconnect materials. To isolate the electronics, standard passivation materials can be used (i.e., thin film coatings). Yet, the interconnects (e.g., wirebonds and bondpads) are often not protected by this wafer-scale passivation. Therefore, other means of protecting these areas from

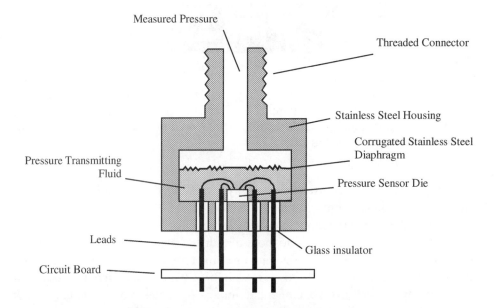

Figure 7.51 Typical secondary diaphragm pressure transducer design.

Figure 7.52 Media compatible pressure sensor based on PFA packaging.

ASSEMBLY AND TEST FOR MEMS AND OPTICAL MEMS 391

harsh environments must be conceived. Fortuitously, the tools of micromachining (e.g., alkaline anisotropic etching) can provide shapes of the silicon die itself that can isolate the interconnect form the harsh environment.

One of the best examples of this is a topside constraint for an absolute pressure sensor, where pressure is applied from the backside of the device. A group at Motorola has created a backside absolute pressure sensor. The advantage of this pressure sensor is that it isolates the corrodable metal areas from the pressure/media with the silicon substrate.[202–206] Figure 7.53 shows examples of this device and its fabrication. Figure 7.54 shows some media exposure results. As can be seen, there is no effect on the output of the device following the initial effect of the test setup.

New SensorNor pressure sensors use a similar construction.[172] In their application, a backside absolute pressure sensor is formed by anodically bonding a constraint to the topside of the bulk micromachined pressure sensor. Pressure is applied from the backside of the die, thus isolating the harsh environment from the wirebonds and other metallization. To isolate the package stress from the device, a softer polymeric material is applied to the top of the constraint die prior to wirebonding.[172] The final product is a two-chip sensor: a bulk micromachined,

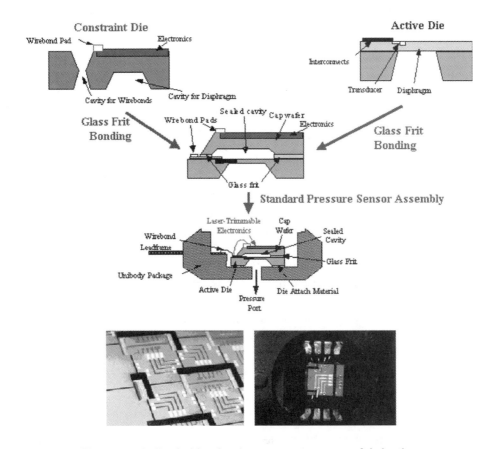

Figure 7.53 Backside absolute pressure sensor fabrication.

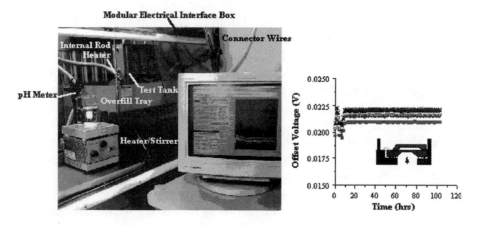

Figure 7.54 Media testing results on the backside absolute pressure sensor.

topside constraint, piezoresistive pressure sensor and a CMOS control IC for manifold absolute pressure, barometric absolute pressure, and tire pressure applications [http://www.sensonor.no/ ca. May, 1999].

Isolating the harsh media to the backside of the pressure sensor (i.e., on the opposite side of the interconnect) is one simple way of creating a media compatible pressure sensor.[255] Similar backside pressure sensor methods have been patented for absolute pressure applications[51,256–258] and differential pressure sensor applications.[259–264] In the previous case, a topside constraint was used to create an absolute pressure reference cavity. In the case of a capacitive pressure sensor, wafer-bonding and backside exposure can also be used to create a media compatible pressure sensor.[265–268]

Electrical feedthroughs is another term for the type of isolation of the interconnects that has been described in the previous example. Najafi et al. at the University of Michigan have been using a variation of this technique to create hermetic packages for integrated sensors.[269,270] They use a glass capsule for hermetic sealing onto a micromachined silicon substrate that acts as a hybrid carrier for other integrated circuits. All of the circuitry is within the hermetically-sealed glass capsule, and the sensor or other electrode is outside of the capsule being exposed to the environment. Therefore, electrical feedthroughs provide a means for carrying the signal from the exposed electrode to the internal circuitry.

The group at Danfoss has suggested using silicon-based pistons to isolate the media from the pressure sensor.[271] This would be constructed by adding a second piece of silicon above the diaphragm (Fig. 7.55). This second piece of silicon would act as a secondary diaphragm. However, instead of a liquid pressure transmission media (as is used with package-level secondary diaphragms), this concept would use a silicon piston to transmit the pressure.

7.3.1.6.4 Through-wafer interconnects.
Several groups have suggested other types of wafer bonding constructions to isolate the sensor interconnects from the

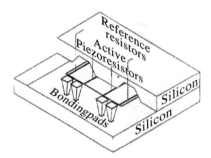

Figure 7.55 Silicon used as a media barrier and pressure transmission mechanical element for a media compatible pressure sensor.[271]

harsh media. An example of approaches to media compatibility using wafer bonding is the anodic bonding work from Hanneborg et al.[272–274] Another method of using the silicon substrate to isolate the device active areas and electrical interconnects from the media is the use of through-wafer interconnections, especially for chemical sensors. An example of a through-wafer interconnect concept on a pressure sensor is shown in Fig. 7.56. Other groups have also used through-wafer interconnects, especially for pressure sensors and chemical sensors.[2,221,262,275–281]

7.3.1.6.5 Selective encapsulation. Selective encapsulation, either on the wafer-level or on the package-level, is another possible technique for creating a media-compatible sensor. For example, Hower et al.[285] have used screen printing to create wells with solid-state ion selective electrodes. The screen printed material creates a dam so that the ion-selective membrane for the sensor can be deposited only within the electrode well. While this application does not create a media compatible sensor, an extension of this screen-printing concept to create a dam which restricts the flow of another material could be used to create a selective encapsulated sensor. Alternatively, Krassow et al.[286] have used package-level photolithography (i.e., shadow masking) to create a selectively encapsulated sensor structure. Finally, Smith and Collins[287] have used wafer bonding techniques to isolate portions of the die with a CHEMFET. Unfortunately, little additional work has been published in this area.

7.3.1.6.6 Packaging material isolation of interconnects. Three approaches have used packaging materials to attempt to isolate the interconnects from the harsh environment. The first approach is a type of selective encapsulation. In this approach the overmolding or potting of the package provides a coating for the interconnect (i.e., wirebonds) while leaving the diaphragm of a pressure sensor exposed to the environment[288–290] (Fig. 7.57). The disadvantages of this approach are that the molding will require a pin to restrict the flow of the mold compound or the potting material from the sensor diaphragm. Placing this pin onto the sensor diaphragm will, inevitably, break the fragile diaphragm. Furthermore, even if

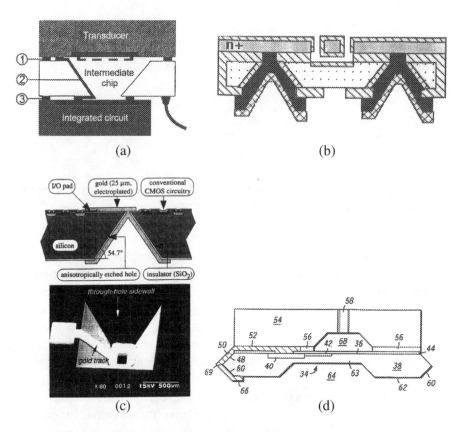

Figure 7.56 Examples of through-wafer interconnects: (a) Bouwstra et al.,[188,190,282] (b) Schmidt et al.,[283] (c) Linder et al.,[189] and (d) [284].

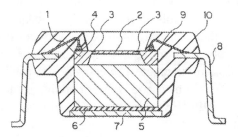

Figure 7.57 Selective overmolding or potting of the wirebond areas on a pressure sensor to create a media compatible package.[290]

the sensor survives the overmolding or potting process, the CTE of the mold compound is very different from the silicon, so the temperature coefficient of the sensor output will be very different than the sensor without the overmolding.

The second approach uses flip-chip bonding and techniques for isolating the flip-chip bond with the die attach material.[138] This could even include adding a

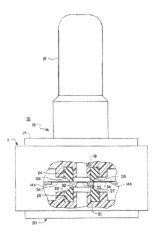

Figure 7.58 Elastomeric seals used by Honeywell to create a media-compatible pressure sensor package.[291]

post-bonding coating to isolate the sensor flip-chip interconnect from the harsh environment. The disadvantage of this type of approach is that for high pressure exposure from the backside of the sensor (i.e., the die attach side), the die attach and flip-chip bond will be tension, which could result in creep of the material over the longterm.

Finally, Honeywell has created a package housing in which two rubber gasket-like materials hold the sensor in place in the package and isolate the interconnect from the media.[291] The package is shown in Fig. 7.58.

7.3.1.6.7 Module or multi-chip package solutions. Multi-chip or module packaging is another possible technique for isolating the sensor chip from other electronics chips. Figure 7.59 shows one example of this approach. There are several module packages being used for sensors today. A variety of module packages are shown in Fig. 7.60. Typically, this type of approach will work for the electronics, but one of the previous approaches must be used for the sensor portion of the multi-chip package.

7.3.1.7 Pressure Interconnection (Ports)

Pressure sensor packaging requires some means of mechanical interconnection. Examples of pressure interconnection or ports are shown in Fig. 7.61.[211] All of the porting options are based on the Motorola unibody package (right side in Fig. 7.43). Ports are adhesively mounted to the unibody package. Adhesives must form a complete seal between the package and the port to minimize pressure leakage through the package. Also, the adhesive must have a similar CTE to the rest of the package to minimize package-induced stress on the sensor from the port attachment.

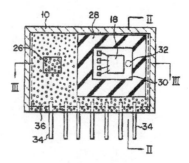

Figure 7.59 An example of a multichip package to isolate the electronics from the harsh media while allowing exposure of the sensor to the harsh environment.

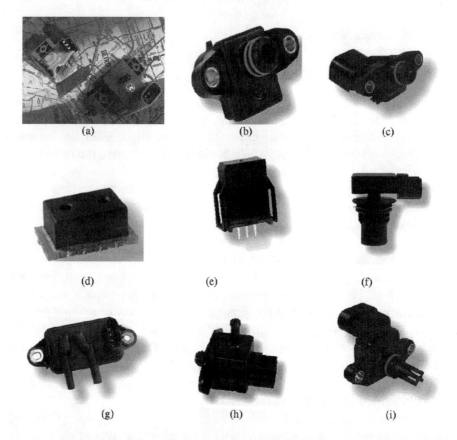

Figure 7.60 (a) A pressure sensor package that is placed in a next level assembly.[292] (b–i) Several examples of module package outlines for pressure sensors in automotive applications [http://www.mot.com/AECS/General/AIEGSensors/index.html, ca. 2/99; and [Schuster, 1997 #159]]. (a–c) Manifold absolute pressure (MAP). (d–e) Barometric absolute pressure (BAP). (f) Gasoline evaporative sensor (EVAP). (g) Delta PFE. (h) Fuel rail. (i) Temperature MAP (TMAP).

7.3.1.8 Second-Level Packaging (Modules)

If we view the micro-processor as the brain (decision-maker) of the electronic system, sensors are the eyes and ears, while actuators are the arms and legs. For efficient functioning, all the components have to fit together within the system requirements. A typical electronic system relies on packaging concepts such as a series of components assembled on a printed wiring board (i.e., PWB). The manufacturing of PWBs has reached a maturity and acceptance level in a diverse range of applications, from low-cost, mass-produced consumer and automotive electronics to high-reliability military and space applications. It is paramount that sensor technology complements the existing production methodologies, such as surface mount and through-hole soldering technologies.

It should be noted that the pin count for the integrated pressure sensor is not very high. For example, even though Motorola's sensor package is designed with 8 pins, the actual number of pins used by the customer is only three. The additional pins are used for the trimming process. Also, the leads are designed with 50 mils width on a 100-mil spacing. For the Surface Mount package (Fig. 7.44), the leads are formed to create a gull-wing shape, while for the Top Piston Fit and Dual Piston Fit package (Fig. 7.44), the leads are trimmed and formed for through-hole solder joint. The size, spacing, and shape of the leads follow a standard industry practice, so no special requirements for pad layout or via hole design are required during the PCB layout. In short, the solder bond pad sizes and solder paste application will be same as other semiconductor components.

Conventional semiconductor components are typically shipped with leads that are solder dipped or tin plated after overmolding. This is not a major issue since the packages are not open. For pressure sensor packages, there is always an opening for the pressure interface. Thus, it is difficult to perform solder dipping or tin plating. Without such a treatment, the underlying Ni layer may not pass the solderability requirement. In such a case, an acceptable solution is to provide a flash of Au on the solderable portion of the leads, which would protect the underlying Ni and meet the solderability requirements. In surface mount assemblies using Sn-Pb solder, the presence of Au is known to form a Au-Sn intermetallic. An excessive amount of Au is likely to cause embrittlment of the solder joint, which will result in lower fatigue life. A solution adopted by Motorola and many other sensor manufacturers is to use a flash of Au on the lead frame, which maintains the solderability to the lead while minimizing the amount of Au in the Pb-Sn solder. In a typical solder joint, this amount of Au on the lead when dissolved in solder is less than commonly acceptable limit of 5%. Another possible solution is Ni/Pd or Ni/Pd/Au plating.

An additional concern for the next-level assembly is the ability to automate, particularly in the surface mount assembly processes. The surface mount assembly process will require a flat surface to facilitate the pick and place operation, using automatic assembly equipment. Furthermore, it is a common practice that after soldering, the PWB goes through cleaning and passivation. The passivation for the PWB is often either dipped coated with silicone or spray coated with acrylic. Thus, the pressure interface has to be designed such that it will permit these operations without adding significant additional processing and cost. In some cases, the package may need a means of holding the sensor in place before and during the solder reflow process, when the leads are unattached. For example, Motorola's Next Generation sensor package is designed for flexibility and ease of assembly at the next level of assembly, be it a PWB or a module (e.g., Fig. 7.60). Several examples of pressure sensor modules are shown in Fig. 7.60. Finally, an alternative to this component-package inside a module-package approach is to simply place the die within the module package itself (without the component package).

7.3.1.9 Stress Sensitivity of Sensor Packages

Sensors in general are designed to sense physical attributes such as externally applied pressure or inertial forces. Often, pressure sensors utilize the piezoresistive property of Si single crystal. Thus, by design, the sensor produces an electrical response that is proportional to the amount of stress at the location of a doped Si resistor.[293]

Most piezoresistive Si pressure sensors produce a full scale output of between 6 and 10 mV/V. For such a sensor, if the package transmits a small amount of stress to the die, either because of externally imposed loads or thermal mismatch, it could result in significant electrical output. For example, if the packaging-induced stress is around 100 psi (~700 kPa), it would result in an electrical output that is around 1 mV/V. It should be noted that this amount of stress is very small when compared with the failure strength of Si, which is above 100 MPa. It can be seen clearly

Figure 7.61 Several examples of porting options on Motorola's unibody package style. Additional examples are shown in Figs. 7.13, 7.14, 7.20, 7.21, and 7.43.

that stress isolation becomes a key for device performance. In most cases, many of the backend assembly techniques and designs, such as die attach, passivation, the housing material, and geometries are driven by the stress isolation requirement. If proper attention is not paid to stress isolation, the device performance can be affected by packaging-induced stresses.

One potential solution to stress isolation for pressure sensors that has been proposed is the use of a stress relief trench or moat around the sensor diaphragm.[216,294] Rusanen et al.[295] have listed the following guidelines for packaging of microphones, which have the same problems of stress isolation as very high sensitivity pressure sensors:

- mount the sensor onto a substrate with a CTE very close to that of silicon (2.3 ppm/C at room temperature);
- use a very low modulus die attach material;[296] and
- minimize the bond line area of the die attach material.

They experimented with bonding of these microphones in three ways: (1) die bond onto silicon; (2) flip-chip onto glass; and (3) flip-chip onto alumina. Their results support these claims. The samples that were flip-chip bonded onto alumina exhibited the largest nonlinearity in performance over temperature and exhibited the largest temperature hysteresis. The samples that were die bonded onto a silicon substrate (using a low modulus die attach material) exhibited the most linear response over temperature (essentially linear between $-40°C$ and $40°C$) and the least temperature hysteresis.

Similarly, Lin et al. modeled pressure sensors with a variety of die attach.[53] In their models, six different techniques were used: (1) alumina/silverglass/30 mil glass; (2) alumina/30 mil glass; (3) alumina/110 mil glass; (4) aluminum nitride/30 mil glass; (5) aluminum nitride/110 mil glass; and (6) alumina/RTV/30 mil glass. A summary plot of shear stress as a function of temperature shows the results. As listed above, the trend is from worst case to best case stress isolation. Similar to the results of Germer[296] and Rusanen,[295] the soft, low modulus die attach material provides the best stress isolation for the pressure sensors.

Work by Germer and Kowalski further describes a method for decoupling the sensor from mechanical package stress by using a soft die attach material.[296] In their description, they also specify that the die attach thickness be greater than 50 microns or the decoupling of stress from the package to the sensor is ineffective. Both offset drift (following temperature cycling) and temperature hysteresis were reported to decrease when die attach thickness was increased from 10 microns to 50 microns. Also, the metal leadframe is designed such that during transfer molding, there is an inner platform on which the sensor is bonded that has minimal connection to the outer frame of the package (Fig. 7.62).

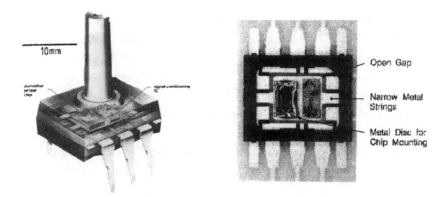

Figure 7.62 Design of a leadframe to minimize the package stress that is observed on a packaged pressure sensor.[296]

7.3.2 Inertial Sensor Packaging

A second example of a commercialized, high volume MEMS physical sensor is an accelerometer. The most prevalent use of accelerometers is in airbag deployment, so automotive specifications are required. Like the pressure sensor case, packaging for accelerometers is an integral part of the device function and reliability, so care must be taken when designing packages for accelerometers. This section is dedicated to showing some of the unique features of accelerometer packaging. A general MEMS packaging discussion can be referred to in Section 7.2 for more details.

Several factors must be considered when designing packaging for accelerometers:

- stiction must be avoided;
- pressure changes can affect the response of accelerometers through damping effects; therefore, hermetic sealing is often desirable;
- orientation of the transducer must be considered (x-, y-, z-, and rotational accelerometers all have applications, and packaging orientation may change depending upon the design of the transducer);
- accelerometers must be able to survive high mechanical shock (to 2000 g and beyond);
- accelerometers must operate in automotive temperature ranges (−40 to 125°C);
- electrostatic shielding is necessary; and
- packaging must have a greater than 10-year lifetime.

As shown in Section 7.2, wafer bonding has been used with accelerometers. For example, in the case of the Motorola accelerometer, glass frit bonding[1,60] is used

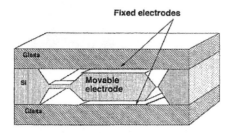

Figure 7.63 An accelerometer structure that uses wafer bonding as part of the transducer.[55]

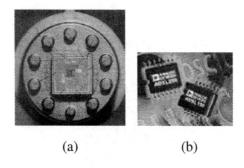

Figure 7.64 The Analog Devices ADXL50 accelerometer in (a) a 10-pin TO-100 metal header[131] or (b) an 8-lead surface mount device [http://www.analog.com/industry/iMEMS/, ca. May, 1999].

at the wafer level to eliminate stiction during assembly and to create a stable pressure environment for the accelerometer during operation over the lifetime of the application (Fig. 7.3).

In other cases, wafer bonding is an integral part of the transducer structure (Fig. 7.63).

Accelerometer packages come in various styles. Analog Devices originally used a 10-pin TO-100 metal header-style package for its ADXL50 accelerometer (Fig. 7.64).

They have also produced accelerometers in 8-lead ceramic dual in-line packages[127] and, more recently, a surface mount package. Motorola and Ford Microelectronics have packaged accelerometers in single-in-line or dual-in-line plastic packages (Fig. 7.65).

EG&G IC Sensors has developed a ceramic leadless chip carrier for accelerometers (Fig. 7.66).

Because there exists a variety of package styles even in a high volume MEMS application like airbag accelerometers, the following section is dedicated to identifying the processes and highlighting the similarities and differences among these different packaging styles.

Figure 7.65 (a) The Motorola accelerometer in a plastic package and (b) the Ford Microelectronics accelerometer in a plastic package.[298]

Figure 7.66 The EG&G IC Sensor ceramic leadless chip carrier.[299]

7.3.2.1 Analog Devices

The process that Analog Devices uses to produce an accelerometer is reviewed by Chau and Sulouff.[127] Figure 7.67 shows the process flow.

The packaging/assembly process starts on the wafer-level following sacrificial layer etch/release with the laser trim/probe setup. ADI uses SiCr thin-film resistors that they believe are more cost effective than EPROM (similar to the process used by Motorola for pressure sensors—see Fig. 7.63). Even the sensitivity is trimmed at the wafer-level by using an algorithm that accounts for circuit gain, resonant frequency, and self-test electrostatic force input.

Following wafer-level trimming, ADI uses a unique upside-down wafer sawing process to minimize the stiction and device damage that would otherwise result from an uncapped accelerometer being exposed to the sawing process.

Then, a modified pick and place tool is used to minimize MEMS device handling during die attach. The dies are mounted in either a 10-pin TO-100 header or a CERDIP (Fig. 7.64). Following this, a proprietary anti-stiction coating is used in the package prior to sealing the packages hermetically. This coating minimizes shock test stiction incidents. Testing is performed with customer automated shak-

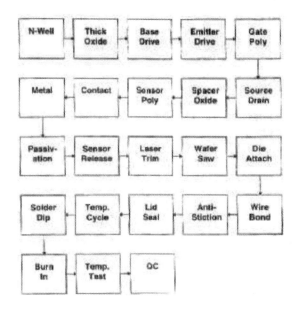

Figure 7.67 The Analog Devices accelerometer assembly process flow.[127]

ers and, although a solution for high-volume final test is in place, "... major efforts are on-going at ADI to develop low-cost high-volume throughput testers ...."[127] Packaging and testing each represent approximately 1/3 of the total product cost; therefore, standardization of these processes with typical IC assembly and test processes is important for cost reduction. In addition, lower cost manufacturing was achieved by establishing a back-end production line in the Far East.

7.3.2.2 Motorola

Motorola uses a different approach for assembly.[126] First of all, the fabrication process is not integrated, so two-chips must be assembled. Secondly, wafer-level, glass-frit bonding is used to create a wafer-level package. This allows standard wafer sawing and plastic packaging processes. It also provides an electrostatic shield for the capacitive accelerometer and can be used to set the pressure of the accelerometer cavity, thereby specifying the damping. Wafers are probed but are not trimmed until in the package because two-chips are not matched until in the package. Both die are bonded onto the leadframe—the g-cell with an elastomeric die attach and the IC with a standard epoxy die attach. The dies are wirebonded to the leadframe, and chip-to-chip wirebonding is performed. Once this is complete, a gel-like material is placed over the g-cell to improve stress isolation. After this material is cured, devices are molded into DIP- or SIP-style packages using a transfer-molding, epoxy process. The next step is to calibrate the device by burning the EPROM, after which devices are leadformed and packed for shipping. The leadforming of the package enables the orientation of the transducer to be varied with respect to the axis of the accelerometer in the car.

7.3.2.3 EG&G IC Sensors

Figure 7.66 shows another approach to accelerometer packaging—a leadless chip carrier.[299] Like the Motorola approach, this is a two-chip package. The package is produced by punching "green" ceramic sheets with pre-patterned via holes. Metal interconnects are formed by screening and printing a refractory metal—tungsten. Eight layers are stacked and sintered at 1500°C to form the package. Die bonding and wirebonding are performed and followed by a lid seal for the package. This sealing process is important because it sets the operating pressure for the accelerometer (at <500 ppm H_2O) in N_2. A 280°C Au/Sn solder seal is used for the hermetic package seal. A unique feature of this package is that it can be mounted flat on a PC board or on its side, which allows the system-level designer flexibility in terms of the orientation of the sensor element. Also, because of the package seal, reliability requirements for automotive and military can be passed.

7.3.2.4 Ford Microelectronics

The accelerometer produced by Ford Microelectronics Incorporated (FMI) is another example of a two-chip solution.[298] The dies are bonded to a leadframe and wirebonded. This includes chip-to-chip wirebonding to connect the two dies. Reuse of existing IC packaging, including automated equipment, standard footprints, and test sockets was emphasized in the development of this package. Stress sensitivity, particle contamination/environmental exposure of the sensor, and die orientation are three important issues that are unique to accelerometer packaging. Stress sensitivity was addressed by selecting a transducer that is relatively insensitive to package stress. Partical contamination and environmental exposure are minimized by wafer-bonding prior to die sawing. Vertical and horizontal mounting capability allows selection of the transducer orientation without changing the die design, through leadforming and molding on standard equipment.

The uniqueness of the FMI packaging approach is the choice of designs for die orientation purposes. Figure 7.68 shows an option for a doubly molded package that, although more costly than a singly-molded device, removed the technical risk of wirebond failure during leadforming to enable die orientation modifications without die design modifications.

Following the first mold operation a 90° leadform is performed before the second mold to create the laterally-sensitive accelerometer package from a z-axis die design. A concern that is present with the doubly-molded package is the possibility of delamination between the two packages, or "popcorning." This was addressed by specifying the surface finish on the inner package to improve the mechanical locking between the two packages.

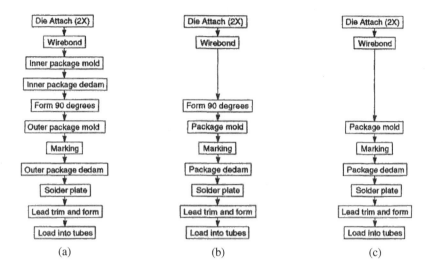

Figure 7.68 A doubly- or singly-molded package for accelerometer transducer orientation flexibility.[298] (a) Double-mold package. (b) Single-mold package. (c) 9-lead SIP.

7.3.2.5 Others

A group Sagem/CNRS/LAAS/ESIEC/LETI[300] has developed another approach to a two-chip accelerometer. In this case, a wafer-bonded and sealed accelerometer is flip-chip bonded onto an ASIC in a vertical-chip packaging technique. Tin-lead bumps are bonded onto a Ti-Ni-Au underbump metallization.

Jet Propulsion Laboratories[301] has used a hybrid package for their space-qualified tunneling tip accelerometer. The hybrid substrate mechanically supports the devices, allows electrical interconnection, and provides heat dissipation. A specially space-qualified conductive epoxy is used to mount the sensors on the alumina substrates in an MCM-style assembly. In addition, JPL has developed a package for its clover-leaf gyroscope.[302] A gold-plated, brass package with a hermetic, solder seal and electrical feedthroughs was used to maintain an initial vacuum of 10^{-6} torr for 2–3 months, which was the specification for space usage.

7.3.3 Optical MEMS Packaging—The Digital Mirror Display (DMD) Example

Probably the highest commercial volume optical MEMS product is the digital mirror display (DMD) from Texas Instruments. The next several sections describe some of the unique packaging challenges with this type of product.[303–305]

The assembly process flow that Texas Instruments uses to manufacture the DMD is shown in Fig 7.69.

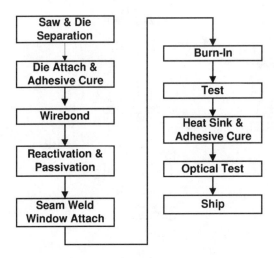

Figure 7.69 DMD assembly process flow.

A more detailed review of the DMD die separation process flow is described in Chapter 5.

7.3.3.1 Assembly

Most IC assembly areas are noncleanroom environments. Since the DMD pixels are vulnerable to ambient conditions, the entire assembly process is performed in a Class 10 (at a 0.3 μm particle size) cleanroom. Once the optical lid is hermetically sealed to the DMD package, the device can be removed from the cleanroom environment after it is helium-leak checked.[304] Although an IC-familiar sequence of assembly operations (i.e., die attach, wire bond, window seal, and final test) is employed for the DMD, the product nevertheless requires special handling and processing. Most assembly equipment has been modified to function within a cleanroom environment. Particulate control is a primary concern in the assembly area. All packaging machines are commercially available semiconductor tools, but each has been modified for low particulate operation. Scrapping a DMD due to particle contamination at this late stage in the process flow is very costly. Figure 7.70 shows an example of particulate contamination and its potential effect on mirror functionality.

Throughout the assembly flow, any exposure to high temperatures during assembly is avoided, if possible. Once the sacrificial layer is removed, the DMD pixels are vulnerable to temperature induced effects (e.g., high temperatures may affect mechanical and optical parametrics). After the individual dies are separated on the saw tape (again, refer to Chapter 5 for dies separation techniques employed on the DMD), the dies are mounted into a ceramic substrate. To avoid damage to the mirrors, a specially designed collet is used to pick and place the dies. This special design prevents any mirror damage when vacuum is applied to the collet

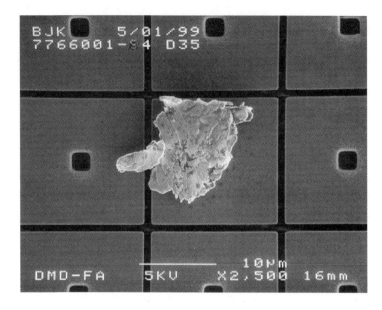

Figure 7.70 DMD pixel with particle defect.

during the pick operation. At the die attach tool, the wafer is in a vertical position. The dies are first removed from the saw tape starting from the bottom of the wafer and progressing upward in a serpentine fashion. This start-from-the-bottom-pick method helps mitigate particle contamination to the upper unpicked die. A die attach adhesive with low outgassing, low thermal cure, and high thermal conductivity qualities was chosen. Adhesive outgassing has been identified as a culprit in mirror stiction. Once the die attach adhesive is cured, a room temperature, high frequency gold wire bonding process is then performed.

Like the wafer portion of a semiconductor process, cleaning steps are used in the assembly process. This cleaning step, however, does not utilize wet chemistry. Instead, a proprietary dry cleaning process has been developed and has proved to be quite effective in the removal of particles from the surface of the mirrors without damage to these mirrors.

Since the DMD is a mechanical component and the DMD pixels make surface contact, stiction is a chief concern. A TI proprietary process was developed to mitigate pixel stiction. This process is a chemical vapor deposition and is performed at the package level just prior to placement of the optical lid. Since the DMD pixel must land millions of times over the life of the product, stiction is a primary reliability focus. Other commercially available MEMS products either do not make surface contact or the surface contact is limited during the life of the product. For instance, MEMS-based accelerometers are typically designed not to make surface contact during use. These accelerometers, however, are sensitive to stiction during shipping and handling, where surface contact may occur under mechanical shock or vibration conditions. Most MEMS devices utilize some type of lubrication strategy, whether it is a thin film deposition, surface roughness treatment, or mechanical and material improvements.

After wire bond, the next step in assembly is the attachment of the window to the substrate with die. The substrate and window are attached utilizing a low-temperature hermetic seam welding process. The attachment is performed in a dry nitrogen environment. This environment helps ensure the DMD pixels will operate in a benign environment. A trace amount of helium is added to the package headspace for leak checking after seam weld. The final assembly operation is the attachment of a heat sink to the backside of the substrate. The heat sink attach adhesive is cured at a relatively low temperature. Like the die attach adhesive, the stud adhesive must possess high thermal conductivity as well as mechanical robustness to ensure the heat sink stud stays in place. Low outgassing qualities are also important to ensure no contamination collects on the surface of the optical lid.

Similar to the processing equipment, the individual package parts must also be particle free. For instance, a mechanical and/or electrical failure occurs if a loose particle on the optical lid falls onto an active DMD pixel. A particle that is not loose on the optical lid can still cause a projection system defect. In this case, an optical lid defect can cast a low contrast defect (e.g., a shadow) on the projection screen (refer to Chapter 5 for more details on this type of defect and its effect on device performance). To avoid this, special handling and attention to particle control is placed upon processing of the optical lid.

7.3.3.2 Package

Packaging is widely considered to be the Achilles' heel of MEMS technology manufacturing. With most, if not all, MEMS devices, the package is a critical component. Typically, equal effort is placed on the package development and the device development. The interface to a system involves electrical and optical connection. Electrical signals that are supplied through the package to the DMD are thus converted to mechanical movement of the DMD pixels. Optical illumination, which is also coupled into the package, reflects off the surface of the DMD pixels and into a projection system. The movement of these reflective DMD pixels allows for the spatial modulation of light: light is steered into or out of a projection lens system. An example of optical coupling is depicted in Fig 7.71.

The DMD package is thus an application-specific design: the entire design was developed internally. Requirements for the optical lid include minimal light losses, optical flatness/parallelism, an optical aperture, and the capability of withstanding a parallel resistance seam welding process that induces mechanical and thermal stresses. The basic elements of the DMD window can be identified in Fig. 7.72.

The substrate on which the DMD device is mounted is an LGA (land grid array) ceramic header product. An aluminum heat sink is mounted on the backside of the ceramic header for thermal heat dissipation. The window assembly has a large, nonround, matched glass-to-metal seal and is postprocessed by polishing and coating the window. The window and substrate for three types of DMD packages are shown in Fig 7.73.

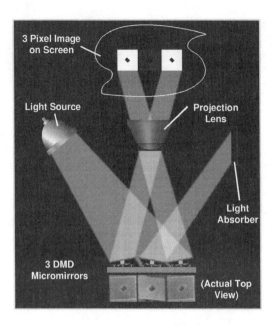

Figure 7.71 Optical schematic of projection operation.

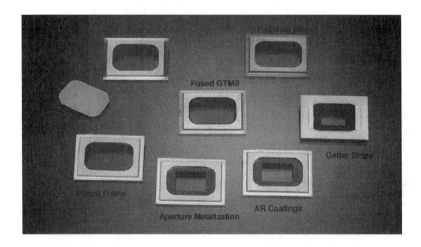

Figure 7.72 Components of the DMD optical window.

The lid assembly consists of two piece parts, a stamped kovar frame and Corning 7056 glass. Packaging material selection, especially matching coefficients of thermal expansion (CTE), is important for a mechanically and environmentally robust package. Since the CTE of the metal and glass are very close, a matched glass-to-metal seal is made by means of a belt furnace with a peak temperature of approximately 1000°C. After the glass-to-metal fuse is complete, the glass of the window assembly is processed through a double-sided grinding and polishing

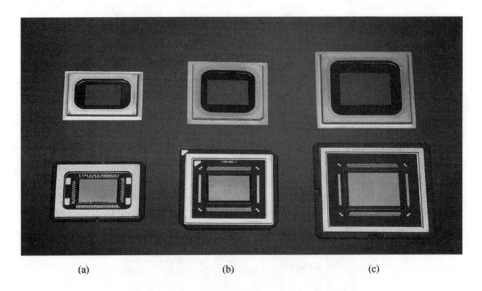

Figure 7.73 DMD Windows and Substrates: (a) 0.7″ SVGA resolution; (b) 0.9″ XGA resolution; (c) 1.1″ SXGA resolution.

operation. The kovar frame is then plated with 2–8 μm of nickel and a minimum of 1–2 μm of gold. It is important that the surface quality of the glass is not damaged when the polished window assembly is exposed to the chemical baths during the plating process.

Two coatings are then applied to the window assembly. First, a low reflectance coating is applied on the surface of the glass that will be on the inside of the package after the window attach process. A photomask is applied and this low-reflectance coating is removed or etched away to form a clear window aperture. To achieve a sharp aperture edge definition (i.e., less than 20 μm protrusions), the etch chemistry and process time used during the process are critical. The purpose of the window aperture is to hide features outside of the DMD active array. Features such as bond wires are blocked from the illumination source. After the aperture coating is etched, the photomask is removed and MgF_2 antireflective (AR) coatings are applied to both sides of the glass. The required average transmittance of the AR coatings is greater than 98% and the reflectance is less than 0.5%. Finally, two getter strips are adhesively attached to the internal region of the window assembly. The purpose of the getter is to control contaminants (specifically moisture). Stiction mitigation is the primary driver for the inclusion of the getter strips: moisture (i.e., capillary attraction) can cause the pixel to stick upon landing. The complete package is shown in Fig. 7.74.

7.3.3.3 Future Packaging Strategy

As is typical with most MEMS devices, the package cost is not insignificant compared to the total device cost. The biggest challenge is to develop a low-cost pack-

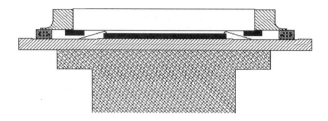

Figure 7.74 Side view of DMD hermetic package.

age that is more aligned with what is typical in the IC industry while still meeting our system-level optical quality goals and specific device requirements. A huge effort is currently underway within the MEMS community to identify a low-cost hermetic package that can be utilized for a large number of MEMS components.

If the DMD is to become a major enabler of new products, a process used to define and successfully implement new packages must be developed. The DMD component packaging strategy is summarized by these five goals:

(1) Reduce the system cost.
(2) Provide differentiated packages.
(3) Improve system robustness.
(4) Reduce the time to qualify new packaging concepts.
(5) Maintain and execute a component packaging road map with key customer inputs.

These goals focus on the need to provide an advantage for the next level of integration and not become myopic about the component package. Although there is overlap among these five goals, each goal has a focus that is useful to emphasize.

7.3.3.3.1 Strategy #1: Reduce the system cost. Although the DMD is only one component of the system, the DMD unit cost must be aggressively reduced. This cost reduction strategy will be implemented with the following tactics:

- reducing device material costs;
- improving back end assembly yields; and
- reducing device assembly time.

The real opportunity for material cost reductions comes with design simplification and increased volumes. The next task is to work with suppliers to determine if the unit cost targets are possible. This exercise will identify barriers to meeting these cost goals which will become the focus of the packaging team—either to solve or to find alternative designs to eliminate these barriers. The result of this focus will be a cost curve showing step decrements in costs due to specific cost-cutting activities.

7.3.3.3.2 Strategy #2: Provide differentiated packaging.
A differentiated package reduces system costs and enables our product to provide features advantageous to our customers. One tactic is to size the DMD to allow the user to miniaturize their product more easily than with a competitive technology. This can be achieved in several ways—by reducing the size of the package or by adding features to the package that are required at the system level (such as alignment aids for DMD integration).

7.3.3.3.3 Strategy #3: Improve system robustness.
The product's reliability is affected by packaging design. Device junction and mirror temperature margins are increased when thermal loads and/or thermal resistance are reduced, both of which are determined by the package design. The maximum mirror temperature is presently set at 65°C, but application needs are identifying opportunities for higher array temperatures. These higher temperatures require larger fans, heat sinks or alternate cooling techniques that cost more, are less efficient and require more electrical power—none of which are conducive to miniaturization or improved system robustness.

The device package is part of the electrical design and can be a source of unwanted electrical noise or increased susceptibility to noise radiated/conducted from cooling fans, color wheels or other high speed circuitry. The package has distributed power and grounds utilizing mutual inductance between high current paths and voltage planes; it also provides isolation between sensitive signals. These considerations become more important as the device operates at higher frequencies along with pin count increases for larger devices.

7.3.3.3.4 Strategy #4: Reduce the time to qualify new packaging concepts.
The packaging strategy requires the development of a process where new designs can be developed and qualified within six months. Of the five strategies, this may be the most difficult to achieve. Relationships with selected suppliers are necessary to reduce the time to obtain products built to alternate designs. A different suite of environmental tests is necessary so that a package design can be evaluated without having to use functional dies. The new capability will need to provide an assembly model shop capability where new devices in new packages are built quickly.

7.3.3.3.5 Strategy #5: Maintain and execute a customer-driven component packaging roadmap.
A rolling 48-month packaging roadmap will allow the organization to focus and track near-term and long-term package development. The goal is to have a major improvement (cost or feature) of the package every six months with improvements tied to specific planned customer product releases. Regular product planning discussions with key customers is desirable to initiate mutually advantageous developments. These relationships can be the source of significant product innovations.

7.3.4 OPTICAL MEMS PACKAGING—FLIP-UP MICROSTRUCTURES USING MICROHINGES

7.3.4.1 Precision Assembly

7.3.4.1.1 Surface micromachining. Before discussing automated and/or self-assembly of MEMS, it is instructive to review briefly how MEMS are realized through surface micromachining with, for example, the Multi-User MEMS Process (MUMPs), which has been used to create self-assembled MEMS.[142] MUMPs is a three-layer polycrystalline silicon (polysilicon) process for prototyping MEMS on a silicon wafer. MUMPs offers three patternable layers of polysilicon and two sacrificial layers of phosphosilicate glass on a base layer of silicon nitride. A top layer of gold is provided as the reflective and/or conductive surface. Table 7.4 provides MUMPs layer thickness and the process is illustrated in Fig. 7.75. The order of the entries in Table 7.4 is consistent with the deposition order of the films on the silicon substrate, with silicon nitride being the first layer. Gold is evaporated on to the device after all other layers have been grown by low pressure chemical vapor deposition. The polysilicon layers and the $<100>$-cut silicon substrate are highly doped with phosphorus (approximately 10^{20} atoms-cm^{-3}) to decrease electrical resistance. After construction, micromachined devices are "released" by etching the sacrificial glass layers in a hydrofluoric acid solution.

After MEMS are released, residual material stresses in a structure composed of gold layer (tensile stress) and Poly-2 layer (compressive stress) may cause the structure to curl slightly into a concave shape.[307] Residual material stress varies somewhat after each production run. For example, for the 16th MUMPs production run, typical peak-to-valley curvature for a 100-μm-wide gold on Poly-2 micromirror was measured as 495 nm.[307] Combining the top and middle polysilicon layers (stacked polysilicon) as support for the gold layer reduced peak-to-valley curvature of a 100-μm-wide micromirror to 140 nm.[307] Peak-to-valley curvature was further reduced to 63 nm by retaining the second oxide layer between the top and middle polysilicon layers (trapped oxide).[307] The second oxide layer is isolated from the etch solution using a via around the edge of the device to connect the top and middle polysilicon layers and seal in the oxide layer.[167] Although structures with

Table 7.4 Structural and sacrificial layers used in MUMPs.[142]

Layer Name	Nominal Thickness (μm)
Nitride (silicon nitride)	0.6
Poly-0 (bottom polysilicon layer)	0.5
1st Oxide (sacrificial layer—phosphosilicate glass)	2.0
Poly-1 (middle polysilicon layer)	2.0
2nd Oxide (sacrificial layer—phosphosilicate glass)	0.75
Poly-2 (top polysilicon layer)	1.5
Metal (gold)	0.5

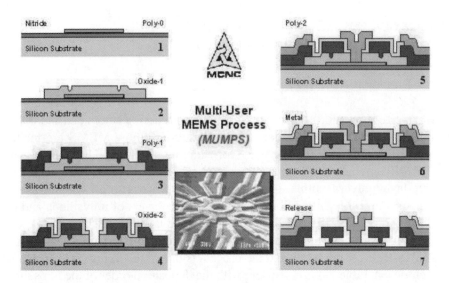

Figure 7.75 Illustration of MUMPs layers through fabrication of a micromotor [after,[142] drawing by Adrian Michalicek, University of Colorado-Boulder].

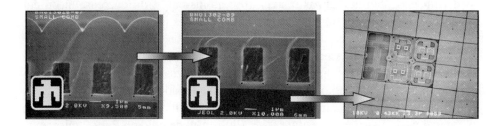

Figure 7.76 Fabrication of piston micromirror array with hidden flexures through Sandia's surface micromachining process ([308] adapted by Adrian Michalicek, University of Colorado-Boulder). Chemical mechanical polishing (CMP) planarization removes topographical effects in uppermost sacrificial oxide layer so that the following polysilicon layer is optically flat.

trapped oxide have the lowest curvature, the use of a via forces the gold layer to be indented further from the edge of the device because the surface over the via is not optically flat. This reduces total reflective surface area of a device such as a micromirror.

In surface micromachining, the thin film layers conform closely to the topology of the previously deposited and patterned layers (see Fig. 7.75). Unless the designer makes sure a layer is flat by controlling the pattern of the layers beneath it, the induced topology can have detrimental effects on the uniformity of the layer and the effective elastic modulus of mechanical structures. In some cases the topology can even trap part of a structure that was intended to move freely. The surface topology can be controlled, however, in more sophisticated fabrication processes where

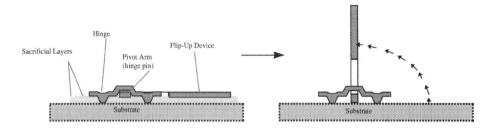

Figure 7.77 Assembly of flip-up optical components using microhinges (after [309]).

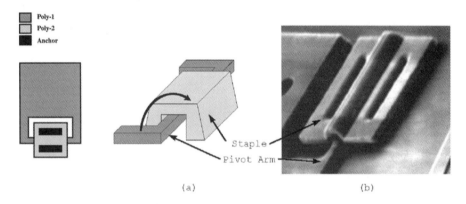

Figure 7.78 (a) Schematic and (b) micrograph of a substrate hinge.[310]

layers are chemically-mechanically polished prior to subsequent layer deposition (Fig. 7.76).[308]

7.3.4.1.2 Microhinges and microlatches for 3D surface micromachined MEMS.
Surface micromachined designs are planar through fabrication and may require some assembly and/or actuation after fabrication. A great number of MOEMS are based on 3D assembly of planar polysilicon structures. The primary components that allow 3D MOEMS are microhinges and microlatches.

Microhinges enable surface micromachined parts to be rotated out of the plane of the substrate as illustrated in Fig. 7.77. This allows fabrication of flip-up components such as plates, mirrors, gratings, and lenses. There are two types of microhinges: the substrate hinge and the scissors hinge. Both hinge designs were originally proposed by Pister et al.[309] A substrate hinge is used to hinge released structures to substrate as shown in Fig. 7.78. It consists of a pivot arm and a staple. The pivot arm is fabricated from the Poly-1 layer. The structure that is to be flipped up off of the substrate is connected at the ends of the pivot arm. The staple is fabricated from the Poly-2 layer, and forms a bridge over the pivot arm. The staple, which is anchored to the substrate, allows free rotation of the pivot arm. Scissors hinges are fabricated by interlocking fingers of two structural layers as shown in

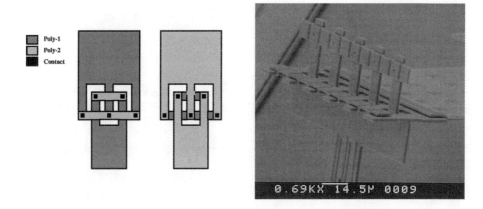

Figure 7.79 Scissors hinges are used to connect two plates while still allowing them to pivot.[310]

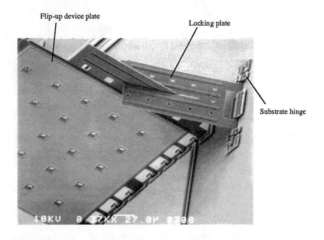

Figure 7.80 Locking of two flip-up plates positioned at a 90 degree angle with respect to each other.[312]

Fig. 7.79. Scissors hinges allow two plates to be connected together while allowing them to pivot at the connection.

Microlatches are used to lock flip-up structures into specific positions. Several locking mechanisms exist. One approach is to use a second flip-up plate positioned at a 90° angle with respect to the first plate (Fig. 7.80).[309] Initially, the first plate is flipped into the desired position. Next, the second plate is flipped up. A notch in the second plate engages with the first plate, thus locking both plates into position. The disadvantage of this design is that both plates must be flipped up and interlocked into position. Another approach of locking flip-up components into position is to use a self-engaging, spring-loaded locking mechanism (microlatch),[311] which is shown in Fig. 7.81. In this approach, when the flip-up plate is raised into position,

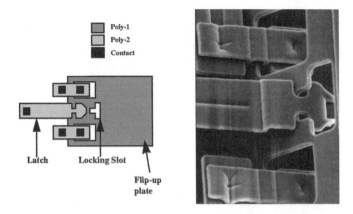

Figure 7.81 A self-engaging locking mechanism (microlatch).[310]

the latch moves along the surface of the plate until it drops into position in the locking slot cut in the plate. The advantage of the self-engaging lock is that it is only necessary to flip-up one plate. This is important in the design of self-assembled microsystems.

7.3.4.2 MEMS Automated Assembly

7.3.4.2.1 Concept of micro-optical bench.
MOEMS are often based on the concept of micro-optical bench.[311] Figure 7.82 illustrates the concept.[145]

7.3.4.2.2 Automated assembly using MEMS mechanisms.
Three-dimensional surface micromachined MOEMS are typically assembled by hand, a time-consuming and delicate process. However, some microactuators have sufficient force to raise and position hinged plates. This eliminates the need for manual assembly or adjustment, thus making batch fabrication and automated assembly feasible. Automated assembly allows deployment and remote assembly of MEMS in the field of operation or readjustment of components to align a system for better performance.

Figure 7.83 shows the automated assembly system for a scanning micromirror.[313] The system consists of three separate parts: a linear assembly motor [Fig. 7.83(a)–(c)], a vertical actuator [Fig. 7.83(d)], and a microlatch [Fig. 7.83(e)]. A lift arm is connected to the drive rod of the linear assembly motor with a hinge. The other end of the lift arm is connected to the flip-up mirrored plate. First, the vertical actuator is used to lift the free end of the flip-up plate off the substrate, forming a triangle with the substrate, the lift arm, and the flip-up plate. The linear assembly motor then drives the lift arm towards the base of the flip-up plate, thus rotating the flip-up plate around its substrate hinges. Finally, the microlatch engages to hold the flip-up mirrored plate in position. If necessary, the linear assembly motor can be reversed to pull the lift arm away from the assembled structure.

A computer-based control system was developed to automate the sequencing of the linear motor during assembly and operation of the micromirror after assembly.[314] A CMOS application specific integrated circuit (ASIC) was designed to handle the current requirements of the thermal actuator arrays in the linear motor (~35 mA) and the scanning micromirror (~15 mA). The computer accepts user inputs for driving the linear assembly motor or micromirror and sends control inputs to the CMOS ASIC via a computer interface card. The ASIC then selects the proper output channel and drives the actuator array. The computer-based sys-

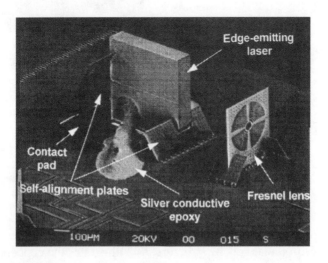

Figure 7.82 Example of edge-emitting semiconductor laser/surface-micromachined MEMS integration.[145]

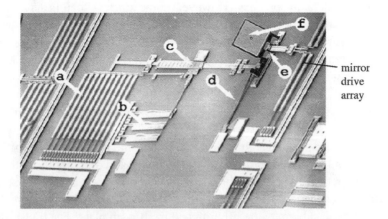

Figure 7.83 Automated assembly system for a scanning micromirror.[313] The various components are: (a) assembly motor drive array, (b) assembly motor push array, (c) linear drive arm, (d) vertical thermal actuator, (e) self-engaging locking mechanism, and (f) scanning micromirror.

tem is highly flexible and has been used successfully to drive manually assembled scanning micromirrors as well as auto-assembled micromirrors. Positioning of the micromirror plate can be controlled either by varying the DC voltage delivered to the micromirror actuator array or by using a pulsed drive signal.[314]

7.3.4.2.3 Automated assembly using stressed cantilevers.

As already mentioned, suspended surface micromachined structures may curl due to residual material stresses. Although typically undesirable in many MEMS structures, the stress-induced curling can be used as a lift mechanism, thus replacing the function of the vertical thermal actuators during automated assembly.

The curling due to residual stress of a MUMPs fabricated cantilever is demonstrated in Fig. 7.84. Prior to release, the cantilever is held flat due to the surrounding oxide. After release, the residual material stresses in the gold (tensile) and Poly-2 (compressive) layers cause the free end of the cantilever to curl upward. The amount of curl or deflection is dependent on the amount of stress and the physical dimensions of the material layers. Poly-2/gold beams fabricated in MUMPs

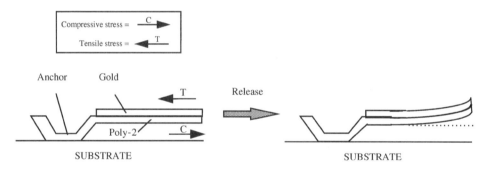

Figure 7.84 Illustration of cantilever curling due to residual material stress.[149]

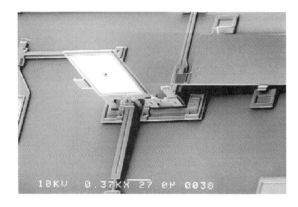

Figure 7.85 Scanning micromirror lifted into locked position by stressed cantilever.[149]

exhibit tip deflection on the order of 20 microns for 300-micron-long beams.[149] Figure 7.85 shows a stressed cantilever used to assemble a scanning micromirror.

Unlike MEMS mechanisms, the stressed cantilever requires no control signals during the assembly process. The stressed cantilever used for lifting represents true *self*-assembly technique as opposed to *automated*-assembly. The availability of releasable, dielectric layers such as silicon nitride with known stress could be of great value in the development of self-assembled MEMS and should be investigated for future MEMS fabrication processes.

7.3.4.2.4 Solder self-assembly for three-dimensional MEMS.

Another method of assembling MEMS has been proposed that uses the surface tension properties of molten solder or glass as the assembly mechanism.[315,316] The solder method uses hinged micromachined plates with specific areas metalized as solder wettable pads. Once the solder is in place it is heated to its melting point, and the force produced by the natural tendency of liquids to minimize their surface energy pulls the hinged plate away from the substrate (Fig. 7.86). Solder is a predominant technology for electronics assembly, and it is being developed for optoelectronic passive alignment.[160] Using solder, hundreds or thousands of precision alignments can be accomplished with a single batch reflow process, and the cost/alignment can be reduced by orders of magnitude. In addition, solder provides mechanical, thermal, and electrical connections.

A fluxless soldering process has been developed for 3D MEMS assembly. Once the solder is deposited, the MEMS are placed in a chamber filled with nitrogen and formic acid gas. The nitrogen gas prevents the solder from oxidizing at high temperatures, and the formic acid gas removes any existing oxide. The chamber is then heated to melt the solder, and the surface tension forces of molten solder lift and assemble MEMS structures. An example of a solder-assembled corner cube is shown in Fig. 7.87.

One way to control precision of solder assembly is through accurate solder volume control. The experimental error resulting from solder volume variation was

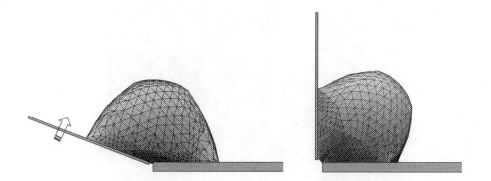

Figure 7.86 Illustration of solder self-assembly of 3D MEMS.

found to be ±2°. Surface energy minimization of the molten solder is the governing principle in determination of the final equilibrium angle of assembled structure.

An alternative approach for precision-demanding applications is to use a self-locking design integrated with solder technology. As already discussed, self-locking mechanisms can be used to control the angle with manual assembly. They can also be used for solder self-assembly as shown in Fig. 7.88. As the molten solder rotates the structure out of the plane of the substrate, a rigid bar attached with a spring rotates with the plate until it slides into a hole or grove in the plate, thus stopping the movement of the plate. The final angle of the 3D MEMS is then a function of design geometry. If designed correctly, the bar would not only stop the plate rotation, but also lock it in position. Angle control accuracy of ±0.3° has been demonstrated using self-engaging microlatches.

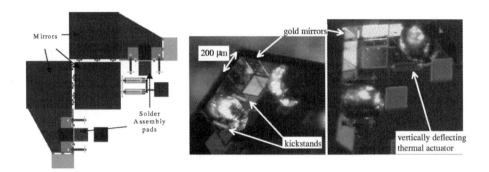

Figure 7.87 Layout and video images, from different perspectives, of a solder self-assembled corner cube reflector. Assembled with 63% Sn/37% Pb, 8 mil diameter solder balls.

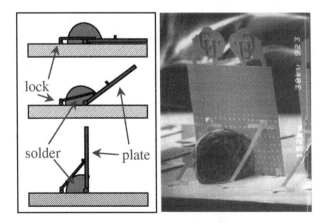

Figure 7.88 Illustration of self-locking. A solder self-assembled polysilicon plate with the microlatches engaged in locking positions.

7.3.5 Microfluidics Packaging

Microfluidics encompasses microfabricated devices that manipulate fluids. Microfluidic devices may perform mixing, separations, reactions, and other operations on liquids. Applications include chemical analysis, combinatorial chemistry, drug discovery, detection of pathogens, DNA sequencing, and drug delivery. Biomedical applications dominate the microfluidics arena, and fluids of interest are typically saline solutions, blood, and complex solutions containing proteins and other organic compounds. For example, the drug discovery process requires screening and testing of thousands of chemicals. With a microfluidic system, the thousands of microfluidic devices on a single chip may conduct chemical experiments with a highly integrated fluidic network. Both the time required and the volume of chemicals expended can be dramatically reduced. Many lab-on-a-chip applications are envisioned, in which conventional analytical equipment would be replaced with a single microfluidic system. The microfluidics area is expanding in concert with advances in genetic and biomolecular sciences, giving rise to a host of new packaging issues. This section identifies main issues in microfluidics packaging and technologies to address them.[20,317–324]

7.3.5.1 Application Requirements

Microfluidic systems based on MEMS technology have unique packaging requirements related to fluid handling. Microfluidics packages typically must address some set of the following criteria:

- chemical compatibility with the media of interest;
- little or no dead-volume;
- wettable surfaces;
- transparent packaging for flow visualization;
- low-cost packaging (disposable devices);
- suitable for autoclaving or purging (reusable devices);
- prevent particle contamination and clogging; and
- fluid inlet and outlet connections, possibly in complex networks.

In order to function properly, microfluidic chips must be supplied with solutions which are pure and fresh, free of air bubbles, particulates, and chemical contaminants. Connections must be designed to allow a rinse fluid to purge through the system, cleansing out old reagent. Packages must have little or no dead-volume, pockets of stagnant fluid where old solutions and contaminants may become trapped. Package surfaces must be clean, inert to the chemical of interest, and wetting so that the fluid can flow properly. The package must allow easy connection to external fluid sources, typically through plastic tubing, although in some tests a fluid sample may be applied directly to the chip. Tubing should be easy to disconnect and reconnect, without creating leaks or air bubbles. Packages must be transparent when optical access is required—for example, to observe fluorescence or in flow cytometry. It is often advantageous to actually see the fluid in the system to reduce human error and for troubleshooting, even if flow visualization is not inherent in the device's function.

7.3.5.2 Materials

Glass is currently the most commonly used material for microfabricated fluidic devices, often in conjunction with wafer-bonded silicon components. Glass exhibits good wetting characteristics, does not react with or contaminate fluids of interest, and permits flow visualization. Plastics such as PMMA are seeing increasing use in microfluidic applications, and due to their lower cost will surpass glass, as the technology for micromachining in plastics improves. Plastic external packages are widely used for microfluidics, due to their low cost, chemical compatibility, and ease of forming into complex shapes such as internal channels and specialized connectors. For example, Affymetrix uses an injection-molded plastic package for their GeneChip products. Silicon is used primarily when greater micromachining precision is required than can be accomplished through wet etching of glass.

7.3.5.3 Fluid Interconnects

A significant problem in microfluidics packaging is making fluid connections to the device. Fluid interconnect technology is important for both monolithic (single-chip) devices and hybrid systems, in which many components such as pumps, valves, and pressures sensors are combined into a larger system.

The most basic fluid connection is formed by inserting a plastic capillary tube into a hole formed in the silicon or glass chip, and then sealing with an epoxy. While adequate for research purposes and some low-volume devices, a better quality and more automated connection technology is needed in most applications. Several researchers are developing fluid interconnect technologies. Desired features include low dead-volume, reusable and interchangeable connections, and high density of interconnections for hybrid fluidic systems. Several researchers, such as Verlee (1996), Jaeggi (1998), and Gonzalez (1997),[317,322,325] have sought to develop a fluid interconnect board, analogous to the printed circuit board. A fluid circuit board contains an array of microchannels, routing fluid through a network of channels to and from externally mounted components, such as valves and pressure sensors. The fluid board may contain chambers for mixing and separation, or electrodes for electrophoresis. For prototyping purposes, a fluid circuit board should allow components to be disconnected and reconnected in different configurations. Variable routing could be accomplished with a network of valves.

Researchers at Abbott Laboratories[317] use both micro- and macrofabrication techniques to assemble fluid circuit boards that bridge the gap between microfluidic components and the macro world. Fluid circuit board assembly begins with multiple sheets of PMMA, which are CNC precision-machined to form fluid channels and vias. The sheets are carefully baked first to remove internal stresses and promote dimensional stability. After machining, the sheets are pressed together at 45 psi and 126°C to form a permanent bond by diffusion. External features, such as connector ports and valve seats, are machined after bonding. Diaphragm valves are formed by sandwiching a thin sheet of mylar between two of the fluid channel blocks. The valve is activated by external pneumatic pressure. Simple optical sensors, consisting of an LED and photodiode pair, determine whether fluid or air is present in a channel. With the proper arrangement of channels, valves, and sensors, a wide range of fluidic operations can be achieved, such as metering, mixing, and incubation. By combining these basic functional elements, larger and more complex fluidic systems can be realized. As a demonstration of a fluid processing instrument using fluid circuit board technology, a functional replica of Abbott's TD_x^{TM} assay system was constructed. The fluid circuit system performed high throughput metering, mixing, manipulation, incubation, and detection functions, with comparable performance to that of a much larger and more complex system, and required only 1/5 of the reagent fluids.

Researchers from Stanford University, Lucas Novasensor, and University of California, Davis demonstrated a fluid circuit board featuring fusion-bonded silicon wafers micromachined by deep reactive ion etching (DRIE), and a pyrex base. With DRIE micromachining, high aspect ratio channels and fluid connectors can be formed into a silicon substrate. By bonding wafers together with features etched on both sides, complex multi-level fluid networks may be formed. Three different types of connections for capillary tubes were demonstrated. For the first type of connector, a round hole is etched through the top wafer. The capillary is inserted and butts against the top of the second wafer, where it meets a channel whose inner diameter matches that of the capillary. This connection minimizes dead volume as the fluid passes seamlessly from the capillary to the silicon fluid board. The capillary is secured in place and sealed with a bead of epoxy around the capillary where it meets the top surface of the top wafer. The second type of connector is similar to the first, except that a cylinder is etched into the top wafer instead of a hole, so that the capillary is placed over a sleeve. This connection promotes a better seal and prevents adhesive from potentially clogging the capillary; however, it introduces some dead volume, since the silicon cylinder is inside of the capillary tube, and the inner diameter of the silicon channel is smaller than the capillary inner diameter. Finally, a third type of connection was introduced, which was the same as the first but used a plastic coupler in place of the adhesive to support the capillary. The plastic coupler was formed by injection molding of polyoxymethylene (POM). The couplers were attached to the silicon board with plastic pegs which fit through vias in the board. The ends of the pegs were then melted to lock the plastic coupler into place. Capillaries were inserted into the coupler with a silicone gasket to promote sealing. The couplers eliminated the need for adhesives but were only leak-tight up to 60 psi, compared with 2000 psi for the adhesive seals. Note that in microfluidic systems, pressure is often not needed to move fluids through the system, since the fluids wick through the channels by capillary action. In addition to forming channels and plastic couplers, pressure sensors were mounted on the board, demonstrating its suitability for hybrid integration of silicon fluidic devices. The pressure sensors were die attached over pressure access ports in the board. The pressure dies were wire bonded to an aluminum metallization layer (0.8 μm thick) on the top of the board.

While the materials and design details for fluid interconnects vary, it is clear that fluid circuit board technology will play a key role in creating commercially viable fluidic systems from the wide range of die-level silicon fluid handling components which have been demonstrated to date.

7.3.6 ACTUATORS

A wide range of actuators has been developed based on micromachining technology. As with sensors, packaging problems for actuators are complicated by the need to both protect the device from, and expose it to the surrounding environment. While microactuators have seen fewer commercial applications than microsensors, packaging technologies used in two notable classes of actuators, microvalves and microswitches, are presented below.

7.3.6.1 Microvalves

Packages for microvalves must allow both electrical and fluid access to the micromachined structure, while protecting the device from the temperature and vibration environment associated with industrial applications. Most commercially available microvalves are packaged in either a TO can adapted from pressure sensor applications, or a custom plastic housing. Redwood Microsystems offers their NC-1500 Normally Closed Valve in a TO-8 can, which is designed to operate in noncorrosive gases over a temperature range of 0 to 55°C. Microvalves from the TiNi Alloy Company are offered in a miniature plastic housing.

Henning et al.[326] from Redwood Microsystems described a microvalve package for refrigeration applications in which a low-cost, easy to assemble package is required. The micromachined valve is comprised of a four-layer pyrex and silicon stack. The base of the package consists of a gold metallized ceramic substrate with vias formed by drilling. The ceramic is brazed to the top surface of copper inlet and outlet tubes. The microvalve die is attached by silicon-gold eutectic bonding to the ceramic substrate. A fill fluid inherent to the valve's operation is added, and the fill port is capped. Finally, gold wire bond connections are made from the chip to the package, and leads are soldered to the ceramic. Silicone is used to encapsulate the wirebonds and leads. This package style provides maximum separation between the fluid and electronic portions of the device, minimizing heat transfer between the thermopneumatic valve and the refrigeration fluid.

Redwood offers their microvalves in a variety of standard packages which address different application areas. While the refrigeration application is based on ceramic, the majority of their packages are metal. The die attach method varies with the application. An important consideration in the die attach process is the compatibility of the application fluid with the die attach material. Other important packaging issues include packaging-induced stresses and thermal expansion matching, particularly since their valve is operated thermally. The trend in package styles is toward surface mountable packages. For semiconductor applications, Redwood offers both traditional VCR connections, and newer surface mount devices with c-seal manifolds, which are suitable for use in larger modules.

For high-performance mass flow control applications, Redwood produces a pressure-based mass flow controller (PMFC), which combines their valve, a pressure sensor, and flow constricting orifice, in a ceramic board.[327] The complete system is precisely calibrated and the calibration parameters are stored in an on-board E^2PROM. Multiple-channel gas panels are made by combining individual PMFC modules.

7.3.6.2 Microswitches

The most common microswitch device is the microrelay, which is expected to play a significant role in future telecommunications applications. Other switching devices include optical switches, pressure switches, temperature switches, and acceleration switches. High reliability is critical in most switch applications. Usually, the package must promote high reliability by providing a clean, hermetically sealed environment in which the switch can operate without oxygen, moisture, or organic vapor contaminants. This precludes plastic packaging unless the device can be sealed at the wafer level.

Researchers at CSEM[328] developed a fully-packaged microrelay. They pointed out the importance of the package in determining the ambient environment for the relay, which is critical for relay contact performance. Their relay is based on a silicon component which is flip-chip bonded to a FeSi electromagnetic substrate. In order to achieve high quality contacts, a well-controlled ambient must be achieved. Suitable ambients for relays include vacuum, nitrogen, or forming gas. In order to achieve a controlled ambient, they form a solder seal during reflow in a vacuum chamber, backfilled with the appropriate gas mixture. First, a fluxless solder joining process is needed to ensure a clean cavity. They use a plasma-assisted dry soldering (PADS) process,[329] whereby the solder is pretreated in SF6 plasma to convert the oxide film into an oxyfluoride which breaks readily during solder reflow, creating a clean surface for joining. The metallization they use for joining the silicon and FeSi chips consists of an under-bump metallization of TiAu, a spacer metal, SnPb (63/37) solder on the bottom chip, and a top-surface metallization of Au on the top chip. After SF6 plasma treatment, a small indentation is made on a portion of the solder sealing ring. A standard flip-chip bonding tool is used to align and pre-bond the devices. The device is placed in a vacuum chamber where a suitable ambient is introduced, and the internal cavity of the device communicates with the chamber ambient through the indentation. The temperature is elevated to 240°C, which reflows the solder and forms a hermetic seal. After sealing the flip-chip cavity, the assembled dies can be attached and wirebonded to a standard leadframe

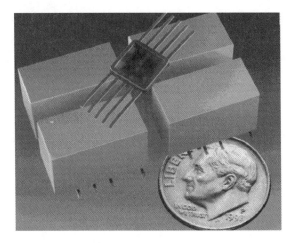

Figure 7.89 Cronos Integrated Microsystems (formerly MCNC MEMS Technology Center) microrelay product.

and molded in a standard package. CSEM uses an SOIC-16 package, with multiple wirebonds per lead to reduce parasitic resistance with the relay resistance.

MCNC/Cronos has recently introduced a microrelay product (Fig. 7.89). Initial shipments are based on ceramic packages, while an injection-molded plastic version is planned for future production.[330] The package is PCM-CIA compliant, and contains from two to eight switches per package. Each switch is electrically tested to ensure that it conforms to specifications. Reliability tests have shown that these devices perform within specifications at over one million cycles. Their robust relay design lends itself well to conventional packaging technology, and does not place unusual requirements on the packaging.

7.4 MEMS AND MOEMS TESTING

7.4.1 MATERIALS TESTING, MODELING, AND MODEL VERIFICATION

With the advent of commercial finite element modeling software, package modeling for MEMS has become more prevalent. However, for any modeling effort, input parameters are very important, as is verification of the final results. Whenever a modeling effort is undertaken, these are the three essential elements for realizing a model that predicts a physical phenomenon successfully. Materials property testing is often performed by using vendor supplied data. Yet, this is not always appropriate. The selection of materials property data, whether from a vendor, taken from the literature, or measured by oneself, represents an assumption to the model that should be stated. In addition, verification of the model with the fabrication of the device that the model was representing is essential to ensure the physical assumptions of the model reflect reality. Pourahmadi has written a review article on modeling for MEMS that discusses these points at much greater length.[331] In this work, Pourahmadi looks at modeling of pressure sensors and accelerometers and shows comparison results of accelerometer output versus finite element results, and pressure sensor temperature coefficient of offset versus sensitivity for actual devices versus finite element analysis. He also presents a method for modeling environmental effects, like shock and vibration, on sensor devices (hence, reliability modeling). The following section shows some work that has been performed to model MEMS devices, with special emphasis on materials testing results and their application to a real device through a mathematical model. Several companies are now working on MEMS modeling tools that combine several tools into one working environment for the MEMS designer. Among the companies are Microcosm's MEMCAD [http://www.memcad.com, ca. 5/99], IntelliSense's IntelliCAD [http://www.intellisense.com, ca. 5/99], Coyote Systems [pbljung@coyotesystems.com, ca. 5/99], Tanner Research [http://www.tanner.com, ca. 5/99], and MEMsCAP [http://memscap.e-sip.com, ca. 5/99].

Dickerson and Ward modeled an accelerometer package in two dimensions and three dimensions.[332] The purpose of the work was to give guidelines for placement of a micromachined device within a package to minimize the impact of packaging stress on device performance. The package was Kovar, the substrate was alumina, the die was silicon, and an epoxy die attach was used. The approach taken was to observe the deflections of the die surface as a function of distance from the center of the package. Both 2D and 3D models were attempted. The conclusions of the work suggest that the 2D model is limited in predicting the absolute deformation of the package. In this case, little is mentioned about the material properties used, nor is there any information on verification of the model with actual devices.

Dougherty et al. performed modeling of the effect of vibration on accelerometer packages.[333] In this work, holographic imagery on various SIP accelerometers was compared to finite element modeling of the devices, including material property assumptions. The method provided a successful use of a verified model that could then be used to reduce the design cycle time for improved versions of the product.

McNeil has also performed modeling on accelerometer packages.[334] In his work, he has modeled the effect of package stress on the micromachined device output by using two separate finite element models. A package model produces displacements as a function of die location which are transformed into the micromachined device model. These displacements are then used in the micromachined device model to predict sensitivity and offset change as a function of room temperature storage drift and external clamping of the package. These data are verified with experimental results of devices following room temperature storage and during external clamping.

7.4.2 TESTING & ELECTRICAL CHARACTERIZATION

Electrical characterization involves measuring electrical output of sensors at a variety of points in the assembly process flow. Electrical measurements are performed on devices when they are in the wafer form (i.e., probe), during the trimming operation (i.e., trim), and after the package is assembled completely (i.e., final test). In addition, electrical measurements are made by a variety of customers on their test equipment.

Several problems exist with sensor test equipment for high volume manufacturing. First, it is often difficult to measure the sensor with a physical stimuli at probe. Correlation among all of the test points that can occur on different pieces of equipment is often difficult to achieve. Some yield loss can be the result of two testers that are not calibrated with each other. Often, "golden units" are used to perform cross-equipment calibration or as experimental controls. Trimming may occur (by necessity) within the assembly process, which limits its ability to calibrate effects of final assembly on the device performance. Also, final test may not be the last step in the assembly/test process. Therefore, outgoing inspection can be limited by the placement of the final test system within the process flow. Finally, little research has been performed on test systems, and the research that has been performed is typically proprietary, often resulting in custom systems for production.

7.4.2.1 Pressure Sensor Test

Pressure sensor probe systems often can only monitor offset (or zero applied pressure output) of the device. Temperature coefficient of offset can also be measured if the probe system is equipped with a hot chuck. However, measurement of the span is difficult because applying a pressure at probe is nontrivial. Gong and White have published a pressurized probe system.[335] Their system employs a vacuum chuck for changing the pressure applied to the sensors. This system will work if a differential, gauge, or backside absolute pressure sensor is being tested. Any frontside pressure application would require additional modifications to the probe system.

Pressure sensor trim systems are a function of the trim process being employed. Laser trimming, which is often used, has some disadvantages because it often cannot be performed at the end of the assembly operation. In a pressure sensor, encapsulation and capping or porting of the sensor occur after trimming. Both can impact the sensor output. Therefore, characterization of the processes after trimming is essential so that the trim targets can be adjusted to compensate. One alternative to performing all of the trimming operation after assembly is to incorporate an initial wafer-level trim by using laser trimming at probe for offset coarse calibration. Electronic trimming (E/EPROM) provides a significant advantage because it can be performed with the final test operation following the completion of assembly.

Pressure sensor electrical characterization and final test equipment combine three control systems: a power supply and a multimeter, an oven and a thermocouple, and a pressure source and a pressure controller (Fig. 7.90). This type of

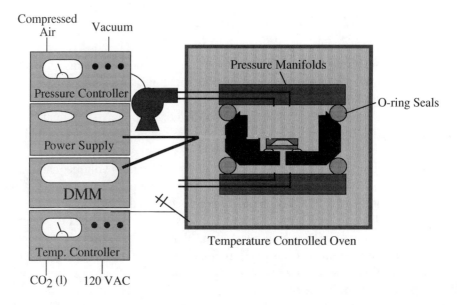

Figure 7.90 A schematic of the pressure sensor test system. This system incorporates three control systems: electrical, pressure, and temperature.

multiple control problem can be difficult to coordinate. In addition, the data generated is complex. Mastery of database usage and sophisticated spreadsheet functions help keep the large amounts of data understandable (and usable).

As has been implied, the test equipment is custom for pressure sensors. A generic control system (voltage, temperature, and pressure) provides the basic tester. To accomodate multiple package styles, custom manifold blocks are used. These blocks are designed to couple the unique device package with a standard test system. However, customers' test systems are often different than the ones used by the manufacturer. Finally, this test equipment, because it is custom-built, is expensive. Adding to the expense is the limited capacity and small, batch processing that are used for testing. Building in a more continuous system, testing fewer devices (because of a better understanding of the variability from a given manufacturing line), and/or standardizing equipment will minimize this throughput bottleneck.

7.4.2.2 Acceleration Test Systems and Self-Test

Accelerometer probe systems, like pressure sensor probe systems, do not employ physical stimuli. Little has been published concerning the possibility of mounting a shaker table with a probe system.

Trimming of accelerometers, again, depends upon the trim method being used. For example, a trim system with EPROM can be performed in concert with final test. Accelerometer final test systems are often more complex than pressure sensor test systems because they involve the use of a shaker table instead of a simple pressure source. In addition, these test systems require an AC stimulus signal, control samples, and care when mounting the samples.

Accelerometers often contain one feature that provides additional testability that is often not found on pressure sensors (in pressure sensors, the feature often requires too high a voltage[336]): self-test. Because accelerometers are used in safety applications, self-test of the transducer is mandatory. Several authors have described their self-test features ([300,337,338], among several others); however, these are typically electrostatic forces that are applied to the transducer via voltage input that are calibrated to test the functionality (i.e., the movement) of the transducer.

7.4.2.3 MOEMS Test Systems

An example of a micro-opto-electro-mechanical device that is in production is the Digital Micromirror Device™.[305] Texas Instruments uses two test processes for its DMDs. Figure 7.70 shows the assembly process for the DMDs. Following the partial wafer sawing process, a functional test is performed before packaging. During this process, typical electrical parametric testing is performed; however, there are not yet any mechanical parametric test structures being used. Furthermore, electrical opto-mechanical testers are custom tools. TI, like other major MEMS and MOEMS manufacturers, employs a test development team to assemble such equipment.

The package-level final test process follows a mechanical burn-in step. The purpose of the burn-in is to mechanically stress each mirror in order to determine whether there are any problems (e.g., stiction) with the mirrors. Also monitored are pixel reflectivity, contrast ratio, and tilt angle. Finally, a projector system test is performed prior to shipment.

7.4.2.4 Microfluidics Test Systems

Testing is a signficant technical hurdle for microfluidic devices. Microfluidic products often require testing of every device prior to shipment. Testing may require expensive reagents to be consumed for each device, and presents a throughput problem. Automated dispensing systems are used to supply fluids to fluidic devices. For one-time use devices, it may not be possible to perform a functional test. Where possible, self-test features incorporated into the device can reduce test costs, extend life, and improve product reliability.

For example, Redwood does a functional test on each of their microvalves, loading the part into a fixture with elastomer seals at the die level. A single test bed provides pnuematic and electrical testing. The specific test varies depending on the application. They want to use known good dies since the packages are expensive. Cronos Integrated Microsystems (formerly the MCNC MEMS Technology Center) also does functional electrical test on each die.

7.5 RELIABILITY OF MEMS

Reliability of MEMS is a relatively unexplored field. Typical MEMS reliability qualifications include the following testing: humidity exposure, temperature cycling, high and low temperature storage, high temperature bias, vibration testing, and electromagnetic interference exposure. Yet, outside of published qualification reports for various MEMS products, there is little in the literature on the reliability of MEMS.

One example of reliability testing on MEMS is the work that has been performed by Brown et al. at Exponent Failure Analysis Associates. Brown et al. have performed experimental work on the reliability of silicon and other micromachining materials by observation of slow crack growth.[339] In this work, a nanoindenter is used to initiate a crack into a test structure. The crack reduces the stiffness of the test, thus changing the resonant frequency at which the test structure vibrates. The resonant frequency is measured and compared against modeling. In addition, it has been observed that the crack growth is very dependent upon the environment. Experimental results show a rapid change in resonant frequency following introduction of humidity into the atmosphere. Subsequent work has also used photolithography to create a notch in a resonant frequency test structure as a crack initiation point.[340]

Another group that has been focusing on reliability of MEMS is Miller et al. at Sandia National Labs. They have been cycling the Sandia microengine through long-term operating conditions, and doing failure analysis and failure mechanism identification.

7.5.1 PRESSURE SENSORS AND MEDIA COMPATIBILITY

Pressure sensor media compatibility is one area in which a considerable amount of reliability work has been performed. The following shows the process for pursing the reliability testing of bulk micromachined pressure sensors. This includes identifying the failure mechanism first, then defining experiments to create accelerated life conditions. This work is used to model the reliability performance of pressure sensor devices.

7.5.1.1 Media Compatibility Failure Mechanisms and Development Approach

Figures 7.91 and 7.92 show examples of one type of failure that has been observed with pressure sensors when exposed to alkaline environments—corrosion.

This section will propose a formal reliability method for characterizing media compatible pressure sensors. Reliability engineering techniques[341] are adapted to the media compatibility of pressure sensors by applying a physics-of-failure reliability approach.[342] It is the intent that this section will: (1) provide motivation for

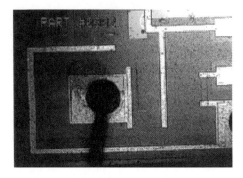

Figure 7.91 An example of corrosion on an experimental, parylene-coated pressure sensor.

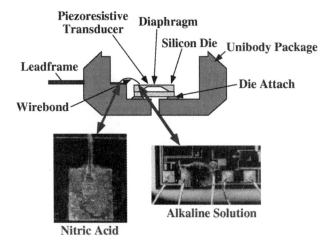

Figure 7.92 Examples of harsh environment-induced failures on pressure sensor. Corrosion is one of the main causes of environmental failures for sensor devices.

addressing this problem; (2) identify the failure mechanisms observed; (3) show examples of using reliability engineering fundamentals to predict lifetime behavior for specific package types in specific media exposure experiments; (4) develop acceleration models for an observed failure mechanism; and (5) compare these results to a customer requirement.

To quantify acceptable long-term media compatibility, a general definition for media compatibility was formulated that can be applied to specific applications: the ability of a pressure sensor to perform its specified electromechanical function over an intended lifetime in the chemical, electrical, mechanical, and thermal environments encountered in a customer's application. All sensors must provide this capability in at least dry, noncorrosive gas media. Specific markets that require additional media compatibility include automotive and/or white goods. Automotive compatible pressure sensors must survive fuel (i.e., organic) exposure in a temperature range from −50 to 150°C for underhood applications. In addition, many are required to withstand exposure to salt water and/or strong acids (e.g., NO_x or SO_x and water), that form electrolytic aqueous solutions. Failure mechanisms are often different for exposure to these two classes of solutions.

White goods applications (i.e., home appliances, like washing machines) most often require exposure to aqueous solutions. For example, a washing machine application requires that the sensor is compatible with an alkaline solution because of the detergents added to tap water.[343,344] One current specification, for a customer who is interested in a water compatible pressure sensor for another white goods application, has been used as a benchmark for our development purposes. The exposure requirements are 2000 hours in a $NaOH/NaHCO_3$ buffered solution[345] with a pH of 11 at 55°C.

Very little data exists for quantifying the media compatibility of a given device.[272–274,346,347] Some of this may be the result of the legality of publishing such data and some may be the result of the difficulty associated with performing such an experiment. Yet, analogous humidity experiments are the standard for quantifying reliability in IC devices. With this in mind, Motorola has been working on: (1) developing experimental test procedures for sensor exposure in these environments; (2) determining the failure mechanisms during exposure testing and modeling these results; and (3) developing a product that has customer applicability.[232]

7.5.1.2 Automotive Environment

In the first set of experiments, various vendor supplied fluorosilicone gel materials used in coating pressure sensor packages were exposed to fuel solutions in a test

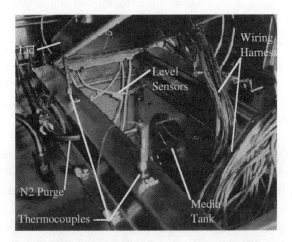

Figure 7.93 Organic solvent (fuel) media exposure system for monitoring pressure sensor offset in situ.

apparatus that was designed as an automated organic solution exposure chamber (Fig. 7.93).

The test apparatus allows in situ offset (i.e., zero applied pressure output voltage) measurements with the ability to provide a supply voltage to the gel-coated devices continuously. An automated mixing and circulating system is used to concoct various test solutions, often per SAE guidelines. For this experiment, reference fuel C with 5% corrosive water and fuel C with 20% ethanol and 5% corrosive water were evaluated. The corrosive water consists of NaCl, Na_2SO_4, and $NaHCO_3$ in DI water. This apparatus is equipped with an external reservoir for temperature control from approximately 10 to 90°C. Most importantly, considerable safety interlocks were designed into the system so that ultimately, if needed, the apparatus could shut itself off unattended. Visual and weight loss measurements were performed before and after these exposure experiments. In addition, manufacturing swelling data has been compared with solubility parameter information on the given solvents to provide an estimate for the solubility parameter of a representative gel.[348]

Pressure sensors with various fluorogels as barrier coatings have been evaluated during exposure to fuel, acid, and combined fuel and acid solutions. The fuels and acids are known to react with the barrier materials.[349] Failure mechanisms expected during the evalution included swelling and dissolution of the polymer, and subsequent corrosion of any exposed metallization. Various fluorosilicone materials were evaluted for each test and lifetime statistics for the various tests computed.

A more general discussion of the observed failure mechanisms with these types of devices is presented elsewhere.[173] Uniform, galvanic, and local corrosion; silicon etching; polymer swelling and/or dissolution; interfacial permeability (e.g., lead leakage[208] and/or adhesive delamination); and mechanical failures (i.e., cracking, creep, fatigue, etc.) all have been observed. In these experiments, polymer swelling and corrosion are observed and modeled in detail.

The swelling of fluorosilicone gels has been analyzed by attempting to determine the solubility parameter of a particular gel. Figure 7.94 shows a plot of manufacturer published swelling data plotted against the solubility parameter of the particular solvent.[348] The solubility parameter is a measure of the interaction energy between species:

$$\delta_i = \left(\frac{\Delta E_i^v}{V_i}\right)^{1/2}, \qquad (7.1)$$

where ΔE_i^v is the energy of vaporization and V_i the molar volume of the species. Because "like dissolves like," a peak in the curve represents the approximate solubility parameter of the polymer. Therefore, the solubility parameter for this particular fluorosilicone gel is between 6 and 8 $(cal/cm^3)^{0.5}$.

In Fig. 7.95, the weight loss, assumed to be caused by loss of low molecular weight species from the solution, is shown for four fluorosilicone gels. The results show that the weight loss for a variety of fluorosilicone gels is not consistent.

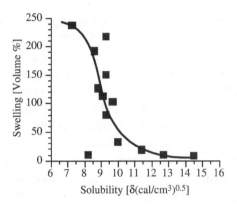

Figure 7.94 Swelling as a function of solubility parameter for a typical silicone gel using material supplier's swelling data following 8 days of exposure at 20°C in various solvents and solubility parameters from Ref. [350].

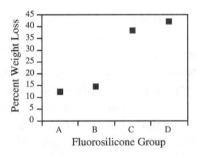

Figure 7.95 Weight loss measurements on four different fluorosilicone gels after 250 hours of exposure to reference fuel C with 5% corrosive water at 65°C.

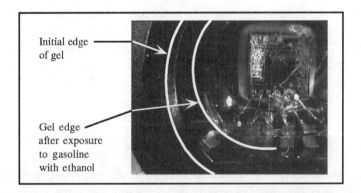

Figure 7.96 Photograph of a pressure sensor device after extended exposure to fuel containing corrosive water, followed by exposure to a strong acid. Evidence of the gel swelling during the test, and corresponding shrinkage after removal from the test media can be seen by the gel retracting away from the sidewall of the package.

Table 7.5 Weibull distribution parameters for fluorosilicone gel coated pressure sensors exposed to fuel and then to nitric acid (pH 1.8 at 85°C).

Fluorosilicone group	Characteristic Life, θ [hours]	$R(t = 48$ hours$)$
A	291	85%
B	51	39%
C	27	16%
D	70	51%

There are at least two potential causes for this result. First, the extent of reaction of the gels varies, so certain gels have more low molecular weight species that will dissolve into solution. Second, the solubility parameters of the polymers are different, resulting in a different degree of dissolution. Also, both of these causes can contribute to the final result. The most significant concern with this variability is the potential for subsequent corrosion that results from the exposure, especially if dissolution (i.e., weight loss) is great enough for metal regions to become exposed (e.g., Fig. 7.96).

A second experiment was performed to address this concern. Automotive applications require that sensors survive both organic solutions (e.g., fuels, oils, etc.) and aqueous solutions (e.g., salt water, NO_x+ water $= HNO_3$, SO_x+ water $= H_2SO_4$, etc.). In this experiment, fluorosilicone gel-filled devices were exposed to fuel for 250 hours and then to nitric acid with a pH of 1.8 at 85°C, which is a standard specification to simulate under-the-hood operation. Failure lifetimes were recorded with in situ offset monitoring. For each group, this lifetime data was assumed to follow a Weibull cumulative failure distribution:

$$R(t) = \exp\left[-\left(\frac{t}{\theta}\right)^{\beta}\right]. \quad (7.2)$$

Statistical likelihood analysis of the lifetime data was used to determine the θ factors (Table 7.5). These results can then be used to compare the performance of devices coated with different fluorosilicone gels. However, this technique can be plagued by the same limitations as traditional corrosion testing (i.e., the difficulty of simulating the field environment with a laboratory experiment). Assuming the test environment produces the same failure mechanism with similar kinetics, these experiments are an invaluable tool for comparing media compatibility schemes and estimating lifetime in a customer application, provided the specifics of that application are known.

7.5.1.3 White Goods Environment

In the second set of experiments, parylene-coated pressure sensors have been used in alkaline exposure experiments. Corrosion under barrier coatings has been observed on several occasions to be the primary failure mode for devices when exposed to electrolytic aqueous solutions. Corrosion under parylene coatings in alkaline solutions was assumed to be analagous to the delamination of paint from metal surfaces and the subsequent corrosion of the metal (e.g., automobile bodies[351,352]). The latter process has been modeled by diffusion of water through the polymer to the metal surface.[352] By solving Fick's First Law of Diffusion, with Ohm's Law as a boundary condition, a "delay time," t, prior to the onset of delamination can be established:[352]

$$t = \frac{\pi b^2}{4D}\left(1 - \frac{V}{4R_0 DFC_0}\right), \tag{7.3}$$

where b is the thickness of the coating, D is the diffusion coefficient of water through the coating, V is the applied voltage, R_0 is the resistivity of the coating, F is Faraday's constant, and C_0 is the concentration of water at the surface of the coating. Our assumption was that delamination occurs prior to the onset of corrosion, similar to the model proposed by Leidheiser et al. (Eq. (7.3) and [352]); therefore, the corrosion failure times observed for these devices should be a function of coating thickness, applied bias, and solution concentration [Eq. (7.3)]. Unfortunately, we know very little about the delamination mechanism for parylene coatings in these environments, so we are assuming that the corrosion failure times are proportional to the delay times in the model proposed by Leidheiser et al.[352]

Therefore, the exposure experiment was designed to examine these variables by using a design of experiments technique.[353,354] Three parylene deposition runs were performed to create sets of devices with three different average parylene thicknesses. A modified Nanometrics Nanospec® program was used to measure thickness. All fabrication, assembly, and surface preparation procedures remained constant within the existing manufacturing tolerances to minimize the effects of these variations on this experiment. Exposure of these devices was performed in individual exposure compartments (Fig. 7.97). All of the closed "cells" were placed in an oven at 55°C. Continuous applied bias (of 5 to 15 Vdc, depending on the device) was applied and offset measurements were performed during the 3022 hours of the experiment.

A second set of media exposure experiments was performed to determine the device lifetime for fluorosilicone gel over parylene C encapsulated sensors. Four separate experiments were used in this evaluation, all of which used a NaOH/NaHCO$_3$ alkaline solution with a pH of 12.5. Three of the experiments are described in detail elsewhere.[175] They include open circuit potential measurements, polarization measurements to determine adhesion strength, and polarization measurements to study the effect of oxygen diffusion on parylene coated

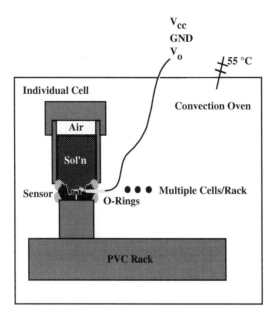

Figure 7.97 Individual compartments for exposure of barrier coated pressure sensors to aqueous solutions. A rack of 64 of these "cells" is placed into an oven at 55°C.

devices.[175] In addition, a media exposure experiment was performed to simulate the customer environment. This experiment used the same NaOH/NaHCO$_3$ (pH = 12.5) solution and apparatus as that shown in Fig. 7.97. Input and output impedance measurements were conducted daily to monitor possible catastrophic corrosion failures. The solution was replenished at approximately 500-hour intervals. All of these devices were biased at 15 Vdc. The customer application is a maximum pH of 11 with a supply voltage of 5 Vdc for at least 2000 hours. Previous work indicated that a conservative acceleration factor of 9× could be used for the corrosion failure mechanisms that are characteristic of this encapsulation system. Therefore, 1% failure rate lifetimes that exceed 225 hours in the environment used for this experiment would be considered "passing." In addition, electrochemical techniques have been used to measure the corrosion rates and give information about the permeability of the coatings.

The use of the open-circuit potential technique and linear polarization has been reported previously.[175] In this work, the use of the linear polarization is described in more detail. Moreover, we introduce chronoamperometry to determine the permeability of the coating to reactants such as water and oxygen and to measure the adhesion strength between the coating and the substrate. In all our measurements a PTFE-block holding 8 devices was used as a test cell. The front side of each device was exposed to the NaOH/NaCO$_3$ solution with no additional applied pressure. Four devices with nominal parylene film thicknesses of 0.5 μm, 1.0 μm, 2.5 μm, and 5.0 μm with fluorosilicone gel and four devices without gel were used.

The actual thicknesses for the gel-filled parts were 0.5 µm, 1.3 µm, 3.68 µm, and 6.14 µm; thicknesses for parylene-only parts were 0.68 µm, 1.33 µm, 3.48 µm, and 6.22 µm.

Linear polarization (linear sweep voltammetry) was performed by varying the working electrode potential at a constant rate of 20 mV/s and measuring the resultant current. The working electrode was always one device lead while the counter and reference electrodes were in the solution (Fig. 7.98). In all our measurements Pt-wire counter electrode and Ag/AgCl reference electrode were used. It is important to note that in the electrochemical measurements, current flows in the solution between the working and the counter electrode and has to overcome the resistance of the coating. In the classical electrical testing of the pressure sensor devices, the two opposing electrodes are two device leads, so there is no contact with the solution and, therefore, no information about the coating. Linear sweep voltammetry was performed in both anodic and cathodic directions. In order to better mimic the actual bias conditions, an anodic (positive) scan was performed on Vcc lead from 0 to 2 V, and a cathodic scan was performed on the ground lead from 0 to −2 V vs. the reference electrode.

Chronoamperometry is another commonly used electroanalytical technique. It is useful for determining diffusion coefficients and for investigating kinetics and mechanisms. We used it in this study as a measure of the permeability of the coating. Chronoamperometry was performed by applying potential steps to the working electrode (device lead) and measuring the current response. In the anodic step, potential to the Vcc device lead was stepped to 2 V vs. the reference electrode. In the cathodic step, potential of −2 V was applied to the ground lead.

Both linear sweep voltammetry and chronoamperometry were performed using an EG&G 283 potentiostat and 270 Research Electrochemistry software. The results are presented in the form of current-voltage $[I = f(E)]$ curves for linear sweep voltammetry and current-time $[I = f(t)]$ curves for chronoamperometry.

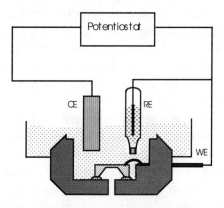

Figure 7.98 Electrochemistry experimental setup. The metallic surfaces on the sensor are the working electrode (WE). A Ag/AgCl reference electrode (RE) and a Pt-wire counter electrode (CE) are used.

The qualification procedure of a media-compatible pressure sensor for a particular white goods customer has been estimated to be 2000 hours of exposure in a NaOH/NaHCO$_3$ buffered solution[174] with a pH of 11 at 55°C. Therefore, an initial set of experiments with parylene-coated sensors was performed for 3022 hours with a continuous supply voltage and in situ offset monitoring. Again, failure distributions are modeled assuming a Weibull cumulative failure function. If all of the Weibull slope parameters are assumed to be 3 [i.e., $\beta = 3$ in Eq. (7.4)], indicating a "wear-out" failure mechanism (e.g., corrosion under a coating), the characteristic lifetimes can be modeled using analysis of variance to develop a linear expression that takes into account the effects of coating thickness, applied bias, and solution concentration [Fig. 7.104(a)]:

$$\begin{aligned}\theta(\text{hrs}) =\ & (15000 \pm 5800) - (400 \pm 340)t_p[\mu m] \\ & +(100 \pm 480)V_{cc}[V] - (1400 \pm 500)C_0[\text{pH}] \\ & -(40 \pm 16)V_{cc}^2[V^2] + (40 \pm 31)t_p \cdot C_0[\mu m \cdot \text{pH}] \\ & +(50 \pm 33)V_{cc} \cdot C_0[V \cdot \text{pH}], \quad R^2 = 0.846.\end{aligned} \quad (7.4)$$

In addition, these distributions can be used to obtain estimates of the 1% failure rate [Fig. 7.99(b)] or the failure rate at 2000 hours [Fig. 7.99(c)].

From these experimental results, the current parylene processing techniques do not produce a device that will satisfy the requirements of this customer. At a pH of 11 and a 5 V supply voltage, the characteristic lifetime [Fig. 7.99(b)] is approximately 450 hours for a 1% failure rate, which was assumed to be our constraint. Devices with fluorosilicone gel encapsulation alone performed similarly to the parylene encapsulated devices.

During these experiments, however, additional devices with a fluorosilicone gel-over-parylene C multilayer coatings were observed to survive 3022 hours of exposure without failure. A subsequent experiment with 37 multilayer coatings with a "worst-case" (i.e., no preparylene-deposition cleaning or adhesion promotion) manufacturing process have been exposed for 500 hours to NaOH/NaHCO$_3$ at pH of 12.5, 55°C, and 15 Vdc without failure. By using a conservative acceleration factor of 9× from Eq. (7.4) (or Fig. 7.99), this represents approximately 4500 hours of exposure in this environment. This estimate assumes that these devices fail with the same mechanism as the parylene encapsulated devices used to develop Eq. (7.4).

An average four-fold improvement in device lifetime was observed when comparing parylene coated devices with parylene coated devices from the same deposition that had fluorosilicone gel placed on top of the parylene. Experimental results, from exposure of devices to NaOH/NaHCO$_3$ solutions with a pH of 12.5 and a temperature of 55°C at 15 Vdc supply voltage, are listed in Table 7.6. As expected for the parylene-only samples, the failure lifetimes increase as a function of thickness. The gel-over-parylene sample results are less dependent upon parylene thickness. The large error values reported are the result of using a small sample size for these experiments. It has been reported that the rate-limiting step for corrosion under polymer coatings is the result of higher diffusion rates at local areas

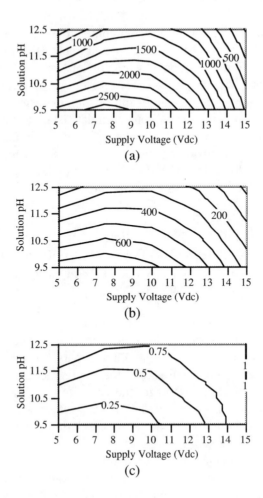

Figure 7.99 Failure analysis for ≈ 5 μm parylene C coated pressure sensors exposed to NaOH/NaHCO$_3$ at 55°C with continuous supply voltage. (a) Weibull characteristic life, θ, as a function of supply voltage and solution pH. (b) Characteristic lifetime for a failure rate of 1% as a function of supply voltage and solution pH. (c) Failure rate at 2000 hours as a function of supply voltage and solution pH.

("effective pinholes"). The longer lifetimes associated with the gel-over-parylene coatings may be the result of two effects: (1) the additional thickness of encapsulant when combining the two materials, and (2) the possibility of low modulus gel filling in any macro defects in the parylene films that would have caused effective pinholes.

The numbers in Table 7.6 are reported for the exposure conditions being used: NaOH/NaHCO$_3$ at pH = 12.5, 55°C, and 15 Vdc. The customer's conditions were a pH of 11 in the same solution at the same temperature with a 5 Vdc supply voltage. To translate these results into those appropriate for the customer, a 9× acceleration factor must be used. Therefore, minimum 1% failure rate lifetimes are on the order of 27000 hours at the customer's specification. This is 13.5×

Table 7.6 Failure times resulting from parylene and gel-over-parylene devices as a function of parylene thickness.

Parylene Thickness (μm)	Results ($t_{1\%}$ in hrs): Parylene only	Results ($t_{1\%}$ in hrs): Gel Over Parylene
0.65 ± 0.025	2 ± 0.3	4000 ± 2000
1.35 ± 0.044	2 ± 0.3	3000 ± 1800
3.49 ± 0.083	500 ± 320	4000 ± 2000
6.1 ± 0.14	1100 ± 510	4000 ± 2100

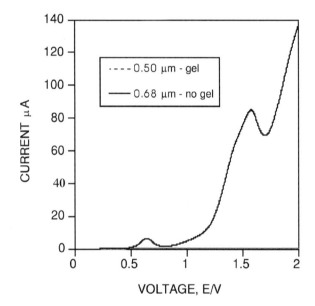

Figure 7.100 Linear Sweep Voltammetry for parylene-only and gel-over-parylene devices.

the customer's lifetime requirements of 2000 hours. The electroanalytical methods were used to rapidly predict the lifetime of devices in the early stages of exposure to alkaline solution. All measurements reported here were taken after the devices were exposed to solution for 36 hours. This "snapshot" of the coating permeability was used to rank different coating combinations and thicknesses, and to predict the relative lifetimes.

In the linear sweep voltammetry measurement, the potential scan from 0 to $+2$ V vs. reference electrode was applied to the V_{cc} lead of the device. The current response in the case of 0.68 mm parylene-only film showed a gradual increase in current and two peaks at approximately $+0.6$ V and $+1.6$ V (Fig. 7.100). In the vicinity of $+2$ V, the anodic evolution of oxygen starts and contributes significantly to large currents.

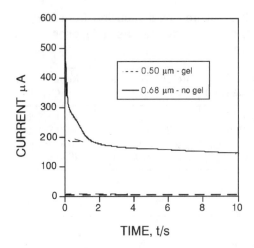

Figure 7.101 Chronoamperometry for parylene-only and parylene-gel films.

While the two peaks have not been assigned yet to specific anodic reactions, it is clear that parylene film alone does not provide adequate barrier protection at this thickness. In contrast, the parylene film of the same thickness with additional gel shows much lower anodic currents and absence of any significant reaction (Fig. 7.100, lower curve).

The anodic scans for the 0.68-μm, 1.33-μm, 3.48-μm, and 6.22-μm parylene-only devices show that the anodic current, which is a measure of the electrochemical activity under the coating, decreases with increasing film thickness, as expected. Furthermore, the two reactions identified for the 0.68-μm parylene-only film are suppressed at thicker films.

The current in the anodic scans for gel-over-parylene devices decreases with film thickness, so that the best barrier protection is provided by the 6.14-μm coating. Compared with parylene-only films, these devices showed currents that were approximately one thousand times lower. Only the 6.22-μm parylene-only film showed behavior similar to 0.5-μm gel-over-parylene coating.

Another comparison of different devices was achieved through the use of chronoamperometry. A typical chronoamperogram is shown in Fig. 7.101 for the case of parylene-only and gel-over-parylene coatings. This figure shows that much larger currents were obtained for parylene-only devices, indicating faster reactant diffusion and higher corrosion rates. After the potential step is applied, the current decreases to a constant value due to diffusion limitations. Plots of the current values at the end of the scan as a function of parylene film thickness reveal that a sharp increase in current occurs at approximately 3 μm, 1 μm for gel-filled devices, and around 3 μm for parylene-only devices.

The linear sweep voltammetry measurements were performed in the cathodic (negative) direction on the ground leads. The two possible reactions in this region involve oxygen reduction and hydrogen evolution. Either one of these reactions can be a depolarizing reaction for the metal dissolution taking place at the anodic

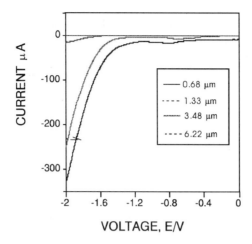

Figure 7.102 Linear sweep voltammetry for parylene-only films.

sites. We found previously that in the biased device, the predominant reaction is the evolution of hydrogen.[175] A comparison of different coating thicknesses and the effect of the gel-over-parylene coating on the diffusion of reactants (water, ions and oxygen) and on the two reactions was performed. Figure 7.102 shows the plot of current vs. potential for all four devices coated with parylene-only film. The peak at approximately −0.6 V for 0.68-μm parylene film is caused by oxygen evolution reaction, while the decrease in current at less than −1 V can be attributed solely to hydrogen evolution reaction.

The comparison of devices with different parylene thicknesses reveals that thicker parylene coating provides better barriers to reactant diffusion and results in lower currents. In particular, the −0.6 V peak current decreases with increased parylene film thickness and the peak disappears for the 3.48-μm parylene film, indicating that the diffusion of oxygen is suppressed. The hydrogen evolution current in the 1–2 V region is also dependent on the parylene film thicknesses, the thicker coating results in lower reaction rates. As in the case of the anodic reaction, the comparison of two devices, one with a 6.22-μm parylene-only film and one with a gel-over-(0.5-μm)-parylene, exhibits similarities in these devices and barrier protection.

The cathodic chronoamperommetry data shows that the cathodic reactions are also diffusion-limited and that the current at −2 V depends strongly on the parylene film thickness and is extremely indicative of the presence of the cooperative gel-over-parylene coating. The best protection is provided by the 6.14-μm gel-over-parylene coating, followed by 3.68-μm gel-over-parylene coating, 1.30-μm gel-over-parylene coating and 0.5-μm gel-over-parylene coating. The parylene-only coatings did not provide adequate protection except for the 6.22-μm coating that exhibits behavior very similar to 0.5-μm gel-over-parylene coating.

Although reliability testing is not used as often to test outgoing devices, it can be a bottleneck because of the length of testing time required. Standard pressure sensor reliability testing performed by Motorola includes several humidity test conditions, high and low temperature storage, temperature cycling, and pulsed pressure temperature cycling with bias (PPTCB).[171] Long term testing of a "standard" distribution of manufacturing devices is used in initial qualification of a device type. For example, multiple wafer/assembly lots would be used for a qualification. Random devices would be extracted from these lots to use in reliability tests. Many of these tests are 1000 hours in duration, so they cannot occur on all outgoing product. When significant changes in the process flow for a given device are made, requalification of the device line is performed to ensure that the change has not affected the level of reliability of the device.

There are a number of limitations to this process. Capacity is often limited because of the length of the tests. Furthermore, because of the wide variety of applications for pressure sensors, some of these tests do not adequately reflect the customer's application. Also, testing is performed to a defined test limit. For example, the devices are defined as "pass" or "fail" depending on electrical performance compared with a specification following the exposure. Because of the evaluation complexity, these experiments are performed ex situ. This may be limiting in the case of a failure mechanism that involves the effect of a packaging material during exposure on the device electrical output (e.g., swelling of a gel in a humid environment). After removal from the test, the effect on the material may disappear, and the failure mechanism may not be observed. Therefore, a concerted effort to do more in situ electrical monitoring is being pursued.

The media compatibility of secondary diaphragm isolated pressure sensors is determined by the diaphragm and housing materials. The most widely used diaphragm isolation material is 316L stainless steel, which can tolerate a wide range of commonly used fluids, such as water, oils, hydraulic fluids, foodstuffs, fuels, solvents, and numerous process chemicals.[355] However, extremely harsh chemicals, such as sulphuric acid, hydrochloric acid, and chloride compounds, will even corrode steel.[355] For the most demanding chemical environments, PTFE (and the mold processable form, PFA) fluoropolymers are the material of choice. PTFE is unaffected by virtually all acids, bases, solvents, and other chemicals. No known material will dissolve PTFE, but it may be degraded by molten salts and strong oxidizers such as chlorine trifluoride. PTFE can be used continously at temperatures as high as 260°C.

In situ testing is the preferred method for all media compatibility experiments that are performed on our pressure sensors.[173–175] Too much information is lost because a failure mechanism is not observed or because an accurate failure time cannot be established with ex situ electrical observation. This type of experimentation is extremely tricky with some of the environments in which pressure sensors are tested. For example, we perform in situ electrical testing of devices in a variety of aqueous (i.e., acidic, tap water, saline, alkaline, etc.) and organic (i.e., fuels, oils, etc.) environments. Aqueous solution testing can be complex because of the potential for corrosion on the device and the test system. Organic solution testing is difficult because of the volatility of the solutions. Safety engineering is essential to minimize the potential flammability problem. In situ testing provides an accurate picture of the failure times, which are the required information for the use of reliability engineering principles. Establishment of failure distributions and acceleration factors has aided product development.

This information also allows a standard for media compatibility to be established.[173] Presently, there is no standard for defining a media-compatible pressure sensor. Admittedly, lifetimes of stainless steel/silicone oil packaged pressure sensors are probably much longer than the customer's intended application lifetimes. However, as more customers find that they cannot use these devices because of their relatively high cost and/or somewhat limited performance,[184] more well-defined standards will be necessary for customers to be able to evaluate low-cost devices. In fact, until some customers see a standard, they may be reluctant to use a low-cost media-compatible pressure sensor. When one considers the potential market for low-cost silicon pressure sensors that are media-compatible (estimated at 50–90% of the total pressure sensor market in 5–10 years), the need for a standard to drive acceptability of these devices is more apparent.

Building these custom test systems is truly a cross-functional acitivity. For example, our testing is limited to atmospheric pressure media exposure testing (see[173–175] for details). The development of a pressurized media test system must include mechanical engineering, polymer engineering, corrosion engineering, chemical engineering, equipment engineering, safety and environmental engineering, and hardware and software engineering. Furthermore, these systems are expensive. They are on the order of several hundreds of thousands of dollars per system. However, the alternative for a single experiment at an external company can be equally expensive, and the learning experience of performing the experiment is lost.

7.5.2 ACCELEROMETERS AND VIBRATION AND SHOCK TESTING

Accelerometers have many of the same reliability requirements as other sensors, including humidity testing, temperature cycling, and high and low temperature storage. In addition, they must also survive mechanical shock testing. As with other MEMS, outside of product qualification reports, little is published on the subject.

At least one published account described how fabrication processes can influence the mechanical shock survivability of an accelerometer. For example, the fabrication of microstructures with a boron diffusion and subsequent removal of this boron-diffused layer have been observed to increase the survival rate in mechanical shock of single-crystal silicon, bulk dissolution fabricated accelerometers by 26.5%.[356]

Reliability of accelerometers has been addressed in some work performed by Jet Propulsion Laboratory for space systems applications.[357] This paper provides the following comparison of the automotive environment and the space environment (Table 7.7). In addition to this comparison and the work that they present concerning qualification of space-based systems, they also make a very important

Table 7.7 Automotive and space environment qualification requirements.[357]

Environmental parameter	Automotive	Space Example: STRV-2
Operating Temperature	−40 to 125°C	−25 to 40°C
Thermal Cycling	>1,000	>5,000
Humidity	Up to 100%	35%–60%
Vibration	15g, 10–200 Hz	Up to 0.4g, 20 to 2000 Hz
EMI Protection	Up to 200 V/m	Up to 70 V/m
Shock	N/A	20g, 100 Hz 2090g, 2–10 kHz
Radiation	N/A	10^6 rads/year
Depressurization	N/A	1 atm to vacuum at 1 psi/sec

point about the cost of commercialization of a MEMS product: "the cost of developing packaging and qualification methods to ensure reliability in the harsh space environment can only be justified for technologies that significantly increase performance or enable new functions not possible with conventional technologies."[357] This is not only true with space applications but with most other applications as well.

7.5.3 MICROFLUIDICS

Microfluidic devices may have high reliability requirements in medical and life-critical applications. For example, if assay chips are used to screen for pathogens in the blood supply, any failure to detect harmful viruses could result in an infected patient. Considering that hundreds of thousands of blood bags are tested each year in the U.S., reliability must be measured in tens or hundreds of millions of readings or more before failure. Likewise, a drug delivery system must exhibit very high reliability, and in the event of failure, should fail in a safe manner, by not releasing the drug, and in life critical situations, provide warning to the recipient so that corrective action may be taken. For one-time use devices, shelf-life becomes important. A disposable device typically must have a shelf life of several years before use. i-STAT Corporation provides a hand-held blood test system based on disposable test cartridges. Their cartridge contains its own calibration fluids, which are released prior to a blood test to calibrate and verify optimal performance of the device just before the blood test is performed.

7.5.4 OPTICAL MEMS

In the example of the Texas Instruments' Digital Micromirror Device™, several specialized reliability tests were developed to test its unique failure mechanisms.[305]

Many of these failure mechanisms are not observed in typical military specifications. Very little additional data has been presented in the literature on the reliability of MOEMS.

7.6 COST MODEL AND SUMMARY

As mentioned at the outset, the cost of packaging, trim, and test for MEMS and MOEMS can represent from 50% to 90% of the cost of a MEMS/MOEMS product; therefore, a significant amount of effort required when developing a product is on the "back-end" of the product as opposed to the "front-end," where much of the research to date has been focused. To illustrate this point, the following simple cost model has been developed.

First, it is assumed that a product mix of three products is being developed (Table 7.8).

Products 1 and 3 are assumed to be in development and will not be produced for another 2 years. Product 1 is assumed to be a CMOS integrated MEMS device (i.e., single chip in a package). Product 3 is assumed to be a 2-chip MEMS device (i.e., sensor and circuit on separate dies within a single package). Product 2 is already in production and is ramping up to high volume production, and it is assumed to be a nonintegrated MEMS device (also single chip in a package).

Table 7.9 shows an estimated die cost for the three products.

Package and test cost is estimated by calculating the labor cost, materials cost, and equipment depreciation. Tables 7.10–7.12 list several assumptions to make these estimates. Table 7.13 shows the package/test cost ("cost less chip"—CLC), and Table 7.14 shows the profit and loss for each product.

A few conclusions can be drawn from this contrived example for cost of a MEMS/MOEMS product. First, CLC can represent a significant amount of the product cost. The values assumed for die size, yields, direct labor salaries, throughput, utilization, materials cost, and equipment cost may not be exact, but they are realistic for MEMS/MOEMS products. Simply by using these values as a basis for the cost model, one can see that the package/test cost represents from 40 to 65% of the total cost for each product. Furthermore, trim and test is often 50% of that CLC value.

Die sizes/wafer cost are being reduced by integration of the MEMS/MOEMS and the control circuitry. Recently, research has opened up several new ways of integration, including MEMS first, mixed processing, MEMS last, and "disintegration" (i.e., low temperature wafer bonding of a MEMS/MOEMS device to a circuit die after fabrication is used for integration).

Table 7.8 Assumed product mix.

Years	200a	200b	200c	200d	200e
Product 1					
Volume (M/yr)	0	0	1	2	4
Volume (K/wk)	0	0	20	40	80
Product 2					
Volume (M/yr)	0.5	1	2	2	2
Volume (K/wk)	10	20	40	40	40
Product 3					
Volume (M/yr)	0	0	1	2	4
Volume (K/wk)	0	0	20	40	80

Table 7.9 Estimated die cost.

Years	200a	200b	200c	200d	200e
Product 1	CMOS Integrated MEMS				
Die size (mils)	140	140	140	140	140
PDPW	2051	2051	2051	2051	2051
Yield	80%	85%	88%	90%	92%
DPW	1640	1743	1794	1845	1876
Wafer cost ($/wafer)	$2,000	$1,700	$1,500	$1,400	$1,250
Wafer cost ($/sq in)	$39.79	$33.83	$29.85	$27.86	$24.87
Die cost ($/die)	$1.22	$0.98	$0.84	$0.76	$0.67
Product 2	Nonintegrated MEMS				
Die size (mils)	125	125	125	105	105
PDPW	2573	2573	2573	3647	3647
Yield	80%	85%	88%	90%	92%
DPW	2058	2187	2251	3282	3337
Wafer cost ($/wafer)	$1,500	$1,300	$1,150	$1,050	$1,000
Wafer cost ($/sq in)	$29.85	$25.87	$22.88	$20.89	$19.90
Die cost ($/die)	$0.73	$0.60	$0.52	$0.32	$0.30
Product 3	CMOS MEMS 2-Chip Product				
Control IC					
Die size (mils)	125	125	125	125	125
PDPW	2573	2573	2573	2573	2573
Yield	80%	85%	88%	90%	92%
DPW	2058	2187	2251	2315	2354
Wafer cost ($/wafer)	$1,340	$1,139	$1,005	$938	$838
Wafer cost ($/sq in)	$26.66	$22.66	$20.00	$18.67	$16.67
Die cost ($/die)	$0.66	$0.53	$0.45	$0.41	$0.36
MEMS chip					
Die size (mils)	75	75	75	75	75
PDPW	7148	7148	7148	7148	7148
Yield	80%	85%	88%	90%	92%
DPW	5718	6075	6254	6433	6540
Wafer cost ($/wafer)	$1,500	$1,300	$1,150	$1,050	$1,000
Wafer cost ($/sq in)	$29.85	$25.87	$22.88	$20.89	$19.90
Die cost ($/die)	$0.27	$0.22	$0.19	$0.17	$0.16
Total silicon cost ($/unit)	$0.93	$0.75	$0.64	$0.58	$0.52

Table 7.10 Estimated direct labor costs.

Tools	Troughput (UPH)	Utilization	People per Operation		Years	200a	200b	200c	200d	200e
					Product 1					
					Volume (M/yr)	0	0	1	2	4
					Volume (K/wk)	0	0	20	40	80
					Product 2					
					Volume (M/yr)	0.5	1	2	2	2
					Volume (K/wk)	10	20	40	40	40
					Product 3					
					Volume (M/yr)	0	0	1	2	4
					Volume (K/wk)	0	0	20	40	80
					Total Number of People					
Laser mark	1000	0.7	1			1	1	1	2	2
Wafer saw	2000	0.7	1			1	1	1	1	2*
Die attach	2000	0.7	1			1	1	1	1	2†
Wire bond	1000	0.7	1			1	1	1	2	3**
Pre-bake	4000	0.7	0			1	1	1	1	1
Encapsulant dispense	400	0.7	1			1	1	2	3	5
Bake	4000	0.7	0			1	1	1	1	3
Cap attach	600	0.7	1			1	1	2	2	3
Leadform/singulate	500	0.7	1			1	1	2	3	4
Trim	100	0.7	1			0	0	4	7	14††
Tri temp test	250	0.7	1			1	1	3	5	7
					Direct labor ($/K)					
					Product 1	$ –	$ –	$0.06	$0.08	$0.09
					Product 2	$1.00	$0.50	$0.09	$0.06	$0.03
					Product 3	$ –	$ –	$0.06	$0.08	$0.09

Assumptions: Hours/week 168; Bisiness Growth/start-up Salary 50 $K/yr; Weeks/Yr 50.

* 2 Saw operations for Product 3; † 2 Die attach operations for Product 3; ** 2 Wirebond operations for Product 3; †† No trim for Product 2.

Table 7.11 Estimated materials cost.

Years	200a	200b	200c	200d	200e
Product 1					
Volume (M/yr)	0	0	1	2	4
Volume (K/wk)	0	0	20	40	80
Product 2					
Volume (M/yr)	0.5	1	2	2	2
Volume (K/wk)	10	20	40	40	40
Product 3					
Volume (M/yr)	0	0	1	2	4
Volume (K/wk)	0	0	20	40	80

Product 1	Material Cost	Units	$/unit
Laser mark	0		
Wafer saw	5	$/wafer	$ 0.00
Die attach	0.05	$/die attach	$ 0.05
Wire bond	0.03	$wirebond	$ 0.03
Pre-bake	5	$/K	$ 0.01
Encapsulant dispense	0.1	$/unit	$ 0.10
Bake	5	$/K	$ 0.01
Cap attach	0.02	$/unit	$ 0.02
Leadform/singulate	5	$/leadframe	$ 0.50
Trim	0		
Tri temp test	0		
Total materials ($/unit)			$ 0.71

Product 2	Material Cost	Units	$/unit	$/unit(*)
Laser mark	0			
Wafer saw	5	$/wafer	$ 0.00	$ 0.00
Die attach	0.05	$/die attach	$ 0.05	$ 0.05
Wire bond	0.03	$wirebond	$ 0.03	$ 0.03
Pre-bake	5	$/K	$ 0.01	$ 0.01
Encapsulant dispense	0.1	$/unit	$ 0.10	$ 0.10
Bake	5	$/K	$ 0.01	$ 0.01
Cap attach	0.02	$/unit	$ 0.02	$ 0.02
Leadform/singulate	5	$/leadframe	$ 0.50	$ 0.50
Trim	0			
Tri temp test	0			
Total materials ($/unit)			$ 0.71	$ 0.71

Product 3	Material Cost	Units	$/unit
Laser mark	0		
Wafer saw	5	$/wafer	$ 0.00[†]
Die attach	0.05	$/die attach	$ 0.10**
Wire bond	0.03	$wirebond	$ 0.06[††]
Pre-bake	5	$/K	$ 0.01
Encapsulant dispense	0.1	$/unit	$ 0.10
Bake	5	$/K	$ 0.01
Cap attach	0.02	$/unit	$ 0.02
Leadform/singulate	5	$/leadframe	$ 0.50
Trim	0		
Tri temp test	0		
Total materials ($/unit)			$ 0.79

Assumptions: Hours/week 168; Business growth/start-up Salary 50 $K/yr; Weeks/Yr 50; Units/leadframe 10.

*After die shrink; [†] 2 Saw Operations for Product 3; ** 2 Die Attach Operations for Product 3; [††] 2 Wirebond Operations for Product 3.

Direct labor salaries are being reduced by moving assembly and test areas offshore. For example, both Motorola Sensor Products Division (for pressure sensors and accelerometers) and Analog Devices (for accelerometers) use manufacturing facilities in the Far East to reduce labor costs.

Table 7.12 Estimated equipment back-end depreciation.

Capital ($M)	200a	200b	200c	200d	200e	Years	200a	200b	200c	200d	200e
Laser mark	0.25	0	0	0.25	0	Product 1					
Wafer saw	0.8	0	0	0	0	Volume (M/yr)	0	0	1	2	4
Die attach	0.25	0	0	0	0	Volume (K/wk)	0	0	20	40	80
Wire bond	0.3	0	0	0.3	0	Product 2					
Pre-bake	0.08	0	0	0	0	Volume (M/yr)	0.5	1	2	2	2
Encapsulant dispense	0.3	0	0.3	0.3	0.6	Volume (K/wk)	10	20	40	40	40
Bake	0.08	0	0	0	0	Product 3					
Cap attach	0.2	0	0.2	0	0.2	Volume (M/yr)	0	0	1	2	4
Leadform/singulate	0.4	0	0.4	0.4	0.4	Volume (K/wk)	0	0	20	40	80
Trim	0.6	0.6	3	2.4	4.2						
Tri temp test	0.6	0	1.2	1.2	1.2						
Total	3.86	0.6	5.1	4.85	6.6						
Contribution/Year*	0.772	0.89	1.912	2.882	4.202						

Tools	Troughput (UPH)	Utilization	Est. Cost ($K)	Number of Tools				
				200a	200b	200c	200d	200e
Laser mark	1000	0.7	250	1	1	1	2	2
Wafer saw	2000	0.7	800	1	1	1	1	1
Die attach	2000	0.7	250	1	1	1	1	1
Wire bond	1000	0.7	300	1	1	1	2	2
Pre-bake	4000	0.7	80	1	1	1	1	1
Encapsulant dispense	400	0.7	300	1	1	2	3	5
Bake	4000	0.7	80	1	1	1	1	1
Cap attach	600	0.7	200	1	1	2	2	3
Leadform/singulate	500	0.7	400	1	2	2	3	4
Trim	100	0.7	600	1	1	7	11	18
Tri temp test	250	0.7	600	1	1	3	5	7

Assumptions: Hours/week 168; Business growth/start-up.

* 5 Year Straight Line.

Table 7.13 Estimated cost less chip (CLC).

Years	200a	200b	200c	200d	200e
Product 1	CMOS Integrated MEMS				
Direct labor	$ –	$ –	$ 0.00	$ 0.00	$ 0.00
Materials & supplies	$ 0.71	$ 0.71	$ 0.71	$ 0.71	$ 0.71
Depreciation	$ 0.08	$ 0.04	$ 0.00	$ 0.00	$ 0.00
Unyielded CLC	$ 0.79	$ 0.76	$ 0.71	$ 0.71	$ 0.71
Yield	75%	80%	83%	84%	84%
Yielded CLC	$ 1.05	$ 0.95	$ 0.86	$ 0.85	$ 0.85
Product 2	Nonintegrated MEMS				
Direct labor	$ 1.00	$ 0.50	$ 0.09	$ 0.06	$ 0.03
Materials & supplies	$ 0.71	$ 0.71	$ 0.71	$ 0.71	$ 0.71
Depreciation	$ 0.08	$ 0.04	$ 0.00	$ 0.00	$ 0.00
Unyielded CLC	$ 1.79	$ 1.72	$ 0.81	$ 0.77	$ 0.74
Yield	75%	80%	83%	84%	84%
Yielded CLC	$ 2.39	$ 1.57	$ 0.98	$ 0.92	$ 0.88
Product 3	CMOS MEMS 2-Chip Product				
Direct labor	$ –	$ –	$ 0.06	$ 0.08	$ 0.09
Materials & supplies	$ 0.79	$ 0.79	$ 0.79	$ 0.79	$ 0.79
Depreciation	$ 0.08	$ 0.04	$ 0.00	$ 0.00	$ 0.00
Unyielded CLC	$ 0.87	$ 0.84	$ 0.85	$ 0.87	$ 0.88
Yield	75%	80%	83%	84%	84%
Yielded CLC	$ 1.16	$ 1.05	$ 1.03	$ 1.04	$ 1.05

Assumptions: 1 Assembly & test line for all three products.

Also, standards are not yet in place for packages, test and trim equipment, and reliability for MEMS/MOEMS. Custom packaging and test will continue to keep depreciation costs high, which will limit the profitability of MEMS/MOEMS devices. In addition, equipment vendors are not likely to be able to afford development of high yield, high throughput, low-cost equipment until standard equipment for MEMS/MOEMS becomes more common.

Yet, as can be seen from the content of this chapter (compared with the rest of this book), there is very little work on MEMS/MOEMS packaging and test. Throughout this chapter, we reviewed the literature that describes the general package/test processes that are being applied to MEMS/MOEMS devices. We showed examples of packaging for pressure sensors, accelerometers, micromirrors, other MOEMS, microfluidics, microvalves, and microswitches. We reviewed the test and reliability literature for MEMS/MOEMS, including an extensive section on media compatibility. Finally, we presented a contrived cost model for a MEMS/MOEMS application to show the importance of packaging and test in the final cost of a MEMS/MOEMS device. Typically, package and test solutions for MEMS/MOEMS have been custom and proprietary. Standardized solutions must become available if MEMS/MOEMS devices are going to meet the expectations of many marketing reports (see Fig. 7.1).

Table 7.14 Estimated profit and loss statement for each product.

Years	200a	200b	200c	200d	200e
Product 1					
Volume (M/yr)	0	0	1	2	4
Volume (K/wk)	0	0	20	40	80
Product 2					
Volume (M/yr)	0.5	1	2	2	2
Volume (K/wk)	10	20	40	40	40
Product 3					
Volume (M/yr)	0	0	1	2	4
Volume (K/wk)	0	0	20	40	80
Product 1					
ASP	$ 7.00	$ 5.50	$ 4.75	$ 4.00	$ 3.50
Die Cost	$ 1.22	$ 0.98	$ 0.84	$ 0.76	$ 0.67
CLC	$ 1.05	$ 0.95	$ 0.86	$ 0.85	$ 0.85
Overhead	$ 0.34	$ 0.29	$ 0.26	$ 0.24	$ 0.23
Total Cost	$ 2.61	$ 2.22	$ 1.96	$ 1.85	$ 1.75
Profit Margin	63%	60%	59%	54%	50%
Product 2					
ASP	$ 4.00	$ 3.50	$ 3.00	$ 2.75	$ 2.45
Die Cost	$ 0.73	$ 0.60	$ 0.52	$ 0.32	$ 0.30
CLC	$ 2.39	$ 1.57	$ 0.98	$ 0.92	$ 0.88
Overhead	$ 0.47	$ 0.33	$ 0.22	$ 0.19	$ 0.18
Total Cost	$ 3.58	$ 2.50	$ 1.72	$ 1.42	$ 1.36
Profit Margin	10%	29%	43%	48%	44%
Product 3					
ASP	$ 7.00	$ 5.50	$ 4.75	$ 4.00	$ 3.50
Die Cost	$ 0.93	$ 0.75	$ 0.64	$ 0.58	$ 0.52
CLC	$ 1.16	$ 1.05	$ 1.03	$ 1.04	$ 1.05
Overhead	$ 0.31	$ 0.27	$ 0.25	$ 0.24	$ 0.24
Total Cost	$ 2.40	$ 2.07	$ 1.92	$ 1.86	$ 1.80
Profit Margin	66%	62%	60%	53%	48%

Assumptions: Overhead 15% of Cost.

REFERENCES

1. L. Ristic and M. Shah, "Trends in MEMS Technology," in *WESCON 96 Conference Proceedings*, pp. 64–72 (1996).
2. S. D. Senturia and R. L. Smith, "Microsensor Packaging and System Partitioning," *Sensors and Actuators* 15, 221–234 (1988).
3. J. Swift, "Rump Session," *Solid-State Sensors and Actuators Workshop* (1998).
4. J. Schuster, "Rump Session," *Solid-State Sensors and Actuators Workshop* (1998).

5. M. Madou, *Fundamentals of Microfabrication*. CRC Press, Boca Raton, FL (1997).
6. R. Allan, "Silicon MEMS Technology Is Coming of Age Commercially," *Electronic Design* 20(January), 75–88 (1997).
7. H. Reichl, "Packaging and Interconnection of Sensors," *Sensors and Actuators* A25–A27, 63–71 (1991).
8. C. Song, "Commercial Vision of Silicon Based Inertial Sensors," paper presented at *Transducers '97, International Conference on Solid-State Sensors and Actuators*, pp. 839–842 (1997).
9. D. S. Eddy and D. R. Sparks, "Application of MEMS Technology in Automotive Sensors and Actuators," *Proceedings of the IEEE* 86, 1747–1755 (1998).
10. G. Beardmore, "Packaging for Microengineered Devices: Lessons from the Real World," in *IEEE Colloqium on Assembly and Connection in Microsystems*, pp. 2/1–2/8 (1997).
11. R. E. Bicking, L. E. Frazee and J. J. Simonelic, "Sensor Packaging for High Volume Applications," in *Technical Digest: Transducers '85*, pp. 350–353 (1985).
12. K. D. Wise, "Integrated Microelectromechanical Systems: A Perspective on MEMS in the 90s," in *Proceedings of the IEEE Micro Electro Mechanical Systems*, pp. 33–38 (1991).
13. *Solid-State Sensors and Actuators Workshop* (1998).
14. *Commercialization of Microsystems* (1998).
15. *Commercialization of Microsystems* (1996).
16. B. Detlefs, "MEMS 1998: Emerging Applications and Markets," System Planning Corporation 19 November (1998).
17. "NEXUS—Results of the Market Analysis, Executive Summary," (1996).
18. R. R. Tummala and E. J. Rymaszewski, *Microelectronic Packaging Handbook*. Van Nostrand, Reinhold, New York (1989).
19. W. H. Ko, "Packaging of Microsensors," in *Microsystems Technologies '94: Proceedings 4th International Conference and Exhibition on Micro Electro Mechanical Systems and Components*, pp. 477–480 (1994).
20. A. Morrissey, G. Kelly and J. Alderman, "Low-Stress 3D Packaging of a Microsystem," *Sensors and Actuators* A68, 404–409 (1998).
21. J. Romig, A. D. P. V. Dressendorfer and D. W. Palmer, "High Performance Microsystem Packaging: A Perspective," *Microelectronic Reliability* 37, 1771–1781 (1997).
22. A. Bossche, C. V. B. Cotofana and J. R. Mollinger, "MEMS Packaging: State of the Art and Future Trends," paper presented at *SPIE Conference on Smart Electronics and MEMS*, pp. 166–173 (1998).
23. M. A. Schmidt, "Wafer-to-Wafer Bonding for Microstructure Formation," *Proceedings of the IEEE* 86, 1575–1585 (1998).
24. W. H. Ko, J. T. Suminto and G. J. Yeh, "Bonding Techniques for Microsensors," in *Micromachining and Micropackaging of Transducers*, C. D. Fung,

P. W. Cheung, W. H. Ko and D. G. Fleming, eds., pp. 41–61. Elsevier, Amsterdam (1985).
25. L. Ristic, *Sensor Technology and Devices*, pp. 524. Artech House, Boston, MA (1994).
26. P. W. Barth, "Silicon Fusion Bonding for Fabrication of Sensors, Actuators and Microstructures," *Sensors and Actuators* A21–A23, 919–926 (1990).
27. C. A. Desmond and P. Abolghasem, "Applications of Wafer Bonding to Non-Electronic Microstructures," in *Proceedings of the 4th International Symposium on Semiconductor Wafer Bonding*, pp. 95–105 (1997).
28. M. Shimbo, K. Furukawa, K. Fukuda and K. Tanzawa, "Silicon-to-Silicon Direct Bonding Method," *Journal of Applied Physics* 60, 2987–2989 (1986).
29. M. A. Huff, A. D. Nikolich, and M. A. Schmidt, "A Threshold Pressure Switch Utilizing Plastic Deformation of Silicon," paper presented at *1991 International Conference on Solid-State Sensors and Actuators*, pp. 177–180 (1991).
30. G.-S. Chung, S. Kawahito, M. Ashiki, M. Ishida and T. Nakamura, "Novel High-Performance Pressure Sensors Using Double SOI Structures," paper presented at *1991 International Conference on Solid-State Sensors and Actuators*, pp. 676–681 (1991).
31. J. A. Folta, N. F. Raley and E. W. Hee, "Design, Fabrication and Testing of a Miniature Peristaltic Membrane Pump," paper presented at *Solid-State Sensor and Actuator Workshop*, pp. 186–189 (1992).
32. L. Parameswaran, V. M. McNeil, A. Huff and M. A. Schmidt, "Sealed-Cavity Microstructure Using Wafer Bonding Technology," paper presented at *7th International Conference on Solid-State Sensors and Actuators* (1993).
33. L. Parameswaran, A. Mirza, W. K. Chan and M. A. Schmidt, "Silicon Pressure Sensors Using a Wafer-Bonded Sealed Cavity Process," paper presented at *8th International Conference on Solid-State Sensors and Actuators, and Eurosensors IX*, pp. 582–585 (1995).
34. C. H. Hsu and M. A. Schmidt, "Micromachined Structures Fabricated Using a Wafer-Bonded Sealed Cavity Process," paper presented at *Solid-State Sensor and Actuators Workshop*, pp. 151–155 (1994).
35. H. Baumann, S. Mack and H. Münzel, "Bonding of Structured Wafers," in *Proceedings of the Third International Symposium on Semiconductor Wafer Bonding: Physics and Applications*, pp. 471–487 (1995).
36. H. Takagi, R. Maeda, Y. Ando and T. Suga, "Room Temperature Silicon Wafer Direct Bonding in Vacuum by Ar Beam Irradiation," pp. 191–196 (1997).
37. T. Suga, Y. Ishii and N. Hosoda, "Microassembly System for Integration of MEMS Using the Surface Activated Bonding Method," *IEICE Transactions on Electronics* E80-C, 297–302 (1997).
38. H. Takagi, R. Maeda, T. R. Chung and T. Suga, "Silicon Wafer Bonding at Room Temperature by Ar Beam Surface Activation in Vacuum," in *Proceedings of the 4th International Symposium on Semiconductor Wafer Bonding*, pp. 393–400 (1997).

39. A. Plößl, H. Stenzel, Q.-Y. Tong, M. Langenkamp, C. Schmidthals and U. Gösele, "Covalent Silicon Bonding at Room Temperature in Ultrahigh Vacuum," in *Materials Research Society Symposium Proceedings*, pp. 141–146 (1998).
40. M. J. Vellekoop and P. M. Sarro, "Technologies for Integrated Sensors and Actuators," *Proceedings of SPIE*, pp. 536–547 (1998).
41. Q.-Y. Tong, W. J. Kim, T.-H. Lee and U. Gösele, "Low Vacuum Wafer Bonding," *Electrochemical and Solid-State Letters* 1, 52–53 (1998).
42. Q.-Y. Tong, G. Cha, R. Gafiteanu and U. Gösele, "Low Temperature Wafer Direct Bonding," *Journal of Microelectromechanical Systems* 3, 29–35 (1994).
43. S. Mack, H. Baumann, U. Gösele, H. Werner and R. Schlögl, "Analysis of Bonding-Related Gas Enclosure in Micromachined Cavities Sealed by Silicon Wafer Bonding," *Journal of the Electrochemical Society* 144, 1106–1111 (1997).
44. H. Henmi, S. Shoji, Y. Shoji, K. Yoshima and M. Esashi, "Vacuum Packaging for Microsensors by Glass-Silicon Anodic Bonding," *Sensors and Actuators* A43, 243–248 (1994).
45. M. Nese, R. W. Bernstein, I.-R. Johansen and R. Spooren, "New Method for Testing Hermeticity of Silicon Sensor Structures," *Sensors and Actuators* A53, 349–352 (1996).
46. F. Secco d'Aragona, T. Iwamoto, H.-D. Chiou and A. Mirza, "A Study of Silicon Direct Wafer Bonding for MEMS Applications," in *Electrochemical Society Proceedings*, pp. 127–134 (1997).
47. M. A. Schmidt, "Silicon Wafer Bonding for Micromechanical Devices," paper presented at *Solid-State Sensor and Actuator Workshop*, pp. 127–131 (1994).
48. V. L. Spiering, J. W. Berenschot, M. Elwenspoek and J. H. J. Fluitman, "Sacrificial Wafer Bonding for Planarization after Very Deep Etching," in *Proceedings of the Third International Symposium on Semiconductor Wafer Bonding: Physics and Applications*, pp. 497–508 (1995).
49. S. Shoji, H. Kikuchi and H. Torigoe, "Anodic Bonding Below 180°C for Packaging and Assembling of MEMS Using Lithium Aluminosilicate-β-Quartz Glass-Ceramic," pp. 482–487 (1997).
50. G. Wallis and D. I. Pomerantz, "Field Assisted Glass-Metal Sealing," *Journal of Applied Physics* 40, 3946–3949 (1969).
51. T. A. Knecht, "Bonding Techniques for Solid State Pressure Sensors," in *Transducers '87*, pp. 95–98 (1987).
52. P. R. Younger, "Hermetic Glass Sealing by Electrostatic Bonding," *Journal of Non-Crystalline Solids* 38–39, 909–914 (1980).
53. Y.-C. Lin, P. J. Hesketh and J. P. Schuster, "Finite-Element Analysis of Thermal Stresses in a Silicon Pressure Sensor for Various Die-Mount Materials," *Sensors and Actuators* A44, 145–149 (1994).

54. N. Ito, K. Yamada, H. Okada, M. Nishimura and T. Kuriyama, "A Rapid and Selective Anodic Bonding Method," paper presented at *8th International Conference on Solid-State Sensors and Actuators, and Eurosensors IX*, pp. 277–280 (1995).
55. T. Sasayama, S. Suzuki, S. Tsuchitani, A. Koide, M. Suzuki, T. Nakazawa and N. Ichikawa, "Highly Reliable Silicon Micromachined Physical Sensors in Mass Production," *Sensors and Actuators* A54, 714–717 (1996).
56. K. B. Albaugh, P. E. Cade and D. H. Rasmussen, "Mechanisms of Anodic Bonding of Silicon to Pyrex® Glass," paper presented at *Solid-State Sensor and Actuator Workshop*, pp. 109–110 (1988).
57. P. Krause, M. Sporys, E. Obermeier, K. Lange and S. Grigull, "Silicon to Silicon Anodic Bonding Using Evaporated Glass," paper presented at *8th International Conference on Solid-State Sensors and Actuators, and Eurosensors IX*, pp. 228–231 (1995).
58. S. Tatic-Lucic, J. Ames, B. Boardman, D. McIntyre, P. Jaramillo, L. Starr and M. Lim, "Bond-Quality Characterization of Silicon-Glass Anodic Bonding," *Sensors and Actuators* A60, 223–227 (1997).
59. I.-B. Kang, M. R. Haskard and B.-K. Ju, "An Assembly and Interconnection Technology for Micromechanical Structures Using a Anisotropic Conductive Film," in *Micromachining and Microfabrication Process Technology II*, pp. 280–287 (1996).
60. S. A. Audet and K. M. Edenfeld, "Integrated Sensor Wafer-Level Packaging," paper presented at *Transducers '97: 1997 International Conference on Solid-State Sensors and Actuators*, pp. 287–289 (1997).
61. M. Waelti, N. Schneeberger, O. Paul and H. Baltes, "Low Temperature Packaging of CMOS Infrared Microsystems by Si-Al-Au Bonding," in *Proceedings of the 4th International Symposium on Semiconductor Wafer Bonding*, pp. 147–154 (1997).
62. L. A. Field and R. S. Muller, "Fusing Silicon Wafers with Low Melting Temperature Glass," *Sensors and Actuators* A21–A23, 935–938 (1990).
63. C. A. Desmond, J. J. Olup, P. Abolghasem, J. Folta and G. Jernigan, "Analysis of Nitride Bonding," in *Proceedings of the 4th International Symposium on Semiconductor Wafer Bonding*, pp. 171–178 (1997).
64. A. G. Pedrine, J. C. De Poorter, L. De Schepper and K. Baert, "Low Temperature Wafer Bonding on Silicon Nitride," in *Microsystems and Microstructures*, pp. 285–292.
65. Y. T. Cheng, L. Lin and K. Najafi, "Localized Bonding with PSG or Indium Solder as Intermediate Layer," paper presented at *MEMS '99: Twelfth IEEE International Conference on Micro Electro Mechanical Systems*, pp. 285–289 (1999).
66. A. Singh, D. A. Horsley, M. B. Cohn, A. P. Pisano and R. T. Howe, "Batch Transfer of Microstructures Using Flip-Chip Solder Bump Bonding," paper presented at *Transducers '97: International Conference on Solid-State Sensors and Actuators*, pp. 265–268 (1997).

67. M. B. Cohn, Y. Liang, R. T. Howe and A. P. Pisano, "Wafer-to-Wafer Transfer of Microstructures for Vacuum Packaging," in *Proceedings IEEE Solid-State Sensor and Actuator Workshop*, pp. 32–35 (1996).
68. M. B. Cohn, Y.-C. Liang, R. T. Howe and A. P. Pisano, "Wafer-to-Wafer Transfer of Microstructure for Vacuum Packaging," paper presented at *1996 Solid-State Sensor and Actuator Workshop*, pp. 32–35 (1996).
69. M. M. Maharbiz, M. B. Cohn, R. T. Howe, R. Horowitz and A. P. Pisano, "Batch Micropackaging by Compression-Bonded Wafer-Wafer Transfer," paper presented at *MEMS '99: Twelfth IEEE International Conference on Micro Electro Mechanical Systems*, pp. 482–489 (1999).
70. J. M. Bustillo, R. T. Howe and R. S. Muller, "Surface Micromachining for Microelectromechanical Systems," *Proceedings of the IEEE* 86, 1552–1574 (1998).
71. C. Mastrangelo, "Thermal Applications of Microbridges," Ph.D. Dissertation, University of California, Berkeley (1991).
72. C. H. Mastrangelo and R. S. Muller, "Fabrication and Performance of a Fully Integrated µ-Pirani Pressure Gauge with Digital Readout," paper presented at *Transducers '91: 1991 International Conference on Solid-State Sensors and Actuators*, pp. 245–248 (1991).
73. L. Lin, K. M. McNair, R. T. Howe and A. P. Pisano, "Vacuum Encapsulated Lateral Microresonators," in *Transducers '93: International Conference on Solid-State Sensors and Actuators*, pp. 270–273 (1993).
74. K. S. Lebouitz, R. T. Howe and A. P. Pisano, "Permeable Polysilicon Etch-Access Windows for Microshell Fabrication," paper presented at *8th International Conference on Solid-State Sensors and Actuators, and Eurosensors IX*, pp. 224–227 (1995).
75. M. W. Judy and R. T. Howe, "Polysilicon Hollow Beam Lateral Resonators," in *Proceedings IEEE Micro Electro Mechanical System Workshop*, pp. 265–271 (1993).
76. D. J. Monk, P. Krulevitch, R. T. Howe and G. C. Johnson, "Stress-Corrosion Cracking and Blistering of Thin Polycrystalline Silicon Films in Hydrofluoric Acid," paper presented at *MRS Spring Annual Meeting*, pp. 641–646 (1993).
77. D. J. Monk, "Controlled Structure Release for Silicon Surface Micromachining," Ph.D. Dissertation, University of California, Berkeley (1993).
78. R. Maboudian and R. T. Howe, "Stiction Reduction Processes for Surface Micromachines," *Tribology Letters* 215–221 (1997).
79. R. Maboudian and R. T. Howe, "Critical Review: Stiction in Surface Micromechanical Structures," *Journal of Vacuum Science Technology* (1996).
80. R. Maboudian, "Surface Processes in MEMS Technology," *Surface Science Reports* 30, 207–269 (1998).
81. N. Tas, T. Sonnenberg, H. Jansen, R. Legtenberg and M. Elwenspoek, "Stiction in Surface Micromachining," *Journal of Micromechanics and Microengineering* 6, 385–397 (1996).

82. J. Y. Kim and C.-J. Kim, "Comparative Study of Various Release Methods for Polysilicon Surface Micromachining," *IEEE* 442–447 (1997).
83. R. L. Alley, G. J. Cuan, R. T. Hose and K. Komvopoulos, "The Effect of Release-Etch Processing on Surface Microstructure Stiction," paper presented at *1992 Solid-State Sensors and Actuators Workshop*, pp. 202–207 (1992).
84. H. Guckel and D. W. Burns, "Fabrication of Micromechanical Devices from Polysilicon Films with Smooth Surfaces," *Sensors and Actuators* A20, pp. 117–122 (1989).
85. C. H. Mastrangelo and C. H. Hsu, "Mechanical Stability and Adhesion of Microstructures Under Capillary Forces—Part I: Basic Theory," *Journal of Microelectromechanical Systems* 2, 33–43 (1993).
86. C. H. Mastrangelo and C. H. Hsu, "Mechanical Stability and Adhesion of Microstructures Under Capillary Forces—Part II: Experiments," *Journal of Microelectromechanical Systems* 2, 44–55 (1993).
87. Y. Yee, M. Park and K. Chun, "A Sticking Model of Suspended Polysilicon Microstructure Including Residual Stress Gradient and Postrelease Temperature," *Journal of Microelectromechanical Systems* 7, 339–344 (1998).
88. F. M. Serry, D. Walliser and G. J. Maclay, "The Role of the Casimir Effect in the Static Deflection and Stiction of Membrane Strips in Microelectromechanical Systems (MEMS)," *Journal of Applied Physics* 84, 2501–2506 (1998).
89. W. Tang, "Electrostatic Comb Drive for Resonant Sensor and Actuator Applications," Ph.D. Dissertation, University of California, Berkeley (1990).
90. L. S. Fan, "Integrated Micromachinery: Moving Structures on Silicon Chips," Ph.D. Dissertation, University of California, Berkeley (1989).
91. Y.-C. Tai and R. S. Muller, "Fracture Strain of LPCVD Polysilicon," paper presented at *Solid-State Sensor and Actuator Workshop*, p. 88 (1998).
92. R. C. Stouppe, "Surface Micromachining Process," U.S. Patent, No. 5,662,771, September 2 (1997).
93. T. Abe, W. C. Messner and M. L. Reed, "Effective Methods to Prevent Stiction During Post-Release-Etch Processing," in *Proceedings MEMS '95*, pp. 94–99 (1995).
94. R. T. Howe, H. J. Barber and M. Judy, "Apparatus to Minimize Stiction During Post-Release-Etch Procesing," U.S. Patent, No. 5,542,295, August 6 (1996).
95. G. K. Fedder and R. T. Howe, "Thermal Assembly of Polysilicon Microstructures," in *Proceedings MEMS '89*, pp. 63–68 (1989).
96. Y. Yee, K. Chun and J. D. Lee, "Polysilicon Surface Modification Technique to Reduce Sticking of Microstructures," paper presented at *8th International Conference on Solid-State Sensors and Actuators, and Eurosensors IX*, pp. 206–209 (1995).
97. R. L. Alley, P. Mai, K. Komvopoulos and R. T. Howe, "Surface Roughness Modifications of Interfacial Contacts in Polysilicon Microstructures," paper

presented at *7th International Conference on Solid-State Sensors and Actuators*, pp. 288–291 (1993).
98. T. A. Lober and R. T. Howe, "Surface Micromachining for Electrostatic Microactuator Fabrication," paper presented at *Solid-State Sensor and Actuator Workshop*, pp. 59–62 (1988).
99. J. Anguita and F. Briones, "HF/H$_2$O Vapor Etching of SiO$_2$ Sacrificial Layer for Large-Area Surface-Micromachined Membranes," *Sensors and Actuators* A64, 247–251 (1998).
100. M. L. Reed and T. Abe, "Method to Prevent Adhesion of Micromechanical Structures," U.S. Patent, No. 5,658,636, August 19 (1997).
101. M. L. Reed and T. Abe, "Method to Prevent Adhesion of Micromechanical Structures," U.S. Patent No. 5,772,902, June 30 (1998).
102. P. R. Scheeper, J. A. Voorthuyzen, W. Olthius and P. Bergveld, "Investigations of Attractive Forces between PECVD Silicon Nitride Microstructures and an Oxidized Silicon Substrate," *Sensors and Actuators* A30, 231 (1992).
103. H. Guckel and J. J. Sniegowski, "Method to Prevent Adhesion of Micromechanical Structures," U.S. Patent, No. 5,013,693 (1990).
104. H. Guckel, J. J. Sniegowski, T. R. Christenson and F. Raissi, "The Application of Fine Grained, Tensile Polysilicon Mechanically Resonant Transducers," *Sensors and Actuators* A21–A23, 346 (1990).
105. J. J. Sniegowski, "Design and Fabrication of the Polysilicon Resonating Beam Force Transducer," Ph.D. Dissertation, University of Wisconsin, Madison, WI (1991).
106. R. Legtenberg and H. A. C. Tilmans, "Electrostatically Driven Vacuum-Encapsulated Polysilicon Resonantors. Part I. Design and Fabrication," *Sensors and Actuators* A45, 57–66 (1994).
107. N. Takeshima, K. J. Gabriel, M. Ozaki, J. Takahashi, H. Horiguchi and H. Fujita, "Electrostatic Parallelogram Actuators," paper presented at *International Conference on Solid-State Sensors and Actuators*, pp. 63–66 (1991).
108. D. Kobayashi, T. Hirano, T. Furuhata and H. Fujita, "An Integrated Lateral Tunneling Unit," in *Proc. IEEE MEMS '92*, pp. 214–219 (1992).
109. C. H. Mastrangelo and G. S. Saloka, "A Dry-Release Method Based on Polymer Columns for Microstructure Fabrication," in *Proc. IEEE MEMS '93*, pp. 77–81 (1993).
110. C. H. Mastrangelo, X. Zhang and W. C. Tang, "Surface Micromachined Capacitive Differential Pressure Sensor with Lithographically-Defined Silicon Diaphragm," paper presented at *Transducers '95—Eurosensors IX: 8th International Conference on Solid-State Sensors and Actuators, and Eurosensors IX*, pp. 612–619 (1995).
111. T. A. Core and R. T. Howe, "Method for Fabricating Microstructures," U.S. Patent, No. 5,314,572, May 24 (1994).
112. M. Orpana and A. O. Korhonen, "Control of Residual Stress of Polysilicon Thin Films by Heavy Doping in Surface Micromachining," paper presented at *International Conference on Solid-State Sensors and Actuators*, pp. 266–269 (1991).

113. F. Kozlowski, N. Lindmair, T. Scheiter, C. Hierold, and W. Lang, "A Novel Method to Avoid Sticking of Surface Micromachined Structures," paper presented at *8th International Conference on Solid-State Sensors and Actuators, and Eurosensors IX*, pp. 220–223 (1995).
114. G. T. Mulhern, D. S. Soane and R. T. Howe, "Supercritical Carbon Dioxide Drying of Microstructures," paper presented at *7th International Conference on Solid-State Sensors and Actuators*, pp. 296–299 (1993).
115. C. W. Dyck, J. H. Smith, S. L. Miller, E. M. Russick and C. L. J. Adkins, "Supercritical Carbon Dioxide Solvent Extraction from Surface-Micromachined Micromechanical Structures," paper presented at *SPIE's 1996 Symposium on Micromachining and Microfabrication*, pp. 225–235 (1996).
116. M. Biebl, G. T. Mulhern and R. T. Howe, "In Situ Phosphorus-Doped Polysilicon for Integrated MEMS," paper presented at *8th International Conference on Solid-State Sensors and Actuators, and Eurosensors IX*, pp. 198–201 (1995).
117. M. R. Houston, R. T. Howe and R. Maboudian, "Effect of Hydrogen Termination on the Interfacial Forces between Polycrystalline Silicon Surfaces," *Journal of Applied Physics* (1996).
118. M. R. Houston, R. Maboudian and R. T. Howe, "Ammonium Fluoride Anti-Stiction Treatments for Polysilicon Microstructures," paper presented at *8th International Conference on Solid-State Sensors and Actuators, and Eurosensors IX*, pp. 210–213 (1995).
119. U. Srinivasan, M. R. Houston, R. T. Howe and R. Maboudian, "Alkyltrichlorosilane-Based Self-Assembled Monolayer Films for Stiction Reduction in Silicon Micromachines," *JMEMS* 7, 252–260 (1998).
120. P. F. Man, B. P. Gogoi and C. H. Mastrangelo, "Elimination of Post-Release Adhesion in Microstructures Using Conformal Fluorocarbond Coatings," *J. MEMS* 6, 25–34 (1996).
121. M. Nishimura, Y. Matsumoto and M. Ishida, "The Method to Prevent Stiction in a Capacitive Accelerometer Using SDB-SOI Structure," in *Technical Digest of the 15th Sensor Symposium*, pp. 205–208 (1997).
122. Y. Matsumoto, K. Yoshida and M. Ishida, "Fluorocarbon Film for Protection from Alkaline Etchant and Elmination of In-Use Stiction," paper presented at *1997 International Conference on Solid-State Sensors and Actuators*, pp. 695–698 (1997).
123. J. R. Martin and Y. Zhao, "Micromachined Device Packaged to Reduce Stiction," WO 98/05935 (World Patent), February 12 (1998).
124. B. P. Gogoi and C. H. Mastrangelo, "Adhesion Release and Yield Enhancement of Microstructures Using Pulsed Lorentz Forces," *Journal of Microelectromechanical Systems* 4, 185–192 (1995).
125. L. T. Inc., *Product Literature* (1999).
126. M. L. Kniffin and M. Shah, "Packaging for Silicon Micromachined Accelerometers," *International Journal of Microcircuits and Electronic Packaging* 19, 75–86 (1996).

127. K. H.-L. Chau and J. Sulouff, R. E., "Technology for the High-Volume Manufacturing of Integrated Surface-Micromachined Accelerometer Products," *Microelectronics Journal* 29, 579–586 (1998).
128. H.-J. Yeh, "Fluidic Self-Assembly of Microstructure and Its Application to the Integration of GaAs on Si," paper presented at *MEMS '94*, pp. 279–284 (1994).
129. H.-J. Yeh and J. S. Smith, "Integration of GaAs Vertical Cavity Surface-Emitting Laser on Si by Substrate Removal," *Applied Physics Letters* 64, 1466–1468 (1994).
130. M. Cohn, R. T. Howe and A. P. Pisano, "Self-Assembly of Microsystems Using Non-Contact Electrostatic Traps," paper presented at *ASME 1995*, pp. 893–900 (1995).
131. T. A. Core, W. K. Tsang and S. J. Sherman, "Fabrication Technology for an Integrated Surface-Micromachined Sensor," *Solid State Technology*, pp. 39–47 (1993).
132. A. K. Henning, "Microfluidic MEMS," paper presented at *1998 IEEE Aerospace Conference*, pp. 471–486 (1998).
133. J. C. Lyke, "Packaging Technologies for Space-Based Microsystems and Their Elements," in *Microengineering Technologies for Space Systems*, H. Helvajian, ed., pp. 131–180. Aerospace Corp., El Segundo, CA (1995).
134. G. Kelly, J. Alderman, C. Lyden, J. Barrett and A. Morrissey, "Microsystem Packaging in 3D," paper presented at *Micromachined Devices and Components III*, pp. 142–152 (1997).
135. H. Han, L. E. Weiss and M. L. Reed, "Micromechanical Velcro," *Journal of Microelectromechanical Systems* 1, 37–43 (1992).
136. B. Shivkumar and C.-J. Kim, "Microrivets for MEMS Packaging: Concept, Fabrication, and Strength Testing," *Journal of Microelectromechanical Systems* 6, 217–225 (1997).
137. F. Mayer, G. Ofner, A. Koll, O. Paul and H. Baltes, "Flip-Chip Packaging for Smart MEMS," paper presented at *Smart Structures and Materials 1998: Smart Electronics and MEMS*, pp. 183–193 (1998).
138. N. Najafi and S. Massoud-Ansari, "Method for Packaging Microsensors," U.S. Patent, No. 5,694,740, December 9 (1997).
139. R. F. Wolffenbuttel, "Microsystems for Multi-Sensory Data-Acquisition," in *ISIE '97: Proceedings of the IEEE International Symposium on Industrial Electronics*, pp. SS146–SS151 (1997).
140. R. Irwin, W. Zhang, K. Harsh and Y. C. Lee, "Quick Prototyping of Flip Chip Assembly with MEMS," paper presented at *RAWCON '98: 1998 IEEE Radio and Wireless Conference*, pp. 293–296 (1998).
141. J. Lyke, "Advanced Packaging Perspectives," briefing presentation slides: Phillips Laboratory, Applied Systems Branch, PL/VTEE, Kirtland Air Force Base, Albuquerque, NM (1994).
142. D. Koester, R. Mahadevan, A. Shishkoff and K. Markus, "Multi-User MEMS Processes (MUMPS) Introduction and Design Rules, rev. 4," July 15 (1996).

143. H. Beutel, T. Stieglitz and J. U. Meyer, "Microflex: A New Technique for Hybrid Integration for Microsystems," in *Proceedings of the 1998 MEMS Conference*, pp. 306–311 (1998).
144. D. J. Warkentin, J. H. Haritonidis, M. Mehregany and S. D. Senturia, "A Micromachined Microphone with Optical Interference Readout," in *Transducers '87*, pp. 291–294 (1987).
145. L. Y. Lin, S. S. Lee, M. C. Wu and K. S. J. Pister, "Micromachined Integrated Optics for Free-Space Interconnections," in *MEMS95*, pp. 77–82 (1995).
146. S. Koh and C. H. Ahn, "Novel Integrated Optical I/O Couplers on MCMs Using MEMS Techniques," paper presented at *Integrated Optics and Microstructures III*, pp. 121–130 (1996).
147. M. A. Chan, S. D. Collins and R. L. Smith, "A Micromachined Biomedical Pressure Sensor with Fiber-Optic Interferometric Readout," paper presented at *Transducers '93: 7th International Conference on Solid-State Sensors and Actuators*, pp. 580–583 (1993).
148. J. Smith, S. Montague, J. Sniegowski, J. Murray and P. McWhorter, "Embedded Micromechanical Devices for the Monolithic Integration of MEMS with CMOS," in *Proceedings of the IEDM*, pp. 609–612 (1995).
149. J. Butler, Ph.D. Dissertation, Air Force Institute of Technology (1998).
150. D. Doane and P. Franzon, *Multichip Module Technologies and Alternatives: The Basics*. Van Nostrand Reinhold, New York (1993).
151. R. Bone, D. Elwell and R. McBride, "Understanding MCM Assembly and Testing," *Solid State Technology* 87–93 (1994).
152. J. T. Butler, V. M. Bright, P. B. Chu and R. J. Saia, "Adapting Multichip Module Foundries for MEMS Packaging," paper presented at *1998 International Conference on Multichip Modules and High Density Packaging*, pp. 106–111 (1998).
153. J. Peltier and W. Hansford, "Flexible Access to MCM Technology Via the Multichip Module Designer's Access Service (MIDAS)," in *Proceedings of the IEEE 1996 Multi-Chip Module Conference*, pp. 86–88 (1996).
154. L. Guérin, R. Sachot and M. Dutoit, "A New Multichip-on-Silicon Packaging Scheme with Integrated Passive Components," in *Proceedings of the IEEE 1996 Multi-Chip Module Conference*, pp. 73–77 (1996).
155. A. Michalicek, personal communication (1999).
156. K. Harsh, personal communication (1999).
157. T. S. McLaren, S. Y. Kang, W. Zhang, T. H. Ju and Y. C. Lee, "Thermosonic Bonding of an Optical Transceiver Based on an 8×8 Vertical Cavity Surface Emitting Laser Array," *IEEE Transacations on Components, Parts, and Manufacturing Technology, Part B* 20, 152–160 (1997).
158. K. F. Harsh, W. Zhang, V. M. Bright, Z. Feng, K. C. Gupta and Y. C. Lee, "Flip-Chip Assembly for Si-Based RF MEMS," in *Technical Digest of 12th IEEE International Conference on MicroElectroMechanical Systems—MEMS '99*, pp. 273–278 (1999).

159. W. D. Cowan and V. D. Bright, "Vertical Thermal Actuators for Micro-Opto-Electro-Mechanical Systems," *Proceedings of SPIE*, pp. 138–146 (1997).
160. Y. C. Lee and Q. Tan, "Soldering for Optoelectronic Packaging," in *Proceedings IEEE Electronic Components and Technology Conference*, pp. 26–36 (1996).
161. Z. Feng, W. Zhang, B. Su, K. Harsh, K. C. Gupta, V. M. Bright and Y. C. Lee, "Design and Modeling of RF MEMS Tunable Capacitors Using Electrothermal Actuators," in *IEEE MTT-S International Microwave Symposium* (1999).
162. W. Daum, W. Burdick, Jr. and R. Fillion, "Overlay High-Density Interconnect: A Chips-First Multichip Module Technology," *IEEE Computer* 26, 23–29 (1993).
163. R. Fillion, R. Wojnarowski, R. Saia and D. Kuk, "Demonstration of a Chip Scale Chip-On-Flex Technology," in *Proceedings of the 1996 International Conference on Multichip Modules*, pp. 351–356 (1996).
164. L. Muller, M. Hecht, L. Miller, H. Rockstad and J. Lyke, "Packaging and Qualification of MEMS-Based Space Systems," in *Proceedings of the IEEE MEMS Symposium*, pp. 503–508 (1996).
165. J. Sweet, D. Peterson, M. Tuck and J. Greene, "Assembly Test Chip Version 04 (ATC04) Description and User's Guide," Albuquerque, NM (1994).
166. N. Hall, "Assembly Test Chips at Sandia National Laboratories," Florida Institute of Technology, Melbourne, FL (1994).
167. J. H. Comtois, "Structures and Techniques for Implementing and Packaging Complex, Large Scale Microelectromechanical Systems Using Foundry Fabrication Processes," Ph.D. Dissertation, Air Force Institute of Technology (1996).
168. L. T. Manzione, *Plastic Packaging of Microelectronic Packages*. Van Nostrand Reinhold, New York (1990).
169. D. S. Soane, "Stresses in Packaged Semiconductor Devices," *Solid State Technology*, 165 (1989).
170. G. Bitko, R. Harries, J. Matkin, A. C. McNeil, D. J. Monk, M. Shah and J. Wertz, "Thin Film Polymer Stress Measurements Using Piezoresistive Anisotropically Etched Pressure Sensors," in *Materials Research Society Symposium Proceedings*, pp. 365–371 (1997).
171. Motorola, *Pressure Sensor Device Data*. Motorola Literature Distribution Centers, Phoenix, AZ (1995).
172. J. B. Nysæther, A. Larsen, B. Liverød and P. Ohlckers, "Structures for Piezoresistive Measurement of Package Induced Stress in Transfer Molded Silicon Pressure Sensors," *Microelectronics Reliability* 38, 1271–1276 (1998).
173. T. Maudie, D. J. Monk, D. Zehrbach and D. Stanerson, "Sensor Media Compatibility: Issues and Answers," paper presented at *Sensors Expo*, pp. 215–229 (1996).
174. D. J. Monk, T. Maudie, D. Stanerson, J. Wertz, G. Bitko, J. Matkin and S. Petrovic, "Media Compatible Packaging and Environmental Testing of Bar-

rier Coating Encapsulated Silicon Pressure Sensors," paper presented at *1996 Solid-State Sensors and Actuators Workshop*, pp. 36–41 (1996).
175. G. Bitko, D. J. Monk, T. Maudie, D. Stanerson, J. Wertz, J. Matkin and S. Petrovic, "Analytical Techniques for Examining Reliability and Failure Mechanisms of Barrier Coating Encapsulated Silicon Pressure Sensors Exposed to Harsh Media," paper presented at *Micromachined Devices and Components II*, pp. 248–258 (1996).
176. S. Petrovic, S. Brown, A. Ramirez, B. King, T. Maudie, D. Stanerson, G. Bitko, J. Matkin, J. Wertz and D. J. Monk, "Low-Cost, Water Compatible Piezoresistive Bulk Micromachined Pressure Sensor," paper presented at *Advances in Electronic Packaging*, pp. 455–462 (1997).
177. M. F. Nichols and A. W. Hahn, "Apparatus for Applying a Composite Insulative Coating to a Substrate," U.S. Patent No. 4,921,723, May 1 (1990).
178. M. F. Nichols and A. W. Hahn, "Article Having a Composite Insulative Coating," U.S. Patent No. 5,137,780, August 11 (1992).
179. M. F. Nichols and A. W. Hahn, "Apparatus for Applying a Composite Insulative Coating to a Substrate," U.S. Patent No. 5,121,706, June 16 (1992).
180. M. F. Nichols, "Flexible and Insulative Plasmalene® Wire Coatings for Biomedical Applications," *Biomedical Scientific Instruments* 29, 77–83 (1993).
181. M. F. Nichols, "Method and Apparatus for Plasma Treatment of a Filament," U.S. Patent No. 5,254,372, October 19 (1993).
182. M. F. Nichols, "The Challenges for Hermetic Encapsulation of Implanted Devices—A Review," *Critical Reviews Biomedical Engineering* 22, 39–67 (1994).
183. G. F. Eriksen and K. Dyrbye, "Protective Coatings in Harsh Environments," *Journal of Micromechanics Microengineering* 6, 55–57 (1996).
184. K. Dyrbye, T. Romedahl Brown and G. Friis Eriksen, "Packaging of Physical Sensors for Aggressive Media Applications," *Journal of Micromechanics Microengineering* 6, 187–192 (1996).
185. R. De Reus, C. Christensen, S. Weichel, S. Bouwstra, J. Janting, G. F. Eriksen, K. Dyrbye, T. Romedahl Brown, J. P. Krog, O. Søndergård Jensen and P. Gravesen, "Reliability of Industrial Packaging for Microsystems," *Microelectronics Reliability* 38, 1251–1260 (1998).
186. T. A. Maudie, D. J. Monk and T. S. Savage, "Media Compatible Microsensor Structure and Methods of Manufacturing and Using the Same," U.S. Patent No. 5,889,211, March 30 (1999).
187. H. Jakobsen and T. Kvisterøy, "Sealed Cavity Arrangement Method," U.S. Patent No. 5,591,679, January 7 (1997).
188. S. Bouwstra, "Stacked Multi-Chip-Module Technology for High Performance Intelligent Transducers," paper presented at *Micromachined Devices and Components II*, pp. 49–52 (1996).
189. S. Linder, H. Baltes, F. Gnaedinger and E. Doering, "Photolithography in Anisotropically Etched Grooves," paper presented at *Ninth Annual International Workshop on Micro Electro Mechanical Systems*, pp. 38–43 (1996).

190. M. Heschel, J. F. Kuhmann, S. Bouwstra and M. Amskov, "Integrated Packaging Concept for an Intelligent Transducer," paper presented at *Smart Electronics and MEMS*, pp. 344–352 (1998).
191. T. Maudie and D. J. Monk, "Pressure Sensor with Isolated Interconnections for Media Compatibility," U.S. Patent No. 5,646,072, July 8 (1997).
192. K. Ryan and J. Bryzek, "Packaging Technology for Low-Cost Media Isolated Pressure Sensors," paper presented at *Sensor 95*, pp. 685–690 (1995).
193. C. P. Wong, "Recent Advances in IC Passivation and Encapsulation," in *Polymers for Electronic and Photonic Applications*, C. P. Wong, ed., pp. 167–220. Academic Press, Boston (1993).
194. V. J. Adams, "Unibody Pressure Transducer Package," U.S. Patent No. 4,655,088, April 7 (1987).
195. M. Summers, D. J. Monk, G. O'Brien, A. McNeil, M. Velez, and G. Spraggins, "Thermal Electric FEA and DOE for a Semiconductor Heater Element," paper presented at *1996 Seventh International ANSYS Conference and Exhibition*, pp. 1.75–1.87 (1996).
196. D. W. Feldbaumer, J. A. Babcock, V. M. Mercier and C. K. Y. Chun, "Pulse Current Trimming of Polysilicon Resistors," *IEEE Transactions on Electron Devices* 42, 689–696 (1995).
197. D. W. Feldbaumer and J. A. Babcock, "Theory and Application of Polysilicon Resistor Trimming," *Sold-State Electronics* 38, 1861–1869 (1995).
198. J. P. Schuster and W. S. Czarnocki, "Automotive Pressure Sensors: Evolution of a Micromachined Sensor Application," in *Proceedings of the 3rd International Symposium on Microstructures and Micromachines*, pp. 49–63 (1997).
199. K. Petersen, "MEMS: What Lies Ahead?" paper presented at *8th International Conference on Solid-State Sensors and Actuators, and Eurosensors IX*, pp. 894–897 (1995).
200. V. I. Vaganov, "Construction Problems in Sensors," *Sensors and Actuators* A28, 161–172 (1991).
201. K. Petersen, J. Brown, T. Vermeulen, P. Barth, J. Mallon, Jr. and J. Bryzek, "Ultra-Stable, High-Temperature Pressure Sensors Using Silicon Fusion Bonding," *Sensors and Actuators* A21–A23, 96–101 (1990).
202. K. Sooriakumar, D. J. Monk, W. K. Chan and K. G. Goldman, "Vertically Integrated Sensor Structure and Method," U.S. Patent No. 5,600,071, February 4 (1997).
203. K. Sooriakumar, D. J. Monk, W. Chan, G. Vanhoy and K. Goldman, "Batch Fabrication of Media Isolated Absolute Pressure Sensors," in *Proceedings of the Second International Symposium on Microstructures and Microfabricated Systems*, pp. 1522–1523 (1995).
204. K. Sooriakumar, D. J. Monk, W. Chan, G. Vanhoy and K. Goldman, "Batch Fabrication of Media Isolated Absolute Pressure Sensors," in *Proceedings of the Second International Symposium on Microstructures and Microfabricated Systems*, pp. 231–238 (1995).

205. K. Goldman, G. Gritt, I. Baskett, K. Sooriakumar, D. Wallace, D. Hughes and M. Shah, "A Vertically Integrated Media-Isolated Absolute Pressure Sensor," paper presented at *Transducers '97: 1997 International Conference on Solid-State Sensors and Actuators*, pp. 1501–1504 (1997).
206. K. Goldman, G. Gritt, I. Baskett, W. Czarnocki, A. Ramirez, C. Brown, D. Hughes, D. Wallace and M. Shah, "A Vertically Integrated Media-Isolated Absolute Pressure Sensor," *Sensors and Actuators* A66, 155–159 (1998).
207. J. Mallon, J. Bryzek, J. Ramsey, G. Tomblin and F. Pourahmadi, "Low-Cost, High-Volume Packaging Techniques for Silicon Sensors and Actuators," paper presented at *IEEE Solid-State Sensor and Actuator Workshop*, pp. 123–124 (1988).
208. D. J. Monk, "Pressure Leakage through Material Interfaces in Pressure Sensor Packages," paper presented at *Sensors in Electronic Packaging*, pp. 87–93 (1995).
209. GE, "VALOX® Property Guide," in *GE Thermoplastic Properties Guide*, September 15 (1993).
210. D. S. Mahadevan, "Effect of Machine Parameters on Polymer Die Attach and a Method to Optimize Them," *Proceedings of SPIE*, pp. 265–272 (1995).
211. R. Frank and V. Adams, "Pressure Sensors Packaged in Plastic," *Electronic Packaging & Production* (December), 62–63 (1986).
212. C. P. Wong, "High Performance Silicone Gel as IC Device Chip Protection," in *Materials Research Society Symposium Proceedings*, pp. 175–187 (1988).
213. G. Bitko, A. C. McNeil and D. J. Monk, "The Effect of Inorganic Thin Film Material Processing and Properties on Stress in Silicon Piezoresistive Pressure Sensors," in *Materials Research Society Symposium Proceedings*, pp. 221–226 (1997).
214. A. J. Polak, "Semiconductor Device Encapsulation Method," U.S. Patent No. 5,357,673, October 25 (1994).
215. A. Polak, "Characterization and Use of Polyfluorosiloxanes in Automotive Applications," in *Materials Research Society Symposium Proceedings*, pp. 281–289 (1992).
216. A. K. Hu and D. H. Green, "Packaging of MEMS Devices," paper presented at *Micromachined Devices and Components*, pp. 273–280 (1995).
217. T. Maudie, D. J. Monk, D. Stanerson, D. Zehrbach and R. Frank, "Packaging and Test Considerations for Sensors," in *Proceedings Sensors Expo Detroit*, pp. 1–9 (1997).
218. T. Maudie, D. J. Monk and R. Frank, "Packaging Considerations for Predictable Lifetime Sensors," in *Proceedings Sensors Expo Boston*, pp. 167–172 (1997).
219. A. Mirza, "Future Directions in Silicon Micromachining for High-Volume Manufacturing," in *Proceedings of the Second International Symposium on Microstructures and Microfabricated Systems*, pp. 33–43 (1995).
220. D. Zehrbach and T. Maudie, "Dry Air and Noncorrosive Gases Only! (Are Silicon Pressure Sensors Really That Finicky?)" *Sensors* 49–52 (1998).

221. W. H. Ko, "Packaging of Microfabricated Devices and Systems," *Matls. Chem. Phys.* 42, 169–175 (1995).
222. D. Walter and R. Frank, "Industrial Control Reaps Benefits of Micromachined Pressure Sensors," *I&CS* 69, 33–36 (1996).
223. W. C. Tang, "Development Process of Automotive Microsensors," *Proceedings of SPIE*, pp. 251–257 (1995).
224. D. Slocum and R. L. Tucker, "Pressure Sensing with Silicon," *Sensors* 20–24 (1989).
225. M. Noble, "Environmental Concerns for Integrated Circuit Sensors," *Measurement + Control* 19, 210–213 (1986).
226. T. Costlow, "Sensor Technology Gains," *Electrical Engineering Times* November 8, 70 (1993).
227. A. S. Voloshin, P.-H. Tsao, A. J. Polak and T. L. Baker, "Analysis of Environment Induced Stresses in Silicon Sensors," paper presented at *Advances in Electronic Packaging*, pp. 489–492 (1995).
228. T. Ito, "Semiconductor Pressure Sensor," Kokai Patent No. Hei 5-223670, August 31 (1993).
229. K. E. Petersen, "Silicon as a Mechanical Material," *Proceedings of IEEE* 70, 420–457 (1982).
230. R. E. Sulouff, Jr., "Silicon Sensors for Automotive Applications," paper presented at *International Conference on Solid-State Sensors and Actuators: Transducers '91*, pp. 170–176 (1991).
231. D. J. Monk, H. S. Toh and J. Wertz, "Oxidative Degradation of Parylene C [Poly(monochloro-*para*-xylylene)] Thin Films on Bulk Micromachined Piezoresistive Silicon Pressure Sensors," *Sensors and Materials*, 307–319 (1995).
232. S. Petrovic, A. Ramirez, T. Maudie, D. Stanerson, J. Wertz, G. Bitko, J. Matkin and D. J. Monk, "Reliability Test Methods for Media-Compatible Pressure Sensors," *IEEE Transactions Industrial Electronics* 45, 877–885 (1998).
233. D. J. Monk and M. Shah, "Thin Film Polymer Stress Measurement Using Piezoresistive Anisotropically Etched Pressure Sensors," in *Materials Research Society Symposium Proceedings*, pp. 103–109 (1995).
234. R. Olson, "Measuring the Stress Component of Conformal Coatings on Surface Mount Devices," paper presented at *International Symposium on Microelectronics*, pp. 236–247 (1986).
235. T. A. Maudie, D. J. Monk and T. S. Savage, "Media Compatible Microsensor Structure by Inorganic Protection Coating," European Patent Application 96104684.4, September 10 (1996).
236. W. Fabianowski, D. Jaffe and R. J. Jaccodine, "Mechanisms of Microelectronics Protection with Polysiloxane Polymers," paper presented at *6th International SAMPE Electronics Conference*, pp. 256–266 (1992).
237. C. P. Wong, "Recent Advances in the Application of High Performance Siloxane Polymers in Electronic Packaging," paper presented at *6th International SAMPE Electronics Conference*, pp. 508–520 (1992).

238. W. F. Gorham, "A New, General Synthetic Method for the Preparation of Linear Poly-*p*-xylylenes," *Journal of Polymer Science* 4, 3027–3039 (1966).
239. W. F. Beach, C. Lee, D. R. Bassett, T. M. Austin and R. Olson, "Xylylene Polymers," in *Encyclopedia of Polymer Science and Engineering*, Vol. 17, 2nd edn., pp. 900–1025. Wiley, New York (1989).
240. W. F. Beach, T. M. Austin and R. Olson, "Parylene Coatings," in *Electronic Materials Handbook: Volume 1: Packaging*, M. L. Minges, C. A. Dostal and M. S. Woods, eds., pp. 789–801. ASM International, Materials Park, OH (1989).
241. W. F. Beach and T. M. Austin, "Update: Parylene as a Dielectric for the Next Generation of High Density Circuits," *SAMPE Journal* 24, 9–12 (1988).
242. C. Christensen, R. de Reus and S. Bouwstra, "Tantalum Oxide Thin Films as Protective Coatings for Sensors," paper presented at *MEMS '99: Twelfth IEEE International Conference on Micro Electro Mechanical Systems*, pp. 267–272 (1999).
243. U. Lipphardt, G. Findler, H. Muenzel and H. Baumann, "Semiconductor Sensor Having a Protective Layer," U.S. Patent No. 5,629,538, May 13 (1997).
244. J. G. E. Gardeniers and N. G. Laursen, "Corrosion of Protective Layers on Strained Surfaces in Alkaline Solutions," *Sensors and Materials* 5, 189–208 (1994).
245. A. F. Flannery, N. J. Mourlas, C. W. Storment, S. Tsai, S. H. Tan, J. Heck, D. Monk, T. Kim, B. Gogoi and G. T. A. Kovacs, "PECVD Silicon Carbide as a Chemically Resistant Material for Micromachined Transducers," *Sensors and Actuators* A70, 48–55 (1998).
246. F. Hegner and M. Frank, "Pressure Sensor Including a Diaphragm Having a Protective Layer Thereon," U.S. Patent No. 5,076,147, December 31 (1991).
247. A. D. Kurtz, D. Goldstein and J. S. Shor, "High Temperature Transducers and Methods of Fabricating the Same Employing Silicon Carbide," U.S. Patent No. 5,165,283, November 24 (1992).
248. C. H. Brown, D. J. Wallace, Jr. and M. F. Velez, "Hermetically Sealed Pressure Sensor and Method Thereof," U.S. Patent 5,454,270, October 3 (1995).
249. S. S. Nasiri, "Media Compatible Pressure Sensor Device Utilizing Self-Aligned Components Which Fit Together Without the Need for Adhesives," U.S. Patent No. 5,880,372, March 9 (1999).
250. K. Ryan, J. Bryzek and R. Grace, "Flexible Miniature Packaging from Lucas NovaSensor Provides Low-Cost Pressure Sensor Solutions to a Wide Variety of Hostile Media Applications," *ECN* 118–120 (1995).
251. K. Kato, "Semiconductor Pressure Sensor Having Double Diaphragm Structure," U.S. Patent No. 5,335,549, August 9 (1994).
252. P. Bauer, M. Frimgerger, L. Pfau, J. Dirmeyer and G. Ehrler, "Protective Diaphragm for a Silicon Pressure Sensor," Patent No. WO 96/03629 (World Patent), July 11 (1995).
253. Y. Yamaguchi, "Silicone Oil-Filled Semiconductor Pressure Sensor," U.S. Patent No. 5,625,151, April 29 (1997).

254. E. L. Sokn, "Pressure Sensor Package and Method of Making the Same," U.S. Patent No. 5,874,679, February 23 (1999).
255. H. Guckel and T. R. Christensen, "Micromachined Differential Pressure Transducers," U.S. Patent No. 5,357,807, October 25 (1994).
256. S. Kobori, K. Yamada, R. Kobayashi, A. Mijazaki and S. Suzuki, "Method for Manufacturing Semiconductor Absolute Pressure Sensor Units," U.S. Patent No. 4,802,952, February 7 (1989).
257. J. Hynecek, W. H. Ko and E. T. Yon, "Miniature Pressure Transducer for Medical Use and Assembly Method," U.S. Patent No. 4,023,562, May 17 (1977).
258. S. Kobori, K. Yamada, R. Kobayaski, A. Miyazaki and S. Suzuki, "Method for Manufacturing Semiconductor Absolute Pressure Sensor Units," U.S. Patent No. 4,802,952, February 7 (1989).
259. M. G. Glenn, R. F. McMullen and D. B. Wamstad, "Pressure Transducer," U.S. Patent No. 4,665,754, May 19 (1987).
260. T. Tominaga, T. Mihara, T. Oguro and M. Takeuchi, "Pressure Sensor," U.S. Patent No. 4,276,533, June 30 (1981).
261. T. Tominaga, T. Mihara, T. Oguro and M. Takeuchi, "Pressure Sensor," U.S. Patent No. 4,287,501, September 1 (1981).
262. A. D. Kurtz and J. R. Mallon, "Media Compatible Pressure Tranducer," U.S. Patent No. 4,222,277, September 16 (1980).
263. T. A. Knecht and R. L. Frick, "Capaticance Pressure Sensor," U.S. Patent No. 4,730,496, March 15 (1988).
264. C. H. Brown and D. L. Vowles, "Pressure Sensor with Stress Isolation Platform Hermetically Sealed to Protect Sensor Die," U.S. Patent No. 5,465,626, November 14 (1995).
265. J. M. Giachino, R. J. Haeberle and J. W. Crow, "Method for Manufacturing Variable Capacitance Pressure Transducers," U.S. Patent No. 4,261,086, April 14 (1981).
266. S. Shoji, T. Nisase, M. Esashi and T. Matsuo, "Fabrication of an Implantable Capacitive Type Pressure Sensor," paper presented at *4th International Conference on Solid-State Sensors and Actuators (Transducers '87)*, pp. 305–308 (1987).
267. M. Esashi, "Micromachining for Packaged Sensors," paper presented at *Transducers '93: 7th International Conference on Solid-State Sensors and Actuators*, pp. 260–265 (1993).
268. H. Kuisma, "Pressure Sensor Construction and Method for Its Fabrication," U.S. Patent No. 4,875,134, October 17 (1989).
269. B. Ziaie, J. Von Arx, M. Nardin and K. Najafi, "A Hermetic Packaging Technology with Multiple Feedthroughs for Integrated Sensors and Actuators," paper presented at *Transducers '93: The 7th International Conference on Solid-State Sensors and Actuators*, pp. 266–269 (1993).
270. T. Akin, B. Ziaie and K. Najafi, "RF Telemetry Powering and Control of Hermetically Sealed Integrated Sensors and Actuators," paper presented at *1990 Solid-State Sensor and Actuator Workshop*, pp. 145–148 (1990).

271. O. Søndergård and P. Gravesen, "A New Piezoresistive Pressure Transducer Principle with Improvements in Media Compatibility," *Journal of Micromechanics and Microengineering* 6, 105–107 (1996).
272. A. Hanneborg, M. Nese and P. Øhlckers, "Silicon-to-Silicon Anodic Bonding with a Borosilicate Glass Layer," *Journal of Micromechanics and Microengineering* 1, 139–144 (1991).
273. A. Hanneborg, M. Nese H. Jakobsen and R. Holm, "Silicon-to-Thin Film Anodic Bonding," *Journal of Micromechanics and Microengineering* 2, 117–121 (1992).
274. M. Nese and A. Hanneborg, "Anodic Bonding of Silicon to Silicon Wafers Coated with Aluminum, Silicon Oxide, Polysilicon or Silicon Nitride," *Sensors and Actuators A* 37–38, 61–67 (1993).
275. A. D. Kurtz and J. R. Mallon, "Methods of Fabricating Transducers Emplying Flat Bondable Surfaces with Buried Contact Layers," U.S. Patent No. 4,208,782, June 24 (1980).
276. S.-H. S. Chen and C. Ross, "Capacitive Pressure Sensor and Method for Making the Same," U.S. Patent No. 5,600,072, February 4 (1997).
277. J. P. Tomase, S.-H. Chen, G. D. Stamm and M. K. Chason, "Hermetically Sealed Interface," U.S. Patent No. 5,173,836, December 22 (1992).
278. C. Liu, M. A. Shannon and I. Adesida, "Side-Wall Feature Definition for Through-Wafer Interconnects Using 45° Mirror Surfaces," paper presented at *Solid-State Sensor and Actuator Workshop*, pp. 25–26. Late-News Poster Session Supplemental Digest (1996).
279. P. Kersten, S. Bouwstra and J. W. Petersen, "Photolithography on Micromachined 3D Surfaces Using Electrodeposited Photoresists," *Sensors and Actuators* A51, 51–54 (1995).
280. K. Matsumi, "Fabrication Process of Semiconductor Pressure Sensor for Sensing Pressure Applied," U.S. Patent No. 5,145,810, September 8 (1992).
281. Y. Hsia, "Silicon Substrate Multichip Assembly," U.S. Patent No. 5,061,987, October 29 (1991).
282. M. Heschel, J. F. Kuhmann, S. Bouwstra and M. Amskov, "Stacking Technology for a Space Constrained Microsystem," paper presented at *MEMS '98: The Eleventh Annual International Workshop on Micro Electro Mechanical Systems*, pp. 312–317 (1998).
283. H. D. Goldberg, K. S. Breuer and M. A. Schmidt, "A Silicon Wafer-Bonding Technology for Microfabricated Shear-Stress Sensors with Backside Contacts," paper presented at *Solid-State Sensor and Actuator Workshop*, pp. 111–115 (1994).
284. T. Maudie and D. J. Monk, "Electronic Sensor Assembly Having Metal Interconnections Isolated from Adverse Media," U.S. Patent No. 5,646,072 (1997).
285. R. W. Hower, R. B. Brown, E. Malinowska, R. K. Meruva and M. E. Meyerhoff, "Study of Screen Printed Wells in Solid-State Ion Selective Electrodes," paper presented at *Solid-State Sensor and Actuator Workshop*, pp. 132–135 (1996).

286. H. Krassow, F. Campabadal and E. Lora-Tamayo, "Photolithographic Packaging of Silicon Pressure Sensors," *Sensors and Actuators* A66, 279–283 (1998).
287. R. L. Smith and S. D. Collins, "Micromachined Packaging for Chemical Microsensors," *IEEE Transactions on Electron Devices* 35, 787–792 (1988).
288. Y. Takahashi, T. Hirose, H. Otani and S. Takemura, "Semiconductor Pressure Sensor for Use at High Temperature and Pressure and Method of Manufacturing Same," U.S. Patent No. 5,333,505, August 2 (1994).
289. M. K. Lam and M. W. Mathias, "Low Cost Wet-to-Wet Pressure Sensor Package," U.S. Patent No. 4,942,383, July 17 (1990).
290. Y. Takahashi, T. Hirose and H. Ichiyama, "Semiconductor Pressure Sensor," U.S. Patent No. 5,207,102, May 4 (1993).
291. M. E. Rosenberger, "Piezoresistive Pressure Transducer with Elastomeric Seals," U.S. Patent No. 4,656,454, April 7 (1987).
292. D. J. Monk and S. M. K., "Packaging and Testing Considerations for Commercialization of Bulk Micromachined, Piezoresistive Pressure Sensors," paper presented at *Commercialization of Microsystems '97*, pp. 136–149 (1996).
293. M. K. Shah, A. C. McNeil, M. D. Summers and B. D. Meyer, "Application of Finite Element Analysis to Predict the Performance of Piezo-resistive Pressure Sensor," paper presented at *1995 ASME International Mechanical Engineering Congress and Exposition*, pp. 79–86 (1995).
294. B. Hälg and R. S. Popovic, "How to Liberate Integrated Sensors from Encapsulation Stress," *Sensors and Actuators* A21–A23, 908–910 (1990).
295. O. Rusanen, A. Torkkeli and J. Vähäkangas, "Packaging Induced Stresses in a Packaged Micro Mechanical Microphone," paper presented at *1998 International Symposium on Advanced Packaging Materials*, pp. 161–164 (1998).
296. W. Germer and G. Kowalski, "Mechanical Decoupling of Monolithic Pressure Sensors in Small Plastic Encapsulations," *Sensors and Actuators* A21–A23, 1065–1069 (1990).
297. F. Shemansky, L. Ristic, D. Koury and E. Joseph, "A Two-Chip Accelerometer System for Automotive Applications," *Microsystem Technologies* 1, 121–123 (1995).
298. W. M. Stalnaker, L. S. Spangler, G. Fehr and G. Fujimoto, "Plastic SMD Package Technology for Accelerometers," paper presented at *1997 International Symposium on Microelectronics*, pp. 197–202 (1997).
299. E. Koen, F. Pourahmadi and S. Terry, "A Multilayer Ceramic Package for Silicon Micromachined Accelerometers," paper presented at *8th International Conference on Solid-State Sensors and Actuators, and Eurosensors IX*, pp. 273–276 (1995).
300. L. Zimmermann, J. P. Ebersohl, F. L. Hung, J. P. Berry, F. Baillieu, P. Rey, B. Diem, S. Renard and P. Caillat, "Airbag Application: A Microsystem Including a Silicon Capacitive Accelerometer, CMOS Switched Capacitor Electronics and True Self-Test Capability," *Sensors and Actuators* A46–47, 190–195 (1995).

301. V. Boyadzhyan and J. Choma, "High Temperature, High Reliability, Integrated Hybrid Packaging for Radiation Hardened Spacecraft Micromachined Tunneling Accelerometer," paper presented at *IEEE International Workshop on Integrated Power Packaging*, pp. 75–79 (1998).
302. T. K. Tang, R. C. Gutierrez, C. B. Stell, V. Vorperian, G. A. Arakaki, J. T. Rice, W. J. Li, I. Chakraborty, K. Shcheglov, J. Z. Wilcox and W. J. Kaiser, "A Packaged Silicon MEMS Vibratory Gyroscope for Microspacecraft," pp. 500–505.
303. M. A. Mignardi, "Digital Micromirror Array for Projection TV," *Solid State Technology* 37, 63 (1994).
304. M. A. Mignardi and T. R. Howell, "Fabrication of the Digital Micromirror Device (DMD)," paper presented at *Commercialization of Microsystems '96*, pp. 208–209 (1996).
305. M. Mignardi, "From ICs to DMD^{TM}s," *TI Technical Journal* 15, 56–63 (1998).
306. K. W. Markus, "The Challenges of Infrastructure—Supporting the Growth of MEMS into Production," paper presented at *Commercialization of Microsystems '96*, pp. 28–32 (1996).
307. D. M. Burns and V. M. Bright, "Designs to Improve Polysilicon Micromirror Surface Topology," *Proceedings of SPIE*, pp. 100–110 (1997).
308. R. Nasby, J. Sniegowski, J. Smith, S. Montague, C. Barron, W. Eaton and P. McWhorter, "Application of Chemical-Mechanical Polishing to Planarization of Surface-Micromachined Devices," in *Technical Digest, Solid-State Sensors and Actuators Workshop*, pp. 48–53 (1996).
309. K. S. J. Pister, M. W. Judy, S. R. Burgett and R. S. Fearing, "Microfabricated Hinges," *Sensors and Actuators* A33, 249–256 (1992).
310. J. R. Reid, "Microelectromechanical Isolation of Acoustic Wave Resonators," Ph.D. Dissertation, Air Force Institute of Technology (1996).
311. M. C. Wu, L. Y. Lin and S. S. Lee, "Micromachined Free-Space Integrated Optics," *Proceedings of SPIE*, pp. 40–51 (1994).
312. J. H. Comtois and V. M. Bright, "Surface Micromachined Polysilicon Thermal Actuator Arrays and Applications," in *Technical Digest: Solid-State Sensor and Actuator Workshop*, pp. 174–177 (1996).
313. J. R. Reid, V. M. Bright and J. H. Comtois, "Automated Assembly of Flip-Up Micromirrors," paper presented at *Transducers '97: 1997 International Conference on Solid-State Sensors and Actuators*, pp. 347–350 (1997).
314. J. T. Butler, V. M. Bright and J. R. Reid, "Scanning and Rotating Micromirrors Using Thermal Actuators," *Proceedings of SPIE*, pp. 134–144 (1997).
315. R. R. A. Syms, "Rotational Self-Assembly of Complex Microstructures by Surface Tension of Glass," *Sensors and Actuators* A65, 238–243 (1998).
316. P. W. Green, R. R. A. Syms and E. M. Yeatman, "Demonstration of Three-Dimensional Microstructure Self-Assembly," *Journal of Microelectromechanical Systems* 4, 170–176 (1995).

317. D. VerLee, A. Alcock, G. Clark, T. M. Huang, S. Kantor, T. Nemcek, J. Norlie, J. Pan, F. Walsworth and S. T. Wong, "Fluid Circuit Technology: Integrated Interconnect Technology for Miniature Fluidic Devices," paper presented at *Solid-State Sensor and Actuator Workshop*, pp. 9–14 (1996).
318. C. Brown, "Common Acrylic Is Ideal Wafer Bonder," *Electronic Engineering Times* November 4, 43 (1994).
319. C. Gonzalez, S. D. Collins and R. L. Smith, "Fluidic Interconnects for Modular Assembly of Chemical Microsystems," paper presented at *Transducers '97: 1997 International Conference on Solid-State Sensors and Actuators*, pp. 527–530 (1997).
320. M. E. Poplawski, R. W. Hower and R. B. Brown, "A Simple Packaging Process for Chemical Sensors," in *Proceedings of the Solid-State Sensor and Actuator Workshop*, pp. 25–28 (1994).
321. T. S. J. Lammerink, N. R. Tas, J. W. Berenschot, M. C. Elwenspoek and J. H. J. Fluitman, "Micromachined Hydraulic Astable Multivibrator," in *Proceedings of the IEEE 1995 Micro Electro Mechanical Systems Workshop*, pp. 13–18 (1995).
322. D. Jaeggi, B. L. Gray, N. J. Mourlas, B. P. van Drieenhuizen, K. R. Williams, N. I. Maluf and G. T. A. Kovacs, "Novel Interconnection Technologies for Integrated Microfluidic Systems," in *Proceedings of the Solid-State Sensor and Actuator Workshop*, pp. 112–115 (1998).
323. P. Gravesen, J. Branebjerg and O. S. Jensen, "Microfluidics—A Review," *Journal of Micromechanics and Microeneingeering* 3, 168–182 (1993).
324. V. L. Spiering, J. N. van der Loolen, G.-J. Burger and A. van den Berg, "Novel Microstructures and Technologies Applied in Chemical Analysis Techniques," paper presented at *1997 International Conference on Solid-State Sensors and Actuators*, pp. 511–514 (1997).
325. C. González, S. D. Collins and R. L. Smith, "Fluidic Interconnects for Modular Assembly of Chemical Microsystems," paper presented at *1997 International Conference on Solid-State Sensors and Actuators*, pp. 527–530 (1997).
326. A. K. Henning, J. Fitch, D. Hopkins, L. Lilly, R. Faeth, E. Falsken and M. Zdeblick, "A Thermopneumatically Actuated Microvalve for Liquid Expansion and Proportional Control," paper presented at *1997 International Conference on Solid-State Sensors and Actuators*, pp. 825–828 (1997).
327. J. S. Fitch, A. K. Henning, E. B. Arkilik and J. M. Harris, "Pressure-Based Mass-Flow Control Using Thermopneumatically-Actuated Microvalves," in *Proceedings of the Solid-State Sensor and Actuator Workshop*, pp. 162–165 (1998).
328. H. A. C. Tilmans, E. Fullin, H. Ziad, M. D. J. Van de Peer, J. Kersters, E. Van Geffen, J. Berqvist, M. Pantus, E. Beyne, K. Baert and F. Naso, "A Fully-Packaged Electromagnetic Microrelay," paper presented at *Twelfth International Conference on Micro Electro Mechanical Systems*, pp. 25–30 (1999).
329. N. Koopman and S. Nangalia, "Fluxless Flip Chip Joining," in *Proceedings of NEPCON WEST*, pp. 919–931 (1995).

330. A. Paultre, "Microrelay Brings MEMS Technology to Commercial Market," *Electronic Products Magazine*, February, p. 28 (1999).
331. F. Pourahmadi, "Review of Modeling Silicon Microsensors and Actuators," *Sensors and Materials* 6, 193–209 (1994).
332. T. Dickerson and M. Ward, "Low Deformation and Stress Packaging of Micro-machined Devices," paper presented at *IEEE Colloquium on Assembly and Connection in Microelectronics*, pp. 7/1–3 (1997).
333. D. Dougherty, Q. Li and M. Shah, "Sensor Package Design Improvements Using Computer Simulation and Experimental Testing," paper presented at *Sensors in Electronic Packaging*, pp. 11–17 (1995).
334. A. C. McNeil, "A Parametric Method for Linking MEMS Package and Device Models," paper presented at *Solid-State Sensor and Actuator Workshop*, pp. 166–169 (1998).
335. S. Gong and M. H. White, "An Automatic Measurement System for Electromechanical Characterization of Silicon Pressure Sensors," *IEEE Transactions on Instrumentation and Measurement* 45, 184–189 (1996).
336. A. Cozma and R. Puers, "Electrostatic Actuation as a Self-Testing Method for Silicon Pressure Sensors," *Sensors and Actuators* A60, 32–36 (1997).
337. H. V. Allen, S. C. Terry and D. W. de Bruin, "Self-Testable Accelerometer Systems," paper presented at *MEMS '89: Micro Electro Mechanical Systems*, pp. 113–115 (1989).
338. T. Olbrich, A. M. D. Richardson and D. A. Bradley, "Built-In Self-Test and Diagnostic Support for Safety Critical Microsystems," *Microelectronics Reliability* 36, 1125–1136 (1996).
339. S. B. Brown, G. Povirk and J. Connally, "Measurement of Slow Crack Growth in Silicon and Nickel Micromechanical Devices," paper presented at *MEMS 93*, pp. 99–104 (1993).
340. S. B. Brown, W. Van Arsdell and C. L. Muhlstein, "Materials Reliability in MEMS Devices," paper presented at *1997 International Conference on Solid-State Sensors and Actuators*, pp. 591–593 (1997).
341. K. C. Kapur and L. R. Lamberson, "Reliability Estimation: Weibull Distribution," in *Reliability in Engineering Design*, pp. 291–341. Wiley, New York (1997).
342. J. M. Hu, "Physics-of-Failure-Based Reliability Qualification of Automotive Electronics," *Communications in RMS* 1, 21–33 (1994).
343. T. Maudie, "Improving Reliability of Electronic Appliance Pressure Sensors," *Appliance* 39 (1995).
344. T. Maudie and J. Wertz, "Semiconductor Sensor Performance and Reliability for the Appliance Industry," paper presented at *International Appliance Technical Conference* (1995).
345. D. R. Lide, *CRC Handbook of Chemistry and Physics*, 75th edn. CRC Press, Boca Raton, FL (1995).
346. T. Maudie, "Testing Requirements and Reliability Issues Encountered with Micromachined Structures," in *Proceedings of the Second International*

Symposium on Microstructures and Microfabricated Systems, pp. 223–230 (1995).
347. A. Nakladal, K. Sager and G. Gerlach, "Influences of Humidity and Moisture on the Long-Term Stability of Piezoresistive Pressure Sensors," *Measurement* 16, 21–29 (1995).
348. F. Rodriguez, *Principles of Polymer Systems*, 2nd edn. Hemisphere Publishing, Washington, DC (1982).
349. L. D. Fiedler, T. L. Knapp, A. W. Norris and M. S. Virant, "Effect of Methanol/Gasoline Blends at Elevated Temperature on Fluorosilicone Elastomers," paper presented at *Society of Automotive Engineers International Congress and Exposition* (1990).
350. E. A. Grulke, "Solubility Parameter Values," in *Polymer Handbook*, Vol. VII, J. Brandrup and E. H. Immergut, eds., p. 519. Wiley, New York (1989).
351. J. Leidheiser, "Mechanisms of De-adhesion of Organic Coatings from Metal Surfaces," paper presented at *Polymeric Materials for Corrosion Control*, pp. 125–135 (1985).
352. W. Wang and J. Leidheiser, "A Model for the Quantitative Interpretation of Cathodic Delamination," in *Equilibrium Diagrams Localized Corrosion*, pp. 255–266 (1983).
353. D. C. Montgomery and E. A. Peck, *Introduction to Linear Regression Analysis*, 2nd edn. Wiley, New York (1992).
354. D. C. Montgomery, *Design and Analysis of Experiments*, 3rd edn. Wiley, New York (1991).
355. C. A. Dostal, *Engineering Materials Handbook*, Vol. 2: *Engineering Plastics*. ASM International, Metals Park, OH (1988).
356. D. G. McIntyre, S. J. Cunningham, J. S. Carper, P. D. Jaramillo, S. Tatic-Lucic and L. E. Starr, "Characterization of the Influence of Fabrication Methods on Microstructure Failure," *Sensors and Actuators* A60, 181–185 (1997).
357. L. Muller, M. H. Hecht, L. M. Miller, H. K. Rockstad and J. C. Lyke, "Packaging and Qualification of MEMS-Based Space Systems," paper presented at *MEMS '96: The Ninth Annual International Workshop on Micro Electro Mechanical Systems*, pp. 503–508 (1996).

CHAPTER 8

MEMS, MICROSYSTEMS, MICROMACHINES: COMMERCIALIZING AN EMERGENT DISRUPTIVE TECHNOLOGY

Steven Walsh
University of New Mexico

Jonathan Linton
New York Polytechnic Institute

Roger Grace
Roger Grace Associates

Sid Marshall
Micromachine Devices

Jim Knutti
Silicon Microstructures Inc.

CONTENTS

8.1 Introduction / 481

8.2 Commercialization of Current M^3 Technology / 483
    8.2.1 Research & development / 484
    8.2.2 Marketing / 484
    8.2.3 Public relations / 485
    8.2.4 Startup funding attraction / 485
    8.2.5 Market research / 485
    8.2.6 Creation of wealth / 485
    8.2.7 Established infrastructure / 486
    8.2.8 Design for manufacturing / 487
    8.2.9 Industry association; standards / 487

8.3 The M^3 Commercialization Process / 488
    8.3.1 The commercialization process—steps 1 and 2: Choosing and sourcing an M^3 technological platform / 492
    8.3.2 Steps 3 and 4: Innovation and market strategies for MEMS / 493

8.4 Market Studies: Their Problems and Value / 494
    8.4.1 Product-market-technology paradigms / 496

      8.4.2   The Nexus MEMS/MST market study / 500
      8.4.3   Related and interlocking European Union efforts / 503

8.5   Technology Roadmap Development for an Emergent Industry / 506
      8.5.1   Nature of the M^3 industry / 506
      8.5.2   Theory and execution of M^3 roadmapping / 507
      8.5.3   Structure of the M^3 roadmap / 507
      8.5.4   Examples of early inputs of the roadmap committee / 509

8.6   Evolving M^3 Infrastructure / 510
      8.6.1   Other infrastructure components / 512

8.7   Conclusions / 513

     References / 514

8.1 Introduction

Microelectromechanical-systems (MEMS), microsystems technology (MST), and micromachines are roughly synonymous terms applied in the United States, Europe, and Japan respectively, to manufacturing technologies that are enabling the development of many exciting microscale or microminiaturized products. For convenience here, the terms are lumped together and labeled M^3. In the parlance of management science, M^3 technologies are:

- exogenous (they originate from outside an existing or familiar frame of reference); and
- disruptive (they are revolutionary rather than evolutionary).

These technologies are creating a second micromanufacturing revolution—the first being microelectronics—as well as a fertile environment for discontinuous innovations that will revitalize existing industries and generate new markets.

M^3 technologies:

- enable the manufacture of products that would otherwise be impossible to produce;
- improve manufacturing efficiencies by as much as an order of magnitude;
- improve critical performance aspects of current products by as much as an order of magnitude; and/or
- improve product quality by reducing component size and number.

Nevertheless, while many firms, large and small, have initiated commercialization activities based on or involving M^3 technologies, and while many governments and national labs also have major M^3 initiatives, commercialization of M^3 technology has proven problematic. This chapter focuses on the underlying reasons for this slow commercialization and discusses industry actions now in progress that could accelerate the commercial adoption of the technology.

Impediments to M^3 commercialization include industrywide differences in nomenclature, differences in manufacturing and marketing infrastructure, and inherent problems that face firms trying to gain competitive advantage in an environment characterized by disruptive technologies and discontinuous innovations.

M^3 technologies generally fall into one of three categories: traditional bulk micromachining, sacrificial surface micromachining, and high aspect ratio micromachining (HARM). The latter includes deep ultraviolet (DUV) lithography techniques and X-ray-based methods such as LIGA (from the German *Lithographie, Galvanoformung, Abformung*, meaning lithography, electroforming or plating, and molding).

These three technology categories are based in large measure on sophisticated processing methods not unlike those used for producing silicon integrated circuit chips, but there are many differences. For example, unlike semiconductor technologies in which nearly all systems are made in silicon, a significant and growing percentage of microsystems utilize alternative substrates such as glass and various plastics. Furthermore, unlike the microelectronics revolution, which generated semiconductor-based products conceptualized primarily in the domain of electrical engineering, M^3 technologies use the skills of many other engineering disciplines.

Firms utilizing M^3 technologies are in the process of making those technologies manufacturable. Many times, a firm's choice of M^3 manufacturing technology, product design, or product application is linked to its historical capabilities. Established firms as well as entrepreneurial startups strive to capitalize on their own inherent competencies. In the case of entrepreneurial firms, the choice may be guided by the background of the founders or by a truly novel manufacturing path.

The absence of a single M^3 manufacturing paradigm underscores the uncertainty that firms must face when searching for competitive advantage in the M^3 arena. The triple nomenclature problem illustrates not only regional preferences in technological approach, but alludes to the emergent nature of the technologies as well. Nevertheless, there is a trait shared in common by all M^3 technologies and microelectronics: They have initiated the emergence of new products that have the potential for profoundly effecting the lives of all human beings. As microelectronics has benefited society in the last half of the twentieth century, so M^3 technologies show the potential for doing this for the society of the new millenium.

The emergent nature of M^3 technologies continues to challenge existing definitions as well as to frustrate suppliers, manufacturers, and M^3 users. Marketers and strategists can be overwhelmed when each day seemingly brings new applications that further push the limits of definitions with which they have become comfortable. But if one is to understand the competitive discourse of this emergent industry, awareness of existing differences in definitions is critical.

The point is illustrated by considering studies that have been made of the growing M^3 market since 1990. Deriving full benefit from these studies requires that one understand the reference point of each. Variations in study definitions can produce wide differences in market estimates. Those who use market analyses must use them prudently and with insight.

Firms involved in commercializing M^3 technology have been confronted with year 2000 market forecasts ranging from US$8 billion[1] to $14 billion.[2] Newer studies have created even wider variations. For example, a 1998 market analysis[3] by an agency of the European Commission (EC), the Network of Excellence in Multifunctional Microsystems (Nexus), predicts a market near US$40 billion by 2003. Considering that the annual market for M^3 products prior to 1995 was less than US$1 billion, all of these studies make two things clear: they predict a significant rate of growth in the M^3 area and they imply an explosion in the number of stakeholder groups.

Industry, government, and academia worldwide are involved in the current commercialization of M^3 technology. Governmental and academic activities often speak to technological research and manufacturing infrastructure requirements, whereas industry is more focused on product development. Firms such as Kulite, Sensym, IC Sensors, MTI, Foxboro ICT, and Novasensor in the United States were among the first to commercialize MEMS devices. More recently, Lucent Technologies, Analog Devices, Texas Instruments, Phillips, and others have launched M^3 entries in existing markets or have generated totally new markets based on M^3 technologies. A noticeable uptick in the entry of large firms in the M^3 arena has occurred in 1998 and 1999.

Recently, the international M^3 community initiated development of a technology roadmap. Since M^3 technologies are highly disruptive, roadmapping them differs greatly from the roadmapping process applicable to more sustaining technologies. The best-known example of a roadmap for a sustaining technology that follows a continuous innovation pathway is the Semiconductor Industry Association's International Technology Roadmap for Semiconductors.

Technology roadmapping in the M^3 arena requires focusing on different paradigms. The design, materials, fabrication processes, generated functionalities, and production methodologies for M^3 will all differ significantly from the counterparts that they are replacing. This was well illustrated when the original $18 mechanical accelerometer device used for automotive airbag triggering was replaced with a microelectromechanical accelerometer (one variation of which now includes on-chip control electronics and costs about $3). Another example of a radically enabling microelectromechanical system is the inkjet printhead. These illustrations underscore the concept that industries that adopt M^3 technologies will dramatically change their product–technology–market paradigms.

Market-pull forces are only now starting to reach the level of technology-push forces for M^3 technology products. M^3 drivers are still more technology than market oriented. Nevertheless, market forces are commanding more and more attention. In the next section, we examine the global state of adoption and commercialization of M^3 technology.

8.2 COMMERCIALIZATION OF CURRENT M^3 TECHNOLOGY

The past five years have witnessed an unprecedented growth of interest in M^3 technologies. It is estimated that more than 600 companies, research institutes, and universities are currently involved with M^3 technologies worldwide. U.S. organizations represent the largest portion of this high-tech international community. Yet with all this interest, a number of issues and barriers are still impeding the commercialization of M^3 technologies by startup companies and well-established firms. Problems are manifested in a variety of business areas, including R&D, marketing, infrastructure support, and venture capital funding.

An M^3 industry "report card" was recently used to assess industry performance in these areas.[4] The 1999 report card is shown in Table 8.1. The accompanying commentary discusses the rating factors and is aimed at highlighting current commercialization barriers.

Table 8.1 M^3 industry commercialization report card (1999).

Subject	Grade
R&D	A
Marketing	C-
Public relations	B
Market research	B-
Design for manufacturing	B-
Established infrastructure	B
Industry association	INC
Standards	INC
Management expertise	C
Startup funding attraction	B
Creation of wealth	B-
Industry roadmap	B-
Profitability	C-
Overall grade	**B-**

8.2.1 RESEARCH & DEVELOPMENT

M^3 R&D activities in the United States were initiated in the 1970s and 1980s by the National Institutes of Health (NIH) and the National Aeronautics and Space Administration (NASA). Worldwide M^3 R&D is currently well funded by a variety of agencies and countries. Vigorous investment by the U.S. Defense Advanced Research Projects Agency (DARPA), the U.S. Department of Energy (DOE), the European Commission's Nexus organization, agencies in Korea and Taiwan, the Micromachines Center and the Ministry of International Trade and Industry (MITI) in Japan, and other government agencies and various private companies worldwide has raised R&D interest in M^3 technologies over the past few years. Increasing attendance at conferences is evident and record attendance was reached at Transducers '97 (Chicago), MEMS '98 (Heidelberg), the Solid State Sensors and Actuators Conference '98 (Hilton Head, North Carolina), the Commercialization of Microsystems '98 conference (San Diego), and Sensors Expo (Baltimore, May 1999).

8.2.2 MARKETING

The insufficiency of successful marketing for M^3 products is undeniable. Foremost among the various marketing problems is the lack of a thorough understanding of market needs and applications requirements. Marketing techniques in firms, even high-tech firms, have been developed for sustaining technologies based on incremental innovation. Most firms utilize market-focused, internally focused, or customer-focused approaches rather than a corporate expeditionary marketing approach—that is, if any approach to M^3 is proactively taken at all. The problem is that traditional marketing techniques simply have not been successful in promoting disruptive technologies. They do not address many basic issues such as a traditional firm's aversion to utilizing a discontinuous innovation-based solution.

8.2.3 PUBLIC RELATIONS

The M^3 industry has received considerable coverage in the academic press. That coverage is now extending to the popular, technical, and business publications sector. However M^3 articles still tend to be overshadowed by other high-tech topics. This is due in part to a lack of marketing expertise in the M^3 industry. Promotion and evangelizing, which could be used productively if an industry association were in place, has been seriously overlooked.

8.2.4 STARTUP FUNDING ATTRACTION

M^3-based companies, until recently, have been considered too risky for most investors, especially venture capitalists. Notable exceptions include TDI, IC Sensors, Novasensor, MTI, Redwood Microsystems, Cronos, TMP, and Standard MEMS. However, most MEMS companies fail to satisfy some of the primary criteria for venture capital and private investor ("angel") funding. Among these are a strong management team with demonstrated previous industry successes (grade = C), a well-defined market-sized growth rate, definition of a target market documented by formal market research (grade = C), and the potential for significant return on investors' capital within the first 5 years of operation (grade = C). Nevertheless, the funding situation for M^3 has been improving as investment funds flow in from nontraditional venture sources.

8.2.5 MARKET RESEARCH

Certifiable, current market data are conspicuously unavailable.[5,6] Of eight frequently cited MEMS market research studies, only the Battelle multiclient study[1] satisfies accepted methodology criteria for a "bottom-up" (supplier of M^3) and "top-down" (user of M^3) study. Now, 7 years later, this 1992 report is dated. The new Nexus[3] and SPC[7] market studies provide excellent insight and are based on, in the former case, over 200 interviews and in the latter case, a nearly equivalent amount of self-reported corporate information. However, these two studies employ different definitions of M^3 technologies.

8.2.6 CREATION OF WEALTH

The creation of personal wealth in the M^3 arena is still a very low visibility phenomenon. It's estimated that fewer than 20 MEMS millionaires currently exist. In

Table 8.2 Prominent M³ mergers and acquisitions.

Startup	M&A (by or into)
Novasensor	Lucas
IC Sensors	EG&G
I.C.T.	Foxboro
Sensym	British Tyre and Rubber
Data Instruments	Next
Silicon Microstructures	Exar → OSI
M.T.I.	Hewlett Packard
Quinta	Seagate

most cases, millionaires have been created not via IPOs (initial public offerings) but rather through mergers and acquisitions of startup firms by larger organizations (Table 8.2).

8.2.7 ESTABLISHED INFRASTRUCTURE

Firms providing infrastructure equipment for M³ technologies have made major investments recently. Companies including Karl Suss, Electronic Visions, STS, PlasmaTherm, and Alcatel produce equipment specifically designed to meet the requirements of M³ manufacturers. Design automation activities are also in high gear. Companies such as Microcosm, Intellisense, Tanner, CFDRC, Coyote Systems, and Memscap/Mentor are developing new software aimed at reducing design iterations, lowering development costs, and reducing time to market.

Test and measurement activities for M³ are being pursued by major producers and research groups, including Analog Devices Inc. and Sandia National Laboratories, and by companies specifically established to provide this service and related equipment, such as ETEK.[8] Ubiquitous M³ processes like Sandia's IMEMS and Summit IV, and MCNC's MUMPS are beginning to be well utilized. The creation of a number of MEMS foundries and specialty fabricators (e.g., Cronos, Sentir, Twente MicroProducts, CSEM, Tronics, MEMS Inc., Standard MEMS, and others) is making it possible for designers and systems manufacturers to put their creations into production and to support mid-to-large-scale production programs. This fabless supplier model enables small-to medium-sized enterprises to become M³ providers. Finally, the recent establishment of capabilities at organizations such as the Jet Propulsion Laboratory (NASA) and Sandia to help M³ companies better understand emerging reliability issues associated with product development and production is an encouraging note.[9]

8.2.8 DESIGN FOR MANUFACTURING

Packaging for M^3 technologies greatly lags behind device development. There are major problems with interconnectivity and packaging that translate directly into cost. It is said that 80% of the cost and value of a current M^3 device resides in the packaging. Papers on packaging are beginning to appear regularly at M^3 technology conferences,[10] suggesting an increase in research efforts in the packaging area. One potential solution is the co-development of devices and packages. Although there are exceptions, generally speaking, interdisciplinary and interdepartmental approaches to packaging are nonexistent.

8.2.9 INDUSTRY ASSOCIATION; STANDARDS

Activity is beginning that could result in the creation of an M^3 industry organization. In the United States, SEMI has taken the lead. In Europe IVAM has promoted MST activities. The Nexus program with its user groups has brought together some 300 private companies, universities, and research organizations under one umbrella. An industry organization could greatly accelerate the commercialization of M^3 technologies. Chief among its benefits would be a formal effort on standards development, an international technology roadmap effort, and the collection of reliable market data.

Industry awareness of issues in M^3 commercialization is growing. Conferences such as the Commercialization of Microsystems'98 (San Diego) and Commercialization of Microsystems: Europe (Dortmund, Germany, July 1999) will continue to provide forums for sharing experiences and stimulating the growth of a healthy industry.

As already noted, M^3 R&D activities are vigorous. They cover a spectrum of technologies, including relays/switches, microfluidics, biosystems, microactuators, photonics, displays, and RF telecommunications areas, to name a few. These efforts will fuel the expected commercialization activities of the next decade (see Table 8.3).[11] Another indicator of increasing commercialization activity is the number of M^3 patents issued. Nagel et al.[12] depict this growth in Fig. 8.1.

For the most part, customers don't care what specific technology is used to solve their applications problems. What they do care about is that the technology solution employed optimally satisfies their requirements in terms of form, fit, function, performance, and price. Better yet, a customer prefers a superior solution based on a known solution base. This problem plagues technology-driven companies, i.e., companies that are developing new technologies and competencies. It is not a problem unique to the M^3 industry. M^3 companies must devote more resources to understanding customer needs through detailed market research that focuses on product-technology paradigms rather than product-market paradigms.

For all of this, many M^3 technology products currently enjoy growing worldwide markets. For example, 1997 production levels reached 25 million units for

Table 8.3 Timetable M³ product evolution.

Product	Discovery	Product Evolution	Cost reduction/ application expansion	Full commercialization
Pressure sensors	1954–1960	1960–1975	1975–1990	1990-present
Accelerometers	1974–1985	1985–1990	1990–1998	1998
Gas sensors	1986–1994	1994–1998	1998–2005	2005
Valves	1980–1988	1988–1996	1996–2002	2002
Nozzles	1972–1984	1984–1990	1990–1998	1998
Photonics/displays	1980–1986	1986–1998	1998–2004	2004
Bio/chemical sensors	1980–1994	1994–1999	1999–2004	2004
RF Switches (RF)	1994–1998	1998–2001	2001–2005	2005
Rate sensors	1982–1990	1990–1996	1996–2002	2002
Chromatography	1975–1980	1980–1985	1990–2000	2001
Microrelays	1977–1982	1993–1998	1998–2006	2006

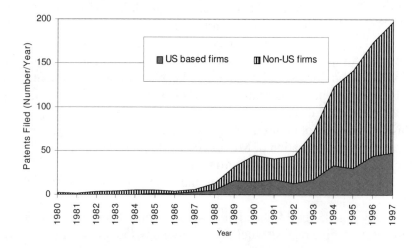

Figure 8.1 Worldwide M³ patent activity (after Ref. 12).

manifold absolute pressure sensors for automotive engine controls, 18 million units for disposable blood pressure sensors for patient monitoring, 18 million units for airbag accelerometers, and 140 million units for inkjet printer heads.[13] However, those in search of the "MEMS killer applications," are still searching.

8.3 THE M³ COMMERCIALIZATION PROCESS

Since the dawn of the industrial revolution, commercialization of new technology has been the source of national wealth. Emergent technologies like M³ have been the wellspring for the wealth-creation process.

Industrial society is now halfway through a third century of accelerating development that has radically changed the world. But the path of new technology commercialization has bifurcated over the last 75 years. One branch has produced products and services that continue traditional streams of technology, while the other branch has been studded with product and service milestones that mark novel and often disruptive technologies.

According to Bower and Christensen,[14] in any given industrial environment, new technologies either *sustain* current manufacturing practices and technological capabilities or *disrupt* those practices and capabilities making them obsolete. A disruptive technology is thus one that introduces something so new and different that it requires an upheaval in existing firm-based manufacturing practices.

Disruptive technologies are closely intertwined with another developmental concept—*discontinuous innovations*. This term may be applied to step-function improvements in current product-market paradigms or to physical and service products that are representative of a new industry or market. Discontinuous innovations thus define new and differing product platforms from which, ultimately, a normal progression of incremental innovations is generated. Discontinuous innovations are frequently based on disruptive technologies, but they can be the products of current sustaining technologies that produce a higher value proposition. Discontinuous innovations often change the way customers utilize products and, in the process, they may redefine an industrial value chain or create a new value chain or market proposition.

Disruptive technologies have been given many different names, but by any name they are increasingly important to individual firms. Moreover, they are no longer the sole domain of entrepreneurial firms. Disruptive technologies change existing product-technology paradigms, often replacing them with manufacturing models that require new technology capabilities. They launch new products and create nascent industries. They stimulate the development of new firm-based competencies. And they are the wellspring of future sustaining technologies.

Firms that embrace disruptive technologies have been most successful when they utilize those technologies to first replace existing products (see Fig. 8.2) and then go on to redefine the value proposition in an industry, thereby repositioning the firm and the industry's value chain. Many firms turn success with a replacement product into a springboard for more aggressive, more value-added, more promising innovation platforms that have a greater impact on the corporate bottom line. This is a learn-by-doing approach[15] that has been described frequently in managerial literature since 1950.

M^3 advances are the emergent phase of a disruptive technology. M^3 is currently long on promise, but the breadth of potential replacement products in its future could radically improve existing product-technology paradigms. Moore[16] noted that disruptive technologies tend to generate discontinuous innovations that then require users/adopters to change their behavior or practices. In terms of commercialization, this raises a question. How can the exceptional but disruptive technological potential of M^3 best be captured and transformed, through discontinuous innovation, into exceptional profit potential?

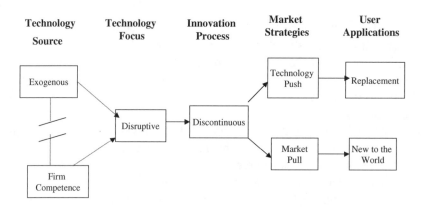

Figure 8.2 The Disruptive technology commercialization model (after Ref. 20).

M^3 technologies are exogenous disruptive technologies that form the basis for large radical[17] or discontinuous innovations.[18] The continuous or incremental innovations of sustaining technologies[19] stand in contrast. Disruptive technologies like those of M^3 often promise strong and early market entry for products, but they are accompanied by a high risk of failure. In large firms, disruptive technologies have traditionally been the responsibility of R&D organizations. In new high-tech firms, they are the responsibility of everyone, especially the founders of the firm.

Can such commercial activities deliver appropriate payoffs to investors? In something of a "Catch-22," firms with broad product lines may find that disruptive technology-based innovation "disrupts" their existing profitable product structure, thereby reducing revenues.

Advances and commercialization of M^3 disruptive technologies can generate financial concerns for both small and large firms. Disruptive technologies, with their unstructured nature and uncertain technological outcomes, make commercialization hard to quantify and therefore hard to justify in financial terms. The promised MEMS "bonanza" is a case in point. It is taking a long time to emerge. In contrast, commercialization of incremental evolutionary or sustaining technologies (e.g., the sub $1000 PC) may find widespread application within months.

Nonetheless, the process of identifying and developing strategically important M^3 disruptive technologies is vitally important to the future viability of technology-based firms. M^3 disruptive technologies will be the birthplace of new corporate technological competencies, the center of evolving and distinctive corporate identities, and the epicenter from which generations of improvements in corporate competencies radiate.

Firms that commercialize innovations based upon disruptive technologies face enormous internal and market-related problems. These are manifested in both external and internal skepticism regarding the value of the technology, in difficulty with manufacturing processes, and in resistance from marketing managers regarding product promotion. Potential users can manifest resistance to product adoption and more significantly, perhaps, to behavioral changes that may be necessary for broader implementation of the technology. One might expect small firms to be more seriously affected by such difficulties, but in fact experience shows that small firms seem to handle them more successfully.

The model presented in Fig. 8.2 (adapted from Walsh and Kirchhoff[20]) for the commercialization of products based on M^3 disruptive technologies begins with a choice of technology type and ends with a user application. This model was influenced by the works of Utterback,[21] Twiss,[22] Frohman,[23,24] Marquis,[25] Rosenbloom,[26] Abernathy and Clark,[27] Moore,[16] Bower and Christenson,[14] and others.

The model of Fig. 8.2 provides a view of major factors affecting the innovation process. The top boxes identify the technology push associated with specific state-of-the-art of technologies and manufacturing processes. In the lower boxes, market pull stems from a demand for capabilities not yet available in existing products.

In advanced technologies, market pull is commonly generated by defense requirements, government regulatory pressures (e.g., environmental, safety, and health); explicit commercial needs (e.g., advanced wireless telecommunications) and unfulfilled market opportunities. Technology push, on the other hand, can result in major innovations that often piggyback on technologies in existing markets,[28] thus decreasing the technological effort required to produce a novel product. For example, this occurred early in the history of the semiconductor silicon industry when the technology for producing silane-based polysilicon feedstock was borrowed from the burgeoning silicone industry. In a similar way, M^3 technologies have borrowed from semiconductor microfabrication.

Market acceptance is a critical element in the commercialization of disruptive technologies. Marketing theorists such as Von Hipple,[29–31] Moore,[16] and Veryzer[32] have codified the connection between innovation and market acceptance. Another consideration in commercialization is the idiosyncratic nature of a firm's approach to the innovation process. This may be displayed positively in the embrace of the discontinuous innovation process.[33]

Finally, there is the nature of an innovation itself. Many researchers have discussed the problems involved with developing radical innovations.[34] In general, as the commercial significance of an innovation increases, the difficulty and duration of its implementation increases.

The commercialization literature is rife with the concept that major change agents play a role in radical commercialization challenges. Two dominant change agents are associated with the disruptive technological commercialization process for M^3. The external agent is proactive government policy. The internal agent is the backing of a product champion. Both can have a significant impact.

Such change agents are powerful because they can decrease the perceived risk that a firm associates with a project. A proactive government policy to fund a selected proposal decreases the monetary risk to an individual firm. An internal product champion often overcomes corporate roadblocks that would otherwise inhibit commercialization of a particular product.

In corporations, product champions are often revered as visionaries, but they may also be feared. These individuals tend to be moderate risk takers. They are not prone to backing bold high-risk projects.[35,36] Product champions often obtain higher than average returns on new product opportunities. On the other hand, when their attempts fail, dire consequences rarely result. The development of the accelerometer device for airbag deployment at Analog Devices provides an excellent example of the positive impact of a product champion, in this case, Dr. Richard Payne.

Until quite recently, M^3 commercialization has progressed at a snail's pace. The model in Fig. 8.2 is useful for identifying causes. Many researchers[21,37,38] have found that the time lag from invention (technological change) to business innovation for radical innovations of all types averages 11 to 18 years. This lengthy period has been remarkably consistent in the twentieth century. Overcoming buyer/user resistance to disruptive technologies involves demonstrating significant cost reductions and/or performance improvements. These induce customers to accept the risks of newness.[37] Interestingly, this does not occur in just a single-product/single-industry market, but rather, in lead user groups in many different markets.[31] Examples include the medical devices industry's utilization of MEMS-based disposable blood pressure sensors and the automotive industry's use, at lead firms such as BMW and Ford, of MEMS-based accelerometers for automotive safety applications.

Schumpeter's[39] "winds of creative destruction" have traditionally moved slowly, but have redefined the way industries conduct business. Schumpeter describes capitalism as an economic system that finds its competitive strength in innovation. His "innovative activity" (creative destruction) is clearly driven by what today we call disruptive technologies. Abernathy and Utterback[40] emphasize this in describing disruptive technologies as those that generate entirely new technology-product-market paradigms that in turn create new markets that initially may be opaque to customers. Opaqueness, they note, constrains customer enthusiasm for changing established behavioral habits. But such enthusiasm is needed if disruptive products are to be accepted and utilized.

To understand the commercial innovation process, one must recognize that in a business sense, innovation is aligned with use. It is not simply technological change. An innovation in a business sense is a new product, process, or service, which gains commercial viability and produces revenue. For this discussion, there is a difference between a technological innovation and a commercial innovation. In discussing commercialization, we use the business sense of the term "innovation."

8.3.1 THE COMMERCIALIZATION PROCESS—STEPS 1 AND 2: CHOOSING AND SOURCING AN M^3 TECHNOLOGICAL PLATFORM

Disruptive technologies not only create sustainable competitive advantage, they provide proprietary markets for evolutionary technologies. Discontinuous innovations lead to changed customer behavior that accepts the advantages of the new disruptive technologies. Once user behavior is changed, the new technology becomes accepted in the customer's organization. As users identify new needs, evolutionary technologies emerge from core competence research to create continuous innovations that satisfy these users. In this way, a technology that was initially disruptive eventually leads to a stream of evolutionary technologies. Evolutionary technologies produce continuous innovations that constantly update and improve current customer-based products. Companies that build on disruptive technologies remain competitive in rapidly changing markets.

Basically, there are three disruptive technology choices for firms embracing M^3 concepts: traditional bulk micromachining, sacrificial surface micromachining, and high aspect ratio micromachining. The oldest—traditional bulk micromachining—has been the dominant production technology for pressure sensors since such sensors were introduced in the 1950s.[1] Many firms have based their products on bulk micromachining. Included among these are the early startups such as Rosemont, Kulite, National Semiconductor, Fairchild, and divisions of large corporations such as General Motors.

HARM technologies such as LIGA, deep UV (laser LIGA), and microplating are becoming more commercially viable. They were developed for the fabrication of high aspect ratio plating molds and provide radical new ways to produce micromachined parts at relatively low cost. The German firm, microParts GmbH, is the world's largest producer of HARM-based products, but the technology is moving to the Pacific Rim. Taiwan's National Science Council is currently funding a National Light Source Project that is chartered to develop HARM-based products. In Japan, a MITI-supported effort called the Micromachine Center has a concentrated program to investigate HARM technologies for its 28 sponsor firms.

Sacrificial surface micromachining was introduced in the 1980s and is the newest M^3 technology. It is the base process for Analog Devices' automotive airbag accelerometers and Texas Instruments' digital light processing (micromirror) technology, as well as a number of other significant commercialization thrusts.

8.3.2 Steps 3 and 4: Innovation and market strategies for MEMS

By definition, according to Schumpeter,[39] innovation implies the commercialization of invention or technological change. Commercialization is widely perceived as the process of bringing a product or service into working applications. This is heavily dependent upon marketing. The model of Fig. 8.2 explores discontinuous market-pull innovations and discontinuous technology-push innovations.

8.3.2.1 Discontinuous Market Pull

This combination of disruptive technology and market strategy may be the lifeblood of true "creative destruction" as described by Schumpeter.[39] The combination is not itself the basis for creative destruction, but it provides a bridge between the old and the new. Disruptive technologies need to evolve over time so that inventions can gradually acquire production infrastructure, e.g., raw material suppliers and process methods. One way to accomplish this is to launch the technology by manufacturing replacements and substitutes for preexisting customer applications. Such replacements are usually easy to sell and require minimal changes in customer behavior. Early applications of M^3 technologies were in products that replaced large and expensive sensing devices. M^3 devices offered equal or better sensing, smaller and more reliable designs, and lower cost. The development and manufacturing of sensing devices still dominates M^3 production in the United States. The previously mentioned automotive airbag triggering accelerometer is a good example of a traditional MEMS device now costing less than $3 that replaced a mechanical device costing far more.

8.3.2.2 Discontinuous Technology Push

While users can be relied upon to pull replacement and/or substitute innovations into the market, they rarely pull major discontinuous innovations into any market. The behavioral changes required by the application of discontinuous innovations invariably create trauma for users, even if the innovation greatly improves performance and/or efficiency. Since customers resist self-imposed trauma, they do not ask suppliers to create discontinuous innovations for them. Discontinuous innovations need to be pushed into markets by technology. Again, medical sensors provide an example. Doctors and patients alike had to be educated to understand the value of the new MEMS disposable blood pressure sensors. In the automotive industry, it took regulatory action to instigate MEMS use for passive restraints.

Numerous small new firms are appearing in the M^3 industry as the technology evolves and expands. M^3 will create major disruptions as these and many other firms push the technology in the future.

8.4 MARKET STUDIES: THEIR PROBLEMS AND VALUE

The good news is that M^3-based product sales are growing rapidly. The bad news is that the emergent nature, regional definitions, and seemingly daily new product unveilings cloud commercialization issues and have caused market researchers to generate widely disparate views of M^3 market growth. Figure 8.3, a compilation of different M^3 market studies, attests to this problem. Some of the difficulty resides in the differences between the definitions of the markets being studied. Accordingly, those using market analyses must use them prudently and with insight.

Some of the studies referenced in Fig. 8.3 are now dated; others used procedures not generally accepted in market research, and still others employed assumptions of uncertain validity, which make their results uncertifiable. Nevertheless, there is value in all of the studies since they focus attention on the emerging M^3 field. Taken together, the lumped data from all studies present sufficient market information to enable one to generate some best guesses on specific M^3 product/market/technology (PMT) paradigms.

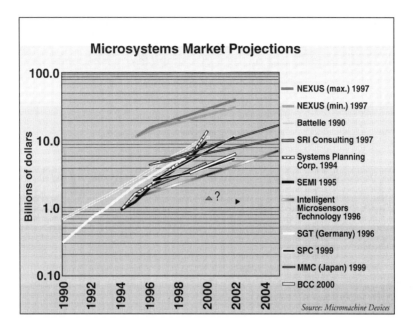

Figure 8.3 M^3 market study comparisons (after Ref. 5).

All the data are of some use, but depending on one's point of view, some studies may provide more appropriate information than others do. To assist those trying to obtain some market perspective based on both existing M^3 market studies and other more informal sources of expert opinion, several things should be kept in mind. Nearly all existing studies and expert opinion depict an industry with a compound annual growth rate (CAGR) of 40% to 60%.

Three points should be clarified here. First, although we have been using the term M^3 for convenience, readers should be reminded that the terms "MEMS," "MST," and "micromachines" are not always directly interchangeable and hence the scopes of the studies that use these terms can vary greatly. Second, the market projections that follow may oversimplify differences of scope in prior studies by using both MST projections based on systems pricing and MEMS projections based on elemental devices. Third, the authors assign no personal value judgment to any set of numbers used in this discussion. The numbers are used for illustrative purposes only.

Examples generally lend credence to M^3 market studies, but they do not entirely substantiate them. The burden is on the market researcher to demonstrate credibility. Uncertainties associated with economic analyses make it critically important that researchers presenting current market study numbers be intimately familiar with market study procedures, with the MEMS marketplace, and with the commercialization aspects of disruptive technologies. Market researchers must be prepared to defend their assumptions, their definitions, their procedures, and their actual market evaluations.

In the examples that follow, M^3 applications are specifically defined in the fields of information technology, communications, aerospace, automotive manufacturing, and medicine and biology. For simplicity, we use only one comprehensive example of an "emerged product" and one best-guess example of a PMT paradigm to illustrate each area.

8.4.1 Product-Market-Technology Paradigms

The tables presented here are meant to offer commercially useful strategic insight. They combine available market study information with much currently existing expert opinion. Each table lists numbers (all based on worldwide production) for six variables in 2 years: the data point year (1996) and the forecast year (2002). The variables examined are:

- the volume range in units of the product application, as determined by various M^3 studies and expert opinion;
- a best-guess MEMS device price (retail) for the product under review;
- an associated MST systems price;
- an indication of whether the MEMS application represents totally new technology or a replacement technology;
- cost drivers, if any;
- performance advantages, if any.

Readers are cautioned against trying to estimate the total size of a particular market segment simply by multiplying the price and volume figures presented here. If this erroneous procedure is used, for example, then the lowest price (variable No. 2) × lowest volume, and the highest price (variable No. 3) × highest volume, will provide grossly exaggerated, unrealistic, and rather meaningless market value ranges.

8.4.1.1 Information Technology (IT)

In IT, the two examples of M^3 product applications are (1) "already emerged" MEMS-based inkjet (IJ) printer heads and (2) "now emerging" MEMS-based data storage devices. Both aim at replacing existing PMT paradigms, and both are making inroads in established markets. The MEMS/MST-based products either already show better performance, lower cost, and a steeper and more accelerated learning curve than the products being replaced, or at least they possess the potential to do so.

Inkjet printheads provide perhaps the best illustration of how MEMS purists and MST proponents differ. The latter regard the entire inkjet cartridge delivery system as a microsystem, whereas the former group counters that only the nozzle of the device is micromachined. Both are correct by their own definitions. However, the definition that one accepts substantially changes the expected market. IJ

Table 8.4 Inkjet printheads.

Variable	1996	2002
Units (millions)	500–100	250–500
MEMS price (US$)	1–4	1
MST price (US$)	50–60	25–50
Technology replaced	Piezo	Laser
Cost advantage	<20%	20%–200%
Performance advantage	Near 10×	1.2× to 10×

Table 8.5 Hard disk trends.

Variable	1996	2002
Units (millions)	100–700	400–2000
MEMS price (US$)	1–10	1–5
MST price (US$)	1–20	1–10
Technology replaced	Macro	Laser
Cost advantage	<20%	20%–200%
Performance advantage	Near 10×	10×

printheads are made by more than eight companies worldwide, including multinational firms such as Canon, Hewlett Packard, and Lexmark. MST now dominates inkjet head manufacturing. MST IJ printheads outperform those made with earlier technologies, and printers that use MST IJ printheads now compete directly with laser printers on performance, but at substantially reduced cost. Printhead sales in 1996 have been placed at 50 million to 100 million units. Some estimates suggest they will reach 500 million units by 2002. Others predict only half that. Volume notwithstanding, what is likely is that up to 95% of future IJ printheads will be produced by MEMS/MST methods. Table 8.4 describes the printhead market.

The second information technology example involves hard disk drives (HDDs) for data storage. Here, head positioning and accelerometer devices are the specific microsystem items discussed. Head positioning is an emerging application of MEMS technology. The economic potential is significant, owing to the ultimate replenishment needs of a large installed equipment base. MEMS-based products offer specific advantages in size, speed, and weight, but in HDDs, competing technologies do exist. With HDDs accelerating along the cost/performance curve, the cost per gigabyte of disk memory is rapidly decreasing. Recent street pricing was about $30 to $45/Gb for IDE drives, which is well ahead of one previous market forecast in 1996 that projected a cost of $50/Gb by 2002. Data access speed has also increased, but at a slightly slower pace. Drive manufacturers are beginning to use MEMS/MST methods. Early applications were directed at the lightweight portable computing market. Table 8.5 reflects MEMS-based HDD market trends.

8.4.1.2 Aerospace

The aerospace industry was the first to use MST since for that industry, MST is an enabling technology. One early application was for aircraft cabin pressure sensors. The factors that are driving the replacement of aerospace macrosystems with MST are clearly weight, performance, reliability, and cost. A key distinction between the lunar rover of the 1970s and the current NASA/JPL Mars rover is enhanced MST utilization in the latter. Table 8.6 defines the established market in just one aerospace product area, pressure sensors. Many corporations produce these systems, including Honeywell, Kulite, and Rosemont.

8.4.1.3 Automotive Applications

The automotive industry was the second to widely utilize MEMS to solve technical problems. Initial applications included the manifold absolute pressure sensor (MAPS) and MST-based nozzles for fuel systems. Today, one of the most widely used automotive MEMS systems is based on the airbag-triggering accelerometer. Other automotive systems now under development include gyros for sensing rollover, vehicle dynamic control (e.g., ABS), and GPS navigation. Both accelerometers and gyros are discussed below. Accelerometer use in automobiles resulted from legislation that initiated an innovation "express." An instant market was created, and with the arrival of MST, macromechanical airbag triggers disappeared almost overnight.

Many companies compete in the accelerometer niche. Among the more prominent are Analog Devices, Bosch, Breed, EGG, Motorola, Rockwell, Sensonor, Siemens, and TRW. Table 8.7 shows an assessment and forward projection for this segment.

Present micromachined angle and angular rate sensing systems (i.e., gyros) are not truly replacements for conventional gyroscopes. The latter never caught on for automotive applications because of their cost and size. On the other hand, MST-based gyros have a significant future in automotive technology, with at least three applications per vehicle now envisioned. Table 8.8 shows projections in this area.

Table 8.6 Aerospace pressure sensors.

Variable	1996	2002
Units (millions)	1–3	1–3
MEMS price (US$)	5–200	4–180
MST price (US$)	5–2000	4–800
Technology replaced	Macro	Macro
Cost advantage	10×	10×
Performance advantage	10×	10×

8.4.1.4 Medicine and Biology

The field of medicine has numerous applications for M^3 products. Among them are body fluid flowmeters, invasive and noninvasive pressure monitors, and patient point-of-care chemical diagnostic and drug delivery devices. There is also a very large market for medical instrumentation and drug discovery and a host of other bioapplications, including those in genomic science and engineering and forensics. Among the many manufacturers in these areas are Hewlett–Packard, i-Stat, IC Sensors, Lucas Novasensor, Perkin-Elmer, and Sentir. Table 8.9 illustrates the market for blood pressure devices, which was initiated by Novasensor.

This section has illustrated how a marketing professional might obtain values inherent in current M^3 market forecasts. To illustrate certain points, we have used data represented in current market studies and in expert opinion. We wish to emphasize again that our use of these data does not represent an endorsement. We acknowledge the difficult problem faced by market researchers who try to accu-

Table 8.7 Airbag accelerometers.

Variable	1996	2002
Units (millions)	20–30	80–150
MEMS price (US$)	5–7	2–3
MST price (US$)	5–15	2–5
Technology replaced	Macro	Macro
Cost advantage	10×	10×
Performance advantage	10×	10×

Table 8.8 Automotive gyros.

Variable	1996	2002
Units (millions)	0.15–0.25	5–40
MEMS price (US$)	25–35	10–20
MST price (US$)	25–35	10–20
Technology replaced	Macro	Macro
Cost advantage	10×	10×
Performance advantage	10×	10×

Table 8.9 Medical pressure sensors.

Variable	1996	2002
Units (millions)	16–20	20–30
MEMS price (US$)	1–10	1–10
MST price (US$)	1–10	1–10
Technology replaced	Macro	Macro
Cost advantage	10×	10×
Performance advantage	Disposable, 1.2–10×	Disposable

rately forecast M^3 markets formed by discontinuous innovations resulting from innovative technologies.

8.4.2 THE NEXUS MEMS/MST MARKET STUDY[41]

In 1996, the Executive Board of the European Commission's Network of Excellence in Multifunctional Microsystems examined a number of the MEMS/MST market studies that had been performed since 1990. The board concluded that none of the studies provided a realistic and quantitative basis for the important MST-related strategic decisions that needed to be made. In the board's opinion, all existing studies either were too restricted to certain devices or technologies, or were essentially insensitive to the economic leveraging effect of implementing microsystems in macrosystems. Moreover, in the board's view, the existing studies basically considered traditional applications while neglecting really novel applications that become possible only through microsystems technology.

To resolve this deficiency, a Nexus task force titled "Market Analysis MST" was established to prepare an applications-oriented, geographically based, in-depth analysis of MST markets through the year 2002. The approach chosen involved:

- extrapolating statistical data for existing products;
- establishing potential market-share estimates for new products based on available data for existing applications;
- making assumptions about the potential economic impact of trends in technology and society;
- obtaining expert opinions on products, technologies, and potential applications.

The committee's work was completed in November 1998 and the report that was issued[3] began with an evaluation of available microstructure products. It considered products already under development that had both a well-defined application and a realistic potential for launch before 2000. The report's value estimates for the size of various market segments were based on then-current MST selling prices. All currency values listed were shown in U.S. dollars (US$). Prices were assumed to be in constant 1997 dollars.

8.4.2.1 Growth Projection

The Nexus report projected a growth in the world microsystems market from a 1996 estimate of $14 billion to $38 billion by the year 2002, a CAGR of 18% (see Fig. 8.3). This forecast has a much greater value than those of earlier studies, primarily because of a valuation procedure that counts all microstructure products (monolithic and hybrid) from a systems point of view. From 1998 to 2002, new products under development will emerge that will contribute to an even larger market growth.

MEMS, Microsystems, Micromachines

Table 8.10 Existing products.

Product	1996 Units (millions)	1996 US$ (millions)	2002 Units (millions)	2002 US$ (millions)
Hard disk drive heads	530	4500	1500	12,000
Inkjet printer heads	100	4400	500	10,000
Heart pacemakers	0.2	1000	0.6	3700
In vitro diagnostics	700	450	4000	2800
Hearing aids	4	1150	7	2000
Pressure sensors	115	600	309	1300
Chemical sensors	100	300	400	800
Infrared imagers	0.01	220	0.4	800
Accelerometers	24	240	90	430
Gyroscopes	6	150	30	360
Magnetoresistive sensors	15	20	60	60
Microspectrometers	0.006	3	0.150	40
TOTALS		**14,330**		**34,400**

Table 8.11 Emerging products.

Product	1996 Units (millions)	1996 US$ (millions)	2002 Units (millions)	2002 US$ (millions)
Drug delivery systems	1	10	100	1000
Optical switches	1	50	40	1000
Lab-on-a-chip (DNA)	0	0	100	1000
Magneto-optical heads	0.01	1	100	500
Projection valves	0.1	10	1	300
Coil-on-chip	20	10	600	100
Microrelays	0	0.1	50	100
Micromotors	0.1	5	2	80
Inclinometers	1	10	20	70
Injection nozzles	10	10	30	30
Anticollision sensors	0.01	0.5	2	20
Electronic nose	0	0.1	0.05	5
Totals		**107**		**4200**

Nexus estimates a threefold growth of MEMS/MST markets based on data for available products. Current principal products and market volumes are listed in Table 8.10. Read-write heads for hard disk drives, inkjet printer heads, and cardiac pacemakers are the product categories with the largest market volume.

Developmental products that have a high probability of being on the market by 2002 are listed in Table 8.11. The rapid progress of microsystems technology makes it likely that many totally new products will appear over the next 5 years. The task force concluded that keeping up with these developments will require continuous monitoring.

Automotive applications were the MST leaders in the early 1990s, but information technology peripherals and biomedical uses will take the lead over the next 5 years. While some studies forecast significant near-term potential in telecommunications products, the Nexus study does not foresee a real breakthrough in this area until 2002.

Integrating silicon micromachining with CMOS technology is a problem that has in many ways been resolved. Both integrated and hybrid solutions have gained in importance. In addition, advances in microfabrication techniques, including laser micromachining, high aspect ratio microreplication based on lithographic patterning, electrodischarge machining (EDM), diamond milling, and other precision mechanical methods have enabled microstructuring, not only in silicon, but also in polymers, metals, and ceramics.

8.4.2.2 Market Leveraging Factors

Since MST products determine the competitiveness of macro products, a leveraging factor of up to 50% must be considered in estimating market sizes. For example, the typical price of an inkjet cartridge is $20 to $40, but the price for the corresponding printing system is $200 to $1000. The Nexus report gave details of

- Product specifications, production quantities, and prices.
- Segments.
- Key factors required for success.
- Applications fields.
- Technological and commercial trends, major MST manufacturers, and buyers.

One of most interesting Nexus activities is its user-supplier club (USC) program. This effort brings together microsystems users (usually systems houses that are implementing microsystems in their products) and microsystems suppliers. The clubs objectives are to facilitate dialogue between users and suppliers; identify opportunities for new products and applications; provide strategic guidance; flag problem areas; discuss producibility, quality, and standardization difficulties; and provide equipment manufacturers with guidance on future requirements. Five USCs have been established (Table 8.12)

Table 8.12 Nexus user-supplier clubs.

USC	Interest area	Coordinator[a]
TE	Telecommunications	R.A. El Fatatry (GEC Marconi, England)
M&BM	Medical and biomedical	G. Legge (Glaxo Wellcome, England)
I&PC	Instrumentation and process control	A Zumstegt (CSEM, Switzerland)
P&M	Peripherals and multimedia	A. Bellone (Olivetti/Balteadisk, Italy)
A&G	Aerospace and geophysics	J. Suski (Schlumberger, France)

[a] As of 1999.

8.4.3 RELATED AND INTERLOCKING EUROPEAN UNION EFFORTS[42]

Traditionally there have been three primary approaches to gaining competitive advantage in an M^3 market place. These are traditional customer-supplier relationships, the development of new customer-supplier relationships, and government-sponsored activities. In the United States, the latter has included efforts sponsored by DARPA, the National Institute of Standards and Technology (NIST), and other federal agencies, as well as state-funded operations. Korea has its K-7 program and in Japan, MITI and organizations such as the Micromachine Center have been the sponsors.

In Europe, M^3 government sponsorship extends to research institutes and development organizations like FZK and IMM in Germany and CSEM in Switzerland. At the European Union (EU) level, there is Europractice, which partners with national governments to sponsor pan-European and localized activities through organizations such as NEXUS, IVAM, VDI/VDE, and others. In this next section, we look at the Europractice–Nexus model.

8.4.3.1 Europractice (www.europractice.com)

Europractice is a foundry/fabrication network funded by the European Commission's Esprit program. It gives small- to medium-sized companies with limited resources access to a wide range of microelectronics technologies, including ASIC and MCM capabilities. The Europractice network consists of five regional European manufacturing clusters. In the microsystems area, each cluster offers unique MST fabrication capabilities. Acknowledging the importance of MST to European industry, the EC committed to a 2-year second-phase Europractice effort as part of the Esprit program. Services were extended in October 1997 both in terms of regional coverage and applications emphasis.

Regional coverage has been improved by the addition of a new manufacturing cluster for the Nordic countries. SensoNor and Sintef in Norway, MIC in Denmark, IMC in Sweden, and VTT in Finland are collaborating to provide design support, manufacturing (using a bulk micromachining process), packaging, and test capabilities. CNM in Spain has been added to the Nordic cluster to improve the access of southern Europe to the technologies offered in this cluster. The new Nordic cluster complements four existing manufacturing clusters now located in Germany, France, the United Kingdom, and Switzerland/Netherlands.

The German cluster, led by Bosch, includes microParts, HL-Planar, FhG (Fraunhofer gesellschaft)-IMS, FhG-ISiT, and GMA.

The French cluster, led by SEXTANT Avionique, includes CEA-LETI, CNRS-LAAS, and Tronics. The latter is a MEMS startup spun off from CEA-LETI. This cluster provides bulk micromachining and SOI surface micromachining for a variety of sensing products.

A new U.K. cluster led by AEA Technology includes Applied Microengineering and Graseby Microsystems, with dissemination activities promoted by the University of Hertfordshire. NMRC in Ireland also participates in this cluster.

Table 8.13 Nexus design-applications competence centers.

Applications area	Lead organization
Automotive and physical measurement systems (including accelerometers and gyroscopes)	Fraunhofer IsiT (Germany)
Bioanalytical and biomedical microdevices, medical, and environmental applications	Fraunhofer IBMT (Germany) IMT (Switzerland)
Process control, machine tools, manufacturing, microactuators, and micromachines	RAL (England)
Microfluidic systems	Pont-Tech (Italy)
Peripherals, telecommunications, and micro-optical microsystems	CEA-LETI (France)
Radiation and imaging sensor-based microsystems, aerospace and scientific instruments	Sintef (Norway) IMEC (Belgium)

A Swiss/Dutch cluster is led by CSEM, with participation by Twente Microproducts (Netherlands) and Holland Signaal (Netherlands).

8.4.3.2 Competence Centers

To expand the role of European research institutes, Nexus initiated the concept of competence centers for development support and application of design skills, including computer-aided design (CAD). The centers (Table 8.13) were established during the second phase of Europractice and are organized around products and applications rather than technologies. Each center aims at performing the role of a fabless design house that can advise microsystems developers on product design and integration and that can choose the most appropriate technology for the task. Nexus anticipates that the competence centers will use the services of Europractice manufacturing clusters or other services outside Europractice.

8.4.3.3 Eurimus

The widely known European Eureka programs are transnational efforts targeted at developing new products, processes, or services having worldwide market potential. Projects are classified as Eureka programs by a panel of national point contacts (NPC) consisting of Eureka organization members. The national administration allocates funding for industry and/or research to each partner of the project. The first strategic Eureka project in the rapidly growing field of microsystems is Eurimus.

The objective of Eurimus is to complement existing European programs by accelerating the growth and rapid commercialization of MST products by European industries. Eurimus has been planned as a 5-year, 400 million EU (~US$418million as of August 1999) effort. It began in 1998. A midterm evaluation is built into the program.

Program participants will include large industrial firms like Bosch, Copreci/Fagor, Daimler-Benz, SEXTANT Avionique, Olivetti, Philips, Schlumberger,

SGS Thomson, Siemens, and Temic. Innovative small- to medium-sized enterprises like CSEM, Electronic Visions, microParts, and SensoNor will also participate, as will applied research institutes including CEA-LETI, CNM, CNR Madess, Fraunhofer ISIT, RAL, and Sintef. Projects will vary significantly in size. A total of 100 to 150 projects will likely be funded by the Eurimus budget. It is anticipated that 50% of the financing will be provided by industry. The other 50% will come from national authorities and the EU. Further funding from financial institutions is also being contemplated.

Nexus has succeeded in establishing an industrial focus on microsystems technologies in Europe. The formation of a Strategic Guidance Group during Nexus IV will increase the focus on areas of development that have both large market potential and particular interest to Europe. In addition, Nexus IV will expand organizational membership to create a truly pan-European organization. It will also concentrate on creating stronger bonds between the user-supplier clubs, increasing interaction among academic groups, and extending coverage provided by its information and communications network.

Specific Nexus objectives include strategic technology workshops, improvements in communications services, benchmarking visits to North America and Asia, and intensive market studies. Nexus also intends to report on approximately 30 organizations in Japan and the Far East. Other objectives include regular written reports to help formulate a European strategic view of MST. Through Nexuspan, Nexus will further integrate Eastern Europe into the overall European MST community.

Current Nexus task forces include:

- The Market Analysis Task Force, which is responsible for generating in-depth analyses of MST markets in Europe based on inputs from the USCs and Nexus members. The Task Force's 1998 report[3] is a strategic guidance document for the MST community that includes a market breakdown, a related growth forecast, and identification of technology and market needs. Regular updates are expected.
- The International Relations Task Force, which is responsible for establishing and maintaining contact with MST-related institutions and authorities outside Europe, such as the U.S. Defense Advanced Research Projects Agency and Japan's Micromachine Center.
- The Eastern Countries Task Force, which maintains relations with the Central and Eastern European MST community through the related network, Nexuspan. This task force and the steering committee of Nexuspan are one and the same.
- The Long-Term Perspectives Task Force, which is associated with the academic working group. This group is taking a 10-year, long-term view of what will be achievable with microsystems.

8.5 Technology Roadmap Development for an Emergent Industry

Current M^3 products reflect the output of a very diverse and still infant community. Industries that adopt M^3 technologies will radically change their product-market-technology paradigms. Since M^3 technologies are at the beginning of their life cycle, an industry overview is not available, and established architectures and processes are only now beginning to emerge.

Market-pull forces are similarly just starting to be felt, and the disruptive nature of M^3 products is still more technologically driven than market driven. Market studies and models support the view that many M^3 products are still fulfilling replacement rather than revolutionary applications. This will change as market forces demand more and more from the technology.

An international industry group has initiated a first-of-its-kind roadmap process to quantify the disruptive M^3 technology base. Like roadmaps in parallel high-technology areas, this document will focus on reducing the corporate risk of embracing M^3 and in turn, increasing the commercialization rate of M^3 technology. The goal of the M^3 roadmap is thus to provide a guide to precompetitive R&D and infrastructure expenditures that will accelerate M^3 commercialization. By providing a consensus assessment of the industry from industry players themselves, as well as an industry information base, the roadmap development will reduce the risks of stakeholder groups interested in entering the worldwide M^3 marketplace. Many interested industry groups and companies seeking to gain competitive advantage from M^3 technologies now support the roadmap activity.

8.5.1 Nature of the M^3 Industry

In the wake of technology advances, many sectors of general industry have identified needs for smaller sized components and systems, reduced use of materials and energy, improved operating performance achieved through tighter tolerances and enhanced functionality, and with systems, "intelligence" that can be realized through integrated electronics. These are the drivers for M^3 technology.

To date, no universal metric has emerged for the M^3 industry, no critical dimension parameter like the one employed by the semiconductor industry to track its progress. However, there are factors in any technology that always influence design, fabrication, and performance. Those factors are the quality-related measures of accuracy and tolerance. At the conference on Commercialization of Microsystems in San Diego (September 1998), participants concluded that a need existed for an M^3 roadmap. They also recognized the challenges that development of a such a roadmap presented. Here are factors that bear on this.

8.5.2 THEORY AND EXECUTION OF M^3 ROADMAPPING

Traditionally, roadmaps serve to integrate business and technology strategies. The outcome of a traditional roadmapping exercise is a description of products and the technologies required to produce those products over short-term (e.g., a 5-year) and longer term (5 to 10 year) periods. Roadmapping requires information about and an understanding of market-product combinations so that product needs can be translated into product specifications that in turn can be converted into technological needs for product manufacture.

However, the process of roadmapping a diverse disruptive technology differs from roadmapping a sustaining technology[43] and focuses on the technology-product paradigm rather than the product-market paradigm. Nevertheless, for M^3, a roadmap need does exist for a market-driven portion that includes the varied views of the M^3 marketplace.

M^3 roadmapping may be performed for technology-product paradigms with short life cycles (e.g., 3–4 years) that characterize some M^3 consumer products, or for technology-product paradigms having a high aggregation level (i.e., a longer anticipated development period) such as optical storage, and thus a potential market scope of up to 10 years. The higher the aggregation level, the further out in time that products can be anticipated.

In M^3 roadmapping, three phases can be distinguished that are related to such questions as,

- Is the objective physically possible?
- Do we want it?
- Is it attractive?
- How can we realize it?

The scope of the roadmap activity is therefore the study of the M^3 technology's current status as well as its long-term trends (i.e., roughly 5–10 years).

8.5.3 STRUCTURE OF THE M^3 ROADMAP

In focusing on the technology-product paradigm rather than the market-product paradigm, a structure for the initial M^3 roadmapping activity emerged from the 1998 San Diego meeting. It addresses the following technological areas:

- simulation, modeling, and design;
- reliability, testing, and metrology;
- packaging;
- assembly;
- cost models;
- glossary;

- IC-compatible manufacturing methods;
- non-IC compatible manufacturing methods;
- potentially IC-compatible manufacturing methods;
- standards;
- integration: monolithic versus hybrid;
- product-technology paradigms;
- commercialization.

In implementing this structure, teams of experts working in each area will provide inputs with respect to the following:

- the current status of the technology;
- developments foreseen in the 3–5 year time frame;
- developments foreseen in the 5–10 year time frame.

A roadmapping activity has been started with support from SEMI. As of September 1999, 143 professionals from 16 countries, representing 113 firms, government agencies, industry organizations, national labs, and academic institutions had become actively involved. A worldwide web-based administrative operation headquartered in The Netherlands has been established to facilitate intra- and interteam communication. Figure 8.4 shows the international roadmap homepage (http://www.roadmap.nl).

Figure 8.4 International M³ roadmap homepage.

8.5.4 EXAMPLES OF EARLY INPUTS OF THE ROADMAP COMMITTEE

Medical instrumentation is just one of the emerging application areas for M^3 technology. In this vein, Wilkinson[44] presented the size trend for catheters and endoscopes that are the vehicles for minimally invasive medical and surgical procedures (Fig. 8.5). Associated with this trend are applications for micromotors, microgrippers, ultrasound scanning, microfiber optical imaging, temperature and pressure sensors, and a variety of biosensors. M^3 technologies that are critical to catheter and endoscope applications are displayed along the x-axis of the figure. Among these technologies are packaging and assembly methods that include wire bonding, chip-on-board, flip chip and film flex technologies, chip-scale packages, multichip technologies, and thinned chips on flex substrates. What this chain of thought illustrates is a logical, hierarchical approach to the construction of just one portion of the M^3 product-technology paradigm.

Figure 8.6 shows a similar projection for DNA analyzers, an important category of emerging instrumentation that will consist of a series of functional subsystems (sample preparation, amplification, hybridization, separation, and detection) that M^3 technologies address directly. Hence a projection of the requirements timetable

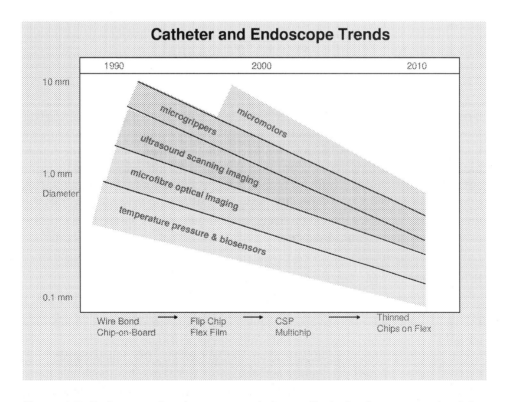

Figure 8.5 Catheter and endoscope trends for medical microinstrumentation (after Ref. 44).

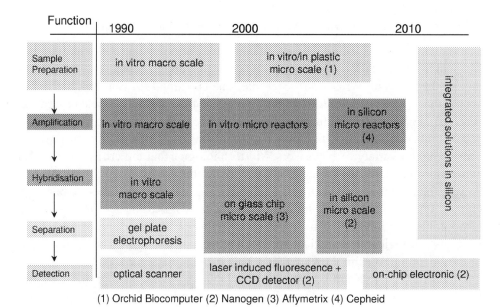

(1) Orchid Biocomputer (2) Nanogen (3) Affymetrix (4) Cepheid

Figure 8.6 DNA analyzer system trends (after Ref. 44).

for DNA analysis applications can lead to a forecast of the supporting M^3 technologies that will be required for these applications.

8.5.4.1 Other M^3 Roadmapping Activities

Nexus, supported by the EC, is also funding a roadmap, but of a somewhat different type. The Nexus roadmap is regional. Focusing on European concerns, it is intended as a product-market-centered document. Elsewhere, many of the companies, especially the larger companies, contributing to the international roadmapping effort maintain internal roadmaps based on the technology-product paradigm and or the product-market paradigm. The product-market paradigm tends to focus on more mature MEMS technical products, such as automotive safety devices, inkjet printer heads, and chemical sensors.

8.6 Evolving M^3 Infrastructure

Infrastructure is one of the fastest-growing elements in the process of commercializing M^3 technologies, yet infrastructure remains a critical bottleneck. To better understand the problem, an infrastructure model (see Fig. 8.7) has been generated that takes the three M^3 production technology groups through a four-stage development process. The model reflects the actual historical growth of M^3 technology. During stage 1, the stage of basic research and nonexistent market channels, managers and technologists who initially investigated M^3 technologies typically forced

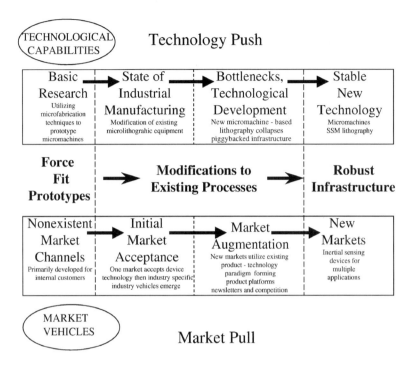

Figure 8.7 Infrastructure model for discontinuous innovations.

a fit between manufacturing and market channels. In lithography, for example, M^3 technologists were required to utilize existing lithographic techniques that were optimized for microelectronics fabrication.

At stage 2, a stage of initial market acceptance of M^3 products, technology-market-product paradigm shifts have begun. At this point, market channels are beginning to widen because high-technology entrepreneurs are entering the industry. Also at this point historically, firms like Novasensor generated highly specific products for special market niches and convinced OEMs to use lightweight throw-away MEMS-based sensors for a variety of applications. Concurrently, modifications of existing manufacturing equipment were beginning to appear. Again, the example cited is in MEMS lithography, where double-sided lithographic aligners constructed from obsolescent tools used for electronics microfabrication emerged to support the manufacture of M^3 products.

At stage 3 of the model—bottlenecks, technological development, and market augmentation—market channels widen still further and awareness of field-tested M^3 products is growing so that OEMs can begin to approach M^3 firms with suggestions for new products. M^3 production begins to be supported by the entry of capital equipment and suppliers of consumables.

Stage 4, the stable new technology and new markets stage, sees a robust infrastructure emerging for some competing techniques. Typically, MEMS production firms are now larger corporations, in contrast to small, high-tech entrepreneurial startups. In the field, the salesman-entrepreneur is being replaced by the salesman-engineer, and the use of strategic market partners to obtain market leverage is becoming common. At this point, a mature infrastructure exists and the behavior of manufacturers, customers, and suppliers is similar to that in many other mature markets.

This four-stage development model can be applied in attempting to understand the infrastructure status of the different M^3 production technologies. For example, at the time traditional bulk micromachining was in the second stage, Novasensor, Motorola, Breed, and other firms had modified traditional manufacturing techniques to produce limited volumes of product. Materials vendors like Okmetic and Siltec, and equipment vendors such as Electronic Visions and Karl Suss, were offering products specifically for M^3 production. The infrastructure model suggests that at this point, traditional bulk micromachining would soon have a robust infrastructure, while sacrificial micromachining, drawing on the vast experience of microelectronics manufacturing, would develop a robust infrastructure before traditional micromachining. Finally, HARM technologies, which at this point are not yet commercially viable, are placed at stage 1 of the model.

8.6.1 OTHER INFRASTRUCTURE COMPONENTS

As noted previously, government support has played and continues to play a role in the development of M^3 production technologies. Again, as noted, the European Commission's five regional Europractice MEMS fabrication service centers facilitate the development of MEMS-based devices and markets in Europe. These service centers are typically corporate–university partnerships that concurrently accelerate the infrastructure development process. In Japan, the Micromachine Center has over 31 corporate members plus university research center involvement. In the United States, efforts like the Advanced Technology Program (ATP) of the National Institute of Standards and Technology and Department of Energy initiatives at locations such as Sandia National Laboratories and several high-energy synchrotron storage ring locations promote development of M^3 applications and technology transfer, and concurrent commercial development for bulk, surface, and LIGA microfabrication.

Information exchanges focused on encouraging the development of M^3 have a history extending back to the early 1990s. A partial list of exchanges that date from the mid-1990s onward includes:

- the IEEE annual MEMS conferences and biennial transducer conferences;
- the European MEMS roundtable at SEMICON/Europa 1994;
- an educational program at SEMICON/West 1994;
- a group of co-sponsored activities between Nexus and SEMI;
- the 1995 MEMS European roundtable at SEMICON Europa;
- a 1995 SEMI Symposium on micromachining and microfabrication;

- a series of SEMI-co-sponsored commercialization conferences in Banff, Canada (1994), Kona, HI (1996), and San Diego(1998);
- a series of annual SPIE symposia dedicated to microsystems fabrication and an active and growing body of technical and trade publications and newsletters.

The number of regional and international cross-disciplinary technical and professional meetings having an M^3 orientation or at least having M^3 special sessions has grown rapidly since 1997.

Another traditional measure of activity in technological fields is patents filed. As shown in Fig. 8.1, since 1993 the rate of M^3 patent filing has increased dramatically.[12] This suggests that inventors or the companies that employ them are increasing activity and perceive their advances as having commercial value.

The latter half of the 1990s has also seen advanced product development and commercialization. Notable examples include Analog Devices' commercial success with its integrated surface micromachined accelerometer for automotive airbag use and Texas Instruments' multiple alliances to commercialize digital micromirror device technology (DMD) for electronic applications. Efforts to raise awareness and involvement in emerging markets, like those based on M^3 production technologies, encourage infrastructure development by increasing customer awareness and acceptance as well as supplier interest and involvement. Surprisingly, in M^3 production technologies and other emerging industries, pioneering firms gain by sharing some proprietary information, owing to the positive effects that shared technical and market information have on the development of infrastructure.

8.7 CONCLUSIONS

Even though M^3 technologies are disruptive, there are at least three areas where mature product-market paradigms have emerged: pressure sensors for automotive and aeronautical applications, accelerometers for automotive safety systems, and inkjet head devices. The latter appeared initially in printers, but their basic operating principle has wide potential in a broad range of industrial systems for accurate placement and/or assembly of materials.

The total dollar volume of the market for microdevices alone is over US$2 billion according to nearly every current market study. Millions of M^3 devices are manufactured and used in microsystems yearly. The market level based on a valuation for complete systems employing microdevices is about 10 times this figure. One must be careful to understand the basis for these valuations, but recent systems estimates for year 2002 market valuations have ranged from US$38 billion to as high as $100 billion.

Numerous new M^3 technology markets are emerging in fields such as biotechnology, medicine, telecommunications, wireless RF applications, and many more.

The serendipitous emergence of M^3 technologies concurrent with industrial developments that can well utilize them, such as the transition to a fiber optic telecommunications base in countries around the world, holds great promise for the M^3 community.

Commercialization efforts focused on M^3 products have intensified noticeably since 1997. Growing global interest virtually ensures that M^3 technology will be widely infused into the overall technology base of twenty-first century society. The path that M^3 commercialization follows in the next two decades will be limited only by the creativity of the M^3 community and the entrepreneurial drive and acumen of its members.

REFERENCES

1. "Micromechanics: Multi-Client Study," Battelle Institute, Frankfurt, Germany, July (1992).
2. "Microelectromechanical Systems (MEMS), an SPC Market Study," Systems Planning Corp., July (1994).
3. "Market Analysis for Microsystems 1996–2002," a Nexus task force report (www. nexus-emsto.com [in Europe], www.delcom.com [in North America]).
4. R. Grace, "Commercialization Issues of MEMS/MST/MICROMACHINES: An Industry Report Card on the Barriers to Commercialization," in *Proc. Sensors Expo*, pp. 299–303 (1999).
5. S. Marshall, "MEMS Market Data: Another Case of Sorry, Wrong Number?" *Micromachine Devices* (Aug.), 6–7 (1997).
6. R. Wechsung, "Wanted: MEMS Market Studies Based on Real Quantitative Data," *Micromachine Devices* (Nov.), 5–6 (1997).
7. "Microelectromechanical Systems (MEMS), an SPC Market Study," Systems Planning Corp. Jan. (1999) (http://memsmarket.sysplan.com).
8. P. Mukai, "Test Strategy Development for MEMS Devices," in *Proc. Sensors Expo*, pp. 487–490 (1998).
9. W. H. Miller, "Reliability: A Hidden Barrier to Successful Commercialization of MEMS?" *Micromachine Devices* (Dec.), 1–4 (1997).
10. K. W. Oh and C. H. Ahn, "Development of an Innovative Flip-Chip Bonding Technique Using Micromachined Conductive Polymer Bumps," paper presented at *Solid State Sensors and Actuators Workshop*, pp. 170–173 (1998).
11. R. Grace, "MEMS: Definitely Smaller, Often Smarter, Faster, & Cheaper," A^2C^2 (July/August), 19–25 (1998).
12. D. J. Nagel, S. J. Walker and D. J. Drosdoff, "Patent Analysis Shows Growth of Microelectromechanical Systems," *Micromachine Devices* (May), 12 (1998).
13. R. Grace, "Making Money with MEMS: How to Become a MEMS Millionaire," paper presented at *Solid State Sensor and Actuator Conference* (1998) (www.rgraceassoc.com).
14. J. L. Bower and C. M. Christensen, "Disruptive Technologies: Catching the Wave," *Harvard Business Review* (Jan.–Feb.), 43–53 (1995).

15. G. Lynn, J. Morone and A. Paulson, "Marketing and Discontinuous Innovation: The Probe and Learn Process," *California Management Review* 38(3), 8–37 (1996).
16. Geoffrey Moore, *Crossing the Chasm*. Harper Business Press, New York (1991).
17. G. Lynn and S. Walsh, "Radical Innovation: Challenges and Insights," in *Proc. Annual Meeting of the Product Innovation Management Society* (1991).
18. J. Morone, *Winning in High Tech Markets*. Harvard Business School Press, Boston (1993).
19. K. B. Clark and R. M. Henderson, "Generational Innovation—Architectural Innovation, the Reconfiguration of Existing Product Technologies and Failures of Established Firms," *Administrative Science Quarterly* 35 (1990).
20. S. Walsh and B. Kirchhoff, "Strategies for HTSF's Embracing Autonomous Disruptive Technologies," paper presented at the *High Technology Small Firms Conference* (1998).
21. J. M. Utterback, *Mastering the Dynamics of Innovation*. Harvard Business School Press, Boston (1994).
22. B. C. Twiss, *Managing Technological Innovation*. Longman, London (1980).
23. A. L. Frohman, "Managing the Company's Technological Assets," *Research Management* 9(Sept.), 20–24 (1980).
24. D. Bitindo and A. Frohman, "Linking Technological and Business Planning," *Research Management* 11(Nov.), 19–23 (1981).
25. D. Marquis, "The Anatomy of Successful Innovations," *Managing Advancing Technology* 1, 35–48 (1969).
26. R. S. Rosenbloom, "Technological Innovation in Firms and Industries: An Assessment of the State of the Art in Technological Innovation," in *A Critical Review of Current Knowledge*, P. Kelly and M. Kranzberg, eds. San Francisco Press (1978).
27. W. J. Abernathy and K. B. Clark, "Innovation: Mapping the Winds of Creative Destruction," *Research Policy* 14, 3–22 (1985).
28. S. Walsh, R. Boylan, A. Paulson and J. Morone, "Core Capabilities and Strategy: Evidence from the Semiconductor Silicon Industry," in *Strategic Integration*, H. Thomas and D. O'Neal, eds., pp. 149–165. Wiley, Cambridge, England (1996).
29. E. Von Hippel, "Users as Innovators," *Technology Review* (Jan.), 30–39 (1978).
30. E. Von Hippel, "Lead Users: A Source of Novel Product Concepts," in *Readings in the Management of Innovation*, M. L. Tushman and W. Moore, eds., 2nd edn, pp. 352–366. Harper Business Press, New York (1986). Also, E. Von Hippel, "Lead Users: A Source of Novel Product Concepts," *Management Science* 32(7) (1986).
31. E. Von Hippel, *Sources of Innovation*. Oxford University Press, New York (1986).
32. R. Veryzer, "Discontinuous Innovation and the New Product Development Process," *Journal of Product Innovation Management* 15(4), 304–321 (1998).

33. G. Lynn, J. Morone and A. Paulson, "Emerging Technologies in Emerging Markets: Challenges for New Product Professionals," *EMJ* 8(3) (Sept.), 23–31 (1996).
34. T. A. Wise, "IBM's $5,000,000,000 Gamble," in *Readings in the Management of Innovation*, M. L. Tushman and W. Moore, eds., 2nd edn. Harper Business Press, New York (1988).
35. M. A. Madique, "Entrepreneurs, Champions, and Technology," in *Readings in the Management of Innovation*, M. L. Tushman and W. Moore, eds., 2nd edn, pp. 79–87. Harper Business Press, New York (1986).
36. M. A. Madique and R. H. Hayes, "The Art of High-Technology Management," in *Readings in the Management of Innovation*," M. L. Tushman and W. Moore, eds., 2nd edn. Harper Business Press, New York (1984).
37. E. Mansfield, *The Economics of Technology*. W. W. Norton, New York (1968).
38. J. Enos, *Petroleum Progress and Profits*. MIT Press, Cambridge, MA (1967).
39. J. A. Schumpeter, *The Theory of Economic Development*. Harvard University Press, Cambridge, MA (1934).
40. W. J. Abernathy and J. M. Utterback, "Patterns of Industrial Innovation," in *"Readings in the Management of Innovation,"* M. L. Tushman and W. Moore, eds., 2nd edn, pp. 25–36. Harper Business Press, New York (1988).
41. "Nexus Market Analysis Task Force Issues Market Study Report on MEMS/MST," *Micromachine Devices* (Oct.), 1–5 (1998).
42. G. Menozzi, "The role of Nexus in European MEMS/MST Development, Parts 1 and 2", *Micromachine Devices* (Sept.), 1–5 (1998); (Oct.), 10–13 (1998).
43. S. Walsh and J. Linton, "Infrastructure for Emerging Markets Based on Discontinuous Innovations: Implications for Strategy and Policy Makers," submitted to *Long Range Planning* (1999).
44. J. M. Wilkinson, "Medical Applications for MST," presented at *Commercialization of Microsystems '99* (www.tfi-ltd.co.uk).

Index

3D beam propagation method (BPM), 238

ABAQUCS, 61
accelerometer, 21, 400
acoustic surface waves (SAW), 270
active equalizer, 324
actuator, 425
actuator design, 95
add/drop multiplexers (ADM), 320
AlN, 257
alternative bonding techniques, 343
amorphous silicon, 273
amplitude detection, 260
angular rate sensor, 83
Anise, 61
anneal process, 234
anneal step, 11
anodic bonding, 340
ANSYS, 61
anti-stiction techniques, 344
antiresonant reflecting optical waveguide (ARROW), 217
arrayed waveguide grating routers (AWGR), 320
ARROW Waveguide Optical Sensors, 279
atmospheric pressure (APCVD), 12

ball bonding, 353
beam splitters, 249
BEMMODULE, 113
bending of the silicon wafer, 232
biaxial stress, 234
BiCMOS, 112
BICMOS processing, 62
bimetallic actuators, 49
Biosensors, 289
birefringence-preserving polarization, 243
bonding, 13
border, 195
boron doping, 11
borosilicate glass (BK-7), 266
boundary-element method (BEM), 136
Bragg Gratings, 284
Bubnov–Galerkin finite-element method, 145
bulk micromachining, 14, 172, 255
buried oxinitride waveguides, 239

calibration, 372
cantilever beam, 259

capacitive pressure sensors, 78
channel waveguide, 212
charge transport, 134
chemical vapor deposition (CVD), 12
Chip-on-Flex (COF) technology, 365
CMOS processing, 62
coefficients of thermal expansion (CTE), 49
comb-drive electrostatic actuator, 256
comparison of LPCVD and PECVD methods, 225
complementary metal oxide semiconductor (CMOS), 112
compressive stress, 232
computer-aided design (CAD), 111
constitutive equations, 126
contrast ratio, 193
control-volume method (CVM), 136
cost model, 449
couplers, 249
coupling efficiency, 251
coupling strategies, 157
creation of wealth, 485
crystal elasticity, 130

DAVINCI, 61
deep reactive ion etching, 33
Defense Advanced Research Projects Agency (DARPA), 3
deformable mirrors, 87
deposition data of LPCVD silicon nitride and silicon oxinitride films, 224
deposition data of PECVD silica, silicon nitride, and silicon oxinitride films, 222
deposition rates for silicon wet oxidation, 220
design failure mode and effect analysis (DFMEA), 74
design for manufacturing, 91
design of accelerometers, 81
design of pressure sensors, 75
design rules, 63
device packaging techniques, 373
die separation, 176, 346
dielectric effects, 132
dielectric permittivity, 132
Digital Light Processing (DLP^{TM}), 170
Digital Micromirror Device (DMD^{TM}), 28, 170
digital mirror display (DMD), 405
direct wafer bonding, 337

directional coupler, 250
discontinuous market pull, 493
discontinuous technology push, 494
discretization Methods, 136
DM, 87
DM modeling, 90
drive-in step, 11
dry oxidation, 219
dynamic thermal scene simulators (DTSS), 157

EDP, 255
eigensolutions, 156
elasto-optic effects, 135, 270
electrical interconnection, 353
electron-beam irradiation, 236
electrostatic, 51
electrostatic microactuators, 256
embedded-strip waveguide, 214
encapsulation, 369
epi-poly, 17
epitaxial (epi) layer, 17
equations of motion, 115
established infrastructure (B), 486
etching, 10
Eureka, 504
Eurimus, 504
Europractice, 503
Evanescent Interaction, 268
evanescent modulation, 281

Fabry–Pérot cavity, 273
fiber-to-waveguide coupling loss, 251
finite-difference method (FDM), 136
finite-element method (FEM), 136
flame hydrolysis deposition, 227
flip chip, 353
flip-chip transfer, 360
FLOTRAN, 61
Fresnel loss, 251

Global, 197
GRIN microlens, 288

HDTV, 170
heat transport thermoelectricity, 136
hermetic seal, 180
hidden hinge, 174
High Density Interconnect (HDI), 363
high-volume production, 73
horizontal "hot wall" LPCVD reactor, 223
humidity resistance test, 230

image, 195
in-use stiction, 344
index contrast, 214
industry association, 487
inertial sensor packaging, 399

inertial sensors, 21
information technology (IT), 496
integration of MEMS devices, 66
IntelliCAD, 61
interferometers, 249
interferometric distance sensors, 287
interferometric or intensity modulation techniques, 275
inverse statement, 148
ion exchange, 236

kinematics of deformation, 127
KOH, 255

Lagrange equation, 115
Lagrangian for crystalline continua, 121
laser irradiation, 236
LIGA, 18, 172
light-actuated micromechanical photonic switch (LAMPS), 315
lightwave networks, 303
linear equation systems, 155
local, 197
local polymerization-driven diffusion reactions, 248
low pressure (LPCVD), 12, 223
luminescence quenching, 289

M^3 commercialization process, 488
M^3 infrastructure, 510
M^3 R&D activities, 484
M^3 roadmap, 506
M^3 roadmapping, 507
M^3 technologies, 481
Mach–Zehnder interferometer, 250
Mach–Zehnder interferometer multiplexers, 243
magnetic actuator, 264
magnetic sensors, 53
manifold absolute pressure (MAP), 75
manifold air temperature (MAT), 75
MAP sensor, 75
market research, 485
marketing, 484
material interface, 124
Materials in micromachining, 254
MCM-ceramic (MCM-C) technology, 357
MCM-deposited (MCM-D) technology, 358
MCM-laminate (MCM-L) technology, 357
MCM-silicon technology, 359
mechanical dissipation, 130
mechanical effects, 127
MEMCAD, 61
MEMS, 333
MEMS acoustic actuators, 57
MEMS and MOEMS testing, 427
MEMS automated assembly, 417
MEMS devices, 5, 302

MEMS modeling, 61
MEMS packaging, 69
MEMS/MOEMS devices, 49
metal leadframes, 349
Michelson interferometer, 250
micro-optoelectromechanical systems (MOEMS), 111
microactuators, 255
microbridge, 277
microcantilever beam, 277
microconveyer belts, 32
microelectromechanical systems (MEMS), 3, 170, 481
microfluidics packaging, 422
microgrippers, 31
microhinges, 415
microlatches, 415
micromachined amplitude-modulated optical sensors, 277
micromachined cantilever beam accelerometer, 278
micromachined deformable mirror (μDM), 87
micromachined integrated optic switches, 261
micromachined nanoprobes, 292
micromachined optical modulators, 267
micromachined optical sensors, 274
micromachining, 4
micromotor, 25
microoptoelectromechanical systems (MOEMS), 210
micropumps, 26
microsensors, 50
microstereolithography, 34
microsystem, 111, 333
microsystem modeling, 112
microsystem simulation, 112
microsystems technology (MST), 481
microtip, 30
mirror planarity, 92
mismatching between the fiber and waveguide modes, 251
Modular, Monolithic Micro-Electro-Mechanical Systems (M^3EMS), 356
modulators, 249
Multi-User MEMS Process (MUMPs), 413
Multichip Module (MCM), 356

nanotechnology, 35
NASTRAN, 61
national point contacts (NPC), 504
near-field detection, 260
near-field scanning optical microscope (NSOM), 327
Nexus, 500
nonlinearities, 153
nonuniformities, 197

optical birefringence, 234
optical bypass switch, 262
optical crossbar switch, 265
optical immunosensors, 291
optical interrogation, 274
optical lithography, 9
Optical MEMS, 334
optical microdevices, 54
optical pressure sensor, 279
optical readout, 276
optical sensors, 50
optical switches, 310
optical switches with active microactuation, 265
optical switches with passive microactuation, 262
optical testing, 181
optochemical sensors, 289
optomechanical pressure sensor, 158
optomechanical sensor, 274
overlap integral, 251
overmolding, 369
oxidation, 9
oxinitride strip-loaded waveguide, 237

P-glass flow (or PSG-flow), 226
package assembly, 179
packaging, 13
Parallel plate PECVD reactor, 221
parallel-plate electrostatic capacitor, 256
partial differential equations (PDEs), 136
passive optical networks (PONs), 314
patterning, 9
PEPPER, 61
Phase Modulation, 268
phosphine, 242
phosphorus doping, 11
phosphorus-doped oxide, 215
phosphorus-doped silica, 226
phosphosilicate glass, 225
phosphosilicate glass waveguides, 242
pick and place, 348
piezoelectric actuators, 57, 257
piezoelectricity, 132
piezoresistance, 57
piezoresistive acceleration sensor, 81
planar lightwave circuit, 263
plasma-enhanced chemical vapor deposition, 220
plasmon resonance, 291
Poisson ratio, 232
polarizers, 249
polycarbonate, 228
polyguide, 249
polyimide, 228
polyimide microplatform, 267

polymer deposition, 228
polymethyl methacrylate (PMMA), 228
polysiloxane, 228, 248
polystyrene, 228
power limiter, 318
pre-molded packages, 349
pressure sensor, 5, 20, 374
propagation loss, 251
PROSIT-ISE, 114
public relations, 485
pure silica, 214
pyroelectricity, 132
PZT, 257

radiation losses, 251
rate constants for wet and dry oxidation of silicon, 220
reconfigurable drop module (RDM), 320
reflectivity, 193
release stiction, 344
reliability, 203
reliability of MEMS, 432
REMESH, 114
residual stress distribution, 276
resonant devices, 57
resonant microsensors, 83
rib waveguide, 212

sacrificial layer, 255
sawing, 177
scaling, 64
scattering loss, 251
scratch drive actuator (SDA), 257
self-test design, 66
SENSIM, 61
shape memory alloys (SMA), 59
shape of microlenses, 164
silicon nitride, 215
silicon oxinitride, 215
silicon oxinitride channel waveguides, 236
silicon piezoresistive pressure sensors, 76
silicon processing, 3
silicon V-groove, 253
silicon wafers, 8
silicon-rich silicon nitrides, 215
single-crystal epi, 17
slab waveguide, 214
smart sensors, 68
solution methods, 152
SPICE, 113
spin-coating technique, 228
splitters, 249
startup funding attraction, 485
stiction, 180, 344
Stoney formula, 232
strain amplitude, 270
strain-elasto-optic coefficient, 270
strain-induced optical birefringence, 282

stress, 128
stress and strain in MOEMS thin films, 98
stress-induced birefringence, 235
stress-induced membrane, 275
strip-loaded waveguide, 214
strong form, 137
structural, 255
structural layer, 255
SUPREM, 61
surface micromachining, 15, 172, 255
surface plasmon resonance, 289
SVGA (Super Video Graphics Adapter), 170
SXGA (Super Extended Graphics Adapter), 171

TAB, 353
tape automated bonding, 353
TE/TM polarization converters, 249
technology CAD (TCAD), 112
tensile stress, 232
TEOS, 226
thermal expansion (CTE), 129, 369
thermal expansion actuators, 49
thermal imager, 157
thermal oxidation, 218
thermo-optic effect, 272
tilt angle, 193
tilt mirror, 304
time integration, 154
total available market (TAM), 334
transducer, 333
trimming, 372
tunneling sensor, 59

user-supplier club (USC), 502

variable optical attenuator, 304
variable optical attenuator (VOA), 306
VLSI, 302

wafer bonding, 336
wafer-scale package, 337
waveguide-cantilever, 294
wavelength division multiplexed (WDM), 243, 305
weak form, 144
wet oxidation, 219
window aperture, 198
wirebonding, 353

X-crosses, 249
XGA, 170

Y-junctions, 249
Young's modulus, 232

ZnO, 257